Edgar Allan Poe, *Eureka*, and Scientific Imagination

Edgar Allan Poe, *Eureka*, and Scientific Imagination

DAVID N. STAMOS

SUNY PRESS

Mathew Brady photograph of Poe, seated, ca. 1845/1848.
Courtesy of the Edgar Allan Poe Museum.

Published by State University of New York Press, Albany

Printed in the United States of America

For information, contact State University of New York Press, Albany, NY
www.sunypress.edu

Production, Jenn M. Bennett
Marketing, Kate R. Seburyamo

Library of Congress Cataloging-in-Publication Data

Names: Stamos, David N., author.
Title: Edgar Allan Poe, Eureka, and scientific imagination / David N. Stamos.
Description: Albany, NY : State University of New York Press, 2017. | Includes bibliographical references and index.
Identifiers: LCCN 2016031441 (print) | LCCN 2016049033 (ebook) | ISBN 9781438463919 (hardcover : alk. paper) | ISBN 9781438463926 (e-book)
Subjects: LCSH: Poe, Edgar Allan, 1809-1849—Knowledge—Science. | Poe, Edgar Allan, 1809-1849—Religion. | Poe, Edgar Allan, 1809-1849. Eureka. | Imagination. | Creative ability. | Science—Philosophy.
Classification: LCC PS2642.S3 S73 2017 (print) | LCC PS2642.S3 (ebook) | DDC 818/.309—dc23
LC record available at https://lccn.loc.gov/2016031441

10 9 8 7 6 5 4 3 2 1

To John Cusack,
and the makers of *The Raven*,
for their wonderfully imaginative *tribute* to Poe,
to the man and his powers of ratiocination,
dedicated here is this—octavo,
for evermore.

Contents

Acknowledgments

I am deeply indebted to my editor at SUNY Press, Andrew Kenyon, and the two scholars he commissioned as anonymous reviewers, to the production editor, Jenn Bennett, and the production team, to the copy editor, John Wentworth, and to the artist. A debt of thanks is also due to a select group of friends and acquaintances, (some of them former students,) either for their inspiration, or for guiding me toward certain readings, or for feedback on certain chapters or parts thereof, namely, Robert H. Haynes, Bernard Lightman, Jason Buccheri, Mauro Buccheri, Julia Tourianski, David Livingstone Smith, Graham McCreath, George Perry, Brandon Fenton, Bobbak Makooie, Andrew Robinson, Karina Chornenka, Neven Sesardić, Susan Chen, Vincenzo Domanico, Kassandra Sharma, Neda Maki, and last but not least, Matthew H. Kelley, Suzanne Underwood Rhodes, William E. Engel, and the other scholars I had the pleasure to meet and converse with at the Fourth International Edgar Allan Poe Conference held in Manhattan at the end of February 2015, where I gave my talk on Poe's "double Dupin."

Sources

For references to and quotations from Poe's writings, I usually refer to the collections listed below rather than the original sources of publication. The year following a work by Poe, for example "The Raven" (1845), indicates the year of its earliest publication or the year of composition if it was unpublished. Pages following a work refer to the following collections:

P&T	Patrick F. Quinn, ed. (1984). *Edgar Allan Poe: Poetry and Tales*. New York: Library of America.
E&R	G.R. Thompson, ed. (1984). *Edgar Allan Poe: Essays and Reviews*. New York: Library of America.
Brevities	Burton R. Pollin, ed. (1985). *The Brevities: Pinakidia, Marginalia, Fifty Suggestions, and Other Works*. New York: Gordian Press.
Letters	John Ward Ostrom, Burton R. Pollin, and Jeffrey A. Savoye, eds. (2008). *The Collected Letters of Edgar Allan Poe*. 3rd ed. New York: Gordian Press.

The following are anthologies I use with varying frequency:

Recognition	Eric W. Carlson, ed. (1966). *The Recognition of Edgar Allan Poe: Selected Criticism Since 1829*. Ann Arbor: University of Michigan Press.
Critical Heritage	I.M. Walker, ed. (1986). *Edgar Allan Poe: The Critical Heritage*. New York: Routledge & Kegan Paul. This is a useful collection consisting primarily of contemporary reviews of Poe's work and obituaries.
Companion	Kevin J. Hayes, ed. (2002). *The Cambridge Companion to Edgar Allan Poe*. Cambridge: Cambridge University Press. A collection of modern essays in the prestigious Cambridge Companion series, but partly anti-Poe.
Context	Kevin J. Hayes, ed. (2013). *Edgar Allan Poe in Context*. Cambridge: Cambridge University Press. A collection rich in background information on the many contexts in which Poe was embedded.

In terms of biographies of Poe, I make use of the following:

Weiss	Susan Archer Weiss (1907). *The Home Life of Poe*. New York: Broadway Publishing. As Susan Talley, she knew Poe during the last three months of his life and also many of those who knew Poe.
Quinn	Arthur Hobson Quinn (1941). *Edgar Allan Poe: A Critical Biography*. New York: D. Appleton-Century Company. In spite of being dated, many Poe scholars still consider this the definitive biography of Poe.
Poe Log	Dwight Thomas and David K. Jackson (1987). *The Poe Log: A Documentary Life of Edgar Allan Poe 1809–1849*. Boston: G.K. Hall & Co. This book is of inestimable value for tracing the chronology of Poe's life on a daily basis.

Silverman	Kenneth Silverman (1991). *Edgar A. Poe: Mournful and Neverending Remembrance*. New York: HarperCollins.
Hutchisson	James M. Hutchisson (2005). *Poe*. Jackson: University Press of Mississippi.
Ackroyd	Peter Ackroyd (2008). *Poe: A Life Cut Short*. London: Chatto & Windus.

Aside from biographies and critical collections, there are many useful monographs and commentaries on aspects of Poe's writings. The ones I make the most use of are the following:

Hoffman	Daniel Hoffman (1972). *Poe Poe Poe Poe Poe Poe Poe*. New York: Paragon House. This is a classic, for better and for worse.
Peeples	Scott Peeples (2004). *The Afterlife of Edgar Allan Poe*. Rochester, NY: Camden House.
Eureka	Stuart Levine and Susan F. Levine, eds. (2004). *Edgar Allan Poe: Eureka*. Champaign: University of Illinois Press. This is an annotated text of Poe's *Eureka* that I use mainly for its annotations.
Critical Theory	Stuart Levine and Susan F. Levine, eds. (2009). *Edgar Allan Poe: Critical Theory, The Major Documents*. Champaign: University of Illinois Press.
Zimmerman	Brett Zimmerman (2005). *Edgar Allan Poe: Rhetoric and Style*. Montreal & Kingston: McGill-Queen's University Press.
Fisher	Benjamin F. Fisher (2008). *The Cambridge Introduction to Edgar Allan Poe*. Cambridge: Cambridge University Press.
Evermore	Harry Lee Poe (2012). *Evermore: Edgar Allan Poe and the Mystery of the Universe*. Waco, TX: Baylor University Press. The author of this book is a professional theologian and a living relative of Edgar Poe's family. This book made a

powerful impression on me and is the catalyst of the volume before you.

McGann Jerome McGann (2014). *The Poet Edgar Allan Poe: Alien Angel.* Cambridge: Harvard University Press.

CHAPTER 1

Prologue

To the few who love me and whom I love . . .

—Edgar Allan Poe[1]

Entrée

This is a book about the nature of scientific imagination, but first and foremost it is a book about Edgar Allan Poe, an American icon, "quite possibly America's most famous literary figure."[2] It is not a book about all of Poe, of course, given the title of the book alone, but about a side of Poe, his science side, a side that has usually been treated either superficially or ignored altogether in Poe studies. Ironically, that side of Poe involves a topic that has usually been treated the very same way in professional philosophy and history of science. The said topic, central to Poe in his lonesome latter years, is none other than scientific imagination itself, the nature of which Poe, in those latter years, applied to the grandest topic of all: the meaning of the Universe, including the meaning of matter, life, terror,—and death. The result was the culmination or *dénouement* of Poe's poetry and fiction and so much more, which in turn shines light back on his works of earlier years, for which he is loved by many. Given the nature and purpose of

1. *Eureka* (Preface). *P&T*, 1259.

2. *Peeples*, ix.

the book before you, then, it is my sincerest desire that it be welcomed with an ardent spirit, to be read by the mind and grasped with the soul, especially by those who love Poe.

Today, of course, those who love Poe are not a few but countless millions, and they are not confined to the English-speaking world (Poe is big in France and Japan, for example).[3] Or rather, his writings are loved. But his writings are such that those who love them cannot help but feel drawn to the author himself, an affinity, even in some a feeling of identity. When most people think of Poe, however, including most Poe people, they usually think of morbid romantic poems such as "The Raven" (1845), perhaps of poems with graves and worms such as "The Conqueror Worm" (1843), or perhaps they think of tales of revenge and terror such as "The Cask of Amontillado" (1846) and "The Pit and the Pendulum" (1842), or of madness such as "Berenice" (1835) and "The Tell-Tale Heart" (1843), maybe even of supernatural horror such as "Ligeia" (1838). Far fewer are aware that Poe literally invented the modern detective mystery, beginning with "The Murders in the Rue Morgue" (1841), or that he helped modernize science fiction by combining minute scientific detail with social commentary and prophetic vision, for example with "The Balloon-Hoax" (1844), or that he wrote comedy and satire, as with "The Angel of the Odd" (1844). Fewer still are aware that Poe was America's first great literary theorist and critic, providing stringent theories on the body and soul of poetry and tales and over a hundred book reviews, or that he wrote essays on a wide variety of topics outside of literary studies, including genius, natural theology, and even a chess-playing automaton, or that he was a pioneer in cryptography and publicly bested all but two of over a hundred attempts designed to stump him.[4] And very few, relatively speaking, are aware that Poe wrote a book containing his own scientific speculations on the Universe, a book he entitled *Eureka*, written and published in 1848, the year before he died at the age of forty, which was an attempt to harmonize the science of his day with his theories of

3. See, e.g., Lois Davis Vines, ed. (1999). *Poe Abroad: Influence, Reputation, Affinities*. Iowa City: University of Iowa Press. Barbara Cantalupo, ed. (2012). *Poe's Pervasive Influence*. Bethlehem, PA: Lehigh University Press.

4. It is no more than a "tantalizing possibility" that Poe himself was the author of the two cryptographs. *Letters*, 320.

poetry and plot so as to provide a grand and panoramic answer to the meaning of the Universe. Still fewer know that in that book he went against much of that science and anticipated at least nine major developments and theories in twentieth-century science, including Big Bang cosmogony.

This book is about that lonesome latter side of Poe, a deeper side that relatively few have explored, let alone from the perspective of philosophy of science. Written in a clear and informative style that should appeal to the general reader, and yet with enough knowledge and theory to challenge the most erudite and stubborn of professionals, this book—*highly* informed in places by the very style and humor of Poe himself, and in more ways than one—is for lovers of Poe, yes, but also for lovers of science, *real* science, and especially for those who are curious about what happens when these two loves, of two seemingly disparate worlds, are amalgamated with a single unity of effect.

Poe was a master of imagination in poems and tales. *Everyone*, of course, knows that. In fact he was arguably second to none, at least compared with his contemporaries in America.[5] The sensational Poe, however, is but the surface Poe. Beneath the surface there is a many-layered Poe, leading ultimately to a very deep Poe, the Poe that most people do *not* know about, and it was by far the more important part of Poe *to Poe*. My thesis is that by studying the deepest part of Poe we are directed to the role of imagination in science (among much else that is related). Poe has something important to say here, and scientific imagination was something that he himself actually exercised to a remarkable degree, such that beyond the grave, haunting us through his printed words, which start once we open the creaking lids of his memetic tombs known as books, his spirit rises up from the pages like a phantasm with vast, sable, overshadowing wings and approaches us with something he desperately wants to communicate—if only we can rise from our sleep paralysis and listen closely to what he is trying

5. As recently pointed out, "one measure of the Poe renaissance is that its 2.2 million likes on his Facebook author site [it is over 3.8 million as I write this] far outstrip the combined popularity of the next ten nineteenth-century American writers." Heyward Ehrlich (2014). "Poe in Cyberspace: To Like, Friend, or Follow? Poe in Social Media." *The Edgar Allan Poe Review* 15 (1), 123.

to say. With unmoving lips he is calling us, trying to stir us from a waking dream of fleeting shadows, murmuring echoes, and unthought-like thoughts. It remains for us to rise from our beds, those lidless and sideless coffins with training wheels, to wipe away the sleep of formal education from our eyes, and to follow the clues he left behind so many years ago, years that now are out of Space—out of Time. That is the purpose of this book. There is an investigation that needs to be done and a mystery to be solved, with a much needed unity of effect.

One would naturally think that whatever the mystery it should surely have been solved by now, given the enormity of Poe scholarship and the fact that Poe died in 1849. Part of the problem, however, a large part in fact, is not only that *Eureka* is very unlike anything else in Poe's corpus and is a difficult read, but that the vast majority of Poe scholars are professors of English and of American literature, with many having accomplished much as writers of fiction and of poetry in their own right. In other words, the problem is that almost invariably they lack the necessary understanding of science and of philosophy of science, from Poe's time to the present, to see what is really going on in *Eureka*. The flipside of *that* problem, of course, is that pathetically few professional philosophers and historians of science have ever bothered to read *Eureka*, given its reputation among the literati themselves as an obscure work (and possibly a hoax) and that it was written, after all, by a mid-nineteenth-century poet and writer of horror, not a bona fide scientist or philosopher or historian of science. Having no motive to read *Eureka* in the first place, they would certainly have no motive to investigate into how *Eureka* connects with the rest of Poe's corpus—let alone into how it connects with, say, the latest research in neuroscience.

Enter yours truly and the book before you. At this point a powerful analogy will hopefully seem useful. Although the Poe community is large, intellectually rich, and is in fact a wonderful world in itself, with numerous books and articles published on Poe every year and two journals (annual and biannual, respectively) entitled *Poe Studies* and *The Edgar Allan Poe Review*, (and of course there is much more, such as the Poe Museum,) one has to recognize that sometimes an outsider can see what the professional insiders have missed, as with the team headed by Luis and Walter

Alvarez in relation to paleontologists in 1980 on the extinction of the dinosaurs. The paleontologists were for some time quite resistant to the impact theory of the Alvarezes, territorially even, but they eventually had to give in due to the evidence. The Poe community, on the other hand, is a remarkably warm and friendly community, (this is my experience,) united, on the whole, by a love for the works and person of a most singular man, such that I anticipate, or at least hope for, a hospitable reception to my interpretation of Poe as the K/T and *Eureka* as the iridium.

The audience for this book, then, is multiple. Those who love Poe for the sake of reading Poe (they are in the millions) will discover a side of Poe that they hitherto had not imagined. Moreover, I have attempted to dig deeper into Poe's thoughts on science and the scientific mind than anyone else, and also to show how those thoughts connect with much of the rest of his corpus, all the while making the book rich enough in small details about Poe's life, (where relevant,) with plenty of new facts, theories, interpretations, interpoelations, and applications,—in fact I have attempted to make this book a veritable *banquet* for those who love Poe,—that I can't imagine a single aficionado coming away feeling disappointed or unsatisfied. Professional Poe scholars should also welcome what is here, for though they may find some of their own interpretations of Poe challenged and even rejected, they will also find much to discuss that is new to them—most of all a Poe from the perspective of philosophy of science. And then we have the professional philosophers and historians of science, who if they pay attention will tend to have their feathers ruffled, too,—this is a Poe book, after all,—for they have routinely pushed imagination outside the boundary of the scientific process, as not a part of it, happy to leave the realm of imagination to artists and others. How could it be that this wordsmith named Edgar, a *Poet*, has anything to teach *them*? But even they, too, should find in these pages something new and exciting, in fact a kind of awakening.

Given that two worlds are brought together here, and vast worlds at that, this necessarily lengthy book requires a certain amount of patience on the part of the reader. In aid of that, I have endeavored to write its pages smoothly, eclectically, analytically, synthetically, even punctually, and of course with some originality. In other words, in keeping with the spirit of Poe, this book was written to produce a degree of pleasure in the reader—which ideally is the state of mind one should have every time one opens its covers, not the sour-faced "but but but" of the critic. This is a *book*; it is sequential; criticisms will hopefully find their answers somewhere in what follows; and ultimately the book needs to be viewed as a whole. My desire is that this book be read first and foremost for pleasure, that this observance will be ongoing to the book's end, and that the experience will be elevated by the book's interactive nature. A balanced application of one's critical faculties is also required, of course, but if the latter becomes predominant, as it will in some, then the overall unity of effect intended for the reader will be utterly and irretrievably—lost.

At this point track 5 of *Tales of Mystery and Imagination* by The Alan Parsons Project, with its mesmerizing riff and closing *punctus contra punctum* fit for a *Maison de Santé*, (the entire album is a magnificent musical tribute to Poe,) should be played in preparation for the following look through.

Chapter Summaries

To jump right into Poe's *Eureka* would be an enormous mistake. It requires careful preparation both to understand and to appreciate the nature of Poe's magnum opus, for not only is the content not easy but the style is quite strange and unique, much like its author. Hence the necessity of the three background chapters.

In Chapter 2 we look at key features of Poe's literary theory, as they developed chronologically, beginning with "Letter to Mr. — —" (1831). We then turn to some of Poe's book reviews, followed by his "A Chapter of Suggestions" (1845), "The Philosophy of Composition" (1846), "The Rationale of Verse" (1848), and finally "The Poetic Principle" (1850). What we shall see is an evolution of insights, particularly in response to Aristotle's *Poetics*, especially Aristotle's theory of a good metaphor, which Aristotle claims is "a sign of genius, since a good metaphor implies an intuitive perception of the similarity in dissimilars." Poe's literary theory is interesting in itself, but the main purpose of Chapter 2 is to better understand how his concepts of plot, poem, unity of effect, and metaphor inform *Eureka*. Along the way we uncover

Poe's distinction between a "tale" and a "story," we at long last unmask the purpose behind Poe's *modus operandi* hoax in "The Philosophy of Composition," which is about how he composed "The Raven" (1845), we hopefully unriddle the ultimate meaning of that poem, (which together with the previous prepares the way for Chapter 3,) and we attempt to unravel the "Prose Poem" paradox posed by Poe's subtitle of *Eureka*.

In Chapter 3 we turn to Poe's theology, which includes his response to the problem of evil and the meaning of life. We examine his view on artistic sensitivity, his view of the world, and his argument for God's existence. The key to understanding the nature of the latter is aesthetic, involving Poe's view on the relation between beauty, discord, and pleasure. Given that the ultimate discord for Poe is death, we introduce the problem of death by way of a mosaic, followed by an examination of Poe's poems "The Conqueror Worm" (1843) and "A Dream within a Dream" (1849). It will be argued that Poe's argument from beauty for God's existence is really an argument from hope, given Poe's view on the relation between beauty and hope, which puts his argument for God's existence in an entirely different category compared with what is normally found in theology and philosophy of religion. In all of this, an examination of Poe's theology proves necessary in order to understand the contrapuntal nature of *Eureka*, which is a harmonic synthesis of his literary theory, his theology, his philosophy of science, and his scientific speculations. Focusing on Poe's theology might seem strange given the central focus of this book,—Poe and the nature of scientific imagination,—but it must be remembered that there are many scientists today (not the majority, since the majority are atheistic or agnostic) whose theological view of the Universe is informed, at least allegedly, by developments in modern science, such as the apparent fine-tuning for life of the cosmological constants. At any rate, one cannot possibly come to understand *Eureka* without understanding Poe's theology.

In Chapter 4 we examine Poe's intellectual background chronologically, for one cannot hope to properly understand *Eureka* until it is viewed as the culmination of Poe's thinking on a lot of matters, including science, which he took very seriously. Specifically, we examine the key features of Poe's formal education, three of his entries in his "Pinakidia" (1836), his controversial authorship of *The Conchologist's First Book* (1839), his series of articles in *Burton's Gentleman's Magazine* entitled "A Chapter on Science and Art" (1840), his criticism of the Bridgewater Treatises in his "Marginalia" (1844), the relation of *Eureka* to Chambers' *Vestiges of the Natural History of Creation* (1844), and the main scientific sources informing *Eureka*. We then finish the chapter by examining Poe's criteria of truth in *Eureka* and whether *Eureka* was another of his clever hoaxes, (an accusation in Poe's time that continues to the present,) all of which needs to be dealt with before one turns to *Eureka* proper.

These three background chapters are interesting in themselves, but they are designed to serve a higher purpose, which in this book is to understand what exactly *Eureka* is about in terms of *process* (Poe's scientific imagination) and *product* (Poe's scientific speculations). Without a strong background in Poe's literary theory, his theology and need to solve the meaning of life, and his understanding of the processes and products of science, one cannot possibly hope to understand let alone appreciate what really goes on in *Eureka*. To simply read *Eureka* without that collective background is a mistake made by almost everyone who has ever read it, which I suspect is why *Eureka* is among the most ignored parts of Poe's corpus, even though he himself regarded it as his magnum opus.[6]

The fundamental point of appreciation is that Poe in *Eureka* anticipated at least nine major theories and developments in twentieth-century science, namely, the rejection of axioms as intuitively true, Big Bang cosmogony, (including the concepts of a primordial atom and an oscillating Universe,) the apparent fine-tuning of the fundamental laws of nature, the nonexistence of laws of nature before the Big Bang, the correct solution to Olbers' paradox, multiverse theory, space–time interdependence,

6. Another mistake is to focus on the mistakes made by Poe in *Eureka*. See, e.g., Peter Swirski (2000). *Between Literature and Science: Poe, Lem, and Explorations in Aesthetics, Cognitive Science, and Literary Knowledge*. Montreal & Kingston: McGill-Queen's University Press. That approach only serves to blind oneself to the bigger picture, which is elaborated in the chapters that follow in the book before you.

matter–energy equivalence, and the rejection of the existence of the material ether. This is the focus of Chapter 5, and it is the core of this book. For each of the topics just listed, I begin with what modern scientists say, using their own language and explaining their ideas, and then in a similar manner I compare that with what Poe says in *Eureka*. This makes for nine striking comparisons, of which there may be more, but certainly these nine will do for our purposes. In all of this every effort is made to avoid the sin that historians call *presentism*, which is reading present ideas into the past without sufficient evidence.

Chapter 5 should awaken us to the realization that Poe had a faculty of imagination far greater and far more diverse than almost all of his admirers ever imagined, and also that the nature of scientific imagination is a topic that calls for serious examination. Accordingly, in Chapter 6 we turn to the field where one would naturally expect to find the answers, namely, philosophy of science, the professional field devoted to the examination of the nature of science. We begin with a brief look at the philosophies of science that Poe was acquainted with, specifically those of John Herschel, John Stuart Mill, and (indirectly) William Whewell. We then turn to the modern scene, specifically logical positivism, logical empiricism, the falsificationism of Karl Popper, the paradigmism of Thomas Kuhn, the new experimentalism, the disunity of science movement, inference to the best explanation, the epistemic virtues and values approach, evolutionary epistemology, and finally contextualist history of science, since the latter also contributes to the field.

What we shall find is that philosophy of science, considered collectively, has pushed the topic of scientific imagination to outside the circumference of the object of its study, as not belonging to the nature of science at all. This is the exact opposite of what one finds in *Eureka*, given that Poe emphasized scientific imagination as *internally* driving science. When the thought first hit me I was absolutely stunned, especially given that the norm today in philosophy and history of science is to conceive of science as a process, not as a product. In fact when the thought hit me, all at once, there came a most deadly nausea over my spirit, and I felt every fibre in my frame thrill as if I had touched the wire of a galvanic battery. At any rate, or volt, the purpose of Chapter 6 is not merely to put Poe's views into an historical and intellectual context by means of comparison and contrast. Instead, and far more important, what we shall see is that Poe, the *artist*, actually had a *philosophy* of science, even a properly *corrective* one. In support of this conclusion, we then turn to two giants of science, namely, Charles Darwin and Albert Einstein, both for their use of imagination in their science and for their expressed views on imagination in science. The chapter ends with an argument by analogy based on the evolutionist Ernst Mayr's view of natural selection combined with my own published work on the foundation of point mutations. The conclusion is that just as mutations and their genesis should be viewed as part of the process of natural selection, (they are routinely viewed as separate from, or preliminary to, that process,) so too should the genesis of theories be viewed as part of the process of science. All in all, the chapter serves as a companion vindication of Poe, further to the scientific anticipations of *Eureka* examined in Chapter 5.

Having established the importance of pursuing an investigation into the nature of scientific imagination, in Chapter 7 we turn to Poe's own theory of scientific imagination. Since Poe wrote no single essay on the topic, we attempt to glean his theory from his various writings. We begin with a close look at what in his time was called "double consciousness," which affords us the opportunity to provide an original interpretation of Poe's poem "Ulalume" (1847), as well as an original theory for why he repeatedly cried "Reynolds" the night he died. We then turn to what was called "mesmeric consciousness," which included the possibility of a connection with the divine. From there we take a detour through a madhouse, looking for possible roles played by madness in Poe's theory of scientific imagination. We then turn to three of Poe's favorite examples of scientific imagination, notably the case of Kepler, followed by three of Poe's examples of failed scientific imagination, notably the case of Newton. These examples afford us the perfect segue into a detailed examination of the nature of Poe's fictional ace detective, Auguste Dupin, including the meaning of his narrator's "double Dupin," which when unravelled is none other than Poe's personification of the scientific mind, complete with a fully developed scientific imagination, (which is the very antithesis of Doyle's Sherlock Holmes). Followed by a close look at what Poe says against deduction and induction, we finish with an examination of what Poe calls "the poetic intellect."

In the final chapter, the Epilogue, (or rather Epiclogue,) we attempt to modernize what Poe had in mind all the while keeping it down to earth. Specifically, we begin with an examination of numerous eureka moments in modern science, examples of "unconscious scientific creativity," drawn from the lives of major scientists such as Darwin, Einstein, and Crick, the "intuitive *leaps*" that are the stuff of what Poe claimed is the main driving force of science. We then search for light on the phenomenon by turning to comparative biography and psychology, including the "ten-year rule." Following a short detour through cognitive science, the interdisciplinary approach to the mind/brain centered on the digital computer as its guiding metaphor, with thinking taken literally as information processing, we then turn to neuroscience, the study of the brain. There we focus on evidence from split-brain patients, including brain hemispheres and dreams, the problem of "metaphor blindness," the neuroscience of metaphorical thinking and of memory, of callosal inhibition, and much more, including an interesting twist on Poe's "double Dupin" as well as his "dream within a dream." We then turn to evolutionary biology, the ultimate explanatory domain in biology, in which we focus on the why of the double brain, on a possible adaptive role for depression in scientific creativity, and on the creativity of terror, with a perfect illustration of "simultaneous, parallel processing" provided by one of Poe's tales. We then finish with an homage to Poe in the form of a thirteen-paragraph mosaic in the first person, the power of Poe condensed, as if he were speaking to us in the here and now. Accordingly, for the full effect, it should be read *viva voce*.

There is still much more that needs to be said in this Prologue, however, before we turn to the chapters proper, which is where we bring Poe *back to life* in accordance with his will.

Discovering Poe

At this point I should like to say something about how I, a mere moth of a philosopher, came to conceive and write the present book. This is a matter of considerable self-indulgence, to be sure, and I hope it will be excused, as I suspect many will have their own interesting tales of how and why they got into Poe.[7]

In what is perhaps typical, I first fell in love with the writings of Poe when I was a child. In looking back, it seems to me that I was primed for it. My childhood, often a happy one, largely due to friends and pets, was too often overcast with the ominous dark clouds known as the fear of dying, either by my father who was possessed by Fiend Intemperance, in particular that demon known as Alcohol, or by my own possession, the demon known as Asthma, with an Isuprel inhaler feebly raised against it as a cross. Many were the nights I lived in terror, fearful I was going to breathe my last. During this time I remember, vividly and palpably to this day, browsing a bookstore in a shopping mall at the age of ten, and seeing on the display desk a singular-looking book, a book with a black bird on its cover. I picked it up, looked at the table of contents, and decided to buy it. (I also bought some storybooks based on TV shows.) At home I read Poe and was in, plain and simple. Like many, I would read Poe, let time go by, read Poe again, let time go by, read Poe again—again—again and even once again at varying intervals. Once you love Poe, you always love Poe.[8]

7. This is a wonderful topic for a book, by the way, a collection of remembrances by Poe lovers, which occurred to me while listening to the Keynote Address given by J. Gerald Kennedy, "Why Poe Matters Now," in which he gave his own remembrance. Further examples can be found in the delightful book by J.W. Ocker (2015). *Poe-Land: The Hallowed Haunts of Edgar Allan Poe*. Woodstock, VT: Countryman Press, 114, 163, 244, 317. This book is an absolute gem for anyone who loves Poe.

8. T.S. Eliot, allowing that "Poe had a powerful intellect," provided a theory for "why the work of Poe has for many readers appealed . . . at the period of life when they were just emerging from childhood." In short, "The forms which his [Poe's] lively curiosity takes are those in which a pre-adolescent mentality delights: wonders of nature and of mechanics and of the supernatural, cryptograms and ciphers, puzzles and labyrinths, mechanical chess-players and wild flights of speculation. . . . in the end the eccentricity and lack of coherence of his interests tire. There is just that lacking which gives dignity to the mature man: a consistent view of life . . . which comes only with the maturing of the man as a whole." Thomas Stearns Eliot (1948). "From Poe to Valéry." In *Recognition*, 212–213. Eliot's bloom has been heralded by others. As for Poe's "wild flights of speculation" and "lack of coherence," the present book serves as a reply. And as for maturity, I prefer Schweitzer to Eliot: "The

When I began university, I acquired another love, no less deep than the first. Even though I initially went to university to get into business school, which was to begin in my third year, in my first year I took a course on ancient Greek and Roman literature (because the electives I wanted were filled). The course included some philosophy readings. Perhaps because of my past already mentioned, and the dominance from my childhood's hour of an old-time religion as well, I was spellbound by Plato's *Apology*, the first work of philosophy I had ever read. Following that course I took as many philosophy electives as I could, and upon getting into business school I shortly thereafter dropped out, lacking the positive appetite for system and regularity, and the ordinary habitudes of my fellow men. I then returned to university as a philosophy major, come what may (an attitude one needs as a philosophy major).

The rest is history, as the saying goes, with much of Madness, and more of Sin, and Horror the soul of the plot. In short, I eventually went back to university for a Ph.D. in philosophy, while already working on a Ph.D. in people at a university called O'Toole's Roadhouse, where I worked with a hockey-player-turned-movie-actor-turned-bouncer by the name of George Finn. The love of philosophy renewed, my interests quickly turned to philosophy of biology, specifically the species problem, determining what a biological species is, which became the topic of my dissertation and of my first book.[9] My attention then turned to Darwin himself, my second book,[10] and to related

epithet 'mature,' when applied to people, has always struck me as somewhat uncomplimentary. It carries overtones of spiritual impoverishment, stunting, blunting of sensibilities. What we usually call maturity in a person is a form of resigned reasonableness. A man acquires it by modeling himself on others and bit by bit abandoning the ideas and convictions that were precious to him in his youth. . . . those who have a deeper experience of life take another tone. They exhort youth to try to preserve throughout their lives the ideas that inspire them." Albert Schweitzer (1965). *The Teaching of Reverence for Life*. Richard and Clara Winston, eds. New York: Holt, Rinehart and Winston, 43–44. In Chapter 6 we shall find further support from another Albert.

9. (2003). *The Species Problem: Biological Species, Ontology, and the Metaphysics of Biology*. Lanham, MD: Lexington Books.

10. (2007). *Darwin and the Nature of Species*. Albany: State University of New York Press. For a condensed version of the basic argument and an

topics of profound interest, most notably the foundation of point mutations in quantum chance,[11] a kind of homage to my biology mentor, a pioneering geneticist, the late Robert H. Haynes. My interests then widened to the implications of evolutionary biology for topics usually thought of as outside the circumference of evolution proper, my third book,[12] which turned my focus to the implications of evolutionary history, both biological and cultural, for the topic of human rights, my fourth book.[13]

Then a curious turn of events occurred, a happy coincidence, what feels now like a synchronicity. The movie *The Raven* (2012) came out in the theaters, starring John Cusack, and I was not disappointed by this wonderful *tribute* to Poe—misunderstood as such by the film critics, who apparently never read much of anything on or by Poe, and who at any rate almost certainly missed the many clever allusions to his life and writings. I watched the movie twice in the theater and immediately thereafter dove, once again, into the world of Poe, the world I had loved since age ten. But this time I wanted to see what was out there in terms of the latest scholarship on Poe. Fortunately, Harry Poe's *Evermore* was fresh off the press, a remarkably well-written and insightful book (except for its *dénouement*, the unwarranted speculation that Poe had a religious revelation a few months before he died) which showed me depths to Poe's thinking that hitherto I had not imagined, especially concerning Poe's magnum opus, *Eureka*. Whatever criticisms I express of various points in Harry Poe's

---

update, see my (2013). "Darwin's Species Concept Revisited." In Igor Ya. Pavlinov, ed. (2013). *The Species Problem: Ongoing Issues*. Rijeka, HR: InTech, 251–280.

11. (2001). "Quantum Indeterminism and Evolutionary Biology." *Philosophy of Science* 68, 164–184. See also my (2010). "Quantum Indeterminism, Mutation, Natural Selection, and the Meaning of Life." In Chérif F. Matta, ed. (2010). *Quantum Biochemistry: Electronic Structure and Biological Activity*. Volume 2. Weinheim, Germany: Wiley-VCH, 837–872.

12. (2008). *Evolution and the Big Questions: Sex, Race, Religion, and Other Matters*. Malden, MA: Blackwell.

13. (2013). *The Myth of Universal Human Rights: Its Origin, History, and Explanation, Along with a More Humane Way*. Denver, CO: Paradigm Publishers.

book, they should not be taken to diminish my high regard for his contribution to Poe studies.

Harry Poe's interpretive approach to Edgar Poe is primarily theological (Harry Poe is a professor of theology, and an actual descendant of one of Edgar Poe's cousins). As a philosopher of science I wanted to bring out a nontheological interpretation of Poe on imagination, since modern science itself is essentially nontheological. Stripped of the theological elements (whatever they are) in Poe's writings, it seemed to me that Poe's insights on the connection between science and imagination are genuine insights and accordingly are worth the effort at bringing to light and submitting to careful examination.

I began thinking of writing a magazine article on Poe, science, and imagination, comparing Poe with the use of imagination by Darwin and Einstein. A little later that idea turned to thoughts of writing a scholarly article. And then later still, literally while driving home from shopping, a couple of hours after I had finished two grueling weeks of checking over the typesetting for my book on human rights and making an index,—I was almost half blind from the process,—the thought suddenly struck me, that with Poe on scientific imagination I have my next book. The division of the book into chapters began forming in my mind, along with what I would do in each chapter. When I got home I wrote it all down, for thoughts are often delicate things and easily lost. It was a remarkable experience, in itself a eureka moment, an act of unforced imagination that filled me with excitement, as if watching a griffin self-assemble right before my very eyes.

In time, during the actual writing of the manuscript, the griffin grew, of course, but beyond my original intention. And it kept growing, to my utter astonishment, from something I intended to be fairly small to something ponderously large. This was not just from the primary source materials of Poe, which are considerable, but moreso from the vast secondary sources that I increasingly found to be relevant, sources that I used as either springboards or foils.

All throughout, be this what it may, I have tried my best *not* to draw loose and weak connections in Poe's thinking, as one sometimes finds in literary criticism (and elsewhere). What some scholars do is draw connections which, albeit fascinating and highly original, have little if any rigor in terms of logic and evidence. This is fine in itself as a matter of self-expression and creativity, but if it is truth outside of ourselves that we are after, including historical truth, then we have got to ground our theories in epistemic virtues and values, given that theories are always underdetermined by the available evidence. For example, one would like to ask some literary critics what they would accept as falsifying a given claim or theory of theirs, a particular connection they have drawn allegedly from the evidence. Silence there and nothing more. In some cases it is the mere drawing of the connection that seems the important matter, like making a constellation and attaching a story to it, rather than asking whether the constellation is real or the story is true. At bottom is the double problem of standards of evidence and underlying motives, the latter basically in some cases the desire to force an ideology on others or more simply to produce a shock or sensation.

Before we begin with the rest of the present book, however, there are two interpretative approaches to *Eureka* that I want to examine as a preliminary. Their focus is on Poe's creativity, but not in the way that I approach Poe in this book. Let us therefore take a brief look at these two approaches before we proceed any further.

Poe's House of Usher

The interpretive focus here is Poe's "The Fall of the House of Usher" (1839), widely regarded as one of Poe's best tales, including by Poe himself.[14] Providing various interpretations of "Usher" from the pens (keyboards, really) of various Poe scholars, Scott Peeples saves the interpretation he clearly prefers for the end of his chapter on "Usher." As he puts it in the *dénouement* of his chapter,

> Reading *Eureka,* in fact, one comes to see the work not so much as a theory of the known universe but as a parallel universe created by Poe; . . . just as *Eureka* is a kind of

14. *Letters*, 450; *E&R*, 870.

> control fantasy in which Poe identifies himself with the creator of the universe, in "Usher" he identifies himself with an artist who has made his house a universe, and then he enacts the artist's fantasy of bringing that dead house to life: . . . Paradoxically, the house comes to life only to collapse and die, but for Poe, the paradox works both ways: the fall of the house gives rise to the story, which "lives" off paradox and other uncanny verbal structures.[15]

There are several problems with this view, one of which is that whether Poe intended *Eureka* as a bona fide theory of the Universe is not going to be determined by examining "Usher." For that we need to get into Poe's biography and into *Eureka* itself, which shall occupy us in Chapters 4 and 5.

Another problem is terminological, which confronts anyone who wants to draw parallels between "Usher" and *Eureka*. One of the key terms in "Usher" is "fissure," which is the undoing of the house.[16] "Fissure" in Poe's day was a well-known physiological and medical term, (as it still is today,) applying to breaches in the skin, and also to the folds of the brain. Corroboration for this interpretation of Poe's use of "fissure" in "Usher" is found in one of his letters to Griswold, in which he affirms that the poem "embodied" in "Usher," "The Haunted Palace," (which was published independently earlier the same year,) was meant "to imply a mind haunted by phantoms—a disordered brain."[17] Moreover, in "Usher" Poe has the narrator state, just before he recalls one of Roderick Usher's "rhapsodies," which is "The Haunted Palace," that "I fancy that I perceived, and for the first time, a full consciousness on the part of Usher, of the tottering of his lofty reason upon her throne."[18] The brain, of course, may be taken to be the "throne" of the mind. Why, then, would the tale in which the poem is "embodied" be about something different than the poem itself?

---

15. Scott Peeples (2002). "Poe's 'Constructiveness' and 'The Fall of the House of Usher.'" In *Companion*, 188.

16. *P&T*, 320, 335.

17. *Letters*, 272.

18. *P&T*, 325.

A further terminological clue is the name "Usher," which quite apparently was taken by Poe from the names of Harriet Ann L'Estrange Snowden Usher and Noble Luke Usher, an older married couple who were friends of and fellow actors with Poe's biological parents, Eliza Arnold Hopkins Poe and David Poe, Jr. While Eliza Poe was sick and dying, Harriet, calling Eliza "Betty," fed Eliza's children "bread soaked with gin," and after the death of Eliza she sometimes added laudanum, to make them "strong and healthy" or to put them to sleep when needed. Her own two children, the last of their line, were orphaned in 1814 and eventually became neurotics.[19] Poe's choice of their surname for his tale was almost certainly, then, neither an accident nor merely for self-amusement (although Poe's writings have plenty of the latter, including the *Mad Trist* in "Usher"). The most likely conclusion, therefore, is the most obvious one and nothing more, that the tale is about the general truth that the state of a person's abode is a reflection of the state of the person's mind. Poe took this general truth and did something artful with it, making the tale about the collapse of a fissured house upon its owner, whose mind was suffering a parallel collapse due to a diseased brain, the one house the reflection of the other—an interesting twist on the cliché that there's no place like home.

The final problem with Peeples' view is chronological. This is because the majority of the main themes found in *Eureka* are also found identical, or nearly so, in Poe's works from around 1844–1845, not around 1839. Specifically, in "The Purloined Letter" (1844) we find, against the smug belief in "self-evident" truth, that "Mathematical axioms are *not* axioms of general truth."[20] In "Mesmeric Revelation" (1844) we find that "there are *gradations* of matter" and that the extreme of "rarity or fineness" is "God," the "ultimate, or unparticled matter," such that God "not only permeates all things but impels all things—and thus *is* all things within itself,"[21] that each of us is God "individualized" by particled matter,[22] that thought, whether we call it God's or ours,

19. *Poe Log*, 14.

20. *P&T*, 681, 692.

21. *P&T*, 720.

22. *P&T*, 723.

is unparticled "matter in motion,"[23] and that pleasure cannot exist without pain.[24] Shortly after in a July 1844 letter to James Lowell we find many of the same ideas, that "The unparticled matter, permeating & impelling, all things, is God. Its activity is the thought of God—which creates. Man, and other thinking beings, are individualizations of the unparticled matter. Man exists as a 'person,' by being clothed with matter (the particled matter) which individualizes him."[25] In "Marginalia" (November 1844) we find that "The Universe is a Plot of God."[26] In "The Power of Words" (1845) we find that "In the beginning *only*, he created," and that "certain operations of what we term Nature, or the natural laws, will, under certain conditions, give rise to that which has all the *appearance* of creation."[27] And in "A Chapter of Suggestions" (1845) we find that "Some of the most profound knowledge—perhaps all *very* profound knowledge—has originated from a highly stimulated imagination. Great intellects *guess* well. The laws of Kepler were, professedly, *guesses*."[28]

All things considered, that "The Fall of the House of Usher" presages *Eureka* fails the test of evidence. The most plausible conclusion is that the house represents the mind and nothing more, whether Poe meant his own mind, overly sensitive minds, the Enlightenment mind, or the human mind in general. What *Eureka*, on the other hand, represents for Poe, is something quite different, as I shall argue in subsequent chapters.

Poe's Poiesis

A more recent interpretation of *Eureka* shares much in common with the interpretation examined above. Much like Peeples, whom he explicitly cites, Jerome McGann, also a Professor of English, sees *Eureka* as "less 'a theory of the known universe [than] a

---

23. *P&T*, 720, 723.
24. *P&T*, 726.
25. *Letters*, 449.
26. *E&R*, 1316.
27. *P&T*, 823.
28. *E&R*, 1293.

parallel universe created by Poe.'" But unlike Peeples, McGann arrives at this conclusion by applying to *Eureka* certain features of Poe's literary theory, especially his poetics (theory of poetry). One feature is the hoax, which he says "is one of the Supreme Fictions of poetic artifice . . . being at once sincere and devious, serious and ironic," such that not only is it a deliberate attempt "to undermine the verification process," but in the context of tales it can be "an especially effective device for putting readers to a test of attention." (On the latter I couldn't agree more.) Hence the imaginary epistle of *Eureka*, dated 2848, is "Poe's farewell salute to the rhetoric of hoax that he had often found so useful for his work." But in *Eureka* this device is especially important because *Eureka* was not meant by Poe as an objective, scientific treatise. It was meant, instead, to be "performative rather than expository," involving the reader as participant, such that "Poe's work assumes that a prosaic exposition will distort poetic truth, tempting readers to make a conceptual rather than an experiential engagement with the issues." One of these issues is Beauty. Hence not only does Poe in the Preface offer *Eureka* as "this Book of Truths, not in its character of Truth-Teller, but for the Beauty that abounds in its Truth; constituting it true," but in his cosmology he has the "Supernal Beauty" of the Universe collapse at the end into what he calls "Nothingness," in line with the death of Lenore in his poem "The Raven" (1845), the point being, as McGann puts it, that the end of *Eureka* is "the moment when a reader understands what Poe is saying: that a thing of beauty—*this* thing of beauty—is not and never can be 'a joy forever.' That is the ultimate meaning of Poe's mortally immortal word 'Nevermore' as well as the sign of the word's pitiless benevolence," that in line with Poe's morbid poetry there is no "compensation for the loss of loved and cherished things . . . beyond a 'mournful and never-ending remembrance.'" Hence *Eureka* is "The work that most fully illustrates and, in illuminating, *explains* Poe's poetics."[29]

But what about the *science* of *Eureka*? Lord Byron, the most notorious and eccentric of the leading figures in the Romantic movement, whose life and poetry was an early influence on Poe, based the flights of his imagination "on the circumstantial facts

29. *McGann*, 217n21, 93, 8, 98–99, 95, 94, 95.

of a lived history," while "Poe's flight of imaginative theory takes off from conceptual facts and scientific information," from "what was called in the nineteenth-century Natural Philosophy." Poe's "flight of imaginative theory," however, is only the "upper current," not the "under current," of *Eureka*. The under current is "not a scientific theory; it is a poetic theory." Poe, we are told, "is not interested in making a scientific claim or building a fact-based scientific model that could be tested as a device for making astronomical discoveries." Instead, given the Romantics, "This entire tradition is important to recall when reading *Eureka*, lest we approach Poe's work as philosophy or, worse yet, as science. It is fundamentally, as Poe said, a 'Prose Poem.'" And as a prose poem the truth of *Eureka* is ultimately "not the truth of science or philosophy or religion, it is the truth of *poiesis*. It is the power to create not *ex nihilo*, but by consistent extravagance, *ex cathedra*." *Poiesis* is derived from the ancient Greek word for "a fabrication or production," and is the root of our word *poetry*. The most that McGann will allow Poe's *Eureka*, "what may be its most startling forecast," is that with Poe's theory of the "intuitive *leap*" of imagination, in criticism of what in *Eureka* he calls the "creeping" and "crawling" methods of "deduction and induction," Poe moves "toward the method of scientific discovery Charles Sanders Peirce would soon call and explain as abduction"—but even so, McGann adds, "poetic abduction unfolds in a discourse of imitation, not a discourse of information."[30]

As Harry Poe aptly puts it, "the history of criticism of Poe is the history of individuals who have imposed their agendas on the body of Poe's work."[31] Denigratory interpretations of Poe and *Eureka* abundantly illustrate this point. And although the interpretive approaches of Peeples and McGann are not exactly uncharitable agendas, they do share a common problem, namely, the problem of unacceptable standards of evidence. What is wanted when dealing with Poe, or the thinking of any important writer for that matter, (aside from having some fun,) is the making of a *case*, (or at least give us some keys,) much like the cases made by top scientists such as Charles Darwin and top lawyers such as

30. *McGann*, 97, 98, 106, 101, 102, 107, 108.

31. *Evermore*, 16.

Vincent Bugliosi—an actual case using the principle of *beyond a reasonable doubt.* For this purpose McGann's critique serves as the perfect segue to the rest of this book. What I shall attempt to show is that Poe actually had a remarkably real (not imitative) scientific imagination, (Chapter 5), that he had an equally remarkable philosophy of science, even a properly corrective one, but it was not abduction, (Chapter 6), and that he had a theory of scientific imagination, (Chapter 7), one that deserves a modern update, (Chapter 8). But before all of that, I shall also attempt to show, as the topic requires, that Poe's literary theory informs *Eureka*, (Chapter 2), but also his theology, which for whatever his morbidity should not be viewed as a theology of despair but as one of hope, (Chapter 3,) and that his intellectual and personal development, along with his biography more generally, needs to be examined in order to fully appreciate *Eureka*, including his background study of science and his position on scientific hoaxes, (Chapter 4). That is what I shall attempt to accomplish in the volume before you. And let me further add, as Poe's ace detective the Chevalier C. Auguste Dupin puts it, "My ultimate object is only the truth."[32]

32. "The Murders in the Rue Morgue" (1841). *P&T*, 420.

CHAPTER 2

Poe's Literary Theory

My first object (as usual) was originality.

—Edgar Allan Poe[1]

Entrée

No one can possibly come to understand, let alone to appreciate, Poe's *Eureka* without knowing Poe's background, not just the biography of his life but the biography of his thought leading up to his magnum opus. That is the purpose of this and the following two chapters. In this chapter we shall examine Poe's literary theory, not completely, but sufficiently for our purposes, and chronologically, leading up to the time of the publication of *Eureka* and a little beyond. This is because Poe's literary theory is not separate from *Eureka*, but informs it from beginning to end.

The Problem

The subtitle that Poe gave to *Eureka* is "A Prose Poem." To this we have to add what he says in the Preface, which must be quoted in full:

1."The Philosophy of Composition" (1846). *E&R*, 20.

> To the few who love me and whom I love—to those who feel rather than to those who think—to the dreamers and those who put faith in dreams as in the only realities—I offer this Book of Truths, not in its character of Truth-Teller, but for the Beauty that abounds in its Truth; constituting it true. To these I present the composition as an Art-Product alone:—let us say as a Romance; or, if I be not urging too lofty a claim, as a Poem.
>
> *What I here propound is true:*—therefore it cannot die:—or if by any means it be now trodden down so that it die, it will "rise again to Life Everlasting."
>
> Nevertheless it is as a Poem only that I wish this work to be judged after I am dead.[2]

All of this seems incredible. Was Poe mad when he wrote this? Or was he sober and sane? And if the latter, what then did he mean? Certainly in the annals of Prefaces, Poe's is unique, totally *alone*. At any rate, I don't think anyone will ever fully unravel what Poe had in mind when he wrote this line by line. On the other hand, it should not be ignored. What Poe says about feeling rather than thinking and about dreams shall be revisited in Chapter 7, for it does seem to make some sense when we look at what he has to say elsewhere about imagination in general and scientific imagination in particular, against pure deduction and pure induction in science. And what he says about truth and beauty also makes some sense, which we shall see in the following chapter, for much of his argument in *Eureka* is an argument from beauty, and in this he in fact is not alone among our own intellectual contemporaries, including philosophers and scientists.

What we shall examine in this chapter is Poe's desire that *Eureka* be judged as a poem. If we hope to understand *Eureka* we need to know what he meant. The problem is that it's a bona fide mystery, because although obviously a work of prose, *Eureka* clearly violates Poe's well-known strictures on poetry. To see this, let us take a tour through Poe's literary theory, considered

2. *P&T*, 1259. It should be noticed that two key terms in the above were used by Poe as early as January 1842, in "Exordium to Critical Notices," a short essay by Poe on literary criticism published in *Graham's Magazine*. The two key terms are "prose-poem" and "*art-product*." *E&R*, 1030, 1031, 1032.

chronologically, a tour that is narrowly focused with the aim of not being complete but sufficient for our purposes, so that we shall be in a much better position to figure out what Poe meant by subtitling *Eureka* "A Prose Poem," and consequently to better understand *Eureka*.

Pleasure, Plot, and Unity of Effect

Near the end of his "Letter to Mr. ——" (1831), later republished as "Letter to B——" (1836),[3] Poe states that "A poem, in my opinion, is opposed to a work of science by having, for its *immediate* object, pleasure, not truth; to romance, by having for its object an *indefinite* instead of a *definite* pleasure, being a poem only so far as this object is attained." So a didactic poem (a teaching poem) is an oxymoron. But what is the nature of the pleasure? Is it ideational or sensual? Apparently it is a combination of both, for "the comprehension of sweet sound is our most indefinite conception. Music, when combined with a pleasurable idea, is poetry; music without the idea is simply music; the idea without the music is prose from its very definitiveness."[4]

It matters little whether Poe "filched" his theory of poetry here from Samuel Taylor Coleridge, as some are quick to point out.[5] What matters for our purposes is that what Poe states constitutes a core thesis in his theory of poetry, subject to little variation onward. And it is important to see why.

For a start, Poe favorably cites Aristotle, who in his *Poetics*, says Poe, "declared poetry the most philosophical of all writing."[6] Poe adds the Greek translation in a footnote, which has led some to believe that Poe's quotation was "taken out of context," since

3. *E&R*, 5–12.

4. *P&T*, 17.

5. *Silverman*, 70. For a nice counterpoise, see *Quinn*, 177. In addition to expressing a lot of praise for Coleridge, Poe only took Coleridge's line about the "*immediate* object" of poetry, as "opposed to works of science," being "pleasure, not truth." Moreover, "That he adopted the definition of Coleridge in part," says Quinn, "is in accord with his later practice to start with an idea from an earlier writer and develop it in his own way."

6. *P&T*, 12.

in *Poetics* (§9) "Aristotle writes that poetry is more philosophical than history, for poetry tends to express the universal, history the particular," as Thompson puts it.[7]

But I suggest this misses the point. Aristotle's many "books" as we have them were not so much written by Aristotle as they were painstakingly compiled (with some slight editing) non-chronologically from his manuscripts and lecture notes and also from lecture notes of his students, almost three centuries after his death by Andronicus of Rhodes, a Greek philosopher living in Rome in the middle of the first century B.C.[8] *Poetics* is no exception. Hence Aristotle's books are not as systematic as one might wish—for one thing, they contain repetitions or "doublets," the same point repeated in different ways and sometimes in different places, with a fair share of inconsistencies and contradictions. Consequently, specialists on Aristotle spend a great deal of time and effort trying to reconstruct Aristotle's claims, concepts, and arguments.

The point I am making, then, is that when one talks about the *context* of Poe's quotation from Aristotle above, one should be talking not so much about the particular passage surrounding the quotation as one should be talking about the particular ideas related to the quotation, ideas found in *Poetics* and also elsewhere. In other words, one should focus more on the *ideational* context of the quotation than on the *textual* context. In the case of Poe citing Aristotle in his *Poetics*, what is of massive importance is what Aristotle says later in that work: specifically, that of all the necessary features of being a good poet, "the greatest thing by far is to be a master of metaphor. It is the one thing that cannot be learnt from others; and it is also a sign of genius, since a good

7. *E&R*, 1493. The exact quotation from Aristotle is, "poetry is something more philosophic and of graver import than history, since its statements are of the nature rather of universals, whereas those of history are singulars. By a universal statement I mean one as to what such or such a kind of man will probably or necessarily say or do—which is the aim of poetry." Jonathan Barnes, ed. (1984). *The Complete Works of Aristotle*. Princeton, NJ: Princeton University Press. All quotations from Aristotle are from the Barnes edition.

8. See Jonathan Barnes (1995). "Life and Work." In Jonathan Barnes, ed. (1995). *The Cambridge Companion to Aristotle*. Cambridge: Cambridge University Press, 6–15.

metaphor implies an intuitive perception of the similarity in dissimilars" (§22).

This is possibly the most profound quotation to be found in the book before you, and it recurs and echoes throughout the remaining chapters, including even the neuroscience section in the final chapter. Its importance is central to understanding Poe and scientific imagination, and we must be as clear as we can on the topic of metaphor right here, right now, before we proceed with the rest of the book. The word *metaphor* ultimately comes from the Greek word *metaphora*, which means "to carry from one place to another." A metaphor, then, is "A figure of speech in which one thing is described in terms of another. . . . A comparison is usually implicit; whereas in simile it is explicit."[9] In the modern world of literary studies, the complaint is commonly made that the word *metaphor* is often used too widely. Strictly speaking, a metaphor is not *literally* true, as evident, for example, in Martin Luther's "A mighty fortress is our God." Nevertheless, whether used widely or strictly, the heart and soul of metaphor is *analogy*, from the Greek words *logos* (meaning "the study of," "proportion," "reason," "rational," "speech") and *ana* (meaning "of each," "bring up"). A metaphor is a kind of comparison between two things, then, between what are called *analogues*, and when an argument is involved, with premises and a conclusion, we have what's called an *analogical argument*. The importance of metaphor and analogy, and the relation between the two, was put beautifully by the psychologist Lawrence Marks, a synesthesia expert, (more on this in Chapter 8,) who I here quote:

> Fundamental to cognitive activity, to the process of thinking, is that we extend the limits of our comprehension. Metaphor is a means to disclose new analogies and equivalences. . . . Metaphor snares independent threads and weaves them into a coherent fabric. Mind follows a metaphorical imperative, seeking out relationships that entwine disparate phenomena and events.

9. J.A. Cuddon (1979). *A Dictionary of Literary Terms*. Rev. ed. Harmondsworth: Penguin Books, 391.

> Metaphor's secret lies in the fact that once an analogy is made between A and B, a whole gamut of associated meanings also become available. Not only is B like A in a certain way, but any and all of A's properties now become fair game to be absorbed into B. And so the process of metaphorization may proceed, through C, D, E, and so on, . . .[10]

Returning to Aristotle, it is important to know that philosophy in his day included science, (this only began to change in the 1830s,) such that someone we might count as a scientist back then was called a *natural philosopher*. In either case, Aristotle clearly believed that the job of the philosopher was to seek out the universals of knowledge, which for Aristotle, contrary to Plato, did not exist statically on their own outside of space and time but instead *in* things. Metaphor, of course, is at the heart not only of poetry but of tales (with an important qualification in the case of the latter, in Poe's view, as we shall see later in this chapter). Metaphor is also, along with simile and more widely analogy, at the heart of philosophy and of science, which ultimately are about knowledge of the world and the Universe.

The fundamental point in all of this is that *both the poet and the scientist do not just simply seek out universals, such as the blackness of all ravens; they also seek out analogical universals, what Aristotle calls "the similarity in dissimilars."* This helps us to understand why in "Pinakidia" (1836), as we shall see in Chapter 4, Poe repeatedly elevates ancient and modern scientists who were also poets and why he seemed interested in everything. Later, near the end of Chapter 6, as part of our vindication of Poe, we shall examine the importance of analogical reasoning in two of the greatest scientists of all time, Darwin and Einstein. Then in Chapter 7 we shall examine why Poe claims that his ace detective, Auguste Dupin, Poe's personification of scientific imagination, is also a poet, we shall examine the importance of analogical reasoning in Poe's criticisms of inductive and deductive reasoning, and we shall then examine Poe's theory of what he calls the "poetic

10. Lawrence E. Marks (1978). *The Unity of the Senses: Interrelations among the Modalities*. New York: Academic Press, 252–253.

intellect," which has analogical reasoning at its core. All of this is absolutely essential for a proper understanding of *Eureka*, both for what Poe did and meant to do there.

Returning to Poe's "Letter to Mr. — —," a second point of influence of Aristotle on Poe is taking it as a "truism" that the end or *telos* of human life is "happiness," which Poe says is "another name for pleasure." "If so," he says, then "the end of every separate part of our existence—everything connected with our existence should be still happiness."[11] What must be remembered is that as part of his formal education Poe intensely studied the classics, including probably Aristotle, and in *Nicomachean Ethics* (I, §§4–5) Aristotle not only states that happiness, or pleasure, is the goal of human life, and that people generally agree on this and only disagree on what happiness is, but all throughout his writings Aristotle displays his belief in *teleology*, the view that virtually everything strives for or has an end, a goal, a purpose, and especially biological wholes and their parts, that not only manmade wholes and their parts exist for a purpose with a function, such as individual chairs and their parts such as legs, but also individual animals and plants and their various parts, and even species. (Aristotle was not only the first biologist, he was also the first ecologist.) Returning to *Poetics*, which is incomplete, as it deals only with epic poetry, tragedy, comedy, and plots, it is to be noticed that Aristotle does argue as a functionalist. Not only is imitation at the core of poetry widely conceived (§1), but humans are the most imitative of animals, learning by imitation since infancy, such that realistic imitation, even of pain, produces pleasure (§4). Moreover, the "tragic pleasure" is produced from evoking pity and fear (§14), and the function of tragedy is catharsis (§6). This is immensely important for understanding *Eureka*, as we shall see in the next chapter.

---

11. *P&T*, 12. Aristotle might have disagreed with Poe's inference about parts as committing the fallacy of division, which is committed when one assumes that what is true of a whole is also true of one or more of its parts (*Sophistical Refutations*, §4). Poe's use of the word "should," however, merits questioning whether Poe actually commits the fallacy, given the profound difference between "is" and "should." At any rate, pointing out the fallacy, as philosophers are wont to do, serves no real purpose in helping to understand Poe's literary theory, and ultimately *Eureka*.

In Poe's "Letter to Mr. — —," however, in the passage quoted at the beginning of this section, Poe claims that teaching, in itself, or instruction, does not immediately produce happiness, but only potentially as a means. Music, on the other hand, produces pleasure or happiness immediately, not mediately. Therefore the goal of poetry, he says, given that music encompasses poetry essentially by virtue of producing pleasurable sounds, should no more be essentially instructional than the goal of music *simpliciter,* which the latter clearly is not. To get a good idea of what Poe is getting at with "the comprehension of sweet sound," take, for example, the last stanza of "To One in Paradise" (1833), which is arguably as good as it ever gets from pen or tongue of any mortal:

And all my days are trances,
    And all my nightly dreams
Are where thy grey eye glances,
    And where thy footstep gleams—
In what ethereal dances,
By what eternal streams.[12]

Poe developed his literary theory not only in essays such as "Letter to Mr. — —" but also in book reviews. An important example is his review of the books of poems by Mrs. L.H. Sigourney, Miss H.F. Gould, and Mrs. E.F. Ellet, published in the January 1836 issue of the *Southern Literary Messenger*. What is of particular interest here is what Poe calls, following the German poet and critic August Wilhelm von Schlegel, the "*unity or totality of interest*," or, in Poe's own words here, the "*totality* of effect," what he would later call the "unity of effect." The reader of a poem, he says, must be able to "include in one comprehensive survey the proportions and proper adjustment of the whole," in other words "the nice adaptation of its constituent parts," so as to gain the "*unique*" pleasure afforded "in the contemplation of the picture *as a whole*." Hence Poe rejected the use of a motto at the head of a poem, whether to indicate the subject of the poem or the proper explanation of the poem, because with the former, "the poem being a mere paraphrase of the motto, the interest is divided

12. *P&T*, 70.

between the motto and the paraphrase," while with the latter "the reader must revert, in mind at least, to the motto for the necessary explanation."[13]

Consider also Poe's review of *Night and Morning* (1841), a novel by Sir Edward Bulwer-Lytton, published in the April 1841 issue of *Graham's Magazine*. Here Poe tells us that a plot is neither, as commonly conceived, (he says Schlegel even fell into this error,) a matter of "simple *complexity*," or even "the greatest involution of incident," nor is a "good plot" one "in which none of the *leading* incidents can be *removed* without *detriment* to the mass." Instead, he says, *plot*, when "properly defined, is *that in which no part can be displaced without ruin to the whole*." Poe is quite strict here. He adds that "It may be described as a building so dependently constructed, that to change the position of a single brick is to overthrow the entire fabric." I should think a house of cards would be a better (though still not perfect) comparison for what Poe has in mind. At any rate, Poe also informs us that his definition, with its concept of "infinite perfection," captures the "unattainable goal" to which the "true artist" aspires.[14] The "unattainable" part, of course, is true unless (and Poe does not say this here) the "true artist" is God, which we have to keep in mind is the true artist in *Eureka*, and who does, according to Poe there, create the perfect plot.

Since Poe is reviewing a novel, he also points out that the object of his review must fail in terms of "the unity or totality of *effect*," since "the mind cannot comprehend at one time, and in one survey, the numerous individual items which go to establish the whole," such that "the pleasure we derive (if any) is a composite one, and made up of the respective sums of the various pleasurable sentiments experienced in perusal." Poe in subsequent writings would elaborate on this point, but suffice it here to add that "the brief tale," he says, is "a species of composition which admits of the highest development of artistical power in alliance with the wildest vigor of imagination,"[15] what is surely the best

13. *E&R*, 876, 877.

14. *E&R*, 148.

15. *E&R*, 152, 153.

description of many of Poe's own tales, at least to my mind and many others.

Another important example of Poe's literary theory found in his book reviews is his second review of Henry Wadsworth Longfellow's *Ballads and Other Poems* (1842), (for which the first review served as prelude,) published in the April 1842 issue of *Graham's Magazine*. This review is important on a number of counts, particularly since Poe used much of it, verbatim in some places, for his final lecture published as "The Poetic Principle" (1850), discussed at the end of this section. Here we find, most notably, Poe's rejection of Longfellow's "*didacticism*," that "he regards the inculcation of a *moral* as essential" to a poem, that "His invention, his imagery, his all, is made subservient to the elucidation of some one or more points (but rarely of more than one) which he looks upon as *truth*."[16] From this alone one might seem required to judge *Eureka* as no poem.[17]

For the rest of the review, there are many elements repeated and amplified in "The Poetic Principle," and I shall only mention some of them here, reserving discussion for later. One is the distinction in "offices" of the mind between "pure intellect, taste, and the moral sense," truth being the goal of the intellect, duty the goal of the moral sense, "Beauty" the goal of taste, which has only "collateral relations" with the other two; to "convey 'the true'" we must be "perspicuous, precise, terse," also "calm, unimpassioned, unexcited," all of which is "the exact converse of the poetical." Also important is Poe's claim that there is a "thirst unquenchable" for "the beauty beyond the grave," which "belongs to the *immortal* essence of man's nature," much as "the desire of the moth for the star," and which is the "first element" of "Poesy."

The "second element" is "the *creation* of Beauty," which combined with the first is "the essence of all Poesy," a term he uses wider than the common meaning, (commonly it means either poetry as a whole or poetic skill,) as an "abstraction . . . applicable

16. *E&R*, 683, 684.

17. In Poe's first review of Longfellow's collection, which I call a prelude to the one under discussion, Poe states that "We do not mean to say that a didactic moral may not be well made the *under-current* of a poetical thesis." His point instead is that Longfellow's "conception of the *aims* of poesy *is all wrong*." *E&R*, 683, 682.

in various moods," a "sentiment" that "may develop itself in Sculpture, in Painting, in Music, or otherwise," which includes "our present business . . . its development in words—that development to which, in practical acceptation, the world has agreed to limit the term ['poesy']." Poe then equates "the Poetical Sentiment" with his wide meaning of "Poesy," which can be rather confusing, until he defines "the Poetry of words" as "the *Rhythmical Creation of Beauty*," which is his meaning of *poetry* proper, such that if beauty is not the ultimate goal of a piece of writing then "what the world terms *prose* may be safely and properly left all else." "The artist who doubts of this thesis," he adds, "may always resolve his doubt by the single question—'might not this matter be as well or better handled in *prose*?' If it *may*, then it is no subject for the Muse."[18]

A month later, in the May 1842 issue of *Graham's Magazine*, in this case his second book review of Nathaniel Hawthorne's *Twice-Told Tales* (1842), (for which again there is an earlier review that served as prelude, and also a third review published much later, in November 1847,) Poe provides further important elements of his literary theory. But I want to return to Aristotle and compare Poe's thinking with Aristotle's. This is because it is sometimes said that Poe really didn't have a coherent literary or aesthetic theory. For example, Rachel Polonsky, in a chapter in *The Cambridge Calumnion to Edgar Allan Poe*, states that Poe provides us with "a fundamentally uncoordinated body of aesthetic theory which ultimately turns away from the possibility of theorizing about art," that there is no "underlying internal logic" but instead "incoherences, eccentricities," some "moments of modest good sense," and perhaps some "intellectual games and hoaxes."[19] Much

18. *E&R*, 685, 684, 686–687, 689. Poe's use of "Poesy" was not unique, as it was virtually the same as that of Coleridge, for example, as Poe well knew in his 1836 review. *E&R*, 183. Much later, in "The Poetic Principle" (1850), Poe would widen his meaning of "Poesy" even a little further, in that "The Poetic Sentiment . . . may develope itself in various modes—in Painting, in Sculpture, in Architecture, in the Dance—very especially in Music—and very peculiarly, and with a wide field, in the composition of the Landscape Garden," to which he adds, "Our present theme, however, has regard only to its manifestation in words." *P&T*, 77. So once again one should not confuse "poesy" and "poetry" in Poe.

19. Rachel Polonsky (2002). "Poe's Aesthetic Theory." In *Companion*, 44. That she includes Poe's literary theory is evident from the fact that much if

the same has often been said of Nietzsche's philosophy, that he provides no system but only shifting and sometimes conflicting perspectives, that it fits postmodernism. The problem is only exacerbated, of course, if one treats the total writings of these authors synchronically rather than, what is surely required, diachronically, for otherwise one cannot possibly perceive a pattern of growth or evolution. It would be like a palaeontologist studying only fossil bones and ignoring strata. In the case of Nietzsche, the perspectival, postmodernist interpretation is plausible until one realizes that Nietzsche did operate from a developing and ultimately unified perspective after all, namely, biology, more specifically physiology and evolution.[20] In the case of Poe's literary theory, I suggest that there is an underlying internal logic also, one that grew and evolved over time in his writings, with an overall coherence in spite of its eccentricities, and that the key to perceiving and understanding it is to view his strictures as a repeated response to Aristotle's literary theory, provided primarily in *Poetics*, which Poe modified over time by his reflections on his own and other people's writings. In fact I suggest that this viewpoint would shed light on many of the literary theories propounded in Poe's day by his contemporaries, but that of course is another topic.[21]

In the case of Poe's second review of Hawthorne's *Twice-Told Tales*, it is interesting to juxtapose Poe's literary theory with Aristotle's, point by point. For my own part, it seems evident that Aristotle's view is ancestral, not only as a matter of history but

---

not most of her chapter is devoted to an analysis of Poe's "The Philosophy of Composition" and "The Poetic Principle." Combined with other critiques in the volume, (most notably, that Poe was a misogynist,) it would appear that the *Companion* is not much of a *compaignon*. At any rate, for a more charitable assessment of Poe's literary theory than one gets from Polonsky, see Stuart Levine's Introduction to *Critical Theory*, ix–xii.

20. See Gregory Moore (2002). *Nietzsche, Biology and Metaphor*. Cambridge: Cambridge University Press.

21. Poe's poetics, as to be expected, had many influences, proximate and remote, notably the German Romantics, such as Schlegel and Schelling, but it is the influence of Aristotle that seems the most important to my mind, and the most overlooked. For the influence of the German Romantics on Poe, see, e.g., Albert J. Lubell (1953). "Poe and A.W. Schlegel." *Journal of English and Germanic Philology* 52 (1), 1–12.

because Poe sometimes explicitly refers to Aristotle and seems familiar with his writings, and of course he had a classical background. Indeed, it often seems, when reading a piece of Poe's literary theory, that he had the ideas of Aristotle readily at hand, even if he did not explicitly refer to them, and that he used them as a springboard for his own views, refashioning them in a more stringent form, and adding what are arguably improvements, most notably the concept of the "unity of effect." There are three themes in particular that I shall focus on here.

First, in *Poetics* Aristotle claims that the length of a poetic work must be sufficient to contain the necessary parts of the work, including a beginning, middle, and end, and that its length should not exceed what can be "taken in by the memory" (§7). Poe is much more strict. In his view, a "prose tale" should take between "a half hour to one or two hours" to read in order to maximally serve its essential effect and purpose. During this time span designed for "one sitting," no "worldly interests" need intervene on the part of the reader, no "external or extrinsic influences" to ruin or diminish the effect, no "weariness or interruption," and "the soul of the reader is at the writer's control." (Think how important and true this is when it comes to a movie.) In the case of a "rhymed poem," the length should be at most "an hour," for the purpose of the poem is to produce in the reader "an exaltation of the soul," a "unity of impression," a soul that is "deeply moved." If either the poem or the tale is too short, the essential purpose cannot be produced, while "a long poem is a paradox" and the "ordinary novel" for the same reason, each depriving the reader of "the immense force derivable from *totality*."[22]

Second, there is what Aristotle calls the "unity of a plot." Just as a "beautiful whole" or a "beautiful living creature" is "made up of parts" (§7), likewise a plot should be a "complete whole," he says, "with its several incidents so closely connected that the transposition or withdrawal of any one of them will disjoin and dislocate the whole." The animal metaphor is an especially interesting one, for just as an animal as a whole has a function or purpose in Aristotle's biology, so too does a poem as a whole in Aristotle's literary theory, even though, in both cases, he allows

22. *E&R*, 572, 571.

for "accidental" (non-essential) features—in the case of an animal, eye color, hair quality, or being bald, etc., and in the case of a poem, "that which makes no perceptible difference by its presence or absence is no real part of the whole" (§8). Moreover, the "perfect plot," he says, "must have a single, and not (as some tell us) a double issue" (§13). Poe again is much more strict. According to Poe, having determined in advance the "unique or single *effect* to be wrought out," the "skilful literary artist . . . then invents such incidents—he then combines such events as may best aid him in establishing this preconceived effect." But there should be no accidents, here, no non-essentials, claims Poe, for "In the whole composition there should be no word written, of which the tendency, direct or indirect, is not to the one pre-established design." "Every word *tells*," he says, "and there is not a word which does *not* tell."[23] And it's not just words. Years later, in "Marginalia" (February 1848), Poe made a related point about punctuation: "a sentence may be deprived of half its force—its spirit—its point—by improper punctuation."[24] It is all like dressing for success. Northrop Frye made a similar point when he wrote, "In ordinary life, as in literature, the way you say things can be just as important as what's said. The words you use are like the clothes you wear."[25] (And again, it's not just the words.) Poe was known for dressing in a very unusual way, (limited pathetically by his wallet, of course,) his appearance always designed to make a certain impression, to indicate a definite persona. As an acquaintance of his put it, Poe was "graceful, and with dark, curling hair and magnificent eyes, wearing a Byron collar and looking every inch a poet."[26] When it came to the way he clothed his ideas, of course, Poe was of the same mind, and always for the designed purpose or "unity of effect."[27]

---

23. *E&R*, 572, 575.

24. *E&R*, 1425.

25. Northrop Frye (1963). *The Educated Imagination*. Toronto: CBC Enterprises. Republished (2002). Toronto: House of Anansi Press, 82.

26. Quoted in *Weiss*, 44.

27. In terms of his writing style alone, Poe over the years has had many detractors among the literati, including the likes of Mark Twain, T.S. Eliot, and Harold Bloom. The definitive refutation of the view of Poe as a bad stylist is *Zimmerman*. In his book Zimmerman shows, with a thoroughness

Speaking as a philosopher, I might add that Poe's strictures mark a profound difference between the philosopher, who much like Aristotle on non-essential animal features allows for digressions, (such as this,) and even disjointed ramblings as to be flopping all over the place, and the true artist, for whom each literary device, such as *hyperbaton* and *expansion*, and every word and phrase, including accessories such as the comma, the dash, and italics,—*much* as each color in a painting, including the kind of brushstroke, and every note and accessory indication in a piece of music, such as staccato or legato,—is chosen and put where it is for a purpose. To the written word one could add footnotes versus endnotes and even the choice of font, as Garamond, for example, is more likely to contribute to agreement than Arial. At any rate, in matters of style alone, philosophers and others could learn a lot from artists like Poe.

Third, and finally, at the end of *Politics* (VIII, §§6–7), in which he provides a discussion on "rhythms," Aristotle states that "music should be studied, not for the sake of one, but of many benefits," one of which is "education," another of which is "purgation," which he says he will return to in his work on poetry, while the third is "intellectual enjoyment" and the fourth is "relaxation" or "recreation after exertion." At the core of them all is "pleasure," for even catharsis, he says, gives "an innocent pleasure to mankind." Poe, on the other hand, although he allows that "Truth is often, and in very great degree, the aim of the tale," including tales of terror, ratiocination, and humor, such that its "products" are "infinitely more numerous, and more appreciable by the mass

---

of evidence and argument that makes his case beyond a reasonable doubt, that Poe was a master of *many* styles, even an inventor, that he used different styles depending on his literary subject and purpose, in each case adapting style to content, so that it is a gross mistake to speak of "Poe's style" (much as it is to speak of Bruce Lee's "self-expression" as limited to any one martial art "style," their respective theories of "unity of effect" a striking example of "the similarity in dissimilars"). Especially useful is Zimmerman's two-hundred page *Catalogue* of three hundred "figures of speech and thought—rhetorical devices," in which each entry receives a learned discussion along with examples from Poe's writings. And *Eureka* itself, it should be added, as Zimmerman shows, involves a variety of rhetorical devices and styles, such as *auxesis*, *climax*, *digestion*, *epic simile*, *hypotaxis*, *iteratio*, and *parrhesia*. *Zimmerman*, 152–153, 171, 185–186, 204–205, 234–235, 242, 274–275.

of mankind" compared with poems, the purpose of a poem, he says, is not truth. This is because "the poem's highest idea" is "the idea of the Beautiful," and, for this, "*rhythm* . . . is an essential aid." A prose tale can be directed toward the "purely beautiful," he admits, but it is not going to happen in tales of "terror, or passion, or horror," and even with tales designed for other effects, the writer of a tale in the pursuit of beauty is "laboring at a great disadvantage," since by its very nature a tale lacks the essential element of the poem, namely "rhythm," so that "Beauty can be better treated in the poem."[28]

It is to be noticed that all of the three main themes just described above were formulated in Poe's mind long before he started thinking seriously and synthetically about the nature of the Universe, and certainly long before *Eureka*. But every single part of Poe's literary theory examined thus far is necessary for understanding *Eureka*, and we are getting closer.

By 1844, as we have seen in the previous chapter in the section on "Usher," Poe was thinking in earnest about the nature of the Universe. In his letter to James R. Lowell, for example, dated July 2, 1844,[29] which is a reply to Lowell's request for a short "spiritual biography" of Poe for *Graham's Magazine*, we find many of the ideas that will be repeated or developed in *Eureka*, for example that "unparticled matter, permeating & impelling, all things, is God," that "Its activity is the thought of God," that "Man, and other thinking beings, are individualizations of the unparticled matter," and that it is "matter (the particled matter) which individualizes him." We also get Poe's theory of poetry in a nutshell, which is that "Music is the perfection of the soul, or idea, of Poetry. The *vagueness* and exultation aroused by a sweet air (which should be strictly indefinite & never too strongly suggestive) is precisely what we should aim at in poetry. Affectation, within bounds, is thus no blemish."

But neither here, nor in "Mesmeric Revelation" (1844) written shortly before it, does Poe yet begin to bring together the three major strands of his thought that are so masterfully interwoven in *Eureka*—the literary, the theological, and the scientific. There

28. *E&R*, 573.

29. *Letters*, 448–451.

is no application of plot to the Universe, no theory of creation or origin, nothing on the purpose of the Universe, no application of Newtonian gravity, no theory of mutuality of adaptation.

It is not until later in that same year, however, in his first installment of "Marginalia" (November 1844), that we can first begin to see the strands coming together, with Poe's criticism of the Bridgewater Treatises for getting wrong the nature of God's design in nature, along with Poe's claim that "The plots of God are perfect. The Universe is a Plot of God."[30] From then on we can see Poe doing increasingly more, developing and intertwining his ideas, even using the language of physics to inform his literary theory. For example, in an obvious elaboration of his definition of *plot* given in his review of Bulwer-Lytton's *Night and Morning*, Poe states, in "A Chapter of Suggestions" (1845), which was written around the same time he wrote his first installment of "Marginalia,"[31] not only that "*Plot* is very imperfectly understood, and has never been rightly defined," but that "In its most rigorous acceptation, it is *that from which no component atom can be removed, and in which none of the component atoms can be displaced, without ruin to the whole*."[32] If the Universe is a plot of God, then it would have to be a plot in its "most rigorous acceptation," which would have to mean that not even a single atom is accidental but plays a role! Poe does not go that far in *Eureka*, at least not explicitly, but he does state that, unlike plots created by humans, "The plots of God are perfect," "The Universe is a plot of God," and throughout the Universe "There is an absolute *reciprocity of adaptation*."[33] What Poe had in mind in using the latter phrase shall be reserved for Chapter 4, when we turn to Poe's comments on the Bridgewater Treatises.

The year 1845 was also the year in which Poe did something very unusual, even for him, which was to write an anonymous book review *of his own book*, the collection of his tales entitled *Tales* (1845). This review is a small goldmine for finding out what Poe thought about particular works of his. But first, let us focus

30. *E&R*, 1315–1316.

31. Pollin, *Brevities*, xxvi.

32. *E&R*, 1293.

33. *P&T*, 1342.

on the title of the collection itself. Why *Tales* rather than *Stories* or *Short Stories*? And, indeed, why not only here but elsewhere does Poe repeatedly refer to his short stories as "tales"? The question is an important one, as Poe always chose his words carefully, so there must be a reason. The first thing to notice is that not every story has a plot. Unlike a story per se, a plot requires a *dénouement*, the final resolution of the various elements or strands in the plot. One can therefore have a never-ending story, but never a never-ending plot. A story *with* a plot, of course, should have a special name, and Poe's own preferred word for that was "tale," refusing to use the word "tale" where there was but a story. Evidence that this interpretation of Poe's usage is correct can be found, for example, in Poe's review of Thomas Moore's *Alciphron, a Poem* (1840), in which he writes, "The design of the story (for plot it has none). . . ."[34] This explains why Poe typically referred to his works of fiction as "tales," not as "stories." He took great care and pride in creating his plots, adapting each part, whether large or small, to the function of the whole. Even his exceptions to the rule are informative. For example, although "Mesmeric Revelation" (1844) is included in his *Tales* (1845), Poe elsewhere, such as in his "Marginal Notes" (1845), refers to it as a "magazine article," adding at the end that "The story is a pure fiction from beginning to end."[35] And in his anonymous review of his own *Tales*, Poe adds that "Mesmeric Revelation" is "evidently meant to be nothing more than the vehicle of the author's views concerning the Deity, immateriality, spirit, &c., which he apparently believes to be true."[36] It was in fact a "vehicle" for ideas that Poe would later develop in *Eureka*. At any rate, "Mesmeric Revelation" can hardly be said in truthfulness to have a *dénouement*, and therefore a plot—unless one considers "he fell back upon his pillow and expired" a *dénouement*, in which case *Lord help your poor soul!*

Returning to his review of *Tales*, Poe, writing anonymously, summarizes "his [Poe's] idea of the perfection of the plot, which he defines as—that, in which nothing can be disarranged, or from

34. *E&R*, 333.

35. *E&R*, 1367.

36. *E&R*, 870.

which nothing can be removed, without ruin to the mass—as that, in which we are never able to determine whether any one point depends upon or sustains any one other." And of course we are told that Poe with his tales has "perfectly succeeded in his perfect aim." We are also told that Poe wrote his tales "backwards." "Most writers," he says, with regard to their plots, write forwards: they "get their subjects first, and write to develope [*sic*] it. The first inquiry of Mr. Poe is for a novel effect—then for a subject; that is, a new arrangement of circumstance, or a new application of tone, by which the effect shall be developed. And he evidently holds whatever tends to the furtherance of the effect, to be legitimate material."[37]

This is of enormous importance for understanding Poe's theology in *Eureka*, the focus of our next chapter, and for Poe's cosmological speculations in *Eureka*, which are the focus of Chapter 5. "Legitimate material" evidently applies to Poe's solution to the problem of evil, of why there is so much pain, suffering, and death in the world if the Universe, considered historically, is a plot of God. And that plots are properly to be written "backwards," with the ultimate effect being determined first in the process of creation, and everything else coming afterward so as to produce that effect, evidently applies to the volition of God in creating the Universe in the first place, the Universe being "a plot of God." What it means is that Poe did not think of the history of the Universe as something written by God aimlessly and on the run, as it were, experimentally, developing the plot throughout the process of writing, a matter of "'exploring' in creation," as the professional biochemist and Anglican priest Arthur Peacocke believes,[38] a matter of "purpose without an exact predetermined plan," as the professional physicist and accomplished theologian Ian Barbour believes,[39] that "mankind's appearance on this planet was *not* preordained," that we are one of many "experiments in progress," as

37. *E&R*, 869, 872, 873.

38. Arthur Peacocke (1985). "Biological Evolution and Christian Theology—Yesterday and Today." In John Durant, ed. (1985). *Darwinism and Divinity*: Oxford: Basil Blackwell, 125.

39. Ian G. Barbour (1997). *Religion and Science*. San Francisco: Harper, 216.

the Catholic cell biologist Kenneth Miller believes.[40] Poe's God, instead, worked it all out beforehand, as a "unity of effect," the end or *telos* being determined first, and then the means, particularly the laws of nature, being planned (though not yet actualized) before the moment of creation of the "primordial Particle," from which the Universe exploded in a big bang.

It is a profound difference in viewpoint on the meaning of the Universe, perhaps the profoundest of all, *whether the Universe is a tale or a story (it could be both, as a series of tales)—or possibly neither.*

The idea that a plot should be written "backwards" did not begin with Poe in 1845, for he had expressed it earlier, as we have seen in his review of Hawthorne's *Twice-Told Tales*, though perhaps not quite as clearly and forcefully. It is an important principle that informs many of his book reviews, for example his second review, published in February 1842, of Charles Dickens' *Barnaby Rudge*, a book originally published in parts, serially, in a weekly periodical. Poe took this book to be an imperfect example of his principle, although it did allow him to predict in his first review (correctly, it turned out, to the consternation of Dickens) the identity of the murderer, such that, as he says in his second review, "if we did not rightly prophesy, yet, at least, our prophecy *should have been* right."[41] In spite of his many criticisms of the details of the plot, as failing his principle, Poe's review was positive enough to state confidently that "Let him [the reader] *re-peruse* 'Barnaby Rudge,' and, with a pre-comprehension of the mystery, these points of which we speak break out in all directions like stars, and throw quadruple brilliance over the narrative."[42]

Oh the temptation to see the last line above as an anticipation of Poe's cosmology in *Eureka*, according to which "the Universe of Stars" is "a plot of God."[43] But we shall resist the temptation as

40. Kenneth R. Miller (1996). *Finding Darwin's God: A Scientist's Search for Common Ground Between God and Evolution*. New York: HarperCollins, 272, 275.

41. *E&R*, 235. According to Tara Moore, "In later editions of the novel, Dickens buried the clues that tipped off Poe." Tara Moore (2013). "Charles Dickens." In *Context*, 280.

42. *E&R*, 233.

43. *P&T*, 1262, 1342.

a nod to the Sons of Temperance. What is important to add is the criticism that immediately follows the passage quoted above: "—a brilliance which a correct taste will at once declare unprofitably sacrificed at the shrine of the keenest interest in mere mystery." There is a difference between a mere mystery and an actual plot, and this *does* inform us of the way Poe viewed the Universe, as not *merely* a mystery. What a plot necessarily has, according to Poe, but not necessarily a mystery, is the adaptation of the parts to the function of the whole. Hence Poe goes so far as to say, at the very end of his review,—what at first might seem excessively sharp, but on further reflection seems justified,—that Dickens "has a *talent* for all things, but no positive *genius* for *adaptation*."[44]

Returning to where we left off with Poe's review of his own *Tales* (1845), Poe's theory of writing "backwards" is repeated in the following year, right at the beginning of "The Philosophy of Composition" (1846). "Charles Dickens," he says, "in a note now lying before me, alluding to an examination I once made of the mechanism of 'Barnaby Rudge,' says—'By the way, are you aware that Godwin wrote his 'Caleb Williams' backwards?'" Poe doubts that this was "the *precise* mode of procedure on the part of Godwin," but he does restate his principle: "Nothing is more clear than that every plot, worth the name, must be elaborated to its *dénouement* before any thing be attempted with the pen. It is only with the *dénouement* constantly in view that we can give a plot its indispensable air of consequence, or causation, by making the incidents, and especially the tone at all points, tend to the development of the intention."[45]

Later in "The Philosophy of Composition" Poe relates in detail how he composed "The Raven" (1845) in conformity with

44. *E&R*, 244. Poe's two reviews of Dickens' *Barnaby Rudge* need to be reread in light of Poe's distinction between a "tale" and a "story" discussed above, as they serve to highlight what Poe really thought of the work. In some places Poe refers to the "tale," in others the "story." To the unwary reader they appear as synonyms. In Poe's hands, however, they are actually a matter of design, not accident. In fact they serve to emphasize when he is making a compliment and when he is making a criticism. This is certainly a topic worth exploring further, especially in comparison with the literary context of his time. But I shall leave the Guggenheim for someone else.

45. *E&R*, 13.

this principle. Some have thought Poe utterly disingenuous on this head, that his reconstruction is a "hoax," recalling to mind "the crazed narrators in some of Poe's tales, whose tone of eerie calm is intended to demonstrate their lucidity and self-control, but arouses only the reader's fear and pity."[46] Some have thought that those who make the hoax criticism "are only superficially correct," that what the critics miss is that Poe "did not have them [the ideas that constitute 'The Raven'] all at once, in the preparation of *one* poem, or in the logical order in which he relates them."[47] Some have thought "Poe's essay on his own *The Raven* is a perfectly accurate account of what he did in that poem, whether he did it on the conscious mental level that the essay suggests or not."[48] For the record, Poe himself in a letter written approximately four months after the publication of "The Philosophy of Composition" called the latter "my best specimen of analysis."[49]

Oh to be Poe.

The height of the problem in "The Philosophy of Composition"—especially for *us*—is Poe's rejection of the Romantic conception of the poet as composing "by a species of fine frenzy—an ecstatic intuition,"—problematic because this is almost precisely how Poe conceives of scientific imagination! And he claims this not only in *Eureka* but also illustrates it in his personification of the scientific mind, the ace detective Auguste Dupin, who first appears in "The Murders in the Rue Morgue (1841), and also in his genius William Legrand in "The Gold-Bug" (1843), as we shall see in Chapter 7. Poe does make it clear in "The Philosophy of Composition," however, that "ninety-nine cases out of the hundred" would have us believe that they compose in the classic Romantic fashion, whereas the reality, he says, if we could "peep behind the scenes," is a process of "vacillating crudities of thought," "painful erasures and interpolations," and "the true purposes seized only at the last moment." These "constitute the properties of the

46. *Silverman*, 296–297.

47. *Quinn*, 440.

48. Northrop Frye (1957). *Anatomy of Criticism: Four Essays*. Princeton, NJ: Princeton University Press, 278. For an extended defense of the conscious level, see *Evermore*, 64–71.

49. *Letters*, 596.

literary *histrio*," he says, the hacks of the literary stage. And what of the remaining one in a hundred? They are like Poe, of course, with a "*modus operandi* by which some one of my own works was put together," from which he selects "The Raven," in which "no one point in its composition is referrible either to accident or intuition—that the work proceeded, step by step, to its completion with the precision and rigid consequence of a mathematical problem."[50]

Perhaps what we have here is a hoax after all, possibly for a purpose in line with Poe's theory of discords in music, of unexpectedness and equality in a poem, of strangeness in proportion to beauty, and of pain in relation to pleasure, (themes we shall focus on below and in the next chapter,) which is to heighten the "unity of effect" of the work. But how could that be *here*? Jerome McGann provides a possible answer, which is that "hoax is primarily a critique and satire upon the truth pretensions of any expository argument, or what Poe in *Eureka* calls 'demonstration.'"[51] Whatever the strength or extent of that proposition, we should hopefully notice that Poe apparently already gave us a clue that should make us think he is not serious about how he composed "The Raven," with no "accident or intuition" involved and that he proceeded "step by step, to its completion with the precision and rigid consequence of a mathematical problem." What should get us thinking is that this doesn't sound human! And that should then get us thinking about what we have seen above in his first installment of "Marginalia" (November 1844), the point (which he would later repeat in *Eureka*) that "The plots of God are perfect. The Universe is a Plot of God." From this one would naturally think that God's *modus operandi* in composing a plot would be perfect, too. And as if to give us yet a further clue, Poe immediately precedes the pair of statements above with "perfection of plot is unattainable in *fact*,—because Man is the constructor."[52] So unless Poe in "The Philosophy of Composition" thinks he is

---

50. *E&R*, 14–15.

51. *McGann*, 93.

52. *E&R*, 1316. Recall Poe's reply to Dickens' claim that "Godwin wrote 'Caleb Williams' backwards," that he doubts that this was "the *precise* mode of procedure on the part of Godwin." Poe's italicization of "precise" only now makes total and perfect sense.

God (enter the principle of charity of interpretation, please) rather than a member of the race of human beings, with their limitations and imperfections, their "fever called 'Living,'" and their glorious unconscious mind, then he can't be serious about how he composed "The Raven." And if this line of interpretation is basically correct, then it is possible if not highly probable that the purpose of Poe's hoax in "The Philosophy of Composition" is to prepare the way for his *ultimate* philosophy of composition, which he would further develop and eventually entitle—*Eureka*.

At any rate, we shall not concern ourselves any further here with this hoax problem. Instead, we shall focus on two differences or additions in "The Philosophy of Composition" compared with Poe's literary theory examined thus far, ones that no one thinks are hoaxes and that are clearly important for understanding *Eureka*. The first concerns what Poe says about beauty and pleasure, that "Beauty is the sole legitimate province of the poem" and that "pleasure . . . most pure, is, I believe, found in the contemplation of the beautiful." What is especially important is what Poe adds to this: "When, indeed, men speak of Beauty, they mean, precisely, not a quality, as is supposed, but an effect—they refer, in short, just to that intense and pure elevation of the *soul—not* of intellect, or of heart."[53] This is very different not only from Plato's concept of the Form of Beauty, existing outside of space and time and perceptible only by the unaided intellect, but from the thinking of the common man. For it follows from what Poe says that whatever beauty we find in people, or animals, or nature, or manmade things, or in the Universe considered as a whole, the beauty is not in what is perceived, it is not what philosophers call a *primary quality*, but is instead a *secondary quality*, like color, for example, a quality evoked in us by both the nature of the object perceived and the nature of us as perceivers, and which we naturally, meaning unconsciously, project onto what is perceived. If the Universe seems beautiful, then, or its laws, or its plot, or yes even Poe's creative God, (which would include ourselves,) it is not because of any beauty in them *per se*, but because of the effect they have on us, which is the combined effect of their nature and ours. One should not take Poe to mean, however, what one often hears, that

53. *E&R*, 16.

beauty is in the eye of the beholder, in the sense of *merely*. Poe has something else in mind, including something about human nature at its finest.

The second point that stands out is that "It by no means follows from any thing said here, that passion, or even truth, may not be introduced, and even profitably introduced, into a poem—for they may serve in elucidation, or aid the general effect, as do discords in music, by contrast."[54]

Certainly death could be said to be one of the many discords in the music of life, arguably the brashest and loudest of all, which brings us back to the central focus of "The Philosophy of Composition," which is "The Raven." How can a poem about the death of a loved one, a grossly premature death at that, possibly evoke a heightened effect of "that Beauty," defined as "the excitement, or pleasurable elevation, of the soul"? Poe seems to provide an answer here, partly in saying that there is beauty, and consequently pleasure, in "*sadness*," in that "Beauty of whatever kind, in its supreme development, invariably excites the sensitive soul to tears. Melancholy is thus the most legitimate of all the poetical tones." In the case of the mournful "student" in "The Raven," Poe tells us that "he experiences a phrenzied pleasure in so modeling his questions as to receive from the *expected* 'Nevermore' the most delicious because the most intolerable of sorrow." Poe even goes so far as to refer to "the human thirst for self-torture," such that, combined "in part with superstition," the student repeats his "queries to the bird as will bring him, the lover, the most of the luxury of sorrow, through the anticipated answer 'Nevermore.'"[55]

54. *E&R*, 16–17. The earliest reference I can find in Poe's writings to discords in music is in his letter to Nathaniel Beverley Tucker dated December 1, 1835. On the topic of versification Poe remarks that there seems to be no essential consensus and his suggested reason is that "music is a most indefinite conception." Poe then goes on to remark that "I especially pride myself upon the accuracy of my ear," that he has established its accuracy "by some odd chromatic experiments," that "to speak technically" about versification would be "without the intervention of any discords," and that "but imperceptibly the *love of these discords* grew upon me as my love of music grew stronger." *Letters*, 115. We shall focus more extensively on Poe's connection between discord and beauty in the following chapter.

55. *E&R*, 16, 17, 19, 24.

This certainly adds an interesting twist to Poe's definition of "poetry" as the rhythmical creation of beauty using words, and it connects with his claim in *Eureka* that pleasure requires pain for its existence (more on this in the next chapter). But still, if Poe's answer does not make much sense to the reader, know that you are not alone. Keep in mind as well that Poe was not by nature morbid, in the sense of having a gloomy disposition and surrounding himself with the paraphernalia of death. Neither was he a lover of pain and suffering, a sadist or masochist. All of this is clear from those who knew him.[56]

Perhaps the problem is contextual. Poe *was* morbid, after all, in basically the same sense that Victorian culture was morbid, and this might help to understand why many today find an excessive morbidity in Poe's poems and tales, why they are not attracted to his writings even though they know of them. (It could also be due to the nagging echoes of the slanderous "Griswold myth.") It can hardly be denied that there is a marked cultural difference concerning death between our time and his, given that Victorian culture, in both Britain and America, was what we today (but not they) would call a morbid culture, with mourning rituals that included

---

56. See, e.g., *Weiss*, 17, 85. The same is found in obituaries by people who knew him well. *Critical Heritage*, 317, 379. As for what Poe claims in "The Philosophy of Composition" is "the most poetical topic in the world," namely, "the death . . . of a beautiful woman" (*E&R*, 19), it is almost certainly not a literal statement but a play on words, the death of a beautiful woman being, as Fisher puts it, "the most *Poe*-etic of all themes." Fisher points out that Poe was fond of wordplay and that with the present example Poe "may have tried enterprisingly to boost his own importance as an author and his sales in the book trade." *Fisher* 9, 109–110. I should think, instead, that Poe's evidently cryptic wordplay here is self-referential given that much in his poetry reflects the fact that his personal history was vividly and tragically marked by the death of beautiful young women deeply beloved by him, beginning with his biological mother, Eliza Poe, followed by the mother of one of his schoolfellows, Jane Stanard, then by his foster mother, Frances Allan, and then by his own wife, Virginia. When Poe wrote "The Philosophy of Composition" (1846), the deaths of the first three were well in his past, while the death of the fourth, looming around the corner, had been rehearsed the year before by Poe in "The Raven" (1845), his "mournful, never-ending remembrance" of Virginia as the lost Lenore. If this interpretation is correct, then I should think it a mistake to take Poe's "most poetical topic" as a key to understanding *Eureka*.

post-mortem photos of the deceased dressed up and posed as if still alive, cemeteries with magnificent architecture and elaborate landscape gardening that served as public parks for picnics and marriage proposals, bizarre behavior considered as tolerable such as opening the tomb of a loved one and lying with, embracing, caressing, talking to, and kissing the rotting corpse, (as Emerson and Griswold each did for their deceased wives,) and much more. How many of us today can relate to *that*?[57]

One has to wonder about the reason for all of this morbidity. James Curl provides a fascinating attempt to shed light on the genesis of "the Victorian Celebration of Death."[58] He looks at the "source in archaeology" of the Romantic movement; at "something like a delight in decay," including biology and crumbling churches and monasteries; at the rise of nonconformity against the established Church; and at the gross overpopulation of urban centers, London in particular, where enormous problems of hygiene, diseases such as tuberculosis, pollution, and poverty created unusually high mortality rates, and where the disposal of the dead (a most grave matter) was remedied in large part by the creation of graveyards separate from churchyards, in other words cemeteries, which freed of the Church were allowed to be developed into elaborate landscape gardens with magnificent architecture, especially suited for mourning and the contemplation of life after death—perhaps the meaning, I suggest, of Poe's "enchanted garden," an interpretation supported by the rest of the poem.[59] But

57. See, e.g., Sarah Rutherford (2008). *The Victorian Cemetery*. Oxford: Shire Publications. Mary Elizabeth Hotts (2009). *Literary Remains: Representations of Death and Burial in Victorian England*. Albany: State University of New York Press. *Hutchisson*, 52. *Silverman*, 217–218. Also, google "Victorian post-mortem photos" or "Victorian post-mortem mourning" and then hit "Images" in the upper-left corner of the screen. Not quite competition for the Palermo Catacombs in terms of creepy, but it would win for morbid comedy if it weren't so real and heartbreaking.

58. James Stevens Curl (2000). *The Victorian Celebration of Death*. Phoenix Mill, UK: Sutton Publishing, xix–36, 269n16. Curl uses the word "celebration" in connection with the words "solemn" and "honouring," not in the sense of "jollification."

59. "To Helen" (1848). *P&T*, 95. E.g., "the upturn'd faces of a thousand / Roses that grew in an enchanted garden" refers to the blood-red faces of the wormy corpses lying supine in their graves.

equally important, if not more, in Curl's view, were the "literary influences," especially the eighteenth-century "Graveyard Poets," most important among them Edward Young, with his internationally popular "The Complaint: or, Night-Thoughts on Life, Death, & Immortality" (1742), in which "the Grave," we are told in the poem, "is Creation's melancholy Vault," and "All, all on earth is *Shadow*, all beyond / Is substance; the reverse is Folly's *creed*"—which adds new meaning, I suggest, to Poe's "being young and dipt in folly / I fell in love with melancholy."[60]

But there is something profoundly missing in Curl's account, which seemingly finds echo in Poe's lament that science "dragged Diana from her car" and its breath "dims the mirror of our joy," entailing "Beyond that death no immortality— / But sleep that pondereth and is not 'to be.'"[61] Traditionally the sting of death—of losing loved ones, and of contemplating one's own mortality—was alleviated by religion, which of course was predominantly Christianity in Victorian culture. But that world, unlike any before it, (including even the ancient Greek and Roman worlds,) was marked by the rise of a powerful mode of explanation, a rival to the hegemony of religion called *science*, a rival that created an underlying anxiety with its materialist implications, an anxiety that haunted the mind even when practiced by religious people. (In Poe's time, unlike ours, science was practiced mainly by the latter, who naturally superimposed a theological interpretation on science.) Combined with the rise of nonconformity against the established Church, the result was something totally unique to the Victorian world: *spiritualism*.

At any rate, the "The Raven" in particular and Poe's writings in general have survived the test of time, and in a way that should

60. "Introduction" (1831). *P&T*, 54. Poe's narrator in "The Premature Burial" (1844), by the way, exclaims at the end of his ordeal that "I thought upon other subjects than Death. . . . I read no 'Night Thoughts'—no fustian about church-yards— . . . In short, I became a new man." *P&T*, 679.

61. "Sonnet—To Science" (1829). *P&T*, 38. "Al Aaraaf" (1829). *P&T*, 49. Diana was a Roman goddess, (Artemis, in ancient Greek,) identified in various ways with the Moon, including the Moon as her chariot. Science, of course, established well before Poe's time that the Moon is not divine or a chariot of the divine but a sphere of matter in a Newtonian orbit around the Earth.

suggest that Poe is not of Victorian America, but of the human soul, that his writings appeal to millions today and throughout the world not simply because they thrill but because they speak to something deep inside us, something that affects us all no matter what our culture. And it is surely that Poe, like all thinking and feeling people sooner or later, was deeply concerned about the problem of death, a problem that is universal to the human condition, the problem of why people, in particular, die, and more deeply the problem of why something so precious as a conscious life, a *person*, all too apparently gets snuffed out, and permanently—a problem captured by Poe with that prophetic word "Nevermore."

Poe not only faced, in his poetry and prose, the problem of death, but as we shall see in the next chapter he tried to make sense of the problem and solve it as best he could, as a double problem: death as the ultimate violation of justice as fairness and desert, and death as the most grotesque deformity of life. Poe provided his answer to both of these problems in *Eureka*. Rather than reject science or wholly embrace it, Poe modified it in association with his theology and literary theory. The latter is the focus of the present chapter, the theology is the focus of the next chapter, and the science is the focus of the two chapters after that. The fundamental point is that *Eureka* should be viewed as a *synthesis*.

But for the moment, let us return to "The Raven" and Poe's claim about melancholy and pleasure. In particular, consider alone the second stanza of that "mortally immortal" poem, with my underscores added, (keeping in mind that my understanding of poetics is not only that of a ninety-ninth-rate poetaster, but also of a servile imitator and literary thief,) and experience "the comprehension of sweet sound" from the content of what is said (notice especially the emphasis on *here*, which does not necessarily preclude *there*) combined with the rhythm and the rhyme, the up from the down:

> Ah, distinctly I remember it was in the bleak December;
> And each separate dying ember wrought its ghost upon the
>   floor.
> Eagerly I wished the morrow;—vainly I had sought to
>   borrow

From my books surcease of sorr<u>ow</u>—sorrow for the lost
Len<u>ore</u>—
For the rare and radiant maiden whom the angels name
Len<u>ore</u>—
Nameless *here* for everm<u>ore</u>.[62]

"The Raven" created a sensation in Poe's time and is one of the best-known and best-loved poems to this day.[63] And it is safe to say that a large part of the reason for this is the way Poe paints in the poem with words. But why would the mourning of a deceased loved one, why would the melancholy, which is Poe's claim, result in the pleasurable elevation of the soul? Whatever the answer, it is surely not to be found exclusively in the idiosyncrasies of culture.

Death is, of course, the ultimate downer, (excuse the pun,) the profoundest leveller, (Richard Overton notwithstanding,) and the massive exception to the claim that biology is statistical

62. *P&T*, 81–82.

63. It is surely therefore all the more remarkable, if Susan Talley is to be believed, that sometime within the last three months of his life Poe said to her, in effect, "*The Raven* was never completed. It was published before I had given the final touches. There were in it certain knotty points and tangles which I had never been able to overcome, and I let it go as it was." *Weiss*, 87. One must always keep in mind, of course, the biographical *conditions* under which Poe composed. But even so, to focus on "blemishes and imperfections" in the poem, as literary critics are wont to do, is to miss (assuming the blemishes and imperfections) the proverbial forest for the trees, its profound *adaptation* and continuing *unity of effect*. For a prime example of this failing, claiming Poe sacrificed "*sense*" for "*sound*," summed as "Poe's carelessness and unscrupulousness in the use of words," see T.S. Eliot (1948). "From Poe to Valéry." In *Recognition*, 209–211, 218. If I didn't know better, I would call him T.S. Idiot. At any rate, for an excellent in-depth analysis that undermines Eliot et al. on Poe's poetry, concluding that "the poems of Edgar Allan Poe develop their meaning by means of their intricate sound design," see Sławomir Studniarz (2014). "Poetry as 'an inferior or less capable Music': Sound and Meaning in 'The Conqueror Worm' and 'To One in Paradise.'" *The Edgar Allan Poe Review* 15 (1), 59–81. See also Sławomir Studniarz (2015). "Sonority and Semantics in 'Annabel Lee.'" *The Edgar Allan Poe Review* 16 (1), 107–125. See also *Zimmerman*, 51, 115, 116, 137–138, 148, 220, 221–222, 224–225, 239–240. For my own part, I think we should be prepared to entertain the possibility that Poe brought out the Hop-Frog in Thomas Stearns, a topic that shall await a later work.

(something Michael Ruse forgot to mention). Our awareness of death comes with being human, and is possibly unique in the animal world. Somewhere in the phylogeny of our species, or genus, diachronically conceived, we passed through a threshold into the awareness of our mortality, a threshold also passed in the ontogeny of each of us as normal human beings (in some ways ontogeny really does recapitulate phylogeny). It is the ultimate loss of innocence, reflected in mythology no less than in reality. As fully conscious human beings, naked in the truth, we are likely to view life, or rather conscious life, or maybe just human life, in the very least our own life, as infinitely precious, and therefore as something that *ought* to be extended indefinitely but clearly is not. Hence we come face to face with the problem of death. Whether we die prematurely or maturely, it's not right, it's not just, there *must* be more to life than the grave. The problem of death reroutes us, then, to the problem of justice, conceived in terms of fairness and desert. *I* do not deserve to *end*. The show *must* go on—and not with me written out of it. And not just me but others like me as well, especially the ones I *love*. Otherwise this world, this universe, is grossly, essentially, and infinitely unjust and malformed.

I doubt that there is any one answer for why people feel a pleasurable elevation of soul (if they do) from mourning the death of a loved one. Certainly religious belief would play a role in many, a conviction about the hereafter. For some, of course, it would have nothing to do with religion at all, but possibly would simply be about the thought that they themselves are still alive, the mourning a sort of discord that heightens the beauty of the music of being alive, hence the "luxury of sorrow." Even more commonly, it could be from a kind of catharsis, the instinctive sense of cleansing that accompanies the release of one's emotional pain. And then perhaps, on a note not here but *there*, an ethereal tone or supernal chord, it could simply be about the elevation of soul that comes from—*hope*.

The hope expressed in "The Raven," however, seems but a "dying ember," for after the hope expressed for the "balm in Gilead," for the "distant Aidenn," and the accusation that what the raven says is a "lie," the young narrator's hope has apparently given up the ghost by the end of the poem, with its "And my soul . . . Shall be lifted—nevermore!"

At any rate, Poe says the unity of effect of the poem, the meaning of the "self-torture" of the narrator and of the raven's "Nevermore," indeed the point and purpose of the poem as a whole, is that of a "*Mournful and Never-ending Remembrance*."[64] We shall return to Poe's meaning in the next chapter, on Poe's theology, for a proper exploration of it does not belong here.

Turning next to "The Rationale of Verse" (1848), published in two parts in October and November 1848 in the *Southern Literary Messenger*, only three and four months following the publication of *Eureka*, it was originally submitted for publication roughly two years prior, and was apparently revised during 1848 prior to its acceptance for publication in August of that year.[65] It is therefore, for all intents and purposes, a work of literary theory contemporary with *Eureka*. But we do not find the same themes examined above. Instead, we find a long and technical treatise on spondees, dactyls, accentuation, long versus short syllables, trochees, scansions, ancient Greek versus modern verse, and so forth. And that is part of the problem. According to the Levines, for example, from the viewpoint of English verse Poe's essay "seems perverse and not very sensible," that "Poe is so slippery one can never be sure," that "Filled with objections to pedantry, it is itself pedantic," that it is "pedantic without being learned," that it is "philosophically slippery," that it "seems almost cranky in tone," that it is "also clever, witty, and sometimes downright funny," and that "Poe's hypothetical history of poetry," which views the evolution of complexity from the simplicity of the spondee, suggests "a philosophical kinship to *Eureka*," given that Poe "repeatedly . . . saw oneness evolving toward complexity."[66] McGann, as another example, (himself

64. *E&R*, 25.

65. *Poe Log*, 671, 702, 715, 749, 760. It has an even earlier history, as "Some few passages" of it, Poe tells the reader in a footnote to the title, was published "about four years ago." The reference is in fact to "Notes Upon English Verse," published in March of 1843. *Letters*, 377–378.

66. *Critical Theory*, 77, 78, 79. The comparison with *Eureka*, by the way, is entirely misplaced in the point given, since in *Eureka*, as we shall see in Chapter 5, Poe did not see the evolution of complexity, whether of solar systems or of life, as occurring in the expansion phase of the Universe, but instead as occurring in the contraction phase, so that he did *not* see "oneness evolving toward complexity," which is the modern view in cosmology.

an expert on nineteenth-century British poetry,) noting that "The difficulties of the essay have been apparent to everyone who has worked with it," sees Poe's essay as a "seriocomic performance," but unlike "The Philosophy of Composition" Poe's sequel "does not succeed in handling that contradiction successfully," given that it is longer and "more tedious and full of convoluted detail" than any other of Poe's critical essays, that "much of it . . . [is] close to nonsense—in my view, often deliberated nonsense," and that "the essay's critical focus on pedants of versification has obscured the positive contributions to prosody that he wanted to make."[67]

At any rate, two points in "The Rationale of Verse" are of especial interest for us. First, what Poe says about metaphysics is particularly relevant for shining some light on *Eureka*. Right at the beginning of "Rationale" Poe refers to "the cloud-land of metaphysics, where the doubt-vapors may be made to assume any and every shape at the will or at the fancy of the gazer."[68] This is clearly a disparaging remark about metaphysics, and yet the key phrase is repeated in *Eureka* as "the Cloud-Land of Metaphysics."[69] "Rationale," however, would seem to help inform Poe's usage in *Eureka*, for the reference certainly seems to be to *a priori* metaphysics, what Poe, as we shall see in Chapter 7, would certainly decry as *bad* metaphysics, for not being grounded in empirical facts. Keeping that in mind, *Eureka* itself is referred to by Poe as a work of metaphysics. Right at the start he says, "I design to speak of the *Physical, Metaphysical and Mathematical—of the Material and Spiritual Universe*," while later in the work he refers to its construction as aiming at "metaphysical order."[70] He also refers to "the kingdom of Metaphysics," without the negative qualifier.[71] So we may safely conclude that not all metaphysics for Poe is bad metaphysics—safely, that is, unless one is among "the doubters for Doubt's sake," that "unprofitable and disreputable

67. *McGann*, 74, 75.

68. *E&R*, 26.

69. *P&T*, 1318.

70. *P&T*, 1261, 1330.

71. *P&T*, 1290. Poe also uses "the nebulous kingdom of Metaphysics," but the qualifier might be a positive one, a play on words, as the passage seems to refer to metaphysics based on the nebular hypothesis of Laplace, whose hypothesis Poe made much of, including that it is true. *P&T*, 1270.

tribe," which includes "the professional questioners" and those "out of Bedlam," the latter more specifically "the mere madmen who deny the Newtonian law of Gravity on which the results of the French mathematicians are based."[72] The important distinction would therefore seem to be the *a priori* / *a posteriori* divide, with *a posteriori* metaphysics being metaphysics based on science, (but of course going beyond it,) hence not a mere "cloud-land" of personal doubt and fancy. And this is in fact consistent with Poe's earlier treatment of the term "metaphysics." In "Pinakidia" (1836), for example, Poe claims that "metaphysics" is misunderstood, as it was not meant by Aristotle to imply any superiority to physics, but merely to indicate "the rational order of study" *after* physics.[73] This important point for understanding Poe is repeated in "Marginalia" (November 1844).[74] Meanwhile, elsewhere, in "The Imp of the Perverse (1845), Poe decries "metaphysicianism" as being "concocted *à priori*."[75] The bottom line is that, when we turn to *Eureka* in Chapter 5, we shall indeed see that although the work ascends into the heavens, so to speak, and goes far beyond physics, Poe made quite a serious effort to ground it in empirical science to what he thought was, and what in fact was, a substantial extent. *Eureka*, then, is ultimately for Poe a work of *a posteriori* metaphysics—i.e., of *good* metaphysics.

Also interesting in "Rationale" is a passage in which Poe heaps contempt on the "book-worm" of "classical scholarship," specifically English theorists who take Homer's *Iliad*, rather than "Nature and common sense," as the "starting point" in erecting rules for modern verse, rules that "contradict each other every five minutes, and for nearly all of which there may be found twice as many exceptions as examples." "The only thing," says Poe, "made out in them is a very magnificent contempt for Leibnitz's [*sic*] principle of 'a sufficient reason.'"[76] One of the fundamental principles of the seventeenth-century German metaphysician Gottfried Leibniz was not only that God is the source of all rational activity

72. *P&T*, 1299, 1301, 1317.

73. *Brevities*, 97.

74. *E&R*, 1328.

75. *P&T*, 826.

76. *E&R*, 32.

in the Universe, but given that God is perfectly rational God must have a sufficient reason for everything he does, including the existence of the Universe itself. One of the interesting consequences of this principle is that no two of anything in the Universe, including atoms, can be 100 percent qualitatively identical, for if they were, then God could not possibly have a sufficient reason to have one of them in one place and the other in another place. Given any two similar things, then, similar from our perspective, there must be at least one inherent difference between them, even if we fail to discern what it is. Known as *Leibniz's Law*, it is to say that there must—*logical* "must"—be a numerical identity (one-and-the-same thing) between any two indiscernible entities. In the few places in which Poe mentions Leibniz, he is usually not positive but negative.[77] Nevertheless, in apparently writing favorably of Leibniz's principle of sufficient reason, we can see Poe's application of the principle not only to his own literary theory, in his strictures on the "unity of effect," but to *Eureka*. If the Universe, as Poe says there, "is a plot of God,"[78] and if "The plots of God are perfect," which is also what he says in the same paragraph, and if, as we have already seen, (and shall see again below,) a plot must have a "unity of effect," meaning that everything in the plot must serve the ultimate effect, such that every item, no matter how small, must serve that function, directly or indirectly, then it follows that everything in the Universe, as God's plot, which must be perfect, serves for the unity of effect intended by God. (Just what Poe took that unity of effect to be, as well as what he took God to be, shall be reserved for the following chapter.) This does not mean that Poe was committed to the absolute rationality of Leibniz's God, and so to Leibniz's Law, (we shall see that he was not,) but we do get a new insight into just how *synthetic* Poe was in *Eureka*.

The last work of Poe's literary theory that we shall examine is "The Poetic Principle" (1850), which is the text of a lecture of the same name delivered by Poe (and very successfully) on three different occasions in Virginia, once in August and twice

77. Pollin, *Brevities*, 148n; see 190n and 272n for Poe's possible secondary source on Leibniz.

78. *P&T*, 1342.

in September 1849,[79] the third occasion occurring on the thirteenth day before he lost his individuality. First published not until August 1850, "The Poetic Principle"—in which, remarkably, "hoax plays no part"[80]—represents, arguably, the peak of Poe's literary theory. Although Poe borrowed much from his earlier writings, particularly from his second review of Longfellow's *Ballads and Other Poems*, discussed above, he brought it all together in this work and achieved a level of expression unrivaled by any of his earlier writings on literary theory.

But it is with regard to *Eureka* that "The Poetic Principle" is most important for our purposes. Although the three lectures in Virginia were post-*Eureka* by more than a year, an earlier lecture entitled "The Poetic Principle" was given in Rhode Island by Poe in December 1848, with his acceptance to an invitation to give the lecture indicated by a letter he had written two months earlier.[81] In one of the most bizarre episodes of his life, during which he seems to have gone off the rails,[82] possibly from cholera or from a drinking binge and lack of food, Poe's written version of the lecture was lost (stolen, to his mind, from his valise) during a trip, with an unintended two-week stay in Philadelphia, sometime during the first two weeks of July 1849, along with another lecture, Poe writing in despair when he was well again that unless the written lectures were recovered he should have to "re-write one of them."[83] "The Poetic Principle" was evidently rewritten, and was probably revised to some degree, in time for the first lecture in Virginia, given on August 17.[84] The history of the work, therefore, brings it much closer to the composition and publication of *Eureka* than one might first suppose—roughly three months between the publication of *Eureka* and Poe's acceptance to give a lecture

79. *Poe Log*, 825–828, 834–836, 840–842.

80. *McGann*, 90.

81. *Poe Log*, 778; *Letters*, 745–746, 706.

82. *Silverman*, 412–419.

83. *Letters*, 824.

84. *Letters*, 826–827. That this lecture was meant by Poe to be similar to the Rhode Island version of December 1848 is corroborated by an advertisement published on July 31, 1849, for Poe's first lecture on "The Poetic Principle" *in Virginia*: "The lecture which Mr. Poe proposes to deliver in Richmond, is that which he gave with such *éclat* a winter or two ago." *Poe Log*, 820.

on what was almost certainly meant to be "The Poetic Principle." The fundamental point is that many of the ideas in "The Poetic Principle" find a remarkable correspondence with, and shine considerable light on, many of the ideas in *Eureka*, (which also, by the way, is a reworking of many of Poe's earlier ideas,) such that we may rightly consider the two works as a complementary pair.

For a start, we find, once again, that a poem should not be too short, the extreme of which is "a mere epigrammatism," since the effect a poem is supposed to produce can never be "profound or enduring" when the poem is brief. "There must," Poe says, "be the steady pressing down of the stamp upon the wax." At the other extreme is the "epic mania" of the "long poem." A long poem, he says, whether epic or not so much, is "a flat contradiction in terms," because the "sustained effort" required by reading or listening to a poem cannot be maintained for longer than half an hour. Accordingly a long poem, such as "Paradise Lost," results in "a constant alternation of excitement and depression," so that the net effect of a long poem is "nullity," the positive and negative canceling each other out.[85] By this definition, then, *Eureka*, at almost one hundred and fifty pages in the original,[86] cannot itself count as a poem, and if still regarded as such would be a "nullity." But the concept of nullity returns in *Eureka*, as we shall see in Chapter 5, as the correlative of Poe's concept of "nothingness," such that Poe's "nullity" helps us to understand Poe's "nothingness," which otherwise is a confusing use of the latter by Poe.

Next, the idea that the main purpose of a poem is to teach particular truths is "the heresy of *The Didactic*," which is "a heresy too palpably false to be long tolerated." Truth, instead, is the goal of reason or intellect, and for this is required "severity rather than efflorescence of language," language that is "simple, precise, terse," involving a "calm, cool, and impassioned" mood. To think that the main purpose of a poem is truth, then, is to be "theory-mad beyond redemption," for it is "attempting to reconcile the obstinate oils and waters of Poetry and Truth." It does not follow, however, Poe adds, that "the lessons of Truth, may not be introduced into a poem, and with advantage; for they may subserve,

85. *E&R*, 75, 73, 71, 72.

86. In modern typeset, it is 102 pages in *P&T* and 106 pages in *Eureka*.

incidentally, in various ways, the general purposes of the work:—but the true artist will always contrive to tone them down." Later in the essay Poe adds that if, "through the attainment of a truth, we are led to perceive a harmony where none was apparent before, we experience, at once, the true poetical effect," but he adds that "this effect is referable to the harmony alone, and not in the least degree to the truth which merely served to render the harmony manifest."[87] *Eureka* is clearly a work of prose, but with a writing style that is hardly "simple, precise, terse." We have seen from one of his letters, on the other hand, that Poe did indeed think of *Eureka*, in large part, as a scientific work on the Universe. Aside from these problems, however, considerable as they are, we do seem to be on more solid ground for considering *Eureka* a "prose poem," for the proposed scientific truths in it could be seen as subservient to the work as a whole, subservient to "render the harmony manifest" in the soul of the reader, the universal *lost chord*, if you will.

In league with what Poe says about truth, neither is the purpose of a poem to "inculcate a moral," and so should not be judged on that basis. The "world of mind" is most obviously divided, says Poe, in accordance with three "*offices*" or "distinctions, . . . Pure Intellect, Taste, and the Moral Sense."[88] According to the Levines, "G.R. Thomson shows that Poe derives his divisions from Immanuel Kant's introduction in *The Critique of Judgment*," while years earlier Northrop Frye claimed that Poe's distinction "has been used in poetics ever since Plato's time," and he claims that Poe's reason for putting Taste in the middle—"because it is just this position, which, in the mind, it occupies"—is an "admirable explanation."[89] What matters more to me is the distinction, not the source, for there must be a point to tracing a source, as tracings can go on seemingly forever. For example, one could trace moral sense theory to Francis Hutcheson, via David Hume and Adam Smith. But what would be the point?

What matters most is what Poe does with the distinction. With regard to Poe's Moral Sense, which includes "Conscience," its purpose, he says, is to inform us of "Duty" and "obligation."

---

87. *E&R*, 75, 76, 78–79, 93.

88. *E&R*, 75, 76.

89. *Critical Theory*, 203n13; Frye (note 48), 243.

Much like truth, morality can play a role in a poem, but only a subservient role and then only if it aids in producing the designed effect of the poem. Certainly it is not essential and has only "collateral relations" with the purpose of a poem, if it has any relations at all.[90]

The faculty of Taste, for Poe, which "holds intimate relations" with the extremes of Pure Intellect and the Moral Sense, is the proper "office" of mind for the poem. Here again we see that "a poem deserves its title only inasmuch as it excites, by elevating the soul." But elevate the soul to what? Certainly not to "deformity" or "disproportion," which is what Taste abhors, but "to the fitting, to the appropriate, to the harmonious—in a word, to Beauty." Poe even claims that there is an "immortal instinct," a "sense of the Beautiful," "deep within the spirit of man." Hence the purpose of a poem is the perception, even the "creation," of "Beauty." "In the contemplation of Beauty," he says, "we alone find it possible to attain that pleasurable elevation, or excitement, *of the soul*, which we recognize as the Poetic Sentiment." Accordingly Poe defines "the Poetry of words as *The Rhythmical Creation of Beauty*." But it is not *physical* beauty that Poe means here. Instead, it is "Supernal Beauty," of which the "Poetic Principle" is "the Human Aspiration." Along with this aspiration, or part of it, is the human "thirst" for "immortality," for "perennial existence."[91] Borrowing again a well-known line from Shelley, (one of his favorite poets,) he says, "It is the desire of the moth for the star." So it is not merely the contemplation of Beauty, after all, "no mere appreciation of the Beauty before us," but instead "a wild effort to reach the Beauty above. Inspired by an ecstatic prescience of the glories beyond the grave, we struggle, by multiform combinations among the things and thoughts of Time, to attain a portion of that Loveliness whose very elements, perhaps, appertain to eternity alone."[92]

These are powerful sentiments, indeed, possibly deriving their power from some sort of religion or religiosity instinct coded

90. *E&R*, 71, 76, 78.

91. Poe vacillates between whether human immortality is a given or an aspiration. Of course it could be both. But as we shall see in the following chapter, he was really torn between the two, such that it would be a mistake to think that Poe *assumed* that humans are immortal.

92. *E&R*, 78, 77.

for in our DNA, but no doubt Poe has expressed himself in language equal to the power of his thoughts. To all of this Poe adds a dimension of sadness, for "we weep then," not, he says, from an "excess of pleasure," but instead "through a certain, petulant, impatient sorrow at our inability to grasp *now*, wholly, here on earth, at once and for ever, those divine and rapturous joys, of which *through* the poem, or *through* the music ['the most entrancing of the Poetic moods'], we attain to but brief and indeterminate glimpses."[93] Hence by "Supernal Beauty" Poe includes a dimension of deformity in the "Poetic Principle."

This returns us to the graveyard. In commenting on "June" by William Cullen Bryant, one of Poe's favorite poems, in which the narrator refers to his friends coming to weep by "my tomb," Poe claims that "The impression left [by the poem] is one of pleasurable sadness," and adds that "this certain tint of sadness is inseparably connected with all the higher manifestations of true Beauty." But unlike his vague suggestions in his discussion on "The Raven" in "The Philosophy of Composition," Poe now confesses that the inseparable pleasure from mourning is "how or why we know not."[94] What is especially important here, at any rate, is the idea of pleasure mixed with pain, beauty with deformity, and "inseparably" so, the rationale for which we shall discover when we turn to Poe's theology in the next chapter.

Returning to Poe's comments on "June," the tomb, of course, takes us back to Poe's "most poetical topic" in "The Philosophy of Composition," which is "the death . . . of a beautiful woman." Poe does not repeat this in "The Poetic Principle," but he does claim that "No nobler *theme* ever engaged the pen of poet," namely, "the unwavering love of woman." Mere passion, he states earlier on, is "the excitement of the heart," and he adds near the end that it is more likely "to degrade, rather than to elevate the Soul." But "Love," he immediately adds, in clear contradistinction to passion, "—the true, the divine Eros— . . . is unquestionably the purest and truest of all poetical themes."[95]

93. *E&R*, 77.

94. *E&R*, 81, 82.

95. *E&R*, 91, 93.

In all of this, the learned might see reflections of the ascent to the vision of Beauty itself expressed in Plato's greatest artistic and mystic masterpiece, *Symposium*, especially given what Poe says about "Beauty," "rapturous joys," and "the true, the divine Eros—the Uranian, as distinguished from the Dionæn Venus."[96] But although strongly suggestive, the comparison falls to pieces once we turn to *Eureka*, for there are no eternal Forms in Poe's imagined reality, no immaterial essences existing outside of space and time, and hence no Beauty existing in and of itself. Instead there is only God, the Creator numerically identical with its Creation, both One and the Same, who created our Universe as a true poet creates a poem or tale, for the purpose of Pleasure alone, but now with a perfect Plot, in which each of our lives is a subplot, with every one of us a moth drawn to the light of the star because we ourselves are made of God-stuff, each a differentiated portion of God, each a part of the Many that longs for the One, for a return to God as Unity.

But enough. Poe was an eclectic synthesizer in the extreme; he was also prone to hyperbole, and that's how one has to understand *Eureka*. Nevertheless, given Poe's literary theory, which is absolutely essential to know in its details if one is to come to understand and appreciate *Eureka*, we still have the problem with which we began this chapter—the problem that *Eureka* itself does not count as a poem in Poe's terms, given its length and lack of versification. So what then, after all, did Poe mean in subtitling *Eureka* "A Prose Poem," which is reaffirmed by Poe's Preface?

Eureka as a "Prose Poem"

The temptation might be to take the subtitle as a sign that *Eureka* was meant as a hoax, for whenever Poe attempted to perpetrate a hoax in his published writings he seems to have always given a hint or sign. For reasons that shall have to await the end of Chapter 4, I must reject that interpretation of *Eureka*. The problem then remains, which has occupied the minds of not a few Poe

96. *E&R*, 93.

scholars: if *Eureka* is not a hoax, what did Poe mean by subtitling it "A Prose Poem"?

For a start, it should be kept in mind that the modern hybrid meaning, a genre that has come to be known as "prose poetry," was not yet defined or in currency in Poe's time, not until Charles Baudelaire put it on the map with his *Petits Poèmes en Prose*, published posthumously in 1869. So we should try to figure out from Poe's own writings what he meant by his subtitle for *Eureka*.

Harry Poe provides us with an interesting suggestion. In his discussion on Poe's "The Domain of Arnheim" (1846), a tale we shall visit in subsequent chapters, in which Poe adds landscape gardening to "the poetic sentiment,"[97] Harry Poe turns to the fact that Edgar Poe calls "his scientific and theological essay, *Eureka*, a prose poem." "It is a poem," says Harry Poe, "as are all scientific theories, in the same way that a garden is a poem. All three cases involve the expression of Beauty."[98] Granted, Poe has his narrator say of Ellison, the supreme landscape gardener in "The Domain of Arnheim," that "In the widest and noblest sense he was a poet."[99] But Harry Poe arguably confuses poetry with poesy here, for in Poe's strict usage, as we have seen earlier in this chapter, poesy is the genus of which poetry proper is a species, along with other species such as painting, music, and sculpture, what Poe calls "various moods" and "modes" of poesy. Moreover, as we have also seen, it is not just the expression of beauty that makes a poem a *poem*, in Poe's mature view, repeated throughout the 1840s, including the time of *Eureka*, but instead the rhythmical creation of beauty using words. This is what differentiates a poem from other species of poesy. As such, then, *Eureka* itself completely fails as a poem *in the literal sense* as understood by Poe, as it is not at all an example of the rhythmical creation of beauty using words. Moreover, recall Poe's review (1841) of Bulwer-Lytton's *Night and Morning*. He states that if the ultimate purpose of the work is not the creation of beauty, then the work "is no subject for the Muse" and "what the world terms *prose* may be safely and properly left all else." So

97. *P&T*, 858.

98. *Evermore*, 77.

99. *P&T*, 858.

that would further explain why *Eureka* is expressed in the mode of prose, but not why it is also a poem.

In terms of length also, *Eureka* seems much too long for humans to take in during one sitting, and so would seem incapable of producing the maximal intended effect. Nevertheless, in the first few pages of *Eureka*, Poe does make the repeated claim that the work, a "survey of the Universe" based on the "general proposition" that "runs throughout this volume"—"*In the Original Unity of the First Thing lies the Secondary Cause of All Things, with the Germ of their Inevitable Annihilation*,"—is designed for an "individual impression" of the Universe, an "*individuality* of impression," a "mental gyration on the heel," as it were, as if one were on top of Mount Ætna and "by a rapid whirling on his heel" could "comprehend the panorama in the sublimity of its *oneness*."[100] I doubt, nevertheless, that most mortals could be expected to read *Eureka* in one sitting, and so according to Poe would miss the effect, but Poe himself was apparently capable of this, and if so was once again pretty much *alone*.[101]

At any rate, at the end of his Preface, as we have seen at the beginning of this chapter, although *Eureka* is clearly a work of prose, Poe says "it is as a Poem only that I wish this work to be judged after I am dead." So we shall have to look elsewhere for the answer we seek.

Jerome McGann also provides a valiant attempt at trying to solve the problem of what we may call *Poe's "Prose Poem" paradox*. Not only does McGann view "'The Philosophy of Composition,' much of 'The Rationale of Verse,' and many of the *Marginalia*" as "already working toward the prose-poetic breakthrough that we call *Eureka*," in that they are "poised at any moment to close the gap between the music of poetry and the music of prose," but he claims that *Eureka* "most fully illustrates and, in illustrating,

100. *P&T*, 1261–1262.

101. As an experiment, I sat down one day and read *Eureka* straight through at a modest pace, to see how long it would take. It took me slightly less than three hours, and during that time I found my mind wandering a little, having to take a brief break to eat, etc. In other words, as Poe put the problem of a long poem in "The Philosophy of Composition" (1846), "the affairs of the world interfere, and every thing like totality is at once destroyed." *E&R*, 15.

explains Poe's poetics," so that he further concludes not once but twice that *Eureka* is Poe's "most ambitious poetical work." As for McGann's reasons, one reason *Eureka*, he says, is "a key document in Poe's poetics" is because it is "performative rather than expository," often using "the first person plural," which marks it as "not just a poem but an oral poem," to which we must remember that it "was initially presented as a lecture." But more than that, *Eureka* is written in a manner that will create in the listener or reader "an experiential engagement with the issues," and "the poem's reader, [is] always a prominent player in Poe's work." Moreover there are mechanisms in *Eureka* that make it interactive, such as "its quietly intimate address" and the words "*now* and *here*, which mark the opening sentences of more than 20 percent of *Eureka*'s 266 paragraphs," and also the word "suppose," so that "The poetic action is not offered as a discovery but as *the conscious representation* of a discovery process that is already understood." *Eureka* is also "suppositious, fanciful, and imaginative," while adhering to Poe's stricture of "consistency," one of "the two most obvious 'poetical' features of *Eureka*," as it is "the measure of the formal Beauty of its Truth." The other obvious feature is Poe's use of the dash—"the bibliographical feature that generates the abundant digressions, emendations, transits, and linguistic conversions that comprise the body of the poem," and that also indicates the "Beauty [that] abounds through the 'leap.'" And then there is the end of *Eureka*, with its view of our contracting universe returning to Nothingness, which is the "poetic moment," a "vision of cosmic 'catastrophe,'" says McGann, just like in "The Raven" with its final "Nevermore," each of which "locates Poe's catastrophe of beauty."[102]

McGann's reasons altogether, however, are not convincing. For a start, calling *Eureka* a "poem" while one is making a case that it is a poem is like a prosecutor referring to the defendant in a murder trial as "the murderer." Until proven guilty beyond

102. *McGann*, 88, 95, 86, 116, 95, 105, 95, 98, 100, 104, 106, 96, 97, 109, 94. Poe himself says the dash is "properly employed" when it represents "*a second thought—an emendation*. In using it just above I have exemplified its use." "Marginalia" (February 1848), 1426. It will be noticed that Poe's *use* of the dash in his corpus greatly exceeds his *definition*. See especially *Zimmerman*, 18, 22–23, 183, 204, 216, 217, 271, 303, 346.

a reasonable doubt, it is a question-begging epithet that should prompt an objection. Moreover, many of McGann's reasons in his argument would lead us to believe, if unchallenged, that Descartes' *Meditations*, for example, is literally a poem, given that it is intimate and interactive with the reader and is imaginative, but those are hardly sufficient reasons for concluding that it is even partially a poem. We should also keep in mind that Poe repeatedly warned against the use of a lot of metaphors in prose (though not, of course, in poetry). As Brett Zimmerman puts it, "I have not found a *great deal* of metaphors in Poe's prose—perhaps because of a certain item in his authorial credo stipulating that one must use this device sparingly."[103] This point serves to further distance *Eureka* as a work of poetry, as it contains relatively few metaphors. Even more important, where in *Eureka* are the rhythms, the rhymes, and the stanzas that are so obviously parts of what are so clearly poems by Poe, such as "The Raven"? And, most especially, where is the satisfaction of Poe's famous definition of poetry as the rhythmical creation of beauty using words that would *literally* make *Eureka* a poem? It's not there. But perhaps the fatal objection to McGann's interpretation of *Eureka* as a poem comes from the conjunction of two of his own statements, which destroys his view like a matter–antimatter explosion. On the one hand, we are told that "Poe's mature poems . . . are not voiced by people . . . but by highly imaginative and anonymous characters," while on the other hand there is "*Eureka*'s first person," the "first person of *Eureka*."[104] And we must emphasize here the blatant fact that *Eureka* is voiced by its author, none other than the non-anonymous "Edgar A. Poe," entered as such "according to Act of Congress, in the year 1848," who begins *Eureka* with a Preface that says, "I offer this Book of Truths," and who continues with numerically the same "I" and "me" throughout the rest of the book (with the exception of the imaginary epistle, of course).

A wider suggestion for why Poe called *Eureka* a poem is given by the Levines in the Introduction to their annotated edition of Poe's *Eureka*.[105] They in fact provide a number of "overlapping

103. *Zimmerman*, 251.

104. *McGann*, 117, 100, 104, 95.

105. *Eureka*, xi–xxvii.

explanations" as suggestions: (i) Poe was following a tradition of treatises on the nature of the Universe that were either in poetic form or were charged with poetic emotion, works "that unify poetic, religious, and scientific approaches to truth," such as Parmenides' poem on Being, Humboldt's *Kosmos*, and J.P. Nichol's *Views of the Architecture of the Heavens*; (ii) Poe claims in *Eureka* that "Poetry and Truth are one" and that "Man cannot long or widely err, if he suffers himself to be guided by his poetical . . . instinct";[106] (iii) Poe claims in *Eureka* that "the Universe . . . is . . . the most sublime of poems";[107] (iv) Poe claims in *Eureka* that the work is constructed "in the hope of thus the better keeping unbroken that chain of *graduated impression* by which alone the intellect of Man can expect to encompass the grandeurs of which I speak," which explains why he did not construct the work using "a merely natural . . . arrangement";[108] and (v) a reviewer of Poe's lecture on the Universe, before Poe seems to have produced the subtitle for *Eureka*, wrote that "The work has all the completeness and oneness of plot required in a poem."[109]

The solution of the Levines is to succumb to a "famous contradiction" of *Eureka*, that Poe, in calling *Eureka* a poem, was contradicting his literary dictum that a long poem cannot exist. Poe's literary theory "is sometimes contradictory, too," they tell us, and "Poe's *arguments* are frequently contradictory throughout his writings," such that "Poe reserves to his poet-scientist the right to be inconsistent, as 'our Keplers—our Laplaces' sometimes are. But he is sure that the universe itself is consistent." Furthermore, "one can prove that Poe is playing private literary games in so many places in *Eureka* that one questions whether to trust anything in the book. Poe played such games throughout his career." The situation is so bad, all told, that "It is impossible ever to be sure when Poe is truly committed to an idea." The summary expression of their view, which would have to include the "famous contradiction" of *Eureka*, is in Stuart Levine's Introduction to their companion volume, on Poe's literary theory, which

106. *P&T*, 1349.

107. *P&T*, 1349.

108. *P&T*, 1330.

109. *Critical Heritage*, 277–278.

is that "Certainly one does not want to waste energy on trying to reconcile the irreconcilable."[110] (This, I suggest, is what Poe can do to a Poe scholar!)

That *Eureka* as a whole was not intended as a hoax is, again, a matter I shall reserve for the end of Chapter 4. Here I shall reply to their view that it is best to give up and accept Poe's apparent inconsistencies and contradictions as just that. This answer is unacceptable for two closely related reasons. First, it lowers the standard of scholarship, which can easily lead to mistakes, in the very least the missing of subtle differences in meaning that can resolve the apparent inconsistencies and contradictions. Second, their solution violates the *principle of charity*, a justly-respected principle of interpretation often taught to students but more often (in my opinion) not applied by professors themselves in their own works. In short, there may indeed have been many reasons why Poe called *Eureka* a poem, but by the principle of charity I am impelled to reject thinking of Poe as inconsistent or contradictory *except as a last resort*, when charitable interpretations have been exhausted.

Accordingly, the solution to which I myself am inclined focuses on one of the possibilities in the Levines' list and views it as subsuming the rest. The core of the solution in fact appeared in an anonymous review of *Eureka*, dated August 1, 1848, in the *New Church Repository*, a periodical devoted to the views of the Swedish scientist, poet, and mystic Emanuel Swedenborg. The reviewer, probably George Bush, suggests that Poe "calls his work a poem, perhaps because, with Madame De Stael, he regards the Universe itself as more like a poem than a machine, and therefore to be treated *poematically*."[111]

This is a remarkably insightful suggestion, a profound insight that I believe provides the key to solving the mystery of Poe's "Prose Poem" paradox. Let us proceed step by step, beginning with the Universe as a machine.

110. *Critical Theory*, 4.

111. *Critical Heritage*, 289. Reverend George Bush was Professor of Hebrew at New York University and proprietor of the periodical. He was also a friend of Poe's, which raises the possibility that he got his understanding of *Eureka* as a prose poem directly from the poet himself.

First, in the eighteenth century, the height of the European Enlightenment, the Universe was literally thought to be a machine, created by God. The idea of the Universe as a machine was beautifully summarized by the character Cleanthes in Part II of David Hume's *Dialogues concerning Natural Religion* (1779), as the foundation of the argument from design for God's existence. It begins as follows: "Look round the world: Contemplate the whole and every part of it: You will find it to be nothing but one great machine, subdivided into an infinite number of lesser machines, which again admit of subdivisions, to a degree beyond what human senses and faculties can trace and explain."[112] No less than William Paley extended this view into the nineteenth century, with his *Natural Theology* (1802), his defense and elaboration of the argument from design, in which he compares the Universe to a watch, which he calls a "mechanism" or "machine," such that "every indication of contrivance, every manifestation of design, which existed in the watch, exists in the works of nature; with the difference, on the side of nature, of being greater and more, and that in a degree which exceeds all computation."[113] Similarly, in Poe's time now, the astronomer John Pringle Nichol, in his *Views of the Architecture of the Heavens* (1838), (the American edition of which Poe drew on for *Eureka*,) refers to "the existing mechanism of the Universe" and "the influence of this vast mechanism over the minute arrangements of its subservient parts."[114]

In all of this, the foundation for the view of the Universe as a machine was the physics of Newton, with his laws of mechanics permeating the Universe. And although Newton was not a *deist*, (in his view God is required to "govern" the Universe, not only, for example, setting planets in their orbits but in maintaining them,) it is the key to the thought of deists, that God created this

---

112. Norman Kemp Smith, ed. (1947). *David Hume: Dialogues concerning Natural Religion*. Indianapolis: Bobbs-Merrill, 143.

113. Matthew D. Eddy and David Knight, eds. (2006). *William Paley: Natural Theology*. Oxford: Oxford University Press, 7, 8, 9, 16.

114. J.P. Nichol (1840). *Views of the Architecture of the Heavens: In A Series of Letters to a Lady*. New York: H.A. Chapin & Co., 114. On this and the next page, Nichol also refers to the Universe as a "volume" and to "the sky" as "splendid hieroglyphics," a theme we shall return to in Chapter 7 in the section on Champollion.

big machine and then stepped back to let it run in accordance with the laws of nature that God built into it. Accordingly, Hume has his critic Philo do everything he can to diminish the comparison of the Universe with a machine, arguing at times that the world is more like a dumb animal or a vegetable than a machine, or if a machine that it has too many defects.

Turning now to Poe, Bush claims that Poe in *Eureka* "regards the Universe itself as more like a poem than a machine." I would go even a step further, to the view that Poe regards our universe as *literally* a poem, or rather that he attempts to make the case for the Universe as such, much as the scientists and philosophers of his time and beforehand attempted to construe the Universe as a machine. As we have already seen, Poe calls the Universe "the most sublime of poems" and "a plot of God."[115] A plot, it is to be noticed, can be contained in a poem, as with "The Raven," its *dénouement* the final "Nevermore." And not only, of course, can a plot be put into words, but Poe described the subject of his lecture on the Universe, the Universe itself, as "a broad text."[116] Moreover, since the Universe is a plot of God, in the context of God's creation Poe reminds his readers of "the *physical power of words.*"[117] So great was his "faith in the *power of words*" that he claimed "I do not believe that any thought, properly so called, is out of the reach of language," including the "evanescence of fancies."[118]

But more important, to make the case for the Universe as a poem with a plot, Poe had to argue against the scientific view of the Universe as a machine. And the only legitimate way he could do that was to take on the science itself. Certainly the received cosmology of his time, with an infinite universe of stars, and either a creation of such by God or a universe with no beginning or end, made the Universe at best a story, not a plot, let alone a poem with a plot. Poe makes the case in *Eureka* as follows, given the cosmology of his time: "We should have been forced to regard the Universe with some such sense of dissatisfaction as we experience in contemplating an unnecessarily complex work

115. *P&T*, 1349, 1342.

116. *Letters*, 647.

117. "The Power of Words" (1845). *P&T*, 825.

118. "Marginalia" (March 1846). *E&R*, 1383–1384.

of human art. Creation would have affected us as an imperfect *plot* in a romance, where the *dénoûment* is awkwardly brought about by interposed incidents external and foreign to the main subject; instead of springing out of the bosom of the thesis—out of the heart of the ruling idea."[119] But the situation would seem even worse, given that the view of the Universe as an essentially static and infinite universe of stars, as an infinite machine made of infinitely lesser machines, the parts perhaps in need of a readjustment now and then, has no point and purpose in itself. At most it is mere scaffolding, with most of the Universe serving apparently no real purpose.

Poe also saw the fundamental weakness in the physics and cosmology of his time. Beginning, I believe, as we shall see in Chapter 5, with the realization that the only real solution to Olbers' paradox is to view the Universe not as infinite but as finite, Poe then, given Newtonian gravity, had to think of the Universe as exploding from a "primordial Particle" of "unparticled matter." Everything else followed from that fundamental insight, the first real expression of Big Bang cosmology. Poe inferred the formation and interdependence of the laws of nature immediately following the Big Bang, he applied Laplace's version of the nebular hypothesis for the formation of stars and planets from gaseous nebulae, and he speculated on the formation of life on planets. As a result, the Universe looks more like a composition than a machine, as a plot with a beginning and a *dénouement*, even as a poem with a Poet God. Add to this picture Aristotle's functionalism, which greatly impressed Poe, but which he made even stricter, as Poe's God is not separate from the Universe (as with Aristotle) but *is* the Universe, for the maximization of his (God's) pleasure. Add now an expanding and contracting Universe, over and over again, "a novel Universe swelling into existence, and then subsiding into nothingness, at every throb of the Heart Divine," a process "renewed forever, and forever, and forever,"[120] making music like a beating heart, music that can be read, and you have a poem, the rhythmical creation of beauty using words, or rather each expansion

119. *P&T*, 1352.

120. *P&T*, 1356.

and contraction of the Universe a poem, followed by another, and so on forevermore.

If this interpretation is correct, then the science of *Eureka*, not just the science of Poe's day, but the science as *corrected* by Poe, which we shall see in Chapter 5, *his* science which made him well ahead of his time, *cannot* be viewed as peripheral to *Eureka*, or as only the "upper current," let alone as a hoax. It must, instead, be viewed as deeply serious and as central to Poe's task, as an essential part of the *synthesis* that is his magnum opus.

We need to understand how deeply revolutionary Poe was in all of this. That the Universe is or is like a plot apparently never occurred to Hume, for example. Even though he was arguably the greatest critic of the European Enlightenment, he was still a part of the Enlightenment and accordingly debated the existence of God in mainly Enlightenment terms. Thinking of the Universe as a plot, on the other hand, and even moreso as a poem, marks an enormous conceptual shift: it is about as big as the difference between the Age of Enlightenment, which exalted reason and intellect, and the Romantic Age, a reaction against the Enlightenment, which exalted intuition, imagination, passion, emotion, and aesthetic experience. Poe lived near the tale end of the Romantic Age, but probably (given his interests) would have considered himself a product of both.

But besides the difference in Ages, what is the fundamental difference between viewing the Universe as a machine and viewing it as a poem with a plot? Bush, it would seem, got us started in the right direction by stating that in the latter case the Universe is "to be treated *poematically*." By way of convergence, the same answer was provided in amplified form by Richard Wilbur in a lecture on Poe delivered on May 4, 1959. "Poe conceived of God as a poet," says Wilbur. "The universe, therefore, was an artistic creation, a poem composed by God. Now, if the universe is a poem, it follows that one proper response to it is aesthetic, and that God's creatures are attuned to Him in proportion as their imaginations are ravished by the beauty and harmony of his creation. Not to worship beauty, not to regard poetic knowledge as divine, would be to turn one's back on God and fall from grace."[121]

121. Richard Wilbur (1959). "The House of Poe." In *Recognition*, 257–258.

Harry Poe adds to this Edgar Poe's concept of "the mutuality of adaptation,"[122] which we shall examine more fully in Chapter 4, in the section on Poe's response to the Bridgewater Treatises, Poe's idea that our relation to nature or the Universe is not merely one-way but two-way, that our senses, for example, are not merely adapted to perceive our environment but that our environment is also adapted for us to perceive it. This is also part of understanding the Universe as a plot, as a poem with a unity of effect, and it marks a further difference between viewing the Universe as a machine (with one-way adaptations) and viewing it as a plot (with two-way adaptations). To Harry Poe's important point I should like to add what we have seen above in "The Philosophy of Composition" (1846), Poe's view that beauty is not a primary but a secondary quality. If the Universe is viewed as a poem with a plot, but not if it is viewed as a machine, its "unity of effect," which is the experience of "Supernal Beauty," would require in Poe's view a "mutuality of adaptation" between the Universe and the person perceiving the Universe. This is because a secondary quality exists only in the mind, as the combined product of the primary qualities of the perceived and of the perceiver.

In all of this, the key notion is that of a "proper response." In the case of a machine, the proper response is, first, to determine if it is in fact a machine, and then, using terminology supplied by Richard Dawkins, to "reverse engineer" it, to determine what each part was made for, including possibly its "utility function," what the whole was made to maximize.[123] There may indeed be other responses, such as "admiration" for the accuracy and skill of the designer, also the experience of "beauty," possibly from contemplating the adaptation or "fitness" of parts to "final causes" or the "symmetry" in design,[124] and possibly even the feeling of "Hope spoken in organ tone."[125] But with a machine the proper response is fundamentally *inference*, with other responses following from that.

122. *Evermore*, 157.

123. Richard Dawkins (1995). *River Out of Eden: A Darwinian View of Life*. New York: BasicBooks, 103–104.

124. Hume's Cleanthes (note 112), 143, 202; Paley (note 113), 15, 102.

125. Nichol (note 114), 115.

With a poem, however, the proper response, at least according to Poe's poetics, is not at bottom inference but *experience*, and more specifically *aesthetic* experience. You take in the poem, you put it in your mouth like wine, or let it fill your head and body like music, and experience what Poe called the "unity of effect." Inference cannot give one this experience; only direct experience can. What the particular unity of effect is, of course, will differ depending on the poem, and also on the person hearing or reading it. But according to Poe every poem properly so called must be composed by the poet with a particular unity of effect in mind beforehand, which essentially must involve beauty. And if the unity of effect has to be inferred or explained to someone, then it probably isn't a good poem in the first place, or even a poem at all,—unless the someone is a philistine to begin with.

A further difference in proper response to a machine and to a poem concerns the maker. In the case of a machine, we argue analogically and infer not only the existence but the nature of the maker based on the effect of the maker, the machine. During the Enlightenment, the argument from design for the existence of God was an argument from analogy, where the analogues are human contrivances and the apparent design in the Universe, including the Universe as a whole. The degree to which the Universe is established to be a machine establishes the probability of the conclusion, the existence of a maker something like us, specifically a divine Watchmaker, vastly greater in intelligence, power, and benevolence than human watchmakers. But the degree to which the Universe is established to be a poem does not stop with the probability of the conclusion, the existence of a divine Poet. This is because the "proper response" to a poem properly so called is what Poe, as we have seen above from "The Poetic Principle" (1850), called "the pleasurable excitement, or elevation, *of the soul*," and a poem does this by virtue of its evocation of Beauty, since the latter is "the province of the poem."[126] But even that is not really the end of it, for the "proper response" to the beauty of a poem is—the topic of our next chapter.

Returning to our original problem, Poe's "Prose Poem" paradox, by the principle of charity alone we must be in the right

---

126. *E&R*, 78.

direction, if not entirely in detail. As a clue to its meaning, *Eureka* itself, if not a hoax, makes no sense as a poem *for humans* for it fails Poe's own demands for what a poem is. But the *subject* of *Eureka*, the entire Universe, or rather *our* universe, does make sense as a poem, if viewed from the *viewpoint of Poe's God* in *Eureka*, the God of which we are an individuated part, such that "The Universe is a plot of God" and "The plots of God are perfect."[127] And it is a plot with *symmetry*, which Poe adds "is the poetical essence of the Universe—*of the Universe* which, in the supremeness of its symmetry, is but the most sublime of poems." Now add Poe's consistency criterion of truth not locally but universally. The result is that "symmetry and consistency are convertible terms:—thus Poetry and Truth are one. . . . *A perfect consistency, I repeat, can be nothing but an absolute truth*." Now add the expansion and contraction of our universe as part of the rhythmical creation of beauty, (the phrase is not to be found in *Eureka*,) "a novel Universe renewed forever, and forever, and forever; a novel Universe swelling into existence, and then subsiding into nothingness, at every throb of the Heart Divine,"[128] each Universe beginning with "Nothingness" (or "Nothing," or "Nihility") as Poe uniquely defines the term,[129] and all of it for the aesthetic pleasure of God, for the "joy of his Existence," the "Divine Will," of which (again) each of us is a part.[130]

This, arguably, is the *purpose* of *Eureka* (or rather its *penultimate* purpose). But to see this, from Poe's eyes, we would have to follow a suggestion made by James Hutchisson,[131] which is that by "unity of effect" Poe had in mind a "double unity": first, "the unity imposed by the writer on the work," which we might call *creation*, and second, "the unity felt by the reader," which we might call *reaction*. With regard to *creation*, whether of a universe or of a tale or poem, everything put in it by the creator should

127. *P&T*, 1342. Both statements are found earlier in "Marginalia" (November 1844), *E&R*, 1316.

128. *P&T*, 1349, 1356–1357.

129. *Letters*, 649. *P&T*, 1276.

130. *P&T*, 1358.

131. *Hutchisson*, 65.

have a purpose that it serves in some way, no matter how small, in order to produce the designed effect, while with regard to *reaction*, whether to our universe or to a tale or poem, the reader, the literary critic, the observer, the grand scientific theorizer, each requires the ability to see the unity of the whole and to experience the designed effect and, inasmuch as that is possible, to see the role of the parts, if not each in particular then in general, toward that unity. And to help each of us have that experience, the experience of the Universe as a poem with a plot, not as God but as ourselves, with our "individual identity," each of us *one* of the "infinite individualizations of Himself,"[132] which is what Poe calls us, Poe created *Eureka*. It is a poem, in the sense that it is designed to produce the very same unity of effect, on a *mini* scale for each of us as individuals, that the Universe itself, as a poem with a plot, is designed to produce for God. This, then, is arguably why Poe, in his Preface quoted at the beginning of this chapter, states not only that *Eureka* is offered "to those who feel rather than to those who think," but that "I offer this Book of Truths, not in its character of Truth-Teller, but for the Beauty that abounds in its Truth; constituting it true." And it is also why he states in that same Preface that "it is as a Poem only that I wish this work to be judged after I am dead." The beauty evoked by Poe's *Eureka* is a *glimpse* of the beauty of the Universe experienced *collectively* by God as Many.

In an important sense, then, Poe's "Prose Poem" is his ultimate metaphor, his ultimate "similarity in dissimilars," to use Aristotle's phrase, arrived at in large part by his study (and correction) of the science of his day, the dissimilars being a manmade poem with a plot and the Universe as a whole. As such, *Eureka* is not *literally* a poem, but *metaphorically* a poem, as a figure of speech. This is because, though not itself a poem, it is enough *like* its subject—the poem with a plot known as the Universe—by virtue of its ability, as Poe says at the very beginning of *Eureka*, to allow the reader "to receive and to perceive an individual impression," to "comprehend the panorama" of the Universe "in the sublimity of its *oneness*."[133]

132. *P&T*, 1359, 1358.

133. *P&T*, 1261.

The "individual impression" made on the reader, however, no matter how beautiful, is not the *ultimate* purpose, not the *end* of the "proper response," to *Eureka*. There is something more, and for that we need to turn from Poe's literary theory to his theology, the topic of our next chapter.

CHAPTER 3

Poe's Theology

Why I was not a priest is a mystery, for I feel I am now a prophet.

—Edgar Allan Poe[1]

Entrée

As mentioned in the previous chapter, no one can possibly come to understand, let alone to appreciate, Poe's *Eureka* without knowing Poe's background, both his life and his thought. In this chapter we examine Poe's theology. This might at first seem odd, for is not the focus of this book scientific imagination? In the case of Poe, however, particularly in the case of *Eureka*, it is both artificial and misleading to separate the scientific imagination from the rest, including his literary theory and his theology, for they are inextricably woven together and cannot rightly be understood apart. To modern minds this might seem a failing, but we have to keep in mind that in Poe's time it was quite common for science to be mixed with theology, and not just in the works of amateur speculators but in the works of what we would call professional scientists as well, following the lead of Newton. It is also true

1. Letter to Mrs. Marie L. Shew, dated June 18, 1848. *Letters*, 678. Given the year 2848, Poe could have added what he wrote years earlier in a book review (January 1842), that "a prophet has no honor in his own land" (Luke 4:24). *E&R*, 1027.

today in many cases, professional scientists mixing their science with theological concepts, especially when it comes to cosmology, in trying to make sense of the Universe. If we want to understand Poe and *Eureka*, then, as well as scientific imagination, we must examine this side of the question.

But let us return to Poe's time and begin by considering three examples of science mixed with theology, each of them by bona fide scientists. The first is by the polymath Sir John Herschel, whose *Preliminary Discourse* (1830) was arguably the first book devoted to the nature of science. In the opening chapter he writes of "the vast machinery" of nature, of the scientist being led "to the conception of a Power and Intelligence superior to his own," one that is "infinite," of science bringing those who add to it and those who study it "nearer to their Creator."[2] Another prime example is the geologist Sir Charles Lyell, who established the case for an old Earth based on uniformitarian and observable processes. Although Lyell excluded cosmological speculation about the origin of the Earth from geology proper, he did not exclude God from the second volume of his *Principles of Geology* (1832), which is devoted to the history of animals and plants. He states, for example, that "We must suppose, that when the Author of Nature creates an animal or plant, all the possible circumstances in which its descendants are destined to live are foreseen." For Lyell this also included God's creation of domestic animals, "expressly to enable them to follow man throughout all parts of the globe."[33]Both of these works are classics, highly influential in their time, and it is important to note that each of their authors viewed science as seeking out *verae causae* (true causes) in nature. Or consider a book that played a central role in informing *Eureka*, *Views of the Architecture of the Heavens* (1838) by the astronomer John Pringle Nichol, which was written as a popular up-to-date account of the state of astronomy. Nichol says, for example, that "we are speaking of the energies of that Almighty Mind, with regard to whose infinite capacity a day is as a thousand years,"

2. John F.W. Herschel (1830). *Preliminary Discourse on the Study of Natural Philosophy*. London: Longman, Rees, Orme, Brown, Green, and J. Taylor, 8, 4, 17.

3. Charles Lyell (1832). *Principles of Geology*. Volume II. London: John Murray, 23, 44.

and he refers to "the fulness of that volume which God has spread before us all, in illustration of his own Infinite Nature!"[4]

To Poe's mind when writing *Eureka*, then, he was probably not doing anything unusual per se by mixing natural science with theology, for the days of separating science from religion, of what today is called *methodological naturalism*, had not yet come. What we shall do in this chapter, accordingly, is examine Poe's theological thinking and how and why it is connected with his scientific speculations. The key to it all, I believe, is Poe's solution to the problem of evil, or more specifically what I call, from an artistic point of view, the problem of ugly.

Poe's Theology and the Problem of Evil

To anyone who reads *Eureka*, one of the most striking statements one can find, and it comes shortly after the three-quarter mark, is that "The Universe is a plot of God."[5] Equally striking are the claims, all stated at the end, that the Universe was created as a "Self-Diffusion" of God, (this is a version of what is known as *pantheism*,) for the purpose of "the joy of his Existence," that everything we call "his creatures" are "really but infinite individualizations of Himself," that everything in the Universe—both "those you designate as his creatures," which "you term animate, as well as those to which you deny life"—is not only alive, (this is known as *hylozism*,) but they are also "conscious Intelligences" (this is known as *panpsychism*) and are sentient, in that "*all* . . . have, in a greater or less degree, a capacity for pleasure and for pain."[6]

Poe was apparently quite serious about this doctrine of universal sentience. In the very least the idea has its precursors in his earlier writings, for example in "The Fall of the House of Usher" (1839), in which Roderick Usher, with his "morbid acuteness of senses," insists upon, "with pertinacity," not only "the sentience

4. J.P. Nichol (1840). *Views of the Architecture of the Heavens: In A Series of Letters to a Lady*. New York: H.A. Chapin & Co., 84, 114.

5. *P&T*, 1342.

6. *P&T*, 1358.

of all vegetable things," but also the sentience of "the kingdom of inorganization," which includes his own house, the family mansion, the "House of Usher."[7] The idea is also to be found in Poe's "The Island of the Fay" (1841), in which the narrator states that "I love, indeed, to regard the dark valleys, and the grey rocks, and the waters that silently smile, and the forests that sigh in uneasy slumbers, and the proud watchful mountains that look down upon all—I love to regard these as themselves but the colossal members of one vast animate and sentient whole . . ."[8]

Equally interesting is that in *Eureka* Poe divides his panpsychism into two kinds: those beings that are conscious "of a proper identity," (whatever that means, although presumably he means a sense of identity as an individual being,) and those beings that are conscious "by faint indeterminate glimpses, of an identity with the Divine Being of whom we speak—of an identity with God." Upon death, as conscious beings of the second kind, "we shall lose our individual identity," our subjectivity, our personal self or *me*, but we should take consolation in the fact that we shall eventually become "blended," an "absorption," along with the "bright stars," indeed with "all other intelligences (that is, of the Universe)," into "the *Spirit Divine*," into the Universe returned unto itself, which is God without any diffusion or differentiation, namely, "the One."[9] (Think, perhaps, of emptying a bucket of water onto a concave hard surface and seeing the water break into thousands of water drops, followed by the drops coalescing back into a single body of water.)

Why would Poe produce such a theology? I suggest the answer is the *problem of evil*, as it is called in theology and philosophy of religion. The problem should be obvious enough *to any thinking*

---

7. *P&T*, 318, 322, 327. In a footnote to "Usher" (*P&T*, 327), Poe provides a list of precursors for the view (though limited to the organic) here ascribed to Usher. But the main connection is not in looking backward but in looking forward, to *Eureka*. This connection is entirely overlooked, by the way, in Scott Peeples' attempt to connect "Usher" with *Eureka*, discussed in the second-last section of Chapter 1, and it is arguably the *only* significant connection, not one of many, contrary to the claim of John H. Timmerman (2003). "House of Mirrors: Edgar Allan Poe's 'The Fall of the House of Usher.'" *Papers on Language & Literature* 39 (3), 227–244.

8. *P&T*, 934.

9. *P&T*, 1358, 1359.

person. If God exists and is all-knowing (omniscient), all-powerful (omnipotent), and all-loving (omnibenevolent), then evil should not exist, for if it exists then God would know about it, would be powerful enough to get rid of it, and would want to get rid of it out of benevolence. So given the existence of evil, then either God does not exist or, by the law of logic known as De Morgan's Theorem, God is either not omniscient, not omnipotent, or not omnibenevolent. The problem applies to what is called *natural evil*, such as parasites, cancer, tornadoes, droughts, and floods, and *moral evil*, (I've never liked that label,) manmade evil such as lies, theft, rape, murder, and genocide. The amount of literature on the problem of evil is overwhelming, with solutions ranging from the soul-developing theology of St. Irenaeus, (evil is necessary for the development in us of moral qualities such as charity, courage, and forgiveness,) the denial of the existence of evil by St. Augustine, (evil does not exist but is merely the absence of good, just as dark is the absence of light, and only God is perfectly good), the view that this is the best of all possible worlds, (Leibniz,) process theology, (God is not perfect but is evolving along with the Universe,) and of course atheism. Each solution is known as a *theodicy*.

Poe surely had a good grasp of the problem of evil, as well as of many theodicies. This is not only because he read widely, but also because, following his move to Fordham, New York, in May 1846, his permanent residence for the remainder of his life, he lived quite close to St. John's College, which had just come under control of the Jesuits. Poe often went for walks with a priest named Reverend Edward Doucet, later President of Fordham University, with whom Poe became good friends and who found Poe "well informed on all matters."[10] One can only imagine the conversations they had, and at times they must have dealt with the big questions of life. They must also have agreed to disagree on much, as Poe was not a religious man in any traditional sense of the term—he didn't attend church, but he did possess a belief in God, an unorthodox one not involving Jesus, ritual, or prayer.[11]

So what, then, was Poe's theodicy? And what argument for God's existence did he employ, if any, to inform his theodicy? To answer these questions we need an intellectual context, so let us go

10. Weiss, 72; *Quinn*, 520.

11. Silverman, 332–333, 341; *Letters*, 692.

back and compare the Enlightenment concept of the Universe as a machine with Poe's Romantic concept of the Universe as a poem with a plot.

During the Enlightenment the prime example of manmade design was the watch, the example used not only by Cleanthes in David Hume's classic *Dialogues concerning Natural Religion* (1779) but also by William Paley in his equally classic *Natural Theology* (1802), both of which we visited in the previous chapter. As we have seen, the "proper response" to a machine such as a watch, even if we have never seen one before, is to reverse engineer it as best we can, to infer its utility function, and to infer the existence and nature of the maker or designer. The argument from design for God's existence draws an analogy between machines such as watches and complexity in the Universe, such as eyes and laws of nature. Given that we know from experience that machines like watches do not come into existence by chance but are always manmade, the more the Universe and its parts are shown to be machines or machinelike the greater the probability that they were created by a maker, and in proportion to their complexity a Watchmaker. As an analogical argument, the argument is not a proof in the strict sense, but at best can only establish a high probability for the truth of its conclusion.

The critic Philo, in Hume's *Dialogues*, attempts to weaken the analogy between manmade design and design found in nature and thereby weaken the most famous argument for God's existence in Hume's time. Paley, in return, without specifically replying to Philo point by point, did in fact attempt to answer his points one by one in his book (and he makes it clear near the end that he did indeed read Hume's *Dialogues*). The debate remained until Darwin came along, with his *On the Origin of Species* (1859), who not only showed *how* design in biological nature, such as eyes and the animals that have them, *could* result from blind materialistic processes, most notably heritable variation combined with generation after generation of the natural selection of favorable variations, but that it *is* in fact the process behind biological design in the world. It would not be an exaggeration to say that modern biology has confirmed, beyond a reasonable doubt, Darwin's theory. (We shall have more on this in Chapter 6.)

One might naturally think that if the Universe is viewed not as a machine but as a poem with a plot, which we have seen in the previous chapter is in fact Poe's view in *Eureka*, then we have an argument for God's existence that is similarly analogical, an argument that at best can achieve a high probability for the existence of a Composer or Poet, though never a proof, the more the Universe is shown to be or to be like a poem with a plot, and one might also naturally think that the argument similarly runs afoul given Darwin, that even though a poem requires a poet, (so far as we know at any rate,) and the beauty produced or evoked by a good poem requires a true poet, it still might in fact be the case that the experience of beauty provoked in us by the natural world or Universe does not require a Poet but can be explained without one, perhaps by evolutionary psychology, and if so *should* be explained without one, as per Ockham's Razor, the principle of parsimony, that as a simple matter of avoiding error one should keep one's explanatory entities to an absolute minimum.[12]

It might also be argued, of course, that if Poe's argument for God as Poet is *not* similarly analogical to the argument from design above, with the Universe as a machine made of smaller machines, then Poe not only shifted the debate quite considerably but was guilty of *begging the question*,—a cardinal sin of argumentation often committed in theology and philosophy of religion,—the error of assuming what he supposedly is trying to prove. The classic example is: God exists (the issue is whether God exists, the position taken is that God exists), because the Bible says God exists and the Bible is literally true. To judge Poe in this way, however, is to be guilty of an equal error, which is to judge Poe as a philosopher, which he was not. Poe was, instead, first and foremost an artist, and though he certainly provided arguments, they are to be judged in a different way, not throwing logic out the door but using the principle of charity of interpretation, which in

12. A long run-on sentence with a lot of *ands*, of course, unlike the production of something beautiful, is arguably evidence of a Bedlamite—unless one is Poe, for then arguably it is evidence of genius. See *Silverman*, 113, on Poe's use of the "copulative ending" to call attention to, e.g., tremor in some of the narrators of his tales.

this case means making the effort to understand his reasoning not as a set of premises leading to a conclusion, the way an analytic philosopher or a scientist would argue, but as a piece of artwork, where the whole, the conclusion, has to be understood as the *effect* of the arrangement of the parts, the premises. To blandly say that Poe assumed the existence of God is to entirely miss Poe's argument, which includes a brushstroke here and there that serve as premise indicators.

Moreover, Poe makes clear throughout his writings, but especially in the imaginary epistle near the start of *Eureka*, written in the year 2848 as a retrospective to his time, that the conscious processes of deduction and induction are not the only roads to truth, contrary to the epistemology of his day, that instead they are methods of "creeping" and "crawling" confined to "paths," that they exclude what is most important of all in the search for truth, the soaring of "eagles," the *educated imagination*,—the imagination of "the *only* true thinkers . . . the generally-educated men of ardent imagination,"—in a word, "intuition," which Poe describes as "the conviction resulting from *de*ductions or *in*ductions of which the processes were so shadowy as to have escaped his consciousness, eluded his reason, or bidden defiance to his capacity of expression."[13] (We shall take a close look at these themes in Chapter 7.) *Eureka*, then, is not to be read or interpreted mainly as a rigorous argument or demonstration, but as an "Art-Product alone," as Poe puts it in the Preface.

So what, then, Dr. David N. Stamos, Genius Literary Interpreter of the Art-world, long forgotten because of his lack of action potential, though known for his morbid digestion, was Poe's *argument* inasmuch as it can be understood as such? I *suggest* (I use the word merely to deflect poignant criticism) that Poe's argument for God's existence is to be understood as the *argument from beauty*, in contradistinction to the *argument from design*, the former indeed found in modern philosophy of religion, (rarely,) an example of which shall be examined below.

13. *P&T*, 1264, 1268, 1269, 1270. Similar ideas are to be found in Poe's "Mellonta Tauta." Apparently submitted by Poe to *Godey's Lady's Book* in January 1848, (*Poe Log*, 717; *Letters*, 646n,)—*Eureka* was published later that year in July,—it was subsequently published in *Godey's* in January 1849 (*Poe Log*, 785). Precisely which is the cause and which is the effect remains an interesting question.

Once again we must return to the question of "proper response." We have seen in the case of Cleanthes in Hume's *Dialogues* and in Paley's *Natural Theology* that the proper response to what is believed to be a machine is *inference*, with the conclusion a matter of *probability*. Hume adds a twist to this, however. In Part II of his *Dialogues* his Cleanthes is very much like Paley years later. But in Part III Hume introduces a very interesting distinction. Apparently as a rearguard action to the criticisms already raised by Philo, Hume has Cleanthes make a distinction between an argument of a *regular* nature, one which is a matter of inference and accordingly is subject to the rules of reasoning, of probability in the case of analogical arguments, and an argument of an "*irregular* nature," (italics mine,) one that is a matter not of logic but of human nature. As Cleanthes puts it, "Consider, anatomize the eye: Survey its structure and contrivance; and tell me, from your own feeling, if the idea of a contriver does not immediately flow in upon you with a force like that of sensation. . . . and it requires time, reflection and study, to summon up those frivolous, though abstruse, objections, which can support infidelity."[14]

Hume scholars call beliefs in Hume's writings based on irregular arguments *natural beliefs*, such as the belief in physical objects, the belief in other minds, and the belief in one's own personal identity, and they debate whether Hume himself held that the belief in an "invisible intelligent power," Hume's wonderful phrase in *The Natural History of Religion* (1757), is also a natural belief. Knowing this makes it enormously tempting to construe Poe in *Eureka* as offering not a regular but an *irregular* argument for God's existence, given Poe's poetics. (This is not at all to suggest that Poe ever read Hume, but to suggest a convergence in thought, with the Universe, in the case of Poe, viewed not as a machine but as a poem with a plot.) Recall from the previous chapter Poe's recurring theme of "unity of *effect*," (italics mine,) that plots and poems are to be composed backwards, starting with the intended effect on the reader or audience, and that everything in the tale or poem should serve to that effect. Recall next that, according to Poe, "Beauty is the sole legitimate province of the poem," that poetry is defined as "the *Rhythmical Creation of Beauty*" using

14. Norman Kemp Smith, ed. (1947). *David Hume: Dialogues concerning Natural Religion*. Indianapolis: Bobbs-Merrill, 155, 154.

words, and that "The artist who doubts of this thesis, may always resolve his doubt by the single question—'might not this matter be as well or better handled in *prose*?' If it *may*, then it is no subject for the Muse."[15] It would seem to follow, then, that if the Universe is *literally* viewed as a poem with a plot, which as we have seen is Poe's expressed view in *Eureka*, and if one takes in Poe's "survey of the Universe," as he says at the very beginning of *Eureka*, as allowing one "to receive and to perceive an individual impression," to "comprehend the panorama" of the Universe "in the sublimity of its *oneness*,"[16] then (to use Wilbur's phrase yet again) the "proper response," whether to *Eureka* or to the Universe properly surveyed, is not *inference* but *experience*, that given human nature the natural, the instinctive, the *proper* response is "that pleasurable elevation, or excitement, *of the soul*, which we recognize as the Poetic Sentiment."[17]

If, however, one thinks this is Poe's intention as the *ultimate* response to *Eureka* or to the Universe as a poem with a plot composed by God, one would be wrong—stone, stone dead wrong. And to see this, we have to return to the problem of evil. Just as the argument from design (whether regular or irregular) for God's existence based on the Universe as a machine faces the quite serious problem of evil, which seems to throw a monkey wrench into it, without any aid from theories like Darwin's, (although Darwin's theory magnifies the problem,) so too the problem of evil threatens Poe's argument (whether regular or irregular) for God's existence based on the Universe as a poem with a plot, evil falling under the more inclusive terms *ugly* or *deformity*. The problems are interconnected: design and the existence of bad design or evil, beauty and the existence of ugly or deformity.

Moreover still, just as theologians and philosophers of religion have each proffered their own solutions to the problem of evil, Poe, too, offers his own solution to the problem of ugly, ugly not only in man but in nature as well. Poe, of course, looked at the Universe *both* from the eyes of the science of his time and from the eyes of the artist and literary theorist that he was. But

15. *E&R*, 16, 688, 689.

16. *P&T*, 1261.

17. *E&R*, 78.

since he is known primarily as the latter, it might be thought that if we are to judge his argument for God's existence *charitably*, and more generally his theology, we should judge it on Poe's own terms as an artist, from the viewpoint of an aesthetic, as an irregular argument. I suggest, however, that this would not be convincing to *thinkers*. I suggest even further, quite remarkably, since it is counterintuitive, that as thinkers if we really want to see the power of Poe's argument, if we really want to understand it and fully appreciate whatever its strengths and weaknesses, then we need to reconstruct it *analytically* and subject it to critical analysis; we need to see it not as an artist would but as a philosopher or scientist would, beginning with what I have called the problem of ugly, which incorporates the problem of evil.

Artistic Sensitivity and Poe's View of the World

To understand Poe's view of the world, we must first understand his view on artistic sensitivity, which in turn plays a central role in his theology and in *Eureka*. A key passage for understanding Poe on this matter is number 22 of "Fifty Suggestions" (1849), which is worth quoting in full:

> That poets (using the word comprehensively, as including artists in general) are a *genus irritabile*, is well understood; but the *why*, seems not to be commonly seen. An artist *is* an artist only by dint of his exquisite sense of Beauty—a sense affording him rapturous enjoyment, but at the same time implying, or involving, an exquisite sense of Deformity or disproportion. Thus a wrong—an injustice—done a poet who is really a poet, excites him to a degree which, to ordinary apprehension, appears disproportionate with the wrong. Poets *see* injustice—*never* where it does not exist—but very often where the unpoetical see no injustice whatever. Thus the poetical irritability has no reference to "temper" in the vulgar sense, but merely to a more than usual clear-sightedness in respect to Wrong:—this clear-sightedness being nothing more than a corollary from the vivid perception of Right—of justice—of proportion—in

> a word, of χαλον. But one thing is clear—that the man who is *not* "irritable," (to the ordinary apprehension,) is *no poet.*[18]

Poe completes the passage in Suggestion 23 by referring to "the appreciation of Beauty and horror of Deformity which we call sensibility."[19] The term "sensibility," by the way, is likewise found years earlier in "The Fall of the House of Usher" (1839), in which the narrator informs us that the "ancient family" of Roderick Usher had been known "for a peculiar sensibility of temperament," one that manifested itself in "many works of exalted art," in "repeated deeds of munificent yet unobtrusive charity," and in "a passionate devotion to the intricacies, perhaps even more than to the orthodox and easily recognisable beauties, of musical science."[20]

The poet, then, as poet, along with artists in general, is more likely to be sensitive to beauty and its correlative deformity, including injustice, (perhaps this is why most artists seem to support left-wing causes, like the animal rights vegetarianism of artists as diverse as Shelley, McCartney, and Zombie,) even when their art itself is deformity, more "clear-sighted" with regard to the Universe and especially the humans in it, than the philosopher or scientist, let alone the common man. Hence the poet, for Poe, sees what others miss, much like bees, to speak in simile, see patterns on flowers that the rest of us cannot, because they have ultraviolet vision. At any rate, that Poe was a bona fide artist is something none but a philistine would deny.

This brings us to Poe's *weltanschauung* or worldview, including the way he viewed nature, the world of humans, and his own life. Did he see beauty all around, was he a Polyanna, wearing rose-colored glasses, or was he well aware of the dark side of life, both in nature, in mankind, and in his own existence? I shall provide only a few illustrative examples, for a *yes* to everything following the "or."

18. *E&R*, 1300–1301.

19. *E&R*, 1302.

20. *P&T*, 318.

Beginning with his own existence, one can often see in Poe's letters not only the self-conception of the disinherited aristocrat,[21] but the idea that his life was one big ripoff after another, that he suffered enormously, and that he did not deserve it. Painful to read (at least for my own part) are his explicit references to his life, such as in his letter to the brother of his late foster mother, in which he writes of his "long & bitter struggle with illness, poverty, and the thousand evils which attend them," or in his only extant letter to Virginia, written in the summer before she died, in which he writes of "this uncongenial, unsatisfactory, and ungrateful life."[22]

Connected with this is his perception of people. Poe benefitted in times of distress from the kindness of friends and even strangers, and knew not only friendship love but romantic love and parental love.[23] But on the whole he viewed the human world as "the hollow heartless world." Confining ourselves to the last three years of his life, Poe wrote of the lack of love in "this cold, dreary world," of "the calm, cold language of a world which I loathe—of a world in which I have no interest—of a world which is *not mine*," of "this cruel, unjust, calculating world," of "calumny however underserved, or however unendurable."[24] And the calumny of people was not only toward others, but also toward themselves, a tendency toward self-harm and even self-destruction, which he collectively termed "the Imp of the Perverse," the impulse, though often resistible, "to do wrong for the wrong's sake," to do wrong "merely because we feel that we should *not*," what he calls "a radical, a primitive impulse—elementary." This is not an *a priori* thesis, Poe has his narrator claim, but instead is based on observation, "of what man usually or occasionally did," and on introspection, as "No one who trustingly consults and thoroughly questions his own soul, will be disposed to deny the entire radicalness [the deep roots] of the propensity in question."[25]

21. *Letters*, 175, 730, 735.

22. *Letters*, 725, 578.

23. With regard to the latter, Poe "always regarded as a mother" his foster mother, Frances Allan, and often called Maria Clemm "mother." *Letters*, e.g., 185, 578, 820, 823. See also his heartfelt poem "To My Mother" (1849). *P&T*, 101–102.

24. *Letters*, 678, 692, 707, 754, 757.

25. "The Imp of the Perverse" (1845). *P&T*, 827, 828.

Nor did Poe believe in the possibility of the innate improvement of mankind. For example, in his autobiographical letter to James R. Lowell, dated July 2, 1844, Poe states that "I have no faith in human perfectibility. I think that human exertion will have no appreciable effect upon humanity."[26] He has his narrator make a similar claim in "The Domain of Arnheim" (1846), in which the narrator says of his remarkable friend Ellison, "In the possibility of any improvement, properly so called, being effected by man himself in the general condition of man, he had (I am sorry to confess it) little faith."[27]

And then there is the world of nature. Poe saw, as many do, great beauty in nature. A very powerful expression of this is to be found in "Morning on the Wissahiccon" (1843), which is more short essay than tale. Here he claims that within the United States there is "a realization of the wildest dreams of paradise," scenes that exceed anything in Britain, for example, as will be judged "by the true artist, or cultivated lover of the grand and beautiful amid the works of God." This is because, unlike in Britain, America has parts that are pure, unspoiled by human traffic or settlement. Consequently they are not tourist spots, since they are not accessible by roads, rails, or riverboats, but ultimately must be accessed "on foot." In "the valley of Louisiana," for example, there is beauty which "No fiction has approached." There, he says, "the most gorgeous imagination might derive suggestions from its exuberant beauty. And *beauty* is, indeed, its sole character." Poe especially focuses on river scenery, which he says has "unquestionably, within itself, all the main elements of beauty, and, time out of mind, has been the favourite theme of the poet." This brings us to the Wissahiccon, a brook west of Philadelphia, which must be seen during the height of the day for best effect, "for the narrowness of the gorge through which it flows, the height of the hills on either land, and the density of the foliage, conspire to produce a gloominess, if not an absolute dreariness of effect, which, unless relieved by a bright general light, detracts from the mere beauty of the scene."[28] In these words one can surely see themes from Poe's

26. *Letters*, 449.

27. *P&T*, 858.

28. *P&T*, 939, 941, 943.

literary theory, specifically "true artist," "beauty," "gloominess," and "effect," along with the implication of God as Poet.

The Argument from Beauty

The natural beauty Poe wrote about spoke to him, as an artist especially, of the existence of God. As he puts it in "Eleonora" (1841), after describing "the Valley of the Many-Colored Grass," with "the river to the mountains that girdled it in," "its exceeding beauty spoke to our hearts, in loud tones, of the love and of the glory of God."[29] A few years later, in "The Domain of Arnheim" (1846), Poe would write of "the most direct and energetic efforts of Nature at physical loveliness," of the "adaptation to the eyes which were to behold it on earth," and of "the august purposes for which the Deity had implanted the poetic sentiment in man."[30]

This is the *argument from beauty* for God's existence, which is not a *proof* that the beauty of nature establishes God's existence. The argument instead is *analogical*: just as beauty in art implies a creator, so too beauty in nature. The argument, by the way, in variation, can be found today in philosophy of religion, although it is not common. Richard Swinburne, for example, one of the top names in the field, and one who takes science more seriously than normally found in that field, provides the argument (as a *regular* argument) in addition to the argument from design. Beginning with the latter, science can explain a lot, he says, but it cannot explain the fundamental laws of nature. For that one needs a different kind of explanation, and since chance won't do, it has to be a personal explanation, and that would mean God. The world has just enough order in it, he says, to allow for the evolution of creatures like us and for personal development in terms of knowledge and character. Design in nature has a similar role to beauty. Since "the experience of beauty" allows us to "develop," "God has reason," he says, "to make a basically beautiful world, although also reason to leave some of the beauty or ugliness of the world within the power of creatures to determine; but he would seem to have

29. *P&T*, 469.

30. *P&T*, 859.

an overriding reason not to make a basically ugly world beyond the powers of creatures to improve. Hence, if there is a God there is more reason to expect a basically beautiful world than a basically ugly one." And, of course, not only "poets and painters," but also "ordinary men down the centuries have admired" the beauty in nature, such that "Who in his senses would deny that here [on Earth and in the skies above] is beauty in abundance? If we confine ourselves to the argument from beauty of the inanimate and plant worlds, the argument surely works."[31]

But the argument surely flounders if beauty after all is merely a projection of our brains onto our perceptual experiences, just as the argument from religious experience for God's existence likewise flounders. Both are arguments from inner experience to outward reality, which cannot be a direct and sure route, given not only common sense but neuroscience and psychology, the latter including evolutionary psychology.[32] But equally important, just as the argument from design needs to provide a satisfactory explanation for the existence of *bad* design in the world, which is part of the problem of evil, likewise the argument from beauty needs to provide a satisfactory explanation for the existence, or at least for the apparent existence, of *ugliness* or *deformity* in the world, both the world of nature and the human world.

Poe, quite remarkably, attempts to provide this in "The Domain of Arnheim" (1846). There he says, "Mr. Ellison did much toward solving what has always seemed to me an enigma:—I mean the fact (which none but the ignorant dispute) that no such combination of scenery exists in nature as the painter of genius may produce." Given any landscape, he says, "there will always be found a defect or an excess—many excesses and defects," such that "no position can be attained on the wide surface of the *natural* earth, from which an artistical eye, looking steadily, will not find matter of offence in what is termed the 'composition' of the landscape."[33] Consequently, Poe places landscape gardening among the modes of *poesy*, the "poetic sentiment," given that it affords

31. Richard Swinburne (1979). *The Existence of God.* Oxford: Oxford University Press, 146.

32. For a good albeit rough approximation, see Edward O. Wilson (1984). *Biophilia.* Cambridge, MA: Harvard University Press, ch. 7 "The Right Place."

33. *P&T*, 859–860.

the opportunity for "the creation of novel forms of beauty," and he has Ellison, with his vast financial resources, be the greatest landscape gardener of all. Ellison's solution, by the way, to the problem of ugly, is not only to point to the matter of perspective, that ugliness of composition in nature might be ugliness from the human point of view but not from another point of view (God or angels); Poe also has Ellison claim, which appeals to Poe's narrator in the tale, that the ugliness in nature is "prognostic of *death*," that "The disturbances were the preparation for his [mankind's] subsequently conceived deathful condition."[34] The idea here fits nicely with what Poe says elsewhere, specifically that "discords in music" may be added to a piece of music to increase its beauty, to heighten the intended effect of the music, which is the elevation of the soul. And keep in mind from the previous chapter that this was stated in Poe's elucidation of "The Raven" in "The Philosophy of Composition" (1846),[35] according to which the mourning of the lost Lenore by the lover results in an elevation of his soul, (if only temporarily,) as does the poem for the reader.

But surely the problem of ugliness in nature is not merely a matter of the "composition" of the scenery in terms of, for example, too few trees here, not enough water there. For surely the ultimate ugly in nature is not compositional at all, in the "arrangement of . . . parts," but in the parts themselves, including processes between them: think of the essentiality in nature of pain, suffering, horror, and death by disease, starvation, accident, predation, parasitism, and the decay of aging. This should be obvious to any *sensitive* and *non-stupid* human being,—one does not need to be an artist,—and in this matter Poe was no Tennyson, who clearly saw nature "red in tooth and claw" and the problem created by it:

> Who trusted God was love indeed,
> And love Creation's final law,—
> Tho' Nature, red in tooth and claw
> With ravin, shriek'd against his creed,—[36]

34. *P&T*, 861.

35. *E&R*, 16.

36. Alfred, Lord Tennyson (1850). "In Memoriam A. H. H.," Canto LV. The edition I use is my own copy of *The Poetical Works of Alfred Tennyson, Poet Laureate* (1870). New York: Harper & Brothers, 114.

The poem from which this stanza is taken, written over a period of seventeen years following the death of Tennyson's close friend, Arthur Henry Hallam, and representing his search for hope, bears at its end the unmistakable mark of Tennyson being "quite excited" by *Vestiges of the Natural History of Creation*, especially by its author's narrative of evolutionary progress.[37] At any rate, the perception of nature "red in tooth and claw" is missing in Poe, that not only in composition or arrangement of parts is there ugliness in nature, but in the very parts themselves and their inter-relations, that this is an ugliness built into the very structure or fabric of nature, a disharmony, a symphonic clash, between the idea of nature as beautiful and the idea of nature as ugly, the latter rebelling against the former. What is ugly to the gazelle might be beauty to the lion chasing and then eating it, of course, but those of our species who find enjoyment in such scenes do seem to forget that the lion is guaranteed to experience the ugly of nature, too.

As are we all.

There are two problems, then, that must be dealt with here, given deformities of nature from an artistic view of life, namely, *pain*, certainly physical pain but also emotional pain and fear, and *death*, the end of a life, a rotting corpse, with the mind (self, soul, spirit, call it what you will) apparently snuffed out like the flame of a candle. Let us see how Poe deals with these deformities, one at a time, and how he thinks they contribute to beauty and pleasure.

The Problem of Pain

Beginning with the fact of pain, the mixture of pain with pleasure is a recurrent theme in Poe's writings. For example, in "The Man of the Crowd" (1840) Poe has his convalescing narrator claim that "Merely to breathe was enjoyment; and I derived positive pleasure even from many of the legitimate sources of pain. I felt a calm but

37. James A. Secord (2000). *Victorian Sensation: The Extraordinary Publication, Reception, and Secret Authorship of Vestiges of the Natural History of Creation.* Chicago: University of Chicago Press, 9–10, 530–532.

inquisitive interest in every thing."[38] Even more strikingly, in "The Premature Burial" (1844) the narrator claims, right in the opening paragraph, that "We thrill, for example, with the most intense of 'pleasurable pain,' over the accounts of the Passage of the Beresina, of the Earthquake at Lisbon, of the Plague at London, of the Massacre of St. Bartholomew, or of the stifling of the hundred and twenty-three prisoners in the Black Hole at Calcutta. But, in these accounts, it is the fact—it is the reality—it is the history which excites. As inventions, we should regard them with simple abhorrence."[39]

The idea immediately above seems abhorrent, given that it is taken to the extreme, but the idea itself that pain can produce pleasure is certainly not new. The Germans have a word that covers much of it: *schadenfreude*, pleasure derived from the pain of others. In English we have *sadism*, pleasure derived from *inflicting* pain on others, and *masochism*, pleasure derived from our own pain. The latter is reserved only for the extreme and does not apply, for example, to the "no pain no gain" motto of athletes, which has endorphins at its core (accordingly athletes distinguish good pain from bad, the latter meaning injury pain). But no matter which words we focus on, no one in their right mind would claim that all pain is productive of pleasure.

And Poe does not seem to claim that in the above. But one can see Poe experiment with solutions to the problem of pain in his writings, ultimately as a theodicy. For example, returning to "The Premature Burial," he has the narrator state that "The true wretchedness, indeed—the ultimate woe—is particular, not diffuse. That the ghastly extremes of agony are endured by man the unit, and never by man the mass—for this let us thank a merciful God!"[40] I suppose we should be thankful that the fallacy of composition is indeed a fallacy, that the braggadocio of Hippias, the ancient Sophist, for example, was misplaced, for thinking that if each of a pair of people is healthy, then the pair itself is healthy, and for all other predicates like sick, just, and young, his reason being what

38. *P&T*, 388.

39. *P&T*, 666.

40. *P&T*, 666.

he called "the continuity of reality."[41] By the same reasoning, if each of a pair of people is in extreme agony, then there are three extreme agonies, not two, the extreme agony of each of the two people plus the extreme agony of the pair as a whole. I suppose the fact that this does not logically follow means that there is much less suffering in the world, for which we should be thankful. But it does not solve the theological problem of pain in the world.

Poe's more usual approach is to view pain as being in some way justified by its production of pleasure. Take, for example, "Mesmeric Revelation" (1844), in which a dying subject is temporarily relieved of his symptoms.[42] In a mesmerized (hypnotic) state in which his "intellectual faculties are wonderfully exalted and invigorated," Mr. Vankirk, in answer to the question, "To what good end is pain thus rendered possible," replies that "All things are either good or bad by comparison. A sufficient analysis will show that pleasure, in all cases, is but the contrast of pain. *Positive pleasure* is a mere idea. To be happy at any one point we must have suffered at the same."[43] In a letter of early 1846, addressed to an unidentified "Gentleman," Poe claims that the "philosophy" or "speculation" in "Mesmeric Revelation"—which anticipates many of the ideas in *Eureka*—"is my own—original, I mean, with myself, and long impressed me. I was anxious to introduce it to the world in a manner that should insure for it attention"—to which he adds, "I believe actual truth to be involved." Whether

41. Robin Waterfield, trans. (2000). *The First Philosophers: The Presocratics and the Sophists.* Oxford: Oxford University Press. The passage in question is from Plato's *Hippias Major* (300e–301e).

42. One of the symptoms is asthmatic breathing, which in the story is relieved immediately when the patient is mesmerized. *P&T*, 719. Did Poe know of this actually happening, (I have seen it accomplished with hypnosis,) or is it yet another application of his imagination? Almost certainly it is the former, as throughout the Victorian era, in both Britain and America, "Mesmerism could be seen almost everywhere and was practiced by individuals of virtually all professions and classes," with "mesmeric infirmaries" for the treatment of pain and the cure of a wide variety of illnesses, and even "mesmeric anesthesia" for operations such as the amputation of a leg. Alison Winter (1998). *Mesmerized: Powers of Mind in Victorian Britain.* Chicago: University of Chicago Press, 16, 138–139, 165.

43. *P&T*, 717, 726.

this includes his theory on the purpose of pain he does not say.[44] But in *Eureka* we do find, right at the end, the claim that God, in diversifying from Unity into the Universe, for the purpose of "the joy of his Existence,"—"just as it *is* in your power to expand or to concentrate your pleasures,"—"feels his life through an infinity of imperfect pleasures—the partial and pain-intertangled pleasures of those inconceivably numerous things which you designate as his creatures, but which are really but infinite individualizations of Himself."[45] Poe even adds that God has a natural right to these "intertangled" pleasures and pains: "*all* these creatures have, in greater or less degree, a capacity for pleasure and pain:—*but the general sum of their sensations is precisely that amount of Happiness which appertains by right to the Divine Being when concentrated within Himself.*" And as one would expect that a single cycle of "creation and radiation, returning into itself" is not going to be enough, Poe's imagination is "guided" by "the law of periodicity" to the "belief"—or rather, "hope"—that the cycle will be "renewed forever, and forever, and forever; a novel Universe swelling into existence, and then subsiding into nothingness, at every throb of the Heart Divine," each cycle involving "a new and perhaps totally different series of conditions,"[46] the infinitely extended series of Universes producing the ideal purpose of the poem, in accordance with what we have seen in Chapter 2, namely, "the *Rhythmical Creation of Beauty*."[47]

---

44. *Letters*, 556. The letter was in response to a request about whether "Mesmeric Revelation" and "The Facts in the Case of M. Valdemar" (1845), also involving mesmerism, were fact or fiction, occasioned by the controversy generated by their publication in London. It should be added that earlier, in a letter dated January 4, 1845, to George Bush, Professor of Hebrew at New York University and also a mesmerist and Swedenborgian, Poe wrote that "You will, of course, understand that the article ['Mesmeric Revelation'] is purely a fiction;—but I have embodied in it some thoughts which are original with myself & I am exceedingly anxious to learn if they have claim to absolute originality, and also how far they will strike you as well based." *Letters*, 474. Once again Poe does not specify the "thoughts."

45. *P&T*, 1358.

46. *P&T*, 1356.

47. Instead of each cycle involving difference, why not have something like Nietzsche's doctrine of Eternal Recurrence, according to which (sans creation) everything in the Universe has been repeated and will be repeated over

Poe's solution to the problem of pain won't do on a number of counts, even if we just keep our focus on pain and ignore for the moment the rest. First, it should be clear to just about everyone that although the experience of pleasure can sometimes be heightened by pain, pain is neither a necessary nor a sufficient condition for the experience of pleasure, let alone that pleasure is always "intertangled" with pain. Quite simply, one can have pleasure without pain. We all know this, just as red, for example, does not require other colors to be experienced. In defense of Poe, it might be said that pain is necessary to *know* pleasure. The argument is often made in philosophy of religion, as a solution to the problem of evil, that pain is necessary to know pleasure, that evil is necessary to know good, that were there no pain or evil in the world then we would never know pleasure or good, and as it is important to know pleasure and good, for moral choice and development, so it is necessary to know pain and evil. The argument, quite simply, is destroyed by the fact, as John Mackie shows using the color analogy, that even if we would need to experience some colors other than red to know that we are experiencing red, it wouldn't take a lot to do it, but only a "minute speck."[48] Perhaps in the case

---

and over again, throughout infinite time? Poe has an interesting answer, not found in *Eureka*, but instead in "Marginalia" (December 1844), in an entry on originality in poems and tales, which is that "The desire of the new is an element of the soul. The most explicit pleasures grow dull in repetition. A strain of music enchants. Heard a second time it pleases. Heard a tenth, it does not displease. We hear it a twentieth, and ask ourselves why we admired. At the fiftieth it enduces ennui—at the hundredth disgust." *E&R*, 1354. The implications of this for *Eureka* are surely interesting: If the Universe is a plot of God, for "the joy of his Existence," and God repeats the process of diffusion and contraction, nothingness returning back to nothingness, each universe followed by another, it follows that each universe must be different, it must be original, to maintain God's pleasure throughout infinite time, (unless God has an imperfect memory,) otherwise God would become not just absolutely bored but absolutely disgusted, maybe even suicidal. Poe might have pondered and then rejected this implication of originality when he came to write *Eureka*, as it would seem to require (unless I am missing something) a different "unity of effect" for each universe in the infinite series. We shall probably never know.

48. John Mackie (1955). "Evil and Omnipotence." *Mind* 64, 200–212. Reprinted in Louis P. Pojman and Michael Rea, eds. (2012). *Philosophy of Religion: An Anthology*. 6th ed. Boston: Wadsworth, 302.

of pain and evil we would need more than a "minute dose" in the world for us to know pleasure and good as we need to know them, but the point remains that the world contains far too much pain and evil for that purpose, both in terms of intensity and quantity, otherwise one could easily turn the argument around, such that pleasure and good exist all the better for us to know, to *really know*, pain and evil.

That the world presents us with *too much* pain, suffering, and death, *too much* ugliness, *too much* discord to provide us with the lost chord, becomes especially evident if we bring in modern science. Looking at life from a *synchronic* view, countless scenes of the horror and gore of predation play themselves out on a moment-by-moment basis in Nature, both predator and prey driven by the merciless tyranny of their shifting and evolving genes. It is the same for the agony of parasitism, the sickening paroxysms of hunger and starvation, and the torture of temperature extremes. And you don't need to be a biologist to know or see this. As the philosopher, theologian, musician, humanitarian, doctor, and Nobel laureate Albert Schweitzer put it, in the context of elucidating his philosophy of *reverence for life*, "Only at rare moments have I felt really glad to be alive. I cannot help but feel the suffering all around me, not only of humanity but of the whole of creation."[49] To all but the callous, the morally deformed, and the unthinking, it should be obvious that the problem of evil is not simply about the tragedy "Man."

Now add to this the *diachronic* view of life presented collectively by modern science. Our planet is not 6,000 years old but roughly 5 billion, with life beginning on it roughly 4 billion years ago, sentient life beginning maybe 600 million years ago, and modern anatomical humans evolving on the scene roughly 200,000 years ago. In other words, not only has most of the world's natural ugliness never come under human purview during our collective lifetime, but diachronically we were not even around for the vast majority of it so as to possibly gain any direct benefit. In other words, given the truths of geology and evolutionary biology, the music of life is filled with dissonance so much and so loud and for

49. Albert Schweitzer (1933). *Out of My Life and Thought*. Antje Bultman Lemke, trans. New York: Holt, Rinehart and Winston, 242.

so long that it seems it could not possibly contribute to any "unity of effect" that was about pleasure.

The ultimate point is that, from a scientific point of view, the pain and suffering in the world does not point to a God worth believing in. This is a common view, especially among professional biologists, and it is worth turning to two illustrative examples of it, both coming from two of the most eminent evolutionary biologists of recent decades.

According to Richard Dawkins, looking at life from an evolutionary point of view, "everything makes sense one you assume that DNA survival is what is being maximized."[50] This is the power of *selfish gene* theory, worked into perfection in his now classic *The Selfish Gene* (1976), that the ultimate function of each of the parts of an organism is not for the happiness of the individual, or even for its survival, let alone that of its species, but for the replication of its genes, that the evolution of everything in life, from amoeba to man, from the capacities for pain and suffering, from the evolution of eyes, sex, language, our sense of purposiveness, you name it, none of it really makes much sense except from a modern evolutionary view, as evolving to better perpetuate the transmission of genes from one carrier to another, that phenotypes evolved and keep evolving for the sake of the genes that, in their genotypes, code for them. Thus, says Dawkins, more summarily, in one of the most powerful lines of our time, "The universe we observe has precisely the properties we should expect if there is, at bottom, no design, no purpose, no evil and no good, nothing but blind, pitiless indifference."[51]

The biologist George C. Williams, famous mainly for his classic *Adaptation and Natural Selection* (1966), puts the point even more strikingly. He defines "God" as "Whatever entity or complex of entities is responsible for the universe being as we find it."[52] After briefly listing facts of nature repeatedly documented

50. Richard Dawkins (1995). *River Out of Eden: A Darwinian View of Life.* New York: BasicBooks, 106.

51. Dawkins (note 50), 133.

52. The same definition, interestingly, was given by Hume in his *Dialogues* (1779). In Part II Philo says, "the original cause of this universe (whatever it be) we call God," while in Part IV he says, "By supposing it [the material

in biology, such as "dysfunctional design features," and of course the usual suspects that everyone knows about, such as "Disease, predators, accidents, and murder," Williams then turns to motherhood itself, specifically the example of infanticide in langur monkeys in northern India: when a male monkey usurps another male monkey for his harem and kills the infants, the nursing mother, after initially resisting the infanticide, "soon starts ovulating. She accepts the sexual advances of her baby's murderer, and he becomes the father of her next child." Since all of this behavior is at bottom genetic, the product of evolved genes, Williams asks rhetorically, "Do you still think God is good?"[53]

One might reply, of course, in the way one often hears, which is to say that God's sense of goodness and justice is not like ours, that God has a very different concept, vastly superior to ours, so that it is a gross mistake to judge God by our standards of goodness and justice. The devastating reply to this line of thinking was well expressed (again) by John Mackie, which is to say, "But then why call him good? Is not this description misleading?"[54] I would take Mackie's point a step further. If "God's goodness is not ours" means something other than human goodness, more exactly something incomprehensible to humans, then it is the theist who does not know what he is talking about, not the critic or atheist, so that it is the theist who must pass over in silence, even suffer the *curse* of silence.

It is difficult to see how Poe's theodicy could deal with all of this. But then it is difficult to see how *any* theodicy, especially those that take evolution seriously, (as they should in our modern time,) any *theistic evolution* that sees God having a guiding hand in evolution, or at least setting the world up for evolution, could satisfactorily deal with this, including the fact, accepted beyond a reasonable doubt by modern science, of utter raw contingency

---

world] to contain the principle of its order within itself, we really assert it to be God." Kemp Smith (note 14), 142, 162.

53. George C. Williams (1996). *Plan & Purpose in Nature.* London: Weidenfeld & Nicholson, 211–212, 215, 217.

54. J.L. Mackie (1982). *The Miracle of Theism: Arguments For and Against the Existence of God.* Oxford University Press, 156.

permeating evolution, both biological and cosmological.[55] But I won't digress. The argument by Poe the artist is faced with a deeply ugly problem—not just surface ugliness, but *structural* or *systemic* ugliness—when it is informed by modern evolutionary biology, and I will leave it at that.

But we are not done with Poe. He admits that natural theology is not a science, that the data of the former is not "palpable" and the reasoning is "purely *analogical*."[56] Apparently in league with this, he rejects the *a priori* attribution of "purposes to God" in favor of an empirical approach, particularly with regard to the behavior of humans, for "If we cannot comprehend God in his visible works, how then in his inconceivable thoughts, that call the works into being?"[57] Poe also admits that "any theological disquisition" begins with "imaginary axioms from which it starts."[58] And in *Eureka*, as we shall see in Chapter 5, he rejects axioms as self-evident or necessarily true. So Poe's theology is really a stab at an answer to the question made much of today, "Why is there something rather than nothing?" His answer is *analogical:* inasmuch as a human would create a work of art for the purpose of pleasure, likewise the Universe was created by God for God's pleasure, that in the original Unity, God, who was also Nothingness as Poe defines it, chose to burst into the "almost Infinite Self-Diffusion" so as to "expand" the experience of pleasure.

Even if pain should be necessary for pleasure, one must question whether the end justifies the means. We might indeed argue that the end, no matter how good, is not justified by such atrocious means, which sacrifices with horror the lives of countless zillions upon the altar of God's supposed *right*. We don't accept that the end justifies the means in the case of wars of aggression

55. For fundamental contingency in biological evolution, see my (2001). "Quantum Indeterminism and Evolutionary Biology." *Philosophy of Science* 68, 164–184. (2010). "Quantum Indeterminism, Mutation, Natural Selection, and the Meaning of Life." In Chérif F. Matta, ed. (2010). *Quantum Biochemistry: Electronic Structure and Biological Activity*. Volume 2. Weinheim, Germany: Wiley-VCH, 837–872. For fundamental contingency in cosmological evolution, see the sources discussed in Chapter 5 §§II, III, and VI.

56. "A Chapter of Suggestions" (1845). *E&R*, 1296.

57. "The Imp of the Perverse" (1845). *P&T*, 826–827.

58. "Marginalia" (April 1846). *E&R*, 1391. "The Literati of New York City" (1846). *E&R*, 1121.

and of terrorist attacks on civilians. As Gandhi put it, (never to be accepted by the Malcolm X types,) you can't get a rose from a weed. So why would we change our view and accept the means when it is on a far bigger scale, for a theodicy that views the hundreds of millions of years of pain, suffering, and killing on this planet alone for the purpose of some higher good designed by God? Moreover, even if pain would be necessary for pleasure, the fundamental fact remains not only that countless creatures have suffered horribly for the sake of God's pleasure, but that they, and we, *die* for it, too—and none of us, in our individuality, by *choice!*

The Problem of Death

This brings us to the second (and more fundamental) problem: the problem of death. From an evolutionary point of view, of course, one might reject at the start that there really is a problem, since one could look at Darwin's main mechanism of natural selection, which is often compared with a sieve, as the creative aspect of death. As Darwin puts it in the *Origin*, "More individuals are born than can possibly survive. A grain in the balance will determine which individual shall live and which shall die,—which variety or species shall increase in number, and which shall decrease, or finally become extinct."[59] What this means is that you, me, and the rest of us are all here ultimately because of billions of years of *creative death*. Every one of our traits, from our knee ligaments and hearts to our eyes and the language capacity in our brains, evolved the way they did because earlier versions of those traits made our ancestors statistically a little more successful in what Darwin called the "struggle for existence," compared with those who had those traits to an inferior degree or not at all.

Given Poe's God, one might call this a *poe*tic conception of natural selection, or *poe*tic justice. But in saying this one would be *gravely* mistaken. This is because there is a subjective side to death, which is what Poe focused on throughout his various works. That focus was on humans, of course, not on animals, which might be justified to some extent by the *apparent* fact that

59. Charles Darwin (1859). *On the Origin of Species by Means of Natural Selection.* London: John Murray, 467.

we, unlike the animals in Tennyson's poem, have the *awareness* of death, of the *finality* of death. This indeed is the problem of ugly in its fullness that Poe dealt with, the ultimate deformity of life, the truly grotesque fact of our existence, that Death presents itself to us not as a temporary sleep or as a metamorphosis corresponding to the two conditions of the worm and the butterfly, but quite apparently as dark—all dark—the intense and utter raylessness of the night without end, of the Night that endureth for evermore, all void, and black, and silent, with Darkness there and nothing more, as the stilly, clear abyss, where total annihilation could be no more, and with no balm in Gilead for the fever called "Living," including the doubly dead in that she died so young, namely, the lost Lenore, that when we die we are not only merely dead, but *really, most sincerely dead*—shadows of shadows embodied in the word "Nevermore."

The problem finds expression in "The Conqueror Worm," what Poe considered to be one of his best poems.[60] If, as Poe claims, the Universe is a plot of God, and God's plots are perfect, then what kind of plot is it? Shakespeare famously claimed, through Macbeth following the death of Lady Macbeth, that "Life's but a walking shadow, a poor player that struts and frets his hour upon the stage and then is heard no more: it is a tale told by an idiot, full of sound and fury, signifying nothing."[61] Poe, in "The Conqueror Worm," as if in reply to Shakespeare, affirms that we are indeed part of a play, a "play of hopes and fears," with "much of Madness, and more of Sin, and Horror the soul of the plot," such that "the play is the tragedy, 'Man,' and its hero the Conqueror Worm."[62] But Poe adds that we are "Mimes, in the form of God on high," an apparent reference to the *imago dei* of Genesis. The problem of death thus expressed, the hope in the form of the *imago dei* materializes, so to speak, in the rising of the dead Ligeia back to life through the body of the newly dead Lady Rowena, in the tale in which Poe embedded the poem, entitled "Ligeia" (1838), which looking back in 1844 he considered to be one of his best tales, even "undoubtedly the best story I

60. *Letters*, 450.

61. *Macbeth*, Act 5, Scene 5.

62. *P&T*, 78, 79.

have written," his "*best* tale" in terms of imagination, as he put in 1846.[63]

The tale is fiction, of course, but the problem it deals with is real enough, finding possibly its most powerful expression in Poe's poem "A Dream within a Dream" (1849), which ends with one of the most achingly plaintive stanzas of all time:

> I stand amid the roar
> Of a surf-tormented shore,
> And I hold within my hand
> Grains of the golden sand—
> How few! yet how they creep
> Through my fingers to the deep,
> While I weep—while I weep!
> O God! can I not grasp
> Them with a tighter clasp?
> O God! can I not save
> *One* from the pitiless wave?
> Is *all* that we see or seem
> But a dream within a dream?

In both works—it is no accident—the description of beauty, the *evocation* and *experience* of beauty, is heightened by discord. The shore with "golden sand" is "surf-tormented." Who among us has not felt drawn by the powerful beauty of such a scene? And Ligeia, we are told by Poe's narrator, was an exquisitely beautiful woman, not only in mind—surpassing in erudition all men and women the narrator ever knew—but in face, such that "In beauty of face no maiden ever equalled her." And yet, says the narrator, before proceeding to describe her face in detail, her features were not of the classical mould, but had a strangeness to them, a strangeness—just like what Poe said elsewhere about discords in music—that heightened her beauty, which he supports by quoting Sir Francis Bacon, who said, speaking generally, "There is no exquisite beauty without some *strangeness* in the proportion."[64]

63. *Letters*, 450, 550, 596. See also Poe's anonymous book review, published in October 1845, of his collection of his own tales, entitled *Tales* (1845). *E&R*, 869, 870.

64. *P&T*, 263.

This, I suggest, is the key to understanding Poe's theology as well as his theodicy, his argument from beauty for God's existence, including our own immortality following the grave. It is not that beauty *proves* the existence of God, even probabilistically—not the beauty of a face, or of nature, or even the elegance of the laws of nature as discovered by science. Neither do we find in Poe the idea that beauty *proves* that we shall survive death. All of this is ridiculous, and it is not Poe. Instead, it is the *effect* that beauty has on our soul. It "elevates the soul," yes, it does so proximately, but ultimately what matters is what beauty conveys.

Beauty and Hope

One of Poe's attempts at solving the problem of evil, including the problem of ugly, is to be found in his "Marginalia" (March 1846). Here Poe doesn't reveal his ultimate solution, but given its explicitly close connection to his literary theory, as an analogy even, what he provides serves as a wonderful introduction to that solution. Quoting once again Bacon, the same quotation we have just seen him use in "Ligeia," that "there is no exquisite beauty without some *strangeness* in the proportions [*sic*]," Poe applies the point to the question of "perfection of rhyme" in a poem. The problem with many poems is that the rhyme at the end of a line creates in the reader or listener an expectation of a similar rhyme in the next line or two, such that "the rhymes were always *anticipated*." The anticipation, however, and its realization, reduces the effect of the poem as properly understood, which for Poe, as we have seen in the previous chapter, is the experience of beauty. (It's like anticipating the end of a movie, which makes me want my money back.) Add to the poem the "element of strangeness," however, "of unexpectedness—of novelty—of originality—call it what we will," one has thereby added to the poem "the *unknown*—the vague—the uncomprehended." The result, the effect on the reader or listener of the poem, if it is done right, is to produce the kind of puzzlement that turns the soul upward, to a higher experience of beauty, to "all that is *ethereal* in loveliness," an elevation of the soul from "the beauty of earth" to "what we dream of the beauty of Heaven." It is like discords in music, to use another of Poe's applications, but it has to be done right. Poe does "not contend for mere *arbitrariness* of rhyme." "In the first place," he says, "we must have equi-distant or regularly recurring rhymes, to form the basis, expectedness, out of which arises the element, unexpectedness, by the introduction of rhymes, not arbitrarily, but with an eye to the greatest amount of unexpectedness."[65] Poe then turns to some of his verses in "The Raven" to illustrate his point.

The relation to the problem of evil is also to be found in this remarkable passage, but in a brief line, as a quick analogy: "as evil cannot exist without good, so unexpectedness must arise from expectedness." What this apparently means is that if good is good, then good with a certain proportion of evil is even better. And what this apparently means more specifically is that if pleasure is good, then pleasure with a certain proportion of pain is even better. In other words still, evil exists in the world, including pain, it is real, and it is necessary, not only because it makes the good in the world, including pleasure, more precious, but also because it raises our awareness, the eyes of our soul, to something higher and more wonderful than this world, to "all that is *ethereal* in loveliness," to "the beauty of Heaven."[66]

But how does it do *that*? If the medium is the message, to use Marshall McLuhan's famous line, the message of beauty for Poe is *hope*. This idea recurs throughout his writings. For example, in his 1836 review of a book written by Joseph Rodman Drake, in which he quotes in full the fifth and sixth stanzas of Percy Bysshe Shelley's "Hymn to Intellectual Beauty" (1816) and tells us that Shelley shows "consciousness of this truth . . . only once," Poe writes that the "Faculty of Ideality" is "the sentiment of Poesy," which although he admits is difficult to define he describes as "the sense of the beautiful, of the sublime, and of the mystical," which is evoked by "the fair flowers, the fairer forests, the bright valleys and rivers and mountains," by "love of the gleaming stars and other burning glories of Heaven," such that "Poesy is the sentiment of Intellectual Happiness here, and the Hope of a higher Intellectual Happiness hereafter." Years later he would express the same idea in a nutshell in his poem "To Helen" (1848), in which

65. *E&R*, 1381.

66. *E&R*, 1381.

he writes of "Beauty (which is Hope)." Similarly in *Eureka*, Poe shows his hand, briefly and subtly, when he puts forward the idea of the repeated expansion and contraction of the Universe and says he is "indulging in a hope."[67] In all of this, Poe gives expression to his fundamental idea that the hope evoked by natural beauty, amplified by discord, speaks to the existence of a supreme Author, or Artist, just as artificial beauty, evoked by a work of art such as a poem or painting, especially with its "unexpectedness," speaks to us of the genius of the artist that created it.

In connection with this hope, at the end of *Eureka*, in a theological flourish, Poe *identifies* each and every one of us with God, such that "each soul is, in part, its own God—its own Creator: in a word, that God—the material *and* spiritual God—*now* exists solely in the diffused Matter and Spirit of the Universe." (The word "solely" is absolutely essential for understanding Poe's theology.) He adds that when we die, we "lose our individual identity," but the "pain of the consideration" ceases, he says, when we think of our future "absorption," when we shall be "blended" with "the bright stars," back "into One." Contra Carl Sagan, then, for Poe we are not only made of star-stuff, we are made of God-stuff, the former proximately, the latter ultimately. At any rate, when we die and lose our personal identity, presumably there will be a long lapse of consciousness until the Universe has collapsed back into the primordial Unity, the "Concentrated Self." But this view is enough, says Poe, for us to "comprehend the riddles of Divine Injustice—of Inexorable Fate. In this view alone," he says, "the existence of Evil becomes intelligible; but in this view it becomes more—it becomes endurable."[68]

This is no better as a solution to the problem of evil than when Poe dealt with the problem of pain. (Here begins the discord.) If *I* die when my body dies, if the *person* or *inner me*, my *subjective self*, is annihilated at death like the snuffing out of the flame of a candle, then whatever *qualitative* identity exists between me and the reconstituted God in the future is not *numerical* identity (one-and-the-same identity), in the very same sense that a *copy* of me is not me. What is required for genuine postmortem existence is

67. *E&R*, 510; *P&T*, 97, 1356.

68. *P&T*, 1357, 1359, 1358, 1357.

continuity of self, such that whatever is good for the *copy* of me in the future is irrelevant as a good for *me*.

This idea is implicit in our legal system, and sometimes explicit in science fiction. If a monozygotic twin of me, or a perfect psychophysical copy of me, commits a crime, *I* am not guilty, the twin or copy is guilty. To think otherwise is to miss the logical difference in meaning when we say, "They are the same" or, synonymously, "They are identical." *Two* things can be the *same*, can be *identical*, in terms of their qualities, but *not* the same, *not* identical, in terms of their number, as they are still two, not one.

Poe is certainly not alone in confusing qualitative identity with numerical identity, what I hereby dub (for lack of a label in the annals of logic) the *same/identical fallacy*. This error in reasoning is in fact a common one, especially in religion and philosophy, and was committed no less than by one of the top philosophers of religion in his defense of the idea of a "resurrection body." In discussing the "Judaic-Christian belief in resurrection" and appealing to the writings of St. Paul, John Hick takes "death" to mean "sheer unqualified extinction" and "resurrection" to mean "God's re-creation or reconstitution of the human psychophysical individual." With these two definitions Hick then provides his three John Smith thought experiments: in the second one, John Smith dies in America leaving behind a corpse, while "at the moment of his death" an "exact replica" appears in India, an exact "psychophysical individual," such that he and his friends think he's one and the same with the John Smith who died in America; in the third thought experiment the replica appears "not in India, but . . . in a different world altogether, a resurrection world inhabited only by resurrected persons."[69] This is great for my replica, but the fallacy discussed above should nevertheless be obvious here, in that the replica *is not me*, but a *copy* of me, so that numerically *I* did not survive my death. In a very different vein, Friedrich Nietzsche made the same mistake with his doctrine of Eternal Recurrence. Near the end of *Die Fröhliche Wissenschaft* (1882), for example, he asks us to believe that "This life as you now live it and have lived it, you will have to live once more and innumerable times more;

---

69. John H. Hick (1990). *Philosophy of Religion.* 4th ed. Englewood Cliffs, NJ: Prentice-Hall, 122, 123, 124.

and there will be nothing new in it, but every pain and every joy and every thought and sigh and everything unutterably small or great in your life will have to return to you, all in the same succession and sequence—even this spider and this moonlight between the trees, and even this moment and I myself. The eternal hourglass of existence is turned upside down again and again, and you with it, speck of dust!"[70] Here the identity is perfectly qualitative, but Nietzsche is just as mistaken in thinking it is also numerical. The fallacy is the same in both cases, however different the moths that committed it.

Poe's solution to the problem of evil is therefore *not* "intelligible," let alone "endurable." Consequently I find no consolation in his idea, no hope from beauty, because I find no good reason to *believe* it. It fails the test of logic and evidence, as arguably do all theodicies—to my mind, at least, and I have studied many, such as that of Boethius, Irenaeus, Augustine, Leibniz, Paley, Hick, Swinburne, Ekstrom, Russell, Tracy, to name some of them. Poe's theodicy is substantially no better or worse than the others. It might only seem worse because of its pantheism, let alone Poe's unique brand of it, which does violence to the parental analogy of the Abrahamic religious tradition, Christianity in particular. Poe, by the way, knew that his thinking would appear filled with "irreverence" as it was "heretical in the extreme."[71] (Here ends the discord.)

What we have to keep in mind is that Poe was principally an artist. And once we try to view his argument from the viewpoint of an aesthetic, it does actually make some sense in a way. The key is to see that his argument from beauty, behind all his hyperbole, with its "*seems, because it is*" and its "a species of proof far surpassing what Man terms demonstration," is really the *argument from hope*, and nothing more.

70. Walter Kaufmann, trans. (1974). *Friedrich Nietzsche: The Gay Science.* New York: Vintage Books, §341.

71. See, e.g., "Mesmeric Revelation" (1844), "The Power of Words" (1845), and *Eureka. P&T*, 722, 822, 1356. Moreover, Poe didn't care. In reply to the theology student John Henry Hopkins, Jr., who in his review of *Eureka* got upset mostly with its theology, Poe didn't care what the critic or others called him—"impious," "pantheist," "polytheist," "Pagan"—so long as it was not "Student of Theology." *Letters*, 690.

As such, Poe's approach bypasses the seemingly endless twisted and tortured defenses of arguments for God's existence, of the cosmological arguments, the design arguments, the arguments from religious experience, for example, and their related solutions to the problem of evil. They are similar to the arguments of modern resurrectionists, who cry along with Paul, "O death, where is thy sting? O grave, where is thy victory?" (I Corinthians 15:55), tempered by "if Christ is not risen, then is our preaching in vain, and your faith is also in vain" (15:17), who then go through the most remarkable argumentative contortions, based on an ancient text by superstitious people, trying to prove beyond a reasonable doubt that the resurrection of Jesus actually happened.

Once one views the argument from beauty for God's existence and our own existence after the grave as the argument from hope, as it surely is in Poe's hands, then it ceases to appear so strange. Beauty gave Poe the only legitimate hope he could find for viewing the Universe as having a plot—as a *tale*, to use the term from his literary theory—with a *dénouement* worthy of his artistic and moral sense. And Poe's focus on hope is surely the way to go for anyone who is intellectually honest and deserves to be called a *thinker*. Let me provide two examples from other thinkers in support, of which there are surely plenty more.

Louis Pojman, late Professor of Philosophy at West Point Military Academy, (Poe's alma mater,) was not only the most successful editor of textbook anthologies in philosophy to date, (literally making millions of dollars for his publishers,) but he was also an accomplished philosopher of religion in his own right—one who went from agnosticism to theism, by the way. In one of his articles, Pojman argues not only that hope is the key to understanding all the theodicies and arguments about God that one finds in religion and theology, rather than logic and evidence, which is to say they are *post hoc*, but that hope is "the proper religious propositional attitude for doubters," (by which he means *thinkers*,) that hope can "serve as a *type of faith* in a religion like Christianity without the belief that the object of that faith exists."[72] The

72. A propositional attitude is an attitude toward a proposition/statement, a sentence with a truth value (true, false, partly true, etc.). *Belief* is a propositional attitude, as in "I believe that the raven hath spoken a lie," and so is

"verific person," says Pojman, as opposed to the "true believer," "recognizes the tragedy of existence," a universe with the proverbial cliff overlooking "a yawning gorge." There is no turning back, for the verific person. The view from the cliff is that of "death and the extinction of all life in a solar system that will one day be extinguished." But there is also a rope that spans the gorge, with a tight-rope walker, who claims he can take you across on his back. You are unsure whether he is "insane or simply overconfident," but you also "have a profound, even *desperate*, hope that he will be successful." Given the problem of the meaning of life, "There is just enough evidence to whet his or her appetite, to inspire hope, a decision to live according to theism . . . as an experimental hypothesis, but not enough evidence to cause belief. So keeping his or her mind open, the hoper opts for the better story, gets on the back of what may be the Divine Tight-Rope Walker, and commits to the pilgrimage."[73]

Given Poe's "We stand upon the brink of a precipice," his "We peer into the abyss," and given what follows,[74] my advice to the committed hoper is: *Don't look down!* At any rate, the upshot of Pojman's approach is that one *can* be without belief in God or gods and still be a theist. One can be a belief atheist *and* a hope theist.

The same is true with regard to life after death, which brings us to my second example. Carl Jung, who was chosen by Freud to be his "successor and crown prince," went his own way into the depths of the mind, creating theories such as the introvert/extravert distinction and the collective unconscious. But he also devoted a large portion of his life as a psychologist to the study of parapsychological phenomena, (a field rejected by Freud as "sheer bosh," much as it is today by psychologists and other scientists,) such as poltergeists, ESP, ghost sightings, and séances, a study that involved many of his own personal experiences. Interestingly, in 1944, at the age of sixty-eight, while in a hospital following

---

hope, as in "I hope that the raven hath spoken a lie." There are, of course, many more propositional attitudes than these two.

73. Louis P. Pojman (1999). "Faith, Hope, and Doubt." In Pojman and Rea (note 48), 556, 558–559.

74. "The Imp of the Perverse" (1845). *P&T*, 829.

a heart attack preceded by breaking his foot, Jung experienced (subjectively, at least) one of the most extraordinary out-of-body experiences ever recorded, not one but a series of visions over a protracted period of time, including a view of Earth from "approximately a thousand miles," a temple with a "black Hindu" sitting "silently in lotus posture upon a stone bench," and a visit to Jung in this temple from his attending physician from the hospital below, communicating in his "primal form" that Jung must return to Earth, which was followed by the doctor's death in real life shortly after Jung was first able to sit up in his bed. Jung devotes an entire chapter to these visions in his autobiography, written in the few years before he died, stating, for example, that "Those inner states were so fantastically beautiful that by comparison this world appeared downright ridiculous" and that "I would never have imagined that any such experience was possible. It was not a product of imagination. The visions and experiences were utterly real; there was nothing subjective about them; they all had a quality of absolute objectivity."[75] One might be tempted to pass this off as a case of bad memory, like exaggerated memory in old fishing stories, given that it was written many years afterward. What has to be recognized, however, is that hallucinations typically do have a subjective quality to them of "absolute objectivity," a topic that has been extensively studied by psychologists and neuroscientists for decades now with relation to diseased, damaged, and altered brain states.[76] But this is all beside the point, which is that, while Jung never feared death following his extraordinary series of experiences, he did *not* (unlike the multitude with far less intelligence, knowledge, and experience) settle into a firm and unshakable belief in life after death. Instead, as he puts it at the end of his autobiography, written shortly before his death in 1961 at the age of eighty-five, "There is nothing I am quite sure about. I have no

75. C.G. Jung (1963). *Memories, Dreams, Reflections.* Aniela Jaffé, ed. Richard and Clara Winston, trans. London: Routledge & Kegan Paul, 333, 152, 270, 271, 273, 275.

76. See, e.g., Oliver Sacks (2012). *Hallucinations.* New York: Alfred A. Knopf. Sacks was a professor of neurology, with a background in theory and practice that began in the 1960s. This book is a must for anyone interested in the topic of hallucinations, both for its fascinating content and its entrée into the vast literature on the topic.

definite convictions—not about anything, really. I know only that I was born and exist, and it seems to me that I have been carried along." As to the meaning of life, "Which element we think outweighs the other, whether meaninglessness or meaning, is a matter of temperament." And then comes the key: "I cherish the anxious hope that meaning will preponderate and win the battle."[77]

The intellectual honesty in the above is absolutely remarkable, all things considered.

Returning to Poe, it was not, of course, religion, philosophy, or parapsychology that served as the basis of his hope. He was essentially an artist, and so accordingly it was beauty (whether of a poem, a woman, natural scenery, stars, laws of nature, etc.) that elevated his soul and gave him hope—hope that there is more to life than what is marked by the grave—hope that suffering has a higher meaning and that death is not the end. This might not be the response to beauty experienced by most people, given that most are not artists, let alone like Poe, but arguably hope, for *most* people by far, (I will not say *all*, given that biology is statistical, excluding the fact of death itself,) is absolutely essential, for without it life becomes unbearable. This is rather obviously the human condition (taken, again, statistically). And Poe, it must be remembered, what is rarely truly appreciated among modern literati, (who need to imagine walking a mile in his shoes,) had a life far harder than almost anyone who has ever read any of his works, (and not entirely of his own making,) a life filled with poverty, pain, hunger, disappointment, failure, death, and despair. Indeed I suggest that hope, for most people who have ever walked this planet, was the only thing that kept them struggling through their miserable existence, the only thing that kept their spirits up in a life marked by hunger, illness, desperation, murder, war, loss of loved ones, fear, terror, and a short life expectancy. Indeed for many, when life is hanging by a thread, often the thread itself is nothing but hope, which when broken or lost spells the end, the one-way door marked EXIT.

Hope indeed seems to be the key at a number of levels. Even at the level of the most mundane, that of material prosperity, one can see hope as a recurring theme in Poe's writings, especially in

77. Jung (note 75), 330.

his correspondence. For example, in a letter dated May 4, 1845, Poe writes of working "14 or 15 hours a day," (he was now one-third owner of the failing *Broadway Journal*, trying to turn it into a success,) adding "I have made no money" and "I am as poor now as ever I was in my life—except in hope, which is by no means bankable." Later, in a letter dated July 22, 1846, roughly six months after losing sole proprietorship of the *Broadway Journal* and roughly six months before Virginia would lose her battle against tuberculosis, Poe writes of "my illness, and how impossible it is for me to put my foot out of the house or indeed to help myself in any way," adding, "Do not let anything in this letter impress you with the belief that I *despair* even of worldly prosperity. On the contrary although I feel ill, and am ground into the very dust with poverty, there is a sweet *hope* in the bottom of my soul." And yet again, in a letter written less than half a year before his end, sometime after May 5, 1849, in which he complains repeatedly about magazines and pay and that "misfortunes never come single," he adds that "all seems to be frustrated" and even more seriously that "my sadness is *unaccountable*, and this makes me the more sad. My life seems wasted—the future looks a dreary blank; but I will struggle on and 'hope against hope.'"[78]

In the bottom of Poe's soul resided another insight into hope, one that found expression in his tales of terror—which in the minds of most lovers of Poe is the quintessential speciality of Poe. In the Preface to his *Tales of the Grotesque and Arabesque* (1840), Poe responds to the criticism, not that his tales are excessively morbid, (for as we have seen in the previous chapter, they were not from the Victorian perspective,) but that his horror is excessively German, that his tales are dominated by "'Germanism' and gloom," as he puts it.[79] Poe replies by stating, "If in many of my

---

78. *Letters*, 504, 592, 796.

79. *P&T*, 129. Germany was the home of Gothic literature. See Alan Brown (2013). "The Gothic Movement." In *Context*, 241–250. Poe states that the "Arabesque" in his tales have "induced one or two of my critics to tax me, in all friendliness, with what they have been pleased to term 'Germanism' and gloom." But it should be noticed that the focus of the criticism in both cases (I can find no more than two) is "Berenice" (1835). The first comes from Thomas W. White, the editor of the *Southern Literary Messenger*, who first published "Berenice" in the March issue of the fledgling *Messenger* but with

productions terror has been the thesis, I maintain that terror is not of Germany, but of the soul." Terror is of the soul, it is part of the human condition, one may even say that the propensity to experience terror is part of human nature (statistically conceived, of course).

And what, one may now ask, does terror have to do with hope? The answer is simple: *just about everything*. What terror triggers, in the soul, in the human psyche, (and not just terror,) is one of the deepest and most powerful survival mechanisms known to man. But there is no better way to get this point across than to look at extreme situations. And for that—*but of course!*—we need not look any further than Poe. Consider, for example, the following passage from "The Pit and Pendulum" (1842), the words of the narrator as he lies supine tied to a bench while the huge pendulum above him, with every swing, inches closer to his chest with its foot-long edge of glittering steel: "It was *hope* that prompted the nerve to quiver—the frame to shrink. It was *hope*—the hope that triumphs on the rack—that whispers to the death-condemned even in the dungeons of the Inquisition."[80] I don't want to ruin the end of "The Pit and the Pendulum" for those who have not yet read it, but observe the role played by hope during the remainder. Observe its role in the narrator not giving up. Observe the actions that it prompts in him. Do that and you will see what I am talking about, what *Poe* is talking about. If I may be so bold as to add to the words of Poe, we all—every one of us—by the very fact of our being alive, are the "death-condemned" in the "dungeons of the Inquisition."[81]

---

grave misgivings. We do not know White's exact words to Poe, but Poe wrote a letter to White, dated April 30, 1835, in which he acknowledges that "Your opinion of it is very just. The subject is by far too horrible, . . ." Poe then goes on to defend "Berenice," mainly in terms of magazine sales. *Letters*, 84–85. The second critic, James E. Heath, wrote an editorial notice of the publication of "Berenice," "Morella," and "Hans Pfaall." The notice is a generally friendly one, although in the paragraph on "Berenice" he wrote, "we confess that we think that there is too much German horror in the subject." *Critical Heritage*, 80. According to Thomas and Jackson, White was "perhaps influenced by James E. Heath" in giving Poe his opinion of "Berenice." *Poe Log*, 149.

80. *P&T*, 501.

81. There is another beneficial effect of terror brought out by Poe. As his narrator puts it at the end of "The Premature Burial" (1844), "out of Evil proceeded

As much as terror is of the soul, then, so also is hope. Without hope, quite simply and powerfully, our ancestors would have been lost, would have given up, would not have tried, would not have made the profound effort to survive, and therefore would probably not have reached adulthood and reproduced. To put it another way, you and I are here, me writing this and you reading this, (what price this book?,) not only because each and every one of our ancestors survived to adulthood and produced at least one child,—a striking thought in itself, emphasizing exponentially our utter contingency and the unbearable lightness of our being,—but they did this, in virtually all cases, because they were powerfully motivated by both terror and hope. In other words, to wax neo-Darwinian, terror and hope are of the soul because they are neural *adaptations*, cognitive/behavioral products coded for in our DNA because they increased survival and ultimately reproduction.

Of course, however much we may understand hope in this way, in the case of Poe and countless others, it does not make his argument from hope for God's existence—in itself, intellectually—any the more *believable*. But it does make it all the more *understandable*. And to this I should like to add, in closing this chapter, possibly the strongest reminder that many an academic needs, (along with being cracked over the head, in the interest of phrenology,) one that flowed with ease from the pen of arguably the greatest philosopher to ever have written in the English language, whose theory of human nature included a moral sense based on sympathy and empathy. I speak of David Hume: "Be a philosopher; but, amidst all your philosophy, be still a man."[82]

Good. . . . My soul acquired tone—acquired temper. I went abroad. I took vigorous exercise. I breathed the free air of Heaven. I thought upon other subjects than Death. . . . I read no 'Night Thoughts'—no fustian about church-yards—no bugaboo tales—*such as this*. In short, I became a new man, and lived a man's life. . . . There are moments when, even to the sober eye of Reason, the world of our sad Humanity may assume the semblance of Hell— . . . Alas! the grim legion of sepulchral terrors cannot be regarded as altogether fanciful—but, . . . they must sleep, or they will devour us—they must be suffered to slumber, or we perish." *P&T*, 679.

82. David Hume (1758). *An Enquiry concerning Human Understanding*. London: A. Millar, §1, ¶6. Originally published in 1748 as *Philosophical Essays concerning Human Understanding*.

CHAPTER 4

Poe's Intellectual Background

> I was astonished, too, at the vast extent of his reading; and, above all, I felt my soul enkindled within me by the wild fervor, and the vivid freshness of his imagination.
>
> —Edgar Allan Poe[1]

Entrée

As stated at the beginning of the previous two chapters, (and which needs to be stated yet again,) no one can possibly come to understand, let alone to appreciate, Poe's *Eureka* without knowing Poe's background, which includes not only the biography of his life but especially the biography of his thought leading up to his magnum opus. The focus of this chapter is Poe's intellectual background, both in terms of his education and his interest in science. In all of this we shall proceed chronologically, for there is an interesting pattern of development to be discerned. We begin with a brief look at Poe's early days, the days of his formal education, focusing on a number of significant highlights. We then look at some of the entries in Poe's "Pinakidia" (1836), which is a set of fillers he wrote when he was the editor of the *Southern Literary Messenger*. After that we turn to the question of Poe's authorship of *The Conchologist's First Book* (1839), a science book on

1. "The Murders in the Rue Morgue" (1841). *P&T*, 400.

seashells. There is a mystery here that begs to be solved, a mystery that should put any Poe lover into a pitiable state of agitation, and accordingly I shall try to solve it. We then look at Poe's "A Chapter on Science and Art" (1840), which is a series of columns published in *Burton's Gentleman's Magazine*. After that we turn to what Poe claims in "Marginalia" (1844) is the fundamental flaw in the reasoning displayed in the Bridgewater Treatises of the 1830s, a set of eight books on natural religion. We then examine the relation of *Eureka* to Robert Chambers' *Vestiges of the Natural History of Creation* (1844). Following that, we turn to an examination of the principal books that informed the scientific background of *Eureka*. We then finish by examining two important matters of veracity, the first concerning Poe's criteria of truth in *Eureka*, and the second concerning whether *Eureka* is another of Poe's hoaxes, which is what many have thought with a velocity accumulating in the inverse proportion of the squares of the distances at which lay the inevitable End.

Poe's Formal Education

What kind of formal education did Poe have? What were his interests? In what did he excel? What did he read, both by choice and by necessity? We shall only take a brief look here, but it will be sufficient to show that Poe did not begin his intellectual life by studying science. That came later, and apparently in a roundabout way, with an unexpected turn. But before that, Poe did display something more, a passionate interest in learning, and seemingly about everything. At any rate, what we shall glimpse here, in Poe's formal education, including remembrances of those who knew him at the time, is a strong foundation for what is the ultimate focus of this book, namely, scientific imagination in the deepest sense of the term. What clearly followed Poe's days of formal education was a genuine interest in science, which had remarkable consequences, most notably the writing of *Eureka*.

We start with the period from June 1815 to June 1820. Beginning at this time, John Allan took his wife, his sister-in-law, and little Edgar, barely aged six and a half, (Poe was born on January

19, 1809,) to Britain in the pursuit of improved business prospects. Although Poe attended two very good private schools there, boarding schools at that, the most notable for our purposes was the second, the Manor House School north of London run by Reverend John Bransby, which Poe attended from late July 1818 until the family left Britain. According to a report recorded much later by a student who had been there, "when he [Poe] left it he was able to speak the French language, construe any easy Latin author, and was far better acquainted with history and literature than many boys of a more advanced age who had had greater advantages than he had had."[2] It is to be noted that just prior to the trip to Britain, Poe had attended a school in Richmond run by William Ewing, in which were taught "The Greek, Latin, French, and English languages, together with Writing, Arithmetic, and Mathematics."[3] Later, in June 1818, John Allan wrote to Ewing from London stating that "Edgar is a fine Boy and reads Latin pretty sharply."[4]

There are two points especially worth noting, aside from the fact that Reverend Bransby is featured in Poe's tale "William Wilson" (1839). First, as Arthur Quinn points out, Bransby had an M.A. from Cambridge University and "was an active student of botany and a member of the Horticultural Society of Stoke Newington." Quinn then remarks, interestingly, that "It is quite within the bounds of possibility that Edgar Poe, noticing how effective 'miscellaneous information' may be when given offhand, took Mr. Bransby as his model later in the acquiring of all kinds of valuable odds and ends of literary and scientific knowledge." Second, Quinn adds that "If the English education was limited in scope, it was probably thorough, and it may have contributed to that quality of Poe's mind which is best described as disciplined."[5]

With the Allans moving back to Richmond in June 1820, Poe then attended the school of Joseph H. Clarke, starting in September 1820. This episode is remarkable in a number of respects for

2. *Quinn*, 71.

3. *Poe Log*, 23.

4. *Poe Log*, 36.

5. *Quinn*, 72, 80.

our purposes. One is the advertisement by Clarke for his school: "the English, French, Latin, and Greek languages are taught in this School," using a method that Clarke claims is superior, and "Arithmetic, Geography, and the Elementary principles of Geometry and Astronomy form also an important appendage to his course of instruction."[6] In later life when Clarke reminisced about Poe, he mentioned not only the particular classics that Poe had studied, that "Poe was a born poet," and that Poe "had no love for mathematics," but also, most notably, that "His ardent and predominant passion seemed to me to be an enthusiastic ardor in everything he undertook."[7] The significance of this reminiscence is that it remained a key feature of Poe's personality right to the end, which is eminently evident in *Eureka*. It is also a key feature of Poe's theory of scientific imagination reconstructed in Chapter 7, which I shall modernize in Chapter 8, which is that scientific imagination in its fullest is not just an educated imagination but a passionate one as well.

That Clarke mentions astronomy in his advertisement is also interesting. There is no record of whether Poe actually learned astronomy in Clarke's school, but what is surely interesting, and one is tempted to connect some dots here, is that John Allan brought from England in their trip back to Richmond a telescope, which found a home on the porch of the Allan home.[8]

From April 1823 to March 1825 Poe attended William Burke's school in Richmond, which Burke in his advertisements referred to as a "Seminary" and as providing preparation for an "honourable entrance into any University in the United States."[9] Poe was a "much more advanced scholar than any of us," according to a remembrance of one his classmates,"[10] but nothing else is notable here.

From February to December 1826, a period of ten months, Poe attended Thomas Jefferson's newly created (and very liberal) University of Virginia in Charlottesville. Poe took two courses,

6. *Poe Log*, 41–42.

7. *Poe Log*, 47.

8. *Quinn*, 93.

9. *Poe Log*, 55.

10. *Poe Log*, 58.

one on ancient languages (Greek and Latin) and the other on modern languages (French, German, Italian, and Spanish). Both of his professors reported in December that Poe "excelled" in his examinations.[11]

Poe's background in languages, emphasized throughout this section, might seem trivial given the purposes of this chapter, but two things must be kept in mind. First, his mastery of French must have played a key role in whatever was his contribution to *The Conchologist's First Book*, discussed below. Second, and more important, Poe's mastery of a variety of languages, both ancient and modern, arguably contributed to his remarkable scientific imagination. This is because, as Northrop Frye puts it, following what "the humanists have always insisted," "you don't learn to think wholly from one language: you learn to think better from linguistic conflict, from bouncing one language off another."[12]

What is also notable from Poe's time at the University of Virginia is his artistic ability. One of his classmates, in a later remembrance, states that Poe would recite poetry to his fellows and "then suddenly a change would come over him & he would with a piece of charcoal evince his versatile genius, by sketching upon the wall of his dormitory, whimsical, fanciful, & grotesque figures, with so much artistic skill, as to leave us in doubt whether Poe in future life would be Painter or Poet."[13]

Apparently no one at that time saw Poe as anything more. But what has to be added to this picture is Poe's use of the university library, (which he reported to have a "fine collection,") namely the books he withdrew that apparently were not required by his courses. Most notably, he borrowed David Hume's *The History of England*, John Lingard's *A History of England*, Charles Rollins' *Historie Ancienne*, William Roberston's *The History of America*, John Marshall's *The Life of George Washington*, two volumes of Voltaire's writings, and Nicholas Gouin Dufief's *Nature Displayed in Her Mode of Teaching Language to Man*.[14] What else

11. *Poe Log*, 73.

12. Northrop Frye (1963). *The Educated Imagination*. Toronto: CBC Enterprises. Republished (2002). Toronto: House of Anansi Press, 72.

13. *Poe Log*, 69.

14. *Poe Log*, 70–73.

Poe would have liked to withdraw is difficult to say, but from what is known it is apparent that he had an avid interest in history, which he clearly would later combine with his growing interest in science, all of which culminated in *Eureka*.

From May 1827 to April 1829 Poe was in the army. This portion of his life does not concern us here, except for one item. From November 1827 to December 1828 Poe was stationed at Fort Moultrie, on Sullivan's Island, which is off the coast from Charleston, South Carolina. The island features prominently in Poe's tale "The Gold-Bug" (1843), whose hero, William Legrand, is a resident "recluse" who was "well-educated," with "unusual powers of mind," and whose favorite pastime was searching for "shells or entomological specimens;—his collection of the latter might have been envied by a Swammerdam [a pioneer on the anatomy of insects]."[15] But the real importance of the island for us consists in the fact that Dr. Edmund Ravenel had a summer residence there. Dr. Ravenel was a physician and the Chair of Chemistry and Pharmacy at the Medical University of South Carolina. He was also an expert on conchology, having published in 1834 the first catalog of shells in America. It is surmised that Poe, who had plenty of spare time while stationed on the island, made the doctor's acquaintance, which somehow contributed to his (Poe's) interest in shells and his subsequent role in the publication under his own name of *The Conchologist's First Book*.[16] We shall return to this quaint and curious volume below in order to determine Poe's precise role behind its words.

From June 1830 to February 1831, Poe was a cadet at West Point Military Academy. His academic studies during this failed attempt at a military career consisted of two courses in the fall semester, in French and in mathematics, at which he excelled, officially being ranked among the best in French for two months straight, also in mathematics for the second of those two months, and he stood third in French and seventeenth in mathematics, in a class of eighty-seven, as a result of the exam in January.[17] This was in the context of an exceptionally rigorous regime that saw most

15. *P&T*, 561.

16. *Quinn*, 130, 277.

17. *Quinn*, 171; *Silverman*, 61.

cadets not make it to graduation, while Poe himself toward the end of the semester was suffering financially because of grossly insufficient funds sent to him by his foster father, was unable to deal with the cold, and as a consequence suffered a bout of ill health.[18] At any rate, Poe must have gained during this short time some interest in military technology, including training in explosives, which was part of his military training.[19] This interest evidently stayed with him, as we shall see in the section on "A Chapter on Science and Art."

Later remembrances by Poe's roommates are also significant. One not only noted that Poe was "an accomplished French scholar, and had a wonderful aptitude for mathematics," but that he "was a devourer of books."[20] All three of these traits, by the way, especially the latter, clearly continued straight through to the time Poe wrote *Eureka*. Another roommate wrote that "Very early in his brief career at the Point he established a high reputation for genius." A few paragraphs later he added, "The studies of the Academy Poe utterly ignored. I doubt if he ever studied a page of Lacroix [*Elements of Algebra*], unless it was to glance hastily over it in the lecture-room, while others of the section were reciting."[21] I suspect that what is being referred to here is not only previous learning as well as intellectual capacity, but also memory, possibly on the verge of what today is known as a photographic memory. This is corroborated by something else this source says, which is that when Poe saw a book of Thomas Campbell's poems on his roommate's desk, he called the poet a "plagiarist" and "turned the leaves over rapidly until he found the passage he was looking for."

But there is more, much more. As William Engel points out, there are reports concerning later in his life that Poe could recite all 108 lines of his poem, "The Raven" (1845), at will, on the spot, on demand. Moreover, Poe had a lifelong interest in mnemonics,

18. Poe can hardly be blamed for dropping out of West Point. Certainly it was not because of academic difficulties, for he had none. For a brief reconstruction of his unique set of reasons, see *Quinn*, 174, and for a more detailed discussion, see *Silverman*, 63–66. For Poe's own reasons stated to his foster father, see *Letters*, 58–62.

19. *Silverman*, 60–61.

20. *Poe Log*, 107.

21. *Poe Log*, 108–109.

the mastery of memory. Evidence for this includes the fact that two of the books included in Roderick Usher's library in "The Fall of the House of Usher" (1839) are books dealing with the art of memory, that Poe in his writings often favorably cites Francis Bacon, whose *Of the Proficience and Advancement of Learning* (1605) includes a comprehensive examination of artificial memory systems, and that Poe wrote a remarkably positive book review of Francis Fauvel-Gouraud's *Phreno-Mnemotechny; or The Art of Memory* (1845).[22] Nothing else explains, one should think, the enormous wealth of references in Poe's mature works, especially in his nonfiction prose, to literary and scientific sources, including quotations that usually have only minor errors, unless he was a master of memory, that he had an enormous memory capacity combined with remarkable recall. Given Poe's typically meager financial means, which included the lack of a sufficient library at home,—and home is where he did the vast majority of his writing (and usually in the evenings),—Poe when he wrote must have relied much more on his memory than on written sources before him.[23] (And it matters little if his sources were more secondary than primary, as they were at all counts massive and multifarious.) This extraordinary ability, by the way, speaks directly to Poe's capacity to exercise what is the ultimate focus of this book, namely, scientific imagination in the fullest sense, which requires the ability to process large quantities of information from disparate and seemingly unrelated sources into a unified picture.[24]

22. William E. Engel (2012). *Early Modern Poetics in Melville and Poe: Memory, Melancholy, and the Emblematic Tradition*. Burlington, VA: Ashgate, 107–110.

23. This view is corroborated by the Levines, who state that Poe "seems often to have quoted passages he had learned by heart." *Critical Theory*, 127n3.

24. This conclusion seems corroborated by Poe himself in one of his entries in "Marginalia" (November 1844), "respecting the greatest amount of erudition attainable by an individual." Poe not only claims, in terms of reading and retention, that "knowledge breeds knowledge, as gold gold; for he who reads really much, finds his capacity to read increase in geometrical ratio," but that "there *are* minds which not only retain *all* receipts, but keep them at compound interest for ever." *E&R*, 1318.

What is also of interest from this period in Poe's life is the publication of his poem "Sonnet—To Science," which was republished in September and October of 1830 in two periodicals, having been originally published in Poe's second book of poems, which appeared in December of the previous year, as the preface to "Al Aaraaf."[25] The poem is important since it characterizes Poe's attitude toward science up to and including his time at West Point. On the surface of it, the poem bears the mark of what C.P. Snow would later, in 1959, call the "two cultures" in academia, what came to be known as the "culture wars" between the sciences and the humanities. According to Arthur Quinn, "Poe already was a reader of scientific works" and yet this poem "is an attack upon Science," largely, suggests Quinn, as part of the Romantic rejection of the "dogmatism of Science."[26] Science, says Poe in the poem, not only "alterest all things" with its "peering eyes," but it is a "vulture, whose wings are dull realities," and it is difficult to love—"how should he love thee? or how deem thee wise"—given, for example, that it has "dragged Diana from her car" and "torn . . . from me the summer dream beneath the tamarind tree."[27] James Hutchisson claims that the poem expresses Poe's fear that science and technology are "destroying their [man's] inner spirituality," that they would "remove mankind's heart and soul."[28]

I want to defend a different reading of this poem, one that views it as neither a warning about man's inner life nor an attack upon the dogmatism of science. If we compare this early poem with *Eureka*, what we shall see is something strikingly in common, which is that science, as *understood* in Poe's time, not only *demythologizes*, to use a twentieth-century term, but it removes *imagination* from the understanding of reality, making reality "dull" and "altered." Nascent in "Sonnet—To Science," Poe would develop this idea and express it fully, years later, in *Eureka*. Science, understood as restricted to the method of induction and/or deduction, is the view of science "creeping" and "crawling" and is grossly incomplete, but when combined with the power of

25. *Poe Log*, 100, 107.

26. *Quinn*, 161–162.

27. *P&T*, 38.

28. *Hutchisson*, 90.

imagination,—the poet's imagination, the ability to perceive "the similarity in dissimilars,"—the combination of which is the true nature of science, science properly understood, science then no longer follows a "path," since like an eagle its true nature is to "soar."[29] Understood as such, "Sonnet—To Science" and *Eureka* are not antipodes, or symptoms of a change in view, but are of a piece, differing only in length and in expression—different stages of an individual view of science much as the metamorphosis of the worm into the butterfly.

"Pinakidia"

In the August 1836 edition of the *Southern Literary Messenger*, of which he was editor at the time, Poe published 172 little pieces on a variety of topics. These served mainly as fillers for incomplete columns. In his Introduction to the "Pinakidia," which is what Poe called the set collectively, Poe claims the entries were taken from "marginal notes" and a "common-place-book" to which he contributed during his many readings.[30] Whether this is true or Poe was amusingly including himself among "some little American pilferers" that he disparages in his Introduction, who didn't rely on primary sources but on secondary sources of erudite miscellanea, is beside the point.[31] What matters, for our purposes, is the content, in illustrating the fact that Poe's interests were vast, even when confined to science, which of course they were not. Three of the entries are of particular interest in relation to *Eureka*.

In Pinakidia 12, for a start, Poe claims that "Theophrastus, in his botanical works, anticipated the sexual system of Linnaeus" and that "Philolaus of Crotona maintained that comets appeared after a certain revolution."[32] The interest here does not at all consist of the fact that Poe's source or sources are unknown. Instead, it resides between the lines, first in the claim that ancient Greeks

29. *P&T*, 1268, 1269.

30. *Brevities*, 2.

31. Pollin briefly examines both possibilities and seems pulled in opposite directions as to Poe's veracity here. *Brevities*, xi, 9.

32. *Brevities*, 17.

anticipated a number of modern theories and discoveries, and second in the fact, certainly known to Poe, that the ancient Greeks did not disparage poetry but held it in high esteem, not only in the culture as a whole but also among their philosophers who rejected the common religion, such as Pythagoras and Aristotle.[33] But this passage is even more interesting than that, for Poe then attributes, to two *poets*,—specifically Luigi Pulci of Florence (d. 1484) and Dante Alighieri (d. 1321) "two centuries before,"—a most remarkable claim, namely, that each "has a passage expressly alluding to a western continent." This, of course, was before Columbus and 1492. Poe's two cases might not in fact involve clear allusions.[34] But his fundamental point is clear. And it is the very same point that permeates *Eureka*, which is that the typical scientific intellect is incomplete, or rather, that it "creeps" and "crawls," while the genuine scientific intellect is in fact the educated poetic intellect, which "soars," and that the scientific mind needs the poetic mind in order to see the big picture and to arrive at the big truths of reality.

Pinakidia 69 is equally interesting, and for much the same reason. Here Poe states that "Darwin is indebted for a great part of his 'Great poem' to a Latin one by De La Croix, published in 1727 and entitled 'Connubia Florum.'"[35] The Darwin here is Erasmus Darwin, the illustrious grandfather of Charles Darwin, a physician, theoretical scientist, and poet, whose popularizing of science in his *The Botanic Garden* (1791), which consists of "The Economy of Vegetation" and "The Loves of the Plants," was written in verse. The former poem consists largely of scientific miscellanea,

33. Poe's examples were well chosen. Philolaus was a Greek Pythagorean, and the Pythagoreans, beginning with the legendary Pythagoras himself, had well-developed theories of music, poetry, and harmony based on number mysticism. As for Theophrastus, sometimes called the "father of botany," he was a pupil of Aristotle's and eventually became his successor, and was equally prolific. Aristotle himself, usually called the "father of biology," was of course the author of *Poetics*, and as we have seen in Chapter 2 he used a biological metaphor for plot and emphasized the importance of metaphor for both poets and philosophers, all of which must have had a profound influence on Poe.

34. Pollin, *Brevities*, 17–18.

35. *Brevities*, 49.

including speculation about the history of the cosmos, while the latter poem is based on Linnaeus' sexual classification of plants. Both of these didactic poems contain evolutionary themes, which became the focus of Erasmus Darwin's *Zoonomia (*1794–1796), which was not written in verse, in which he proposed a theory of evolution akin to what Lamarck would later provide in his *Philosophie Zoologique* (1809). Poe's reference to Erasmus Darwin is almost certainly not because of the latter's evolutionism.[36] Instead, it is apparently because Erasmus Darwin combined science and poetry, and as Poe says was "indebted" to (pilfered from?) another writer who combined science and poetry, namely, Demetrius de la Croix, in his *Connubia Florum* (1727), (*The Marriage of the Flowers*,) in which he anticipated Linnaeus' sexual system of classification. At any rate, the bottom line is that Pinakidia 69 is of a piece with Pinakidia 12, which are both of a piece with *Eureka* in the said respect.

Finally, in Pinakidia 151 Poe comments on Empedocles' physics, which consists of four elements and two forces. Of the latter, Poe writes, "What are these but attraction and repulsion?"[37] As Burton Pollin points out, this is interesting as it seems to anticipate an important item in Poe's physics in *Eureka*, according to which, as we shall see in the next chapter, matter is nothing but attraction and repulsion. But the passage is important for another reason, given that Poe's source on Empedocles, which Pollin quotes, states that "The Greeks, men of subtile and inquisitive genius, went further [than the Egyptians and Hebrews in physics] and sometimes guessed right enough, though very rarely. Empedocles, for example, . . ."[38] The idea of guessing right, combined with a background in the physics of the day and a culture that exalts poetry,—Empedocles' *On Nature*, by the way, was written as a poem,—makes Poe's use of this item hardly a coincidence, not only because of what we have seen in Pinakidia 12 and 69, but because of Poe's theory of the educated imagination that figures in so many of his works and that reached its culmination in *Eureka*.

---

36. Pollin, *Brevities*, 49–50.

37. *Brevities*, 93.

38. *Brevities*, 93–94.

The Conchologist's First Book

Just prior to April 20, 1839, a book was published in Philadelphia entitled *The Conchologist's First Book: or, A System of Testaceous Malacology*.[39] The subtitle is much longer, but the curiosity is not the subtitle; instead, it is the name of the author: Edgar A. Poe. Written primarily as a primer, this book on seashells (*conchology*) and the soft-bodied animals that made them (*malacology* collectively) was the only book written by Poe to go into a second edition within his lifetime. Sold out in two months, a second edition appeared, with corrections and additions and again with Poe's name, just before September 11,[40] although the cover page lists the publication date as 1840.

Concurrent with this curiosity, which we are about to examine in full, is a letter Poe wrote, dated July 14, 1839, to George W. Poe, a distant relative on his father's side (specifically, their fathers' fathers were brothers, sons of John Poe, who was born in Ireland and emigrated to America with his parents, ending up in Baltimore). In this letter Poe states: "There can be no doubt, I think, that our family is originally German—as the name indicates. It is frequently met with in German works on Natural History, and a M. Poe is now living in Vienna who has much reputation as a naturalist. The name there is spelt with an accent thus, Poé, and is pronounced in two syllables, Po-a. As far back, however, as we can trace our immediate progenitors they are Irish."[41] According to the editors of Poe's letters, "The full resources of the New York Public Library fail to verify Poe's statements about the origins of the family name. 'Poe' is not German, nor does this posit a French accent on a terminal 'e.' Moreover, no bibliography or index of Germanic scientists indicates a 'naturalist' or scientist of that name. . . . Poe is obviously trying to 'gentrify' himself through somewhat peripheral and inflated forebears and relationships."[42] On the other hand, there might be an actual connection. This is because a little research on the Internet led me to an Orlando M.

39. *Poe Log*, 260–261.

40. *Poe Log*, 269.

41. *Letters*, 183.

42. *Letters*, 186.

Poe (1832–1895) of Ohio, who served as an engineer and captain in the Civil War. His family tree, interestingly, traces to Germany, specifically to his great-great-grandparents Catherine and George Jacob Pfau, who emigrated from Germany to Maryland. Their sons were the first to Anglicize their surname to Poe, which they probably derived from their father's father, George La Poer.

This only adds to a greater mystery: did Poe, the master of poetry, tales, and literary criticism, really write a scientific work on seashells and the organisms that made them? It is such an anomaly in Poe's corpus that it rarely receives more than a brief mention. And when it does receive some attention, even though the book was well received by reviewers following the publication of the first and second editions, it is usually because of the suspicion of plagiarism, which began apparently seven years after its first publication. In a letter dated February 16, 1847, Poe replies to his friend George Eveleth, who in his letter to Poe informed him that the *Philadelphia Saturday Evening Post* had published an accusation of plagiarism against Poe with regard to his conchology book. Poe was incensed. He wrote, "What you tell me . . . surprised me. It is the first I heard of it. . . . I assure you that it is *totally* false. . . . I wrote it, in conjunction with Professor Thomas Wyatt, and Professor Mc Murtrie [sic] of Ph^a^—my name being put to the work, as best known and most likely to aid its circulation. I wrote the Preface and Introduction, and translated from Cuvier, the accounts of the animals etc. All school-books are necessarily made in a similar way. The very title-page acknowledges that the animals are given 'according to Cuvier.'"[43] Poe immediately continues: "This charge is infamous and I shall prosecute for it, as soon as I settle my account with the 'Mirror.'" The date of the slanderous publication in the *Post* was March 14, 1846.[44] Poe already had a defamation lawsuit going against the *New York Evening Mirror*, which in its June 23 issue of the same year printed an article by Thomas Dunn English in which English accuses Poe of forgery and viciously attacks his morals and sanity.[45] In a letter to Eveleth dated March 11, 1847, we find that Poe won the lawsuit,—"The vagabond, at

43. *Letters*, 619.

44. *Poe Log*, 629.

45. *Quinn*, 504–506.

the period of the suit's coming on, ran off to Washington for fear of being criminally prosecuted. . . . and 'the Mirror' could not get a single witness to testify one word against my character,"—being awarded a quite handsome sum of $225.00 plus costs. But Poe decided not to litigate against the *Post*, because "I fear that according to the law technicalities there is nothing 'actionable' in the Post's paragraphs—but I shall make them retract by some means."[46]

In reading this, I am reminded of Montresor's family motto given in Poe's "The Cask of Amontillado" (1846), *Nemo me impune lacessit*, which Montresor uttered shortly before he walled up the unfortunate Fortunato, sealing his fate.[47] It lacks the full-bodied flavor, to be sure, of *Sic gorgiamus allos subjectatos nunc*, the family motto of the Addams family, but it will certainly do as a motto against Poe's enemies and detractors, then—and now.

Looking at modern biographies of Poe, we find, for example, James Hutchisson state that Poe "in the winter of 1838 ghost-wrote a piece of hackwork for Thomas Wyatt, a British professor and lecturer, called *The Conchologist's First Book*. Poe put together this geology [*sic*] textbook by cribbing the information from a similar book published five years earlier. Poe added an introduction, and some library research may have gone into his descriptions of different animal species, but on the whole, the book was a work of plagiarism."[48] On the other hand, according to Kenneth Silverman, "Just two weeks after Virginia's death, a Philadelphia newspaper had arraigned him [Poe] for committing one of 'the most remarkable plagiarisms.' This empty charge concerned Thomas Wyatt's *Conchologist's First-Book*, the textbook to which his name had been attached as a copyright convenience eight years earlier."[49] In the one biography, which is a defense of Poe, we find the plagiarism affirmed, while in the other, a cold and unfriendly biography of Poe, (which gets the chronology with Virginia's death wrong,)

46. *Letters*, 626.

47. *P&T*, 851.

48. *Hutchisson*, 73.

49. *Silverman*, 334–335.

we find the plagiarism rejected as an "empty charge." I find it all *odd*.

Anyway, according to the editors of Poe's letters, "The body of the work . . . is largely a paraphrase of Wyatt's *A Manual of Conchology* (New York: Harper & Brothers, 1838), and Poe's own contribution is a close paraphrase of Thomas Brown's *The Conchologist's Text-Book* (Glasgow: Archibald Fullarton, 1833) and, as he says, a translation from Cuvier."[50] And then in the biography by Arthur Quinn we find that "The curious may find a close parallel between Poe's Introduction and Brown's, and the pictures of the shells are also copied. The 'Explanation of the Parts of the Shells' is verbatim from Brown. The bulk of the book is a paraphrase from Wyatt's *Conchology*, with Wyatt's consent." Quinn then gives what he believes is the best explanation of the whole matter:

> Wyatt had published a Conchology with Harpers in the previous year which proved too expensive [for students]. Harpers declined to bring it out in a cheaper form, so Wyatt decided to have a book prepared which he could sell in connection with his lectures and Poe put his name on the title page for a consideration [money]. He did, apparently, translate the description of the animals from Cuvier as he announced on the title-page. It was not entirely a piece of hack work, for Poe had had plenty of opportunity to talk with Dr. Edmund Ravenel, an eminent conchologist who lived during Poe's army service on Sullivan's Island. The book remains, however, out of the current of Poe's creative work.[51]

There is yet more. In *The Poe Log*, Thomas and Jackson quote from a remembrance by John Gould Anthony recorded in a letter he wrote dated July 7, 1875:

> Wyatt's Manual of Conchology, . . . being somewhat above the means of beginners and too high also for a text-book, it

50. *Letters*, 620.

51. *Quinn*, 277.

> was soon apparent that a smaller and less costly work was needed and the author was beset to make an abridgment which could be sold for $1.50 and contain all that was actually needed, but no abridgment could be published without consent of the house that held the copy right of the larger work and they would not spoil the sale of their book by issuing a cheaper one so soon after its publication. So the only way was to get up the abridgment and have it published with the name of some irresponsible person whom it would be idle to sue for damages, and Poe was selected for the scape goat . . . . The *facts* I had from Mr. Wyatt himself who was then lecturing on Conchology and using the abridgement as a help in that business—I think too he had them for sale at his lectures.[52]

Later in their book Thomas and Jackson quote from William Whitelock, who claims to have received a presentation copy of *The Conchologist's First Book* from Thomas Wyatt who inscribed it "From the author." In a remembrance published on April 7, 1881, Whitelock states that

> the Professor informed me he had prepared the work, but paid Poe $50 for the use of his name. This naturally led him to speak of the poet, whose neighbor he was in Philadelphia—the sickness of his wife, his pecuniary straits at times, and his assistance in enabling him to bridge these over. He alluded to him in the kindest manner, and while conceding to the poet a brilliant genius, attributed his troubles to a want of thrift and prudence in his domestic affairs.[53]

As everyone knows who deals with the intimate details of biographies, remembrances by persons recorded decades after the fact have to be taken with a large grain of salt (historians and lawyers know this better than anyone). Both of the above remembrances, after all, claim that Poe was used only for his name. But

52. *Poe Log*, 259.

53. *Poe Log*, 390.

the problem even deepens, given that the publisher issued a third edition in 1845 with Poe's name removed from the cover and title page and only his initials retained at the end of the Preface,[54] and that Wyatt, in a book of his published in 1846, listed himself as the author of several earlier works, including *The Conchologist's First Book*.[55]

Indeed, the more deeply one delves into the details of Poe's life, the more it seems that practically everything about him is shrouded in mystery, from minor questions to major ones. For example, where were Poe's love letters to Elmira intercepted by her father? Was it Charlottesville or Richmond? What is the relation of the so-called Mathew Brady daguerreotype of Poe (reproduced on the cover of the book before you) to the November 1848 "Ultima Thule" daguerreotype of Poe? Is the former a modified version of the latter? And, of course, what happened to Poe during the five days he was missing before he was hospitalized and died? Moreover why, in a state of delirium, did he repeatedly yell "Reynolds" during the last few hours of his life?

In the case of Poe's relation to *The Conchologist's First Book*, a matter mixed with fact, fiction, and calumny, there is a mystery that begs to be solved and I want to take a shot at it here, especially as it provides a crucial and unappreciated link in understanding Poe's concept of scientific imagination that underlies *Eureka*. Ironically, a substantial advance toward solving the mystery of Poe's authorship of *The Conchologist's First Book* was provided by another Gould, other than the one quoted above, namely, Stephen Jay Gould, the late Harvard professor of paleontology and famous popularizer of science, (mainly evolutionary biology,) in a chapter entitled "Poe's Greatest Hit," originally published in the July 1993 issue of *Natural History* magazine, to which he contributed a regular column. I say "substantial advance," for I do not believe that Gould solved the mystery. Accordingly, I shall first summarize the main features of his argument in two parts and then provide what I think is missing.

Looking at the fourth edition of Brown's book, which was published in 1836, Gould finds no reason to doubt that Poe wrote

54. *Poe Log*, 608.

55. *Poe Log*, 678.

the two-page Preface, and concludes that only three paragraphs, roughly a quarter of the four-page Introduction, were borrowed from Brown's Introduction. The ten-page "Explanation of the Parts of Shells," however, was almost entirely borrowed, "with a few minimal changes to simplify terminology and eliminate British parochialisms." Not good for Poe. Of the twelve pages of diagrams, or plates, they were "lifted *in toto*, jot for jot and tittle for tittle, from Brown."[56] Now it's really looking bad for Poe, and Gould provides no excuses for him.

Poe's own defense, of course, is given in his letter to Eveleth, in which he says "*All* school-books are necessarily made in a similar way. The very title-page acknowledges that the animals are given 'according to Cuvier.'" Granted, this does not excuse the lifting from Brown, but no doubt a lot of unacknowledged lifting did go on in his time, given that international copyright laws did not exist until 1891.[57] Moreover, to add to Poe's defense, he might not have been the one who did the lifting (all things considered, I think it was more likely Wyatt). But even if it was Poe, he may have rightly perceived that there is a vast difference in the unacknowledged lifting of material from one science book to another and the unacknowledged lifting of *original* material. The former is a matter of information or knowledge being passed around as part of the public domain, and as we shall see in the following section, in the first paragraph of "A Chapter on Science and Art" (1840), Poe expressed his belief that scientific knowledge "is the property not of any individual . . . but of mankind." The unacknowledged lifting of *original* material, on the other hand, is quite another matter, as scientists typically want credit for their discoveries, theories, experiments, and inventions, largely as part of their legacy. In art it is even more important, as originality is the very heart and soul of artists (at least the top ones). Hence the charges against Poe for plagiarism in *The Conchologist's First Book* would likely seem to Poe (if he indeed was guilty) a far cry from, and in a very different category from, the charges of plagiarism

56. Stephen Jay Gould (1995). *Dinosaur in a Haystack: Reflections in Natural History*. New York: Harmony Books, 179, 178.

57. Michael J. Everton (2013). "Literary Piracy." In *Context*, 189.

that he often levelled against others in the literary field, including English but most notably Longfellow.

Returning to the lifting of the plates, this is where things get even more interesting. The first four plates illustrate parts of shells, while the remaining eight plates illustrate whole shells. The difference with those eight plates is that they are in reverse order. Brown explicitly followed the ordering of the French naturalist Lamarck, who arranged organisms in accordance with the ladder-like concept of the Great Chain of Being, but starting from the top down. In Poe's book, however, the ordering is from the bottom up, starting with the plate with the least complex shells and finishing with the plate with the most complex shells, along with rearrangements of the figures on each plate. Does this illustrate a sense of evolutionary progressionism on the part of Poe? Gould does not go there, but leaves the reader dangling.[58] As for the roughly 120-page body of the book, describing each genus, Poe follows the order presented in Wyatt's *Manual of Conchology*.

Then comes the most interesting part. In his Preface to the first edition, Poe states that "the ruling feature" of the book is "that of giving an anatomical account of each animal, together with a description of the shell which it inhabits." Brown did not do that in his book. Instead, he almost exclusively described only the shells.[59] This is also, says Gould, how Linnaeus based his classification of molluscs.[60] In fact Gould says this was the common

58. It should be noted that near the end of the Preface to the first edition Poe states that "No attention has been given to the mere *History* of the subject; it is conceived that any disquisition on this head would more properly appertain to works of ultimate research, than to one whose sole intention is to make the pupil acquainted, in as tangible a form as possible, with *results*. To afford, at a cheap rate, a concise, yet sufficiently comprehensive, and especially a well-illustrated school-book, has been the principal design."

59. To be more specific, in Brown's book each genus is described in terms of its generic characters, followed by a description of the type species, focusing almost exclusively on the shells in both cases, with the geographical location of the type species added. In Poe's book, each genus is described in two parts—the animal, then the shell—followed by the list of known living and fossil species belonging to the genus, but with no descriptions of any of them.

60. Gould (note 56), 180. On the cover page Brown states that his book embraces "the Arrangements of Lamarck and Linnaeus."It should be added that not only Linnaeus but also Lamarck classified molluscs according to

approach of the time, repeated by Mary Roberts in her *The Conchologist's Companion* (1834) and also notably by Thomas Wyatt in his *Manual of Conchology* (1838).[61] So where did Poe get his descriptions of the animals? "Poe and Wyatt," says Gould, turned to "the work of Europe's greatest anatomist, Georges Cuvier." Given Poe's strict education in ancient and modern languages, in particular his fluency in French, Wyatt's choice of Poe was no accident but "good judgment," so that Poe is to be taken at his word in his letter to Eveleth quoted above, that "I wrote the Preface and Introduction, and translated from Cuvier, the accounts of the animals," a point reflected on the cover page, which states "In Which the Animals, According to Cuvier, are Given With the Shells." Poe justifies this synthetic approach, says Gould, in the Preface, in which he claims that the etymology of the word "Conchology" does not properly exclude "Malacology," so that, contrary to popular usage, "there is no good reason why a book upon *Conchology* (using the common term) may not be malacological as far as it proceeds." Hence, "they [Poe and Wyatt] promoted a good and innovative idea—elevating conchology from artificial description to integrative biology." This is the redeeming feature of *The Conchologist's First Book*, says Gould, it "defined the central, and surely admirable, feature," which Wyatt put to profitable use by selling copies of the book at his public lectures.[62]

The first problem with Gould's argument is the inclusive "Poe and Wyatt." Why not give Poe the credit for the idea? After all, Wyatt's *Manual of Conchology* lacked the "good and innovative idea." Moreover, Wyatt was not a professional naturalist, but a

---

their shells. See Peter F. Stevens (1994). *The Development of Biological Systematics*. New York: Columbia University Press, 422n14. As Stevens puts it, Lamarck thought "there were too few of the animals known to base a classification on them. Also, if the classifications of molluscs were to be based on the animals inside the shells, the collections of the shells themselves that were the pride and joy of many cabinets would become useless from the point of view of the study of natural history."

61. Gould (note 56), 180–181. In Wyatt's book, interestingly, although he divides his species into four classes compared with Brown's five, he has far more subdivisions for each class, and describes most of the species in each genus. But notably his book is like Brown's in that his descriptions are confined almost exclusively to the shells.

62. Gould (note 56), 182, 180, 183, 182.

British immigrant who settled in Philadelphia only to become an amateur naturalist. The "M.A." that follows his name cannot be traced to any university, and his title of "Professor" was apparently a self-appointed label, as he lacked a university affiliation—little wonder, then, that Gould "had never heard of Wyatt before writing this essay."[63]

The second problem concerns Poe's acknowledgment at the end of the Preface, his "great indebtedness to the valuable labours, as well as private assistance, of Mr. Isaac Lea, of Philadelphia. To Mr. Thomas Wyatt, and his late excellent *Manual of Conchology*." Isaac Lea was an acquaintance of Poe's, a publisher in Philadelphia whom Poe solicited back in 1829 to publish his long poem "Al Aaraaf." Lea was also a highly-respected conchologist and geologist (he would later receive an LL.D. from Harvard in 1852). Wyatt also acknowledges the help of Lea in the Introduction to his *Manual of Conchology*. The problem is that in his letter to Eveleth, quoted above, in which Poe replies to the charge of plagiarism, he states that "I wrote it [*The Conchologist's First Book*], in conjunction with Professor Thomas Wyatt, and Professor Mc Murtrie [*sic*] of Ph^a^." Instead of Lea, then, Poe names Mc Murtrie. Granted, this letter was written roughly eight years after Poe had worked on *The Conchologist's First Book*, and Poe sometimes exhibited an inaccurate memory in his letters (including the misspelling of names), while 1847 was a horrible year for him, his "most immemorial" to date (more on this below). But the information he gives in this letter cannot be so easily dismissed. This is because Wyatt, in his book *A Synopsis of Natural History*, which was published *only a month after the first edition* of *The Conchologist's First Book*,[64] acknowledges at the end of the Introduction "Dr H. M'Murtrie, and his excellent translation of Cuvier." This relatively short book, which is a translation of a book by C. Lemmonier with additions from the works of Cuvier and some others, embraces "Human and General Animal Physiology, Botany, Vegetable Physiology and Geology." Some have thought that Poe might have lent a hand to the work; some have even claimed

63. Gould (note 56), 185.

64. *Poe Log*, 261.

that he did.[65] But there is nothing more than suggestive evidence in the matter.[66] The shock to the system is that Henry M'Murtrie was a Philadelphia medical doctor and amateur naturalist who translated a book entitled *Cuvier's Animal Kingdom*, which was published in 1834. Georges Cuvier was the most celebrated naturalist of his time, especially with the publication of his massive *Le Règne Animal* (*Animal Kingdom*) in five volumes from 1817–1830, which was translated into English in sixteen volumes from 1827–1835, with many other translations as abridgments. As Peter Stevens points out, as early as 1795 Cuvier "rejected the value of mollusc shells in classification—they might be the basis of a special classification, but the 'grand system' would entail dissections of the organisms inhabiting the shells."[67]

I hesitate to follow the trail of evidence here, but Poe's letter and Wyatt's acknowledgment might mean that Poe did not translate from Cuvier after all. It certainly means that there was nothing particularly original about describing both animal and shell in *The Conchologist's First Book*. They were merely importing a novel idea from France, no more American than French (or rather Belgian) fries.

Cuvier adamantly rejected evolution, and even broke the animal part of the Great Chain of Being into four discrete branches (*embranchements*). But contrary to Harry Poe we cannot look to the reordering of the plates in *The Conchologist's First Book* and suggestively claim that "Twenty years before *On the Origin of Species*, Poe's book was the first American text to study animals by beginning with the simplest and moving to the [less?] simple."[68] This is because there is no indication that Poe was thinking of

---

65. E.g., *Fisher*, 6.

66. There is a cordial letter from Poe to Wyatt, in which he gives his feedback on the drawings for the book. *Letters*, 266–267. But since the letter is dated April 1, 1841, the reference cannot be to the first edition. Poe also reviewed the book. *Poe Log*, 259, 264–265. On the other hand, in a letter dated September 4, 1838, Poe turns down a financial offer from a publisher to write a book review, because "I have two other engagements which it would be ruinous to defer." *Letters*, 177. Was he referring to *Conchologist* and *Synopsis*? One can only wonder.

67. Stevens (note 60), 66.

68. *Evermore*, 140.

"temporalizing" the ladder-like Great Chain of Being, replacing the synchronic Great Chain with a diachronic Great Chain, let alone a Tree of Life. (Nor can we see it of a piece with *Eureka*, the Universe going from simplicity to complexity beginning with the Big Bang.) Instead, as professed in the subtitle of *The Conchologist's First Book*, the author simply wanted to bring "The Whole Up, As Accurately as Possible, to the Present Condition of the Science." In this sense Poe can be seen, at best, to be a *little like* Darwin in his classificatory (and non-evolutionary) work on barnacles, undertaken painstakingly from 1846–1854, for which he was awarded the Royal Medal of the Royal Society in 1853. As Darwin complained in a letter of 1849, "In not one large genus of Cirripedia has *any one* species been correctly defined: . . . Literally not one species is properly defined: not one naturalist has ever taken the trouble to open the shell of any species to describe it scientifically, & yet all the genera have ½ a dozen synonyms. . . . The subject is heart-breaking."[69] Or again, "I find that the anatomy of the Cirripedia has been most imperfectly done; nearly all the most striking features in their organization having been overlooked.—Their classification is likewise a perfect chaos, as must be the case until the whole body of every species be examined, as I am now doing."[70]

To see the real value of *The Conchologist's First Book*, then, we have to look in a very different direction. What I suggest is the most important lesson of all, with whatever was Poe's contribution to the book, is that it appears to be of a piece with Poe's later development of the ace detective, beginning with "The Murders in the Rue Morgue" (1841), namely, C. Auguste Dupin, the original model for Arthur Conan Doyle's Sherlock Holmes, for whom no stone is to be unturned in a murder investigation, no matter how insignificant or unrelated it might first appear. As Poe has Dupin put it in the sequel, "The Mystery of Marie Rogêt" (1842–43), "experience has shown, and a true philosophy will always show, that a vast, perhaps the larger portion of truth, arises from the

69. Frederick Burkhardt and Sydney Smith, eds. (1988). *The Correspondence of Charles Darwin, Volume 4, 1847–1850*. Cambridge: Cambridge University Press, 206–207.

70. Burkhardt and Smith (note 69), 213.

seemingly irrelevant. It is through the spirit of this principle, if not precisely through its letter, that modern science has resolved to *calculate upon the unforeseen*."[71] This marks a profound difference between Dupin and the Prefect of Police that Poe emphasized, and also between good and bad science, as he has Dupin emphasize here, which is a topic we shall return to in Chapter 7. It is also, not coincidentally, a major part of the key to understanding the genius of Darwin, who brilliantly calculated "upon the unforeseen," as we shall see in Chapter 6. The important point is that with *The Conchologist's First Book*, if not before, Poe became clued in to the importance of thoroughness and completeness in any investigation or treatment of knowledge.

"A Chapter on Science and Art"

From June 1839 to May 1840 Poe served as the "Assistant Editor" for a failing monthly periodical based in Philadelphia, *Burton's Gentleman's Magazine*. Poe was in effect the editor, as William Burton, the proprietor and editor in name, a British comic actor who had come to Philadelphia to establish himself as a literary man, couldn't resist the stage, now that he had Poe, such that he busied himself pursuing his passion of acting and eventually opened in Philadelphia what he called the National Theatre. Poe's addition to *Burton's* received generally positive mention in other periodicals, and the periodical itself received generally positive reviews during Poe's editorship, although one complained of "a palpable want of tact in the manner in which it [the periodical] has been gotten up."[72] Poe turned *Burton's* around into a profitable venture, contributing many of his own artistic pieces, as well as reviews and "fillers," many of which were unsigned. One of the latter was a series entitled "A Chapter on Science and Art," which appeared in the March, April, and May issues of the 1840 volume.[73] Given the content of the series and its inclusion during

71. *P&T*, 534.

72. *Poe Log*, 266.

73. The Edgar Allan Poe Society of Baltimore (2008) – Works – Miscellania – "A Chapter on Science and Art." There is a fourth installment in the June

Poe's tenure, there is no good reason to doubt that Poe was the author.[74]

Of especial interest are the first two paragraphs of the series in the March issue. They are about the eventual founding of the Smithsonian Institution in 1846:

> A writer in the last Southern Literary Messenger proposed the endowment, at Washington, of a *Central School of Natural Science*, as the best disposition of Mr. Smithson's bequest. This academy should have, it is suggested, a perfect apparatus, good cabinets, and the rudiments of a library, to be increased as means would permit. The institution should commence its operations of instruction at the point where our highest universities close—none of which profess, in mathematics, for example, to carry their pupils beyond a general acquaintance with the principles of the Calculus. In natural philosophy itself, the university deficiency of apparatus is a lamentable drawback upon the utility of our colleges. They have no facilities for the conduct of our young men farther than the mere vestibule of the temple. Yet we, above all people of the earth, have the most need of the highest physical instruction, prefaced and aided by the profoundest analytical science. We are, beyond all other nations, a nation of physical wants, means, and opportunities—this not less from the character of our population, than from the extent and general nature of our territory. The entire spirit of the age, too, tends rather to physical than to moral investigation. We want means for the immediate development of all our powers and resources. It may be said, moreover, in favor of physical knowledge, that it is the property not of any individual, or of any people, but of mankind. *All* are interested in its pursuits; its profits *all* share; and herein consists its great superiority to mere literature; for whose advancement,

---

issue, but except for the first entry, which is on the production of air during the putrefaction of animal bodies, it is not normally ascribed to Poe.

74. *Quinn*, 295–296; *Poe Log*, 292.

> indeed, we have already abundant means—whose guidance and control may be safely left to the press.
>
> . . . In respect to the designs of the testator, no doubt should be entertained. He meant to found a College for the advancement of *Science*. His whole life is a plain commentary upon this intention—and this intention should, in the present case, be made a paramount law.

One can see in this a strong lament from Poe about his own formal education, specifically the lack of science education in university. But what is really striking is what Poe says at the end of the first paragraph. When I first read it, I immediately thought of the late, great Carl Sagan, who was not only a first-rate astronomer but one of the greatest popularizers of science, especially with his highly-acclaimed *Cosmos* video series. As Sagan puts it in one of his books, "A proclivity for science is embedded deeply within us, in all times, places, and cultures. It has been the means for our survival. It is our birthright."[75] The spirit here is of a piece with Poe.

Returning to Poe's "Chapter," what we find is a great variety of reports, with some critical comments. On inventions, we find Poe discuss developments in balloon travel, some of which he rejects as impossible; a new method of engraving; a proposed new air and steam engine, which Poe thinks is highly doubtful; improvements made in daguerreotype photography; a new method of copying things made out of metal (like buttons) using galvanism; a method of giving soft stone the look and hardness of marble; a pneumatic engine of which Poe is highly doubtful; the supposed invention of roller skates; the making of an automatic railroad crossing gate; the development and testing of a bomb cannon; and a new kind of paper mill that can convert rags to paper. On theories and experiments, Poe comments on the cause of the flash from the muzzle of an air gun; he favorably discusses a theory that explains "red rain" as droppings from swarms of butterflies; and he favorably discusses a new theory on thunder. On wage issues, he comments on engravers making more than writers who make more than

75. Carl Sagan (1996). *The Demon-Haunted World: Science as a Candle in the Dark*. New York: Random House, 317.

binders. And, finally, on discoveries, Poe discusses the conjectured discovery of a new planet by count Decuppis, which he would later question in number 36 of his "Fifty Suggestions" (1849),[76] he discusses the discovery of subterranean rooms and caverns in the vicinity of the pyramids, and also the discovery of the foundation of worn-away pyramids, which makes him doubt the standard chronology.

When one surveys the variety of interests that Poe displays in "A Chapter on Science and Art," and then one considers his many book reviews and articles and works such as "Al Aaraaf" (1829) and "Scheherazade" (1845), one gets the impression that Poe *was interested in everything*.[77] And when that thought dawned on me, I was immediately carried back in memory to my years as a Ph.D. student in philosophy. I remembered having a conversation about something (I don't remember what) with one of the philosophers on my Ph.D. committee, Joseph Agassi, the most famous of the students of the philosopher of science Karl Popper, (more on them both in Chapters 6 and 7,) who later broke away from the master and became a first-rate philosopher in his own right. I asked Professor Agassi if he's interested in something in particular (again, I don't remember what) and he replied by saying, "I'm interested in everything." When he said that, I was taken aback, at first because it seemed quite arrogant, and then because it made no sense to me at all. Why would someone, I thought to myself, be interested in "everything"? Is that even possible?

Agassi's claim seemed ridiculous to me at the time. But over the passage of many years and the gradual progress in my intellectual development, I came to realize that the higher one ascends in any particular topic the more one is able to see its connections with other topics, (*if* one has wide-ranging interests,) allowing most importantly for something we have seen in Chapter 2—which is to be "a master of metaphor," as Aristotle put it, "the greatest thing by far" and "a sign of genius," by which he meant "an intuitive

76. *E&R*, 1305.

77. Poe was possibly being self-referential when, during the same time as "A Chapter on Science and Art," he has his narrator say, in "The Man of the Crowd" (1840), "I felt a calm but inquisitive interest in every thing." *P&T*, 388.

perception of the similarity in dissimilars." The attitude of being *interested in everything*, then, is the ultimate scholarly attitude and a key feature of scientific imagination in its fullest, as it enables one to see connections between seemingly unconnected topics, topics on the surface that seem irrelevant to one another. It is precisely in being *not* interested in everything that the fault lies. Truly great minds have a sense of curiosity, among their many other contributing attributes, that is unrelenting and thereby fruitful, that by its very nature sees things that other minds miss, what are lesser minds, minds that don't care to know. These other minds, which constitute the vast majority, easily get bored, uninterested, and even impatient with the learning of others unless it is up their alley, (and even then they are often closed). They are parochial minds, ones that suffer from myopia, unfulfilled minds that might achieve some sort of recognition and acclaim in their specialized fields but that overall are destined for underachievement and relative mediocrity. Truly their nature is their fate, the maker of their lot and destiny in terms of legacy.

The Bridgewater Treatises

In 1829 the dying eighth earl of Bridgewater, as part of his will, donated £8,000 for the publication of works "On the Power, Wisdom, and Goodness of God, as manifested in the Creation" (the title is much longer). Accordingly, from 1833 to 1836 eight works (first editions) were published using this endowment, among them William Whewell's *Astronomy and General Physics Considered with Reference to Natural Theology* (1833), Sir Charles Bell's *The Hand, its Mechanism and Vital Endowments as Evincing Design* (1833), and Peter Mark Roget's *Animal and Vegetable Physiology Considered with Reference to Natural Theology* (1833). Collectively the eight works are known as the Bridgewater Treatises, and like William Paley's *Natural Theology* (1802) they were written to demonstrate the existence and glory of God purely on the basis of the evidence of adaptation and design in the natural world.

In his "Marginalia" (November 1844) Poe claims that "All the Bridgewater treatises have failed in noticing *the great* idiosyncrasy in the Divine system of adaptation:—that idiosyncrasy which stamps the adaptation as Divine, in distinction from that which is the work of merely human constructiveness. I speak of the complete *mutuality* of adaptation."[78] An example that illustrates Poe's point about "human constructiveness" is the chair, which is designed for humans so they can sit, but humans are not designed so that chairs can benefit in some way. In the Bridgewater Treatises, ideas of one-way adaptation are indeed ubiquitous, such as Whewell's claim that plant cycles are adapted to the solar year, which allows plants to survive and reproduce, or Bell's claim that the human hand was designed for making and using tools and weapons so that man could have dominion over the world.

So Poe's point is that the authors of the famous Bridgewater Treatises have misunderstood the nature of adaptation, particularly as enshrined in nature by God, that they have missed possibly the most important evidence of "the Power, Wisdom, and Goodness of God, as manifested in the Creation." But what exactly did they miss? With one-way adaptations, says Poe, "The effect does not re-act upon the cause." In other words, the causal relationship is and remains one-way. But with two-way adaptations, or mutual adaptations, the causal relationship is two-way, with each cause of the other being also the effect of the other, such that, as Poe puts it, "we can never (abstractedly, without concretion—without reference to facts of the moment) decide which is which." Hence, says Poe, with divine productions "There is an absolute reciprocity of adaptation, for which we seek in vain among the works of man."

---

78. *E&R*, 1315. Poe claims that his notes for "Marginalia" were taken from "the volumes of my library," from the margins on which he recorded his thoughts "freshly—boldly—originally—with *abandonnement*—without conceit." He then tells us, at the end of his Introduction, that "nonsense [is] the essential sense of the Marginal Note." *E&R*, 1309–1311. The irony of that statement should not be lost on the reader. At any rate, as Pollin points out, "The books owned and retained by Poe are now very rare, probably because he omitted his name and sold most of them." *Brevities*, 109. Poverty, after all, was the norm during Poe's adult life. We should therefore wonder if Poe ever read even one of the Bridgewater Treatises. However, he did in fact write a positive review of Roget's two-volume work that was published in the February 1836 issue of the *Southern Literary Messenger* and he did in fact read an anonymous article on the Bridgewater Treatises published in the *Quarterly Review* two years earlier. *Eureka*, 140n133.

According to Stuart and Susan Levine, Poe "is either being prescient or concealing his source in regard to an aspect of what is now termed coevolution; this is applied to the interdependence between plants and animals, enabling their continued living and reproducing together. Following a suggestion of BRP [Burton R. Pollin], we checked a number of histories of evolutionary theory without finding a reference to the concept before Poe." They even go so far as to say that, although Poe's example falls short of Poe's concept, "Our consultant in the history of biology (Professor Emeritus John A. Weir, University of Kansas) . . . agreed with us that Poe's insight was nevertheless impressive and very unusual for the period—'profound,' he called it."[79]

It is difficult, however, to see how Poe could have been *prescient* (or otherwise) with regard to the concept of coevolution when there is no *evolution* in his concept as presented in "Marginalia" (November 1844).[80] The fundamental problem is that Poe did not give us a concept of evolution at that time, of any kind, so there is no point searching for evolutionary precursors. What he did give us, instead, is a concept of *coadaptation*, which logically is not wedded to evolution at all and conceivably could be found in early creationist texts. (Neither did Poe give us a concept of evolution later in *Eureka*, as we shall see in Chapter 5.)

According to Harry Poe, Edgar Poe's "thought anticipates Charles Darwin's reflections on the adaptation of organisms, but without Darwin's philosophical predisposition to naturalism."[81] But did Poe really do that? In the *Origin* (1859), right at the beginning, Darwin criticizes theories of evolution other than his own for failing to explain what he calls "coadaptation," one of his examples being "the misseltoe, which draws its nourishment from certain trees, which has seeds that must be transported by certain birds, and which has flowers with separate sexes absolutely

79. *Eureka*, 158n217.

80. *Coevolution* is an "umbrella term," covering not only the evolution of symbiotic relationships between species but also the evolution of arms races, the evolution of resource partitioning between species, and parallel evolution. Scott Freeman and Jon C. Herron (1998). *Evolutionary Analysis*. Upper Saddle River, NJ: Prentice Hall, 389. Douglas J. Futuyma (1998). *Evolutionary Biology*. 3rd ed. Sunderland, MA: Sinauer Associates, 220, 539–541.

81. *Evermore*, 145.

requiring the agency of certain insects to bring pollen from one flower to the other."[82] For Darwin, none of this could be *scientifically* explained except by natural selection operating on heritable variations over tens of thousands and ultimately hundreds of millions of years.

Did Poe have something like this in mind? Darwin never claimed that all examples of adaptation in nature are examples of coadaptation. Poe, however, seems to think so, for he begins his entry with "*the great* idiosyncrasy in the Divine system of adaptation." But let us leave that aside for the moment and try to gain some clarity here. Poe's language of cause and effect and reacting or not reacting upon each other is nothing short of confusing. For clarity we need to distinguish between *adapted to* and *adapted for*. In the case of termites, a popular example of symbiosis known as *mutualism*, the termites and the protozoa of the genus *Trichonympha* in their intestines are adapted *to* each other *for* the purpose of turning wood into food, the termites finding the wood and chopping it into little bits, (something the protozoa cannot do), the protozoa digesting the cellulose in the wood, (something the termites cannot do,) each getting calories and nutrients from the combined process. In the case of orchids and moths, one of Darwin's favorite examples, the flowers of the orchids are adapted to attract the moths for the purpose of fertilization, while the proboscis (the tubular mouth) of the moth is adapted to the structure of the orchid so as to extract the nectar, which serves as its food. So certain was Darwin of his discovery and of the ubiquity of the phenomenon that he predicted the existence of a moth in Madagascar with a proboscis "capable of extension to a length of between ten and eleven inches," given that he had been sent several specimens of an orchid found only in Madagascar, *Angraecum sesquipedale*, with "nectaries eleven and a half inches long."[83] Although

82. Charles Darwin (1859). *On the Origin of Species*. London: John Murray, 3. On the next page Darwin explicitly criticizes the "author of the 'Vestiges of Creation'" for failing to explain "the coadaptations of organic beings to each other and to their physical conditions of life." We shall have more on *Vestiges* below.

83. Charles Darwin (1862). *On the Fertilization of Orchids by Insects*. 2nd ed. (1877). London: John Murray, 162–163.

ridiculed at the time for the prediction, Darwin's theoretical moth was in fact discovered in 1903, twenty-one years after Darwin's death, and given the name *Xanthopan morganii praedicta*.[84]

I'm not so sure that you don't find something like this in the world of manmade things, a mutuality of adaptation, for example with hammers and nails, or cans and can openers. At any rate, did Poe really have the basics of Darwin's idea? As a general rule, if one is not sure of a person's concept then one should look at their examples. In the same entry in "Marginalia" Poe does provide an example for his concept. He points out that the human body, in order to maintain its temperature in polar climates, requires fat, what he calls "train-oil" (the actual term used for oil from blubber). He then continues: "In polar climates, the sole food afforded man is the oil of abundant seals and whales. Now, whether is oil at hand because imperatively demanded?—or whether is it the only thing demanded because the only thing to be obtained? It is impossible to say."

This is hardly a clear example, and no matter how it may be clarified it is not at all like Darwin's examples of coadaptation, where each species requires and is adapted to the other. In fact Poe's example looks like a hoax. But then Poe repeats it in *Eureka*, virtually word for word, following his discussion on what he calls "Divine *adaptation*,"[85] which he later in *Eureka* calls "that absolute *reciprocity of adaptation* which is the idiosyncrasy of the Divine Act—stamping it divine."[86] Moreover, both passages, the one examined in "Marginalia" and the one here in *Eureka*, are followed by the same paragraph, virtually word for word, in which Poe applies his concept of mutuality of adaptation to plot: "we should aim at so arranging the incidents that we shall not be able to determine, of any one of them, whether it depends from any one other or upholds it. In this sense, of course, *perfection* of plot is really, or practically, unattainable—but only because it is

84. David Quammen (2004). "Was Darwin Wrong?" *National Geographic* 206 (Nov.), 14–15. I cite this source not only for the beautiful two-page picture of the moth, but also for the words: "Such mutual adaptation—the moth to the flower, the flower to the moth—is called coevolution."

85. *P&T*, 1341–1342.

86. *P&T*, 1354.

a finite intelligence that constructs. The plots of God are perfect. The Universe is a plot of God."[87]

In fairness to Poe, he arguably does indeed express the concept of coadaptation in his writings as a feature of nature, (though not necessarily before Darwin,[88]) but his imagination failed when it came to supplying clear examples in nature. Poe was not, after all, a naturalist. But he does have a clear example that pervades *Eureka* and helps us to understand it, an example that connects with our discussion on Poe and beauty in the previous chapter. The example, interestingly, is made explicit in "The Domain of Arnheim" (1846), in which Poe has the narrator explain Ellison's view on nature's beauty, which is that (i) "the Deity had implanted the poetic sentiment in man" and (ii) nature was made by the Deity as an "adaptation to the eyes which were to behold it [nature] on earth."[89] Here we clearly see the concept of coadaptation, especially when we recall from Chapter 2 that Poe thought of beauty not as a primary quality in things but as a secondary quality produced in us, the sum product of the primary qualities in the constitutions of things and the primary qualities in the constitutions of our sensory organs and minds. Specifically, Poe's idea is that *we are adapted to nature and nature is adapted to us*, all for the purpose of beauty, which is for the purpose of pleasure, God's pleasure, of which (God) we each are a part.

In developing the concept of mutuality of adaptation, then, against the concept of adaptation found throughout the Bridgewater Treatises, Poe was not exactly anticipating a key concept that was already brewing in Darwin's mind and that Darwin would later make much of, the concept of coadaptation that is now a mainstay of modern biology, residing peacefully in the home of the ecological concept of symbiosis, or more specifically mutualism. Instead, with his concept of mutuality of adaptation Poe was developing a concept that proved central to his *argument from*

87. *P&T*, 1342.

88. The concept of coadaptation is found in Darwin's unpublished *Essay of 1844* and, as a proto-concept, in his unpublished *Essay of 1842*. Paul H. Barrett and R.B. Freeman, eds. (1986). *The Foundations of the Origin of Species*. New York: New York University Press, 6, 87. One would like to know if either Darwin or Poe had an earlier source or if we have a case of convergent intellectual evolution here. As far as I know, no research exists either way.

89. *P&T*, 859.

hope, the core of his theology as discussed in the previous chapter. He was also contributing to modern literary theory. But even more, with his literary theory sufficiently developed, combined with his understanding of science up to his time, along with the moth in his heart, (Shelley's, not Darwin's, "the desire of the moth for the star,") Poe anticipated a major development of twentieth-century science, specifically the concept of matter–energy equivalence, as we shall see in the next chapter. In Poe's hands, attraction and repulsion simply are matter, and both are mutually adapted for the unity of effect of the Universe. And what this indicates is the central theme of this book,—namely, the extraordinary power of the eclectic, educated, holistic, and all-searching imagination.

Vestiges of the Natural History of Creation

If years could be envious, from a sense of legacy, many in Poe's time would want to be 1844, the year of publication of a very important book, *Vestiges of the Natural History of Creation*. The author of the work attempted to bring together the latest facts and theories from the sciences of the day—astronomy, cosmology, physics, chemistry, geology, paleontology, biology, embryology, anthropology, phrenology, economics—and synthesize them into a grand narrative of the evolution of the Universe, notably the evolution of stars and planets and the evolution of biological species, (but *not* the origin of the Universe). The author used the nebular hypothesis of Laplace, for example, but distanced himself from Lamarck's particular brand of the evolution of species. In all of this, the view was one of law-based evolution, or unfolding,—hence "the development hypothesis," as the author later called it,—a theory of progressionism with God as lawmaker and "inferior animals" constituting "only less advanced types of that form of being perfected in ourselves." The author even went so far as to claim that our current world is "a stage in a Great Progress" and that there is "some greater phenomenon, the rest of which was yet to be evolved."[90]

---

90. James A. Secord (1994). *Vestiges of the Natural History of Creation and Other Evolutionary Writings*. Chicago: University of Chicago Press, [207], 184, 385.

Originally published in London in October of 1844 and then in New York shortly afterward, in 1845, (by Wiley and Putnam, who published the third London edition of February 1845,) the book caused a "sensation" throughout the English-speaking world. As James Secord puts it in his book devoted to the reading and reception of *Vestiges*, it "was more controversial than any other philosophical or scientific work of its time."[91] The sensation was caused by mainly two factors: its promotion of the evolution of species and its thoroughgoing anonymity of authorship. *Vestiges* went through fourteen editions and included two editions of a sequel, *Explanations*, mainly for the purpose of dealing with new developments in science and with reviews and criticisms of *Vestiges*.[92] Today when we think of life and the Universe (if we are indeed among the genuinely educated) we cannot help but think of both in terms of change from simple to complex, rather than in terms of creation and stasis. "How did evolution gain this pivotal role in the public arena?," asks Secord. "The answer," he says, "turns out to have little to do with Darwinian biology or Big Bang astronomy. Instead, the critical period is the first half of the nineteenth century, and the turning point is the response of readers to *Vestiges*."[93]

A comparison of *Vestiges* with *Eureka* is important on a number of counts. *Vestiges* was published anonymously, by Robert Chambers, a Scottish journalist and publisher who had an abiding interest in science and who attempted to bring the sciences of his day together into one unified historical narrative of the Universe. He was not what we would call a "professional scientist," (or "naturalist," or "natural philosopher," the terms still used in his day,) but would properly be called an "amateur scientific theorist." And yet, as Secord rightly points out, the professional/amateur distinction "is to assume the existence of an acknowledged class of professional scientists and prejudge precisely the question of authority at issue." (Indeed, the word "scientist" was only gradually starting to be used, as science was only starting to become,

91. James A. Secord (2000). *Victorian Sensation: The Extraordinary Publication, Reception, and Secret Authorship of Vestiges of the Natural History of Creation.* Chicago: University of Chicago Press, 1.

92. Secord (note 90), ix, xxxiv.

93. Secord (note 91), 2.

and then ever so gradually, a distinct profession.) Poe, likewise, was not what we would call a professional scientist, but an amateur theorist, with a similar abiding interest in science. And like Chambers, Poe in *Eureka* attempted to frame a grand narrative of the Universe using the science of his day.

Second, as Secord points out, "*Vestiges* was not a bungled attempt to produce 'professional' science, but a skilled intervention in some of the great public debates of the nineteenth century."[94] *Eureka*, too, with its combined use of science, literary theory, and theology, was arguably a "skilled intervention in some of the public debates of the nineteenth century." In fact it can be viewed as an attempt to resolve the Romantic response to the Enlightenment, the "culture wars" that continue to this day in the form of the humanities versus the natural sciences.[95]

Third, it is interesting that Chambers, in the final chapter of *Vestiges*, states that "The book, as far as I am aware, is the first attempt to connect the natural sciences into a history of creation."[96] Poe, too, claimed originality with *Eureka*, stating that "I do not know a treatise in which a survey of the *Universe*—using the word in its most comprehensive and only legitimate acceptation—is taken at all." "The nearest approach," he says," is "the 'Cosmos' of Alexander Von Humboldt." Poe's main criticism of the work, however, is not only that it is confined to the "merely physical universe," but that the "multiplicity" of Humboldt's "particular points" lacks, even "precludes," what Poe calls "all *individuality* of impression."[97] In other words, it lacks what Poe in his literary theory calls a "unity of effect," what on the first page of *Eureka* he says is the ultimate purpose of the book, which is to produce in the mind of the reader "an individual impression."[98]

94. Secord (note 90), xliv.

95. Compare, e.g., John S. Nelson et al., eds. (1987). *The Rhetoric of the Human Sciences: Language and Argument in Scholarship and Public Affairs.* Madison: University of Wisconsin Press. Noretta Koertge, ed. (1998). *A House Built on Sand: Exposing Postmodernist Myths About Science.* Oxford: Oxford University Press.

96. Secord (note 90), 388.

97. *P&T*, 1262.

98. *P&T*, 1261.

Fourth, each magnum opus was written following a mental breakdown, Chambers "during a recovery from a mental breakdown produced by the constant strain of writing for the *Journal*,"[99] Poe during his recovery following the death of his beloved Virginia (more on this in the final section below). (This feature underlying their highly *synthetic* works shall be especially examined in Chapter 8.)

Fifth, equally interesting is how both Chambers and Poe judged criticism coming from those we today would call "professional scientists." In a letter to a confidant dated July 26, 1845, Chambers wrote that "The condemnation so liberally dealt in by the scientific men must surely spring in great measure from inferior feelings, not from offended philosophy. It seems so spiteful. If, however, these men had generally any comprehensive ideas of science, they could not, I think, blind themselves to the harmony of the opus with nature in at least her general character."[100] Poe says something very similar in a letter to a friend, dated February 29, 1848. Poe had recently given his lecture, "The Universe," in New York City on February 3, 1848, and in response to at least one review of *Vestiges* he evidently read, and perhaps anticipating reviews of his own work, he wrote that "One thing is certain; that the objections of *merely* scientific men—men, I mean, who cultivate the physical sciences to the exclusion, in a greater or less degree, of the mathematics, of metaphysics and of logic—are generally invalid except in respect to scientific *details*. Of all persons in the world, they are at the same time the most bigoted and the least capable of using, generalizing, or deciding upon the facts which they bring to light in the course of their experiments. And these are the men who chiefly write the criticisms *against* all efforts at generalization—denouncing these efforts as 'speculative' and 'theoretical.'"[101] Chambers went on to publish a reply to criticisms entitled *Explanations: A Sequel* in two editions, the first in December 1845, and to issue no less than ten updated editions of

99. Secord (note 90), xxiii. The *Journal* was *Chambers's Edinburgh Journal*, the highly successful weekly he owned and operated with his brother.

100. Secord (note 90), xxxiii. The criticism seems to be repeated, though greatly watered down, in *Explanations*, 176, 179.

101. *Letters*, 659.

Vestiges during his lifetime (he died in 1871).[102] Poe, on the other hand, turned "The Universe" into *Eureka*, penning alterations and additions following its publication, but he came to his tragic end less than fifteen months post-publication, and so could not do anything else about it.

All of this should make one wonder whether Poe had read a copy of *Vestiges*, especially given the "sensation" that *Vestiges* created throughout the English-speaking world. Keeping in mind that international copyright laws did not yet exist, *Vestiges* was republished in the United States, according to the editors of Poe's letters, quite a number of times while Poe was alive—"Harper, 1847; Wiley & Putnam, 1845, 1846, and 1846–1848"—and was reviewed in the same journal, the *American Review*, that published Poe's "The Raven" (1845).[103] Perhaps not surprisingly, then, at least one reviewer of Poe's lecture on the Universe made the comparison, claiming that Poe was indebted to *Vestiges*,[104] while at least two reviewers of *Eureka* alluded to similarities.[105] A friend of Poe's sent him the said review of "The Universe." This is the same friend that Poe replied to in his letter quoted from above, dated February 29, 1848. In that letter Poe states that "'The Vestiges of Creation' I have not seen; and it is always unsafe and unwise to form opinions of books from reviews of them. The extracts of the work which have fallen in [*sic*] my way, abound in inaccuracies of fact:—still these may not materially affect the general argument."[106]

Poe's claim about judging books by book reviews is surely ironic, (unless, of course, Poe meant to exempt himself from the comment,) given that Poe himself was a prolific reviewer of books. That he read at least one review of *Vestiges* comes through clearly, while the reference to "extracts" indicates that there may have been more. At any rate, Poe gives the clear impression that the extracts did not impress him. But was Poe deliberately playing down *Vestiges*, perhaps even lying? Certainly Poe's denial of ever having seen a copy of *Vestiges* has not been well received.

102. Secord (note 90), [217–219].

103. *Letters*, 660.

104. *Poe Log*, 723.

105. *Critical Heritage*, 280, 288.

106. *Letters*, 658–659.

According to the editors of Poe's letters, for example, "It seems difficult to believe that Poe would not have made even the slight effort needed to locate a copy of this prominent book, especially one which dealt with a subject clearly of considerable interest to him. Although it is possible that he was familiar with the book only from reviews, as he suggests, Poe's denial has not been entirely convincing to readers and scholars."[107]

Poe, it needs to be said, was not above proffering outright lies, so he may have been simply lying in his letter when he denied ever seeing *Vestiges*.[108] At any rate, what we should really be focusing on is how different Poe's *Eureka* is from Chambers' *Vestiges* in both character and content.

As to character, Chambers, with the exception of promoting the evolution ("development") of species, to which he thought science would eventually come around, wanted to be as consistent as possible with established science: "My sincere desire in the composition of the book was to give the true view of the history of nature, with as little disturbance as possible to existing beliefs,

107. *Letters*, 660.

108. Robert Scholnick makes an interesting case for the view that Poe was somewhat disingenuous regarding *Vestiges*. Robert J. Scholnick (2013). "*Eureka* in Context: Poe, the Newspaper, the Lyceum, and Cosmic Science." In Richard Kopley and Jana Argersinger, eds. (2013). *Poe Writing/Writing Poe*. New York: AMS Press, 31–50. Poe's "The Power of Words" (1845), he says, "would seem to respond directly to the raging *Vestiges* controversy." But the evidence for "strong resonances" is not convincing to my mind, given that, for example, it was a common view of Enlightenment deists that God created "in the beginning *only*," that scientists such as John Herschel in *Preliminary Discourse on the Study of Natural Philosophy* (1830) and Charles Lyell in *Principles of Geology* (1830) emphasized creation by natural law, and that the controversial experiments in 1836 by Andrew Crosse on the creation of life using electricity were discussed by many, such that the discussion of Crosse's experiments by Chambers followed by Poe is likely the result of a common cause rather than the former influencing the latter (especially given that Poe read widely). Similarly, that *Vestiges* "is the work that bears the most immediate comparison with *Eureka*" is a claim based on weak circumstantial evidence, such as that the authors of both works were enthusiastic readers of Nichol's *Views of the Architecture of the Heavens* (1838), or that Poe says very little about biological evolution in *Eureka* since because of *Vestiges* "so familiar had the concept become," or that Poe used the word "development" with regard to the latter. None of this evidence would stand up in a court of law, and surely the dissimilarities between the two works far outweigh whatever their similarities.

whether philosophical or religious."[109] He wanted to synthesize the sciences into a grand scheme, with the evolution of species being only a relatively small part of the harmony of that scheme, a developmental and progressive view of virtually everything in the Universe. He was also concerned not to upset the religious establishment.

Poe, however, although he drew on the science of his day, expressly developed theories (or more accurately hypotheses) that went *against* the scientific orthodoxy of his day, as we shall see in the next chapter, not to mention the religious orthodoxy (as we have seen in the previous chapter). And Poe was quite intentional about this, stating right at the outset of *Eureka* that "I shall be so rash, moreover, as to challenge the conclusions, and thus, in effect, to question the sagacity, of many of the greatest and most justly reverenced of men."[110] In challenging the scientific orthodoxy alone, which is arguably the biggest difference between the two works, Poe anticipated at least nine theories and developments in twentieth-century science. Were it one or two, of course, one could easily pass it off as lucky coincidences. But with *at least nine* the problem of credibility swings toward the skeptic. Chambers' *Vestiges*, of course, can boast no similar outcome, and accordingly pales in comparison.

Poe's *Eureka*, moreover, went *against* the progressionism that pervades *Vestiges* and the Victorian psyche, shared even more by the American psyche. This is because the Universe, as Poe viewed it, was in a period of collapse and return to the original Unity.[111] Little would be more shocking to Chambers and to his readers.[112]

109. Secord (note 90), 388.

110. *P&T*, 1261.

111. Scholnick states that for both authors God created and then "retreated to the sideline" and that for Poe "the process of cosmic development would be reversed and that there would be an eventual collapse." Scholnick (note 108), 46, 47. But as we have already seen in the previous chapter, Poe's God does not retreat to the sideline, and in the next chapter we shall see that Poe's theory of cosmic and biological development is that it is occurring within the *contraction* phase of the Universe, a phase that is directed toward "Nothingness."

112. Poe's anti-progressionism is evident in a number of his writings. In "The Colloquy of Monos and Una" (1841), for example, a dialogue set in the afterworld a few centuries after the global conflagration, Monos says the world "grew infected with system and abstraction," such that "huge smoking cities

Chambers viewed *Vestiges* as his magnum opus, which it was, for he rode it for the rest of his life, claiming even, following the success of Darwin's *Origin* beginning in 1859, that *Vestiges* not only anticipated Darwin's work but contained basically the same ideas.[113] Poe, too, viewed *Eureka* as his magnum opus, but he struck a very different chord. As he wrote in a letter to the behind-the-scenes hero of his adult life, his "ever beloved mother, and dearest, truest friend," his aunt Maria Clemm, the mother of his cousin-wife Virginia, (Poe was in a severely ill and depressed state, but almost prescient,) "I must die. I have no desire to live since I have done 'Eureka.' I could accomplish nothing more."[114] That was on July 7, 1849, twelve months following the publication of *Eureka*, and precisely three months before he lost his individuality in Washington College Hospital in Baltimore.

---

arose, innumerable. Green leaves shrank before the hot breath of furnaces. The fair face of Nature was deformed as with the ravages of some loathsome disease." The viewpoint is not just anti-West, as Monos adds that "the Earth's records had taught me to look for widest ruin as the price of highest civilization," and he explicitly refers to ancient China, Assyria, Egypt, and Nubia. Only "one or two of the wise among our forefathers," he adds, "had ventured to doubt the propriety of the term 'improvement,' as applied to the progress of our civilization," and they lived "amid the scorn of the 'utilitarians.'" *P&T*, 449–457. Similarly in "Some Words with a Mummy" (1845), Count Allamistakeo, a 5,050-year-old Egyptian mummy (almost 5,800 years old by birth) discovered with intact organs and resuscitated by use of the galvanic battery, refutes his interlocutors on point after point, who believe that the world has progressed since the time of ancient Egypt. This leaves the narrator the next morning "penning these memoranda for the benefit of my family and of mankind." After his shave and coffee he is about to get himself embalmed, for "The truth is, I am heartily sick of this life and of the nineteenth century in general. I am convinced that everything is going wrong. Besides, I am anxious to know who will be President in 2045." *P&T*, 805–821.

113. Chambers' three published responses to the publication of Darwin's *Origin* are reprinted in Secord (note 90), [209–214]. Remarkable are his claims in his appendix to the eleventh edition of *Vestiges* (1860) that Darwin's "book, in no essential respect, contradicts the present," that "it expresses substantially the same general ideas," and that "the difference seems to be in words, not in facts or effects." Remarkable also is that the eleventh edition of *Vestiges* was published so as to match the *Origin* in size and color of the cover.

114. *Letters*, 820.

We can only wonder what more Poe *would* have accomplished had he lived. With Chambers we know. With Poe we can only speculate.

Poe's Scientific Sources for *Eureka*

What were Poe's sources for his understanding of the contemporary science of his day? Did he rely mainly on primary or secondary sources? What did he read? And how reliable is his presentation in *Eureka* of contemporary scientific facts and theories? Let us begin with the last of these questions. According to Stuart and Susan Levine, (and it must be kept in mind that they used at least two scientific consultants,) "Poe is very good at briefing lay readers on the state of knowledge in several fields. . . . [which] was a difficult job, because in 1848 there was a great deal of 'late-breaking news.'" Again, *Eureka* contains "skill and accuracy as scientific summary." Again, "Poe's summary of scientific matters is in general excellent; he understands most available astronomical theory. His summaries of the arguments of different scientists, as the notes indicate, are very competent." And yet again, "*Eureka* is very successful as a brief introduction to good mid-century knowledge of the cosmos."[115]

I see no reason to argue against this. So if Poe was not writing off the top of his head, what did he read? What were his sources for the latest scientific knowledge and theories?

Eureka is dedicated to Alexander von Humboldt. The first two volumes of his *Kosmos* were published in 1845 and 1847, respectively. The work quickly became quite popular in Britain and the United States, with the English translation of the first volume appearing in the same year as its original. (The rest of the history of the work does not concern us, as it is post-*Eureka.*) In *Kosmos* Humboldt attempted to unify the sciences into a single framework, and it is no doubt because of Humboldt's effort at the big picture, at times passionate and poetic, that Poe dedicated *Eureka* to Humboldt. Poe also states in *Eureka* that *Kosmos* is the "nearest approach" to *Eureka*, but his main criticism is that

115. *Eureka*, xx, xxii.

Kosmos treats the Universe "*not* in its individuality but in its generality."[116] This, of course, means that *Kosmos* fails at producing a "unity of effect," a deficiency Poe attempts to rectify with *Eureka.* Poe elsewhere makes references to Humboldt in the text, in one place quoting from *Kosmos* in English, providing the original German in a footnote, after praising Humboldt's "generalizing powers."[117] So Poe apparently did, then, have at least part of the work in front of him.[118]

Another important scientific source for *Eureka* comes from John Pringle Nichol, Regius Professor of Astronomy at the University of Glasgow. His *Views of the Architecture of the Heavens,* first published in 1837, was followed by a second edition in 1838, which contained twenty-three plates and which was the edition republished in New York in 1840.[119] In this book Nichol explains for the general reader the contemporary state of astronomy, with a special focus on the nebular hypothesis of Laplace and the revelations that came from the large telescope of Sir William Herschel, his forty-foot reflector with a four-foot diameter, which gave Herschel unprecedented observations of stellar nebulae and allowed him to estimate stellar distances. Nichol followed this book with *The Phenomena and Order of the Solar System* (1838), which was republished in New York in 1842. In this book Nichol begins with the history of thought on the solar system and then discusses modern theory, including Kepler's three laws, Newton's theory of gravity, and the general characteristics of the Sun and planets. Poe explicitly refers to Nichol three times in *Eureka* and quotes from *Architecture* twice, such as when he criticizes Nichol for calling gravity an "ultimate principle," since Newton refrained from speculating on the principle behind the law and Poe believes that he himself has found it.[120] Poe also makes a number of allusions to Nichol and his works, including quotations from Nichol's enormously popular lecture series on astronomy given in New York

116. *P&T*, 1262.

117. *P&T*, 1346–1347.

118. *Eureka*, xiv, 118n9, 148n178, 150–151n179, 155–156n209, 163–164n232, 171.

119. Poe may well have read the American edition shortly after its publication, for he lists "Dr. Nichol" among Dupin's "meditations" in "The Murders in the Rue Morgue" (1841), which is prefaced by the observation of "the vast extent of his [Dupin's] reading." *P&T*, 403, 400.

120. *P&T*, 1289.

City (which, unfortunately for Poe in terms of competition, was given at the same time as Poe's lecture on the Universe given on February 3, 1848, also in New York City).[121]

Any shortlist of Poe's scientific sources must also include the son of Sir William Herschel, namely, Sir John Herschel, who was considered in his time the greatest living scientist and the greatest authority on the nature of science.[122] Charles Darwin certainly treated him as such, both for his writings and in visiting him in person, and Poe's respect for Herschel was not much different. In *Eureka* Poe quotes from Herschel's *A Treatise on Astronomy* (1833), which was republished in Philadelphia in 1834 and for many years thereafter, which Poe follows with a criticism.[123] It is possible that Poe first read Herschel's book shortly after it was published, for in his "The Literati of New York City" (1846), in his entry on Richard Adams Locke, Poe claims that an American edition of Herschel's *Treatise* informed his own "story" (hoax) about a balloon trip to the Moon, "Hans Pfaall" (1835).[124] At any rate, there are also many allusions to Herschel's writings in *Eureka*,[125] including Herschel's classic on the nature of science, *Preliminary Discourse on the Study of Natural Philosophy* (1830).[126] Poe also discusses in *Eureka* the telescope of Sir William Herschel and his calculations on interstellar distances.[127]

There are other scientific sources that Poe used. For example, he either read Kepler's *Harmonice Mundi* (1619) or secondary sources that quoted from it.[128] The same is true for Laplace's *Mécanique Céleste* (1798) and Comte's *Traité Philosophiqie d'Astronomie Populaire* (1844).[129] Poe was familiar with the

121. *Eureka*, 135n75, 135n76, 141n137, 141n138, 141n140, 142n149, 143n155, 144n157, 144n160, 146n163, 148n173, 152n193, 161n221, 162n229, 162–163n230.

122. Walter F. Cannon (1961). "John Herschel and the Idea of Science." *Journal of the History of Ideas* 22 (2), 215–239.

123. *P&T*, 1345–1346, 1347–1348. Poe was likely using the 1835 Philadelphia edition. *Eureka*, 162n227.

124. *E&R*, 1215–1216.

125. *Eureka*, 118n7, 135n73, 146n165, 149n178, 150n179, 150n180, 158n217.

126. *P&T*, 1290. *Eureka*, 136n77.

127. *Eureka*, 156n213.

128. *Eureka*, 127–128n24, 151n190.

129. *Eureka*, 136n77, 140n134, 140–141n135, 147n170, 144n161, 146n163.

experiments on motion and gravity conducted in the late 1700s by Nevil Maskelyne and then by Henry Cavendish.[130] He was aware of the method of measuring the speed of light developed by Otto Struve.[131] And he examines and then rejects the hypothesis of Johann Mädler that there is a "non-luminous vast central orb" in the center of our galaxy, around which our Sun and the other stars in our galaxy revolve.[132]

But we have seen enough. Supposing that Poe's "every week was a desperate struggle of the scrivening pen against starvation," Daniel Hoffman skeptically wonders "when, and how" Poe could "find or make the time to read, much less to master, these complex and mathematical discourses."[133] But the fact remains that he did, with the proof in the pudding, so to speak. At any rate, the point of the exercise in this section is to see that Poe was not a hack when it came to understanding the science of his day. Instead, his was a sufficiently educated imagination, combined with a capable memory and an intellect well suited to extending the science of his day into novel speculations, the most remarkable products of which we shall examine in the following chapter. Even the Levines have to admit that in *Eureka* "The scientific speculation and theorizing are consistently intelligent and no more rhetorically overblown than other comparable statements of the era."[134]

Poe's Criteria of Truth

In *Eureka*, Poe provides a criterion of truth near the end of the imaginary epistle that opens the book. As Poe puts it in the mouth

130. *Eureka*, 132–333n62, 146n163.

131. *Eureka*, 156n210.

132. *Eureka*, 160n220, 161n221. Poe clearly blew his chance to anticipate the modern concept of the black hole, much like Einstein blew his chance to anticipate the Big Bang theory. This is evident especially in Poe's footnote on the matter: "I must be understood as denying, *especially*, only the *revolutionary* portion of Mädler's hypothesis. Of course, if no great central orb exists *now* in our cluster [galaxy], such will exist hereafter. Whenever existing, it will be merely the *nucleus* of the consolidation." *P&T*, 1346n.

133. *Hoffman*, 275.

134. *Eureka*, xx.

of the writer of the epistle, 1,000 years ahead of Poe's time, the real road to truth, (as opposed to induction or deduction,) "the broadest, the straightest and most available of all mere roads—the great thoroughfare," is what he calls "the majestic highway of the *Consistent*." As he puts it, "*a perfect consistency can be nothing but an absolute truth*." Again, as he puts it in the same paragraph, focusing now on "our Keplers" and "our Laplaces," "their theories are merely corrected—reduced—sifted—cleared, little by little, of their chaff of inconsistency—until at length there stands apparent and unencumbered *Consistency*—a consistency which the most stolid admit—because it *is* a consistency—to be an absolute and unquestionable *Truth*."[135]

What Poe expresses here might be construed as an instance of what is known as the *coherence theory of truth*, a theory of truth opposed to its main competitor the *correspondence theory of truth* (according to which a statement is true if it corresponds to reality). According to the coherence theory of truth, a statement is true if it coheres with, in other words is consistent with, a body of statements accepted as true as a whole. The coherence theory is associated with thinkers such as Hegel and Bradley, while the correspondence theory is associated with thinkers such as Tarski and Popper and is the common-sense theory, the theory writ large in modern science.

The obvious criticism of the coherence theory is that a statement can perfectly cohere with a large body of statements that is nothing but a fiction, such as the galaxy of *Star Wars* or the world of *Harry Potter*. But there is no evidence to believe that Poe was employing the coherence theory of truth, that truth is *merely* a matter of coherence with a large body of statements. His examples of Kepler and Laplace indicate that he was combining the correspondence theory with the coherence theory, that a statement is true, such as that the Universe originated out of a Big Bang as it is called today, when it is perfectly consistent with all the evidence we have of the Universe—subject, of course, as he indicates above, to modification when new evidence is acquired, and is perfect when there is no more evidence to be acquired. Poe in this regard was arguably on solid ground. As the philosopher Walter

135. *P&T*, 1269.

Kaufmann puts it beautifully, in his section on theories of truth, "All correspondence . . . is known through coherence: we have no second sight to see whether appearance and reality correspond, and if we would know whether a proposition is true we must see whether it is consistent with what else we know, with our other experiences."[136]

Poe probably thought he was doing nothing strange here, in spite of his way of putting it. (Many would say it is exaggerated or hyperbolic, but I would say it is artistically expressed.) After all, in Nichol's *Views of the Architecture of the Heavens* (1838), a book, as we have seen in the previous section, that Poe surely studied, Nichol seems to employ the same criterion of truth as Poe's. He says, with regard to the nebular hypothesis of Laplace, which is based mainly on Newtonian gravity and observations of nebulae using telescopes, "The Cosmogony has every mark of truth: its roots are seen in the Heavens, and they appear to go through every nook and alley of solar and planetary arrangements, not only explaining them, but comprehending their variety, and deducing the whole from one grand principle." Nichol contrasts Laplace's theory with "the Cosmogony of Buffon," his "idea that planets were chips struck off the sun by the collision of comets!" "Not one of the fundamental conditions of our system's planets," he says, "could be explained by this wild and reckless imagination, whereas Laplace's bold and brilliant *induction* (may I not now so name it?) includes and resolves all! The theory is so beautiful and perfect, . . ."[137] Laplace's theory is perfectly consistent with everything observed and known, while Buffon's theory falls far short of the mark. Laplace's theory is therefore held to be true of reality, while Buffon's is taken to be false. One can see here a common view in science, which may be connected with the epistemic value of inference to the best explanation that we shall examine in Chapter 6, which is devoted to theories of the nature of science. In the case of Poe and *Eureka*, the focus is not simply on the origin of our solar system, but on the origin of our universe, and indeed

136. Walter Kaufmann (1958). *Critique of Religion and Philosophy.* New York: Harper & Row, 74.

137. J.P. Nichol (1838). *Views of the Architecture of the Heavens.* 1840 ed. New York: H.A. Chapin, 100–101.

of other universes, Poe attempting to apply the same principles and criterion of truth. We see the same idea today, in the debate over the apparent fine tuning of the cosmological constants, some thinking a rational mind behind the constants is the most consistent with the facts and observations, others disagreeing.

In *Eureka*, however, Poe employs another criterion of truth, which is not so easily defended, which may be called the *artistic criterion of truth*. It finds one of its clearest expressions in Poe's works in "The Domain of Arnheim" (1846). In this story (not tale) Poe has the narrator elaborate the theory of truth and beauty of the genius landscape artist Mr. Ellison, which is that the truth about beauty is "felt" by a genuine artist, not reasoned, but being felt it is "no affectation or chimera." The certainty felt is on the same level as mathematics: it is objective. As Poe's narrator puts it, "The mathematics afford no more absolute demonstrations than the sentiment of his art yields the artist." The genuine artist "not only believes," says the narrator, "but positively knows, that such and such apparently arbitrary arrangements of matter constitute the true beauty." Moreover, "he is confirmed in his instinctive opinions by the voice of all his brethren," including the "remedy of the defective composition." This "instinctive" knowledge, adds the narrator, involves reasons that "have not yet been matured into expression" and "It remains for a more profound analysis than the world has yet seen, fully to investigate and express them."[138]

In *Eureka* we find repeated use of the same criterion of truth. For example, in the Preface we find Poe refer to "this Book of Truths" and "the Beauty that abounds in its Truth; constituting it true." Later in the book, in discussing the nebular hypothesis of Laplace, Poe states that "From whatever point we regard it, we shall find it *beautifully true*. It is by far too beautiful, indeed, *not* to possess Truth as its essentiality—and here I am profoundly serious in what I say."[139]

138. *P&T*, 860. "The Domain of Arnheim" (1846) is a rewrite of "The Landscape-Garden" (1842), which Poe in a letter referred to as a magazine "article," just as he did with "Mesmeric Revelation" (1844) in later letters. *Letters*, 353, 453, 473, 556. "Article" is certainly the better term, as neither of the two works are properly tales. See also *E&R*, 870.

139. *P&T*, 1259, 1312.

Elsewhere in *Eureka*, near the end, Poe combines the two theories of truth. Of "the poetical essence of the Universe—*of the Universe* which, in the supremeness of its symmetry, is but the most sublime of poems," he adds that "symmetry and consistency are convertible terms" and "thus Poetry and Truth are one." And to emphasize the point, he says "*A perfect consistency, I repeat, can be nothing but an absolute truth*."[140]

Consistency and beauty are themes that we shall examine in Chapter 6, the chapter devoted to Poe and philosophy of science. But before we turn to Chapter 5, the next chapter, devoted to the science of *Eureka*, we need to examine and resolve the most important preliminary matter of all, and with a consuming curiosity that pervades our soul.

Was *Eureka* a Hoax?

We have just seen Poe in *Eureka* emphasize symmetry as "the poetical essence of the Universe." In the very same passage, however, he also emphasizes a *sense* of symmetry. He says, "in fact, the sense of the symmetrical is an instinct which may be depended on with an almost blindfold reliance." And again: "We may take it for granted . . . that Man cannot long or widely err, if he suffer himself to be guided by his poetical, which I have maintained to be his truthful, in being his symmetrical, instinct," which he likewise calls "the analogical, symmetrical or poetical instinct of Man." Poe adds that it should not be "superficial" symmetries "of forms and motions" that guide one here, but rather "the really essential symmetry of the principles which determine and control them," a "symmetry of *principle*" which Poe himself, of course, applied in *Eureka* not only to the Universe as it is, to cosmology, but to cosmogony as well, to the history of the Universe, a Big Bang universe "renewed forever, and forever, and forever . . . at every throb of the Heart Divine."[141] The overall idea, then, is that there is a "*mutuality* of adaptation," as Poe put it in criticism of the Bridgewater Treatises, a coadaptation, between the deep symmetry

140. *P&T*, 1349.

141. *P&T*, 1349, 1350, 1352, 1356.

of the Universe (both synchronic and diachronic) and the sense or instinct of symmetry in humans (which like all instincts would have to be a matter of degree, the highest, for Poe, being found in great artists and in great scientists alike).

What is yet to be fully appreciated in Poe studies, however, is that this emphasis on symmetry in *Eureka* was not entirely new with Poe, but was rooted far back in his life, in the very least to his teens, and in an interpersonal form. Evidence for this conclusion is both physical and *outré*, namely, a very recently discovered poem that Poe wrote when he was seventeen. I came across this poem by an odd accident, for my own part by far the oddest accident of all, one which could not have been anticipated, to be sure. It was during a week I spent three summers ago in Charlottesville, for a conference on counterfactuals held at the University of Virginia. A few blocks away from the bed and breakfast where I had been staying, my '92 Jeep YJ Renegade broke down in front of a stately old mansion on Belmont Avenue, southeast of the campus, the mansion looking rather the worse for wear. In the driveway the owners were having a garage sale. While waiting for CAA I decided to browse the tables, eventually noticing way at the back an oblong chest of wood. Inside the chest was a small old trunk, and inside the trunk were some dusty books and an oblong box, on the lid of which was inscribed a name, Mrs. Bullitt Fitzhugh. Inside the box I found various papers and letters, as well as a scrap of parchment appearing not as others were but as alone. Crumpled and soiled, it had a handwritten poem on it, entitled "To Elmira" and signed "Edgar A. Poe." I could scarcely believe my eyes—let alone fathom the extent of my good fortune, a treasure of incalculable value, all from a series of accidents and coincidences. Fearing that my excitement would be palpable to others, I put on my best poker face and haggled the old gentleman, perhaps seventy years of age, down from $7.00 and bought it for $5.00 (which is all the cash I had on me). He called the piece a "determination of non significance." Apparently he had never heard of Poe! Days later when I got home I compared the handwriting with some published examples of Poe's handwriting, the earliest I could find. The writing before me was clearly by the same hand, although in comparison it seemed unusually careful and refined. When I saw the letter I immediately knew that it was written for Elmira

Royster, Poe's teenage sweetheart and first fiancée, with whom he was reunited in 1849, apparently getting engaged in the few weeks before his untimely death, for which he even joined the Sons of Temperance to satisfy her. From the date of the letter, April 1, 1826, I soon surmised that the poem was written six-and-a-half weeks after Poe had arrived at the University of Virginia to begin his studies, the place from which Elmira's high-born father was intercepting Poe's letters to his daughter, chilling and killing their engagement. Years later, long after Poe's death, she would claim that "Poe never addressed any poems to me."[142] Here, for the very first time, after all these years, are the *facts* of the affair, conclusive proof that she was wrong, proof published at long last:

> When on my life I ponder
> And half doth seem my soul,
> In my heart I think of you
> And long to be made whole.

When I first announced this remarkable find, believing that the extraordinary case will create a *stir* to some purpose, the rush for the "sole paper which had the news," was something beyond even the prodigious; and, in fact, if (as some assert) Poe *did* not absolutely pen the rhyme recorded, it will be difficult to assign a reason why he *should* not have penned it—especially given *the novelty of the idea* and its fancy-exciting and reason-repressing character.

The profound sense of symmetry expressed in this remarkable poem brings us to the question of whether *Eureka* was intended as a hoax. Certainly several of Poe's contemporary reviewers thought it literally a hoax. One claimed that "The mocking smile of the hoaxer is seen behind his [Poe's] grave mask," which the reviewer followed with the compliment that "If Mr Poe is not a philosopher he is the most adroit of mimics." Another called it "a scientific hoax of the higher order which few men are capable of executing more cleverly." Yet another reviewer claimed that with *Eureka* Poe "has egregiously hoaxed—not his readers—but himself."[143] These

142. *Quinn*, 91.

143. *Critical Heritage*, 281, 285, 292.

ani*mad*versions were not without some foundation. For Poe wrote science fiction in which he took great pleasure in fooling people into believing it was not fiction at all, using a high degree of detail interspersed with accidents, such as with what would come to be called "The Balloon-Hoax" (1844) and with "The Facts in the Case of M. Valdemar" (1845). Even today, among Poe scholars, who collectively know Poe's writings better than ever before, one can find some who seriously entertain the possibility that *Eureka* was intended as a hoax. Take, for example, Stuart and Susan Levine in the Introduction to their annotated text of *Eureka*, according to whom not only was Poe fond of playing literary games, (there is no question they are right about that,) but "one can prove that Poe is playing private literary games in so many places in *Eureka* that one questions whether to trust anything in the book."[144] And then there is the feminist scholar who recently put it to me that *Eureka* is "total bullshit."

Given the importance of this question, much like stress testing a metal, I shall here endeavor to take the hoax interpretation of *Eureka* a few steps further. For a start, when you look at Poe's tales, you find that one of his amusements was to take a common, everyday phrase and play it to the extreme as the basis for a wild tale. He did this, for example, with "Loss of Breath" (1832). How many times have you heard the phrase, "I am out of breath," or "I have lost my breath"? Well, in the tale this literally happens to the narrator, he literally lost his breath on the morning after his wedding while he was grasping his wife by her throat, screaming "Thou wretch!—thou vixen!—thou shrew! . . . thou witch!—thou hag!—thou whipper-snapper!—thou sink of iniquity!—thou fiery-faced quintessence of all that is abominable!" Just as he was about to "launch forth a new and more decided epithet of opprobrium," to his astonishment he found, upon losing his breath, that he was "alive with the qualifications of the dead—dead with the propensities of the living—an anomaly on the face of the earth—being very calm, yet breathless." Having, I might add, what is every asthmatic's dream,—literally not needing to breathe, for which I would join the Sons of Temperance if I thought it would work, and would even get engaged,—a series of misadventures then begins,

144. *Eureka*, xviii.

such as being rendered unconscious from sitting between "two gentlemen of colossal dimensions" during a stage coach ride. The doctor "having applied a pocket-mirror to my mouth, and found me without breath," the narrator was then thrown off the coach and subsequently vivisected by a surgeon, then later "interred in a public vault," from which he managed to escape, and so on.[145]

Poe did the same with "The Man That Was Used Up" (1839). A common expression in Poe's day was to say something or so-and-so is "used up," which meant in Poe's business that a person was severely and thoroughly attacked in print,[146] or more generally among the public that a topic or person was overworked or overused.[147] In Poe's tale the man that was highly praised by everyone, as in "he's the man," Brevet Brigadier General John A. B. C. Smith, who distinguished himself in the Indian wars, especially against the Bugaboo and Kickapoo, turns out to be "a large and exceedingly odd looking bundle of something" who required his valet to "slip on" his prosthetic parts, specifically arms, legs, shoulders, bosom, wig, teeth, an eye, and palate.[148]

Yet another example is the phrase "love at first sight," which Poe took as the springboard for "The Spectacles" (1844), a tale in which the twenty-two-year-old narrator, who is notoriously nearsighted, absolutely refuses to wear glasses for the reason that he is "youthful and good-looking" and that when it comes to wearing glasses there is "nothing, indeed, which so disfigures the countenance of a young person." He is also a "devoted admirer of the women." One evening at the opera with a friend he notices a woman who captures his attention. It is "love at first sight." He pursues her for weeks, always coming across her coincidentally in places with poor lighting, eventually getting her to marry him at 2:00 in the morning, only to find out by daybreak, with the aid of her opera double eye-glass, and to his astonishment and horror, that not only is the woman an eighty-two-year-old hag, beautiful in poor light by the aid "of pearl-powder, of rouge, of false hair, false teeth, and false *tournure*, as well as of the most skilful

145. *P&T*, 151, 152, 155, 158.

146. *Peeples*, 25; *Brevities*, 486.

147. *Hutchisson*, 90; *Brevities*, 486.

148. *P&T*, 311, 314, 315.

modistes of Paris," but that she is also his long-lost "great, great, grandmother."[149] Oh the perils of myopia!

Might not *Eureka* also be part of this *genre*? Is not "Eureka!" a common-enough word and "eureka moment" a common-enough phrase? Perhaps *Eureka*, then, is an elaborate burlesque with a foot in not one but in two genres, one the genre of the scientific hoax, to which Poe contributed substantially, for example with "The Facts in the Case of M. Valdemar," the other the genre that he possibly invented, beginning with "Loss of Breath."

Another consideration that lends itself to the hoax interpretation of *Eureka* concerns a feature of Poe's literary theory that, unless I am mistaken, has hitherto gone unnoticed. In Poe's anonymous book review of his own *Tales* (1845), a collection of some of his more popular fiction, he writes at the end that "A writer must have the fullest belief in his statements, or must simulate that belief perfectly, to produce an absorbing interest in the mind of his reader. That power of simulation can only be possessed by a man of high genius. It is the result of a peculiar combination of the mental faculties. . . . It is possessed by Mr. Poe, in its full perfection."[150] Poe made a similar comment a few years earlier in his high praise of Daniel Defoe, that "the author of Crusoe must have possessed, above all other faculties, what has been termed the faculty of *identification*—that dominion exercised by volition over imagination which enables the mind to lose its own, in a fictitious, individuality. This includes, in a very great degree, the power of abstraction; . . ." What is especially interesting is that Poe calls Defoe a "genius" and *Robinson Crusoe* a "literary performance."[151] These comments, along with Poe on himself above, recall to mind what today is known as *method acting*, in relation to which the label "genius" is sometimes applied, as in the case of Marlon Brando, who was schooled in method acting mainly by Stella Adler at New School's Dramatic Workshop, also a little later by Elia Kazan at his Actors Studio.[152] Poe's parents,

149. *P&T*, 618, 619, 620, 640.

150. *E&R*, 873.

151. *E&R*, 202, 201. See also Poe on Shakespeare. *E&R*, 273.

152. Stefan Kanfer (2008). *Somebody: The Reckless Life and Remarkable Career of Marlon Brando*. New York: Alfred A. Knopf, x, 36–37, 78–79, 126. Marlon

of course, were actors in American theater, his mother one of the most famous of her time. The common denominator of the variations that go by the name of method acting is that acting is not so much pretending as working yourself into believing your lines, (if you did not already believe them,) a kind of self-deception, possibly by employing experiences and emotions from one's past. The roots of method acting may indeed run deep, not just to Constantin Stanislavski in Russia, and I venture to suggest that Poe was well aware of those roots, (excluding, of course, any knowledge of Stanislavski, who was born in 1863,) which he interestingly applied to writing, joining great writing and great acting as two species of a genus, the genus of method acting, with *theatrical* method acting one of the species, and *literary* method acting the other—the latter affording a possible window of insight into Poe's poems and tales, and maybe even *Eureka*, too, with its recurring hyperbolic certainty.

In spite of all of the above, I shall argue here that the balance of evidence should leave one with the opposite conclusion, that Poe took *Eureka* very seriously indeed, not on the whole as a hoax—even though, for example, near the beginning it contains extracts from an epistle on the nature of science that "seems to have been written in the year two thousand eight-hundred and forty-eight," which was found in a corked bottle floating in the *Mare Tenebrarum*, (Sea of Darkness, the Atlantic Ocean as referred to by the geographer Al-Idrisi in the twelfth century,) a body of water "little frequented in modern days unless by the Transcendentalists and some other divers for crotchets."[153] (As we shall see in Chapters 5 and 7, Poe makes some very serious and profound points in that epistle, although clearly, not only from prefacing *Eureka* with it but from his writings and his life as a whole, he seems to have been the type of person who never lost an opportunity to have some fun.)

First, there is the time of composition. Poe's cousin-wife, Virginia, thirteen-and-a-half years his junior and his wife for close

---

Brando (1994). *Brando: Songs My Mother Taught Me*. New York: Random House, 78–85.

153. *P&T*, 1263. For the full meaning of Poe's *Mare Tenebrarum*, see *P&T*, 1399n, *Eureka*, 119n11.

to eleven years, (based on their second marriage shortly before her fourteenth birthday, each with her mother's blessing,) died on January 30, 1847, at the tender age of twenty-four, after a five-year battle against tuberculosis. Poe was devastated, and his writing at first ground to a halt, resulting in the least productive year of his literary life. Truly, for all the suffering that Poe endured throughout his lifetime,—all the severe disappointments and the recurring bouts of deep poverty, driving him to drink and worse, (he was never technically an alcoholic,) and keeping in mind that he died on October 7, 1849,—the year 1847 had to have been his most *immemorial* year (discounting, of course, his early childhood).[154] Almost six weeks after Virginia had met her destiny with the Conqueror Worm, Poe wrote of "Being still too unwell to leave my room," of a "sorrow so poignant as to deprive me for several weeks of all power of thought or action."[155] Of the few poems he

154. Poe's poem, "Ulalume—A Ballad," which in the opening stanza includes the lines "It was night, in the lonesome October / Of my most immemorial year," was composed "CA. October" of 1847. *Poe Log*, 705–707. Officer T.S. Eliot accuses Poe of using the word *immemorial* improperly, which "according to the Oxford dictionary, means: that is beyond memory or out of mind; ancient beyond memory or record: extremely old." He says "sound and sense must cooperate; in even the most purely incantatory poem, the dictionary meaning of words cannot be disregarded with impunity." Thomas Stearns Eliot (1948). "From Poe to Valéry." In *Recognition*, 210. Rather than a violation of the dictionary, however, an "irresponsibility towards the meaning of words," as Eliot also puts it, I suggest that Poe's use of *immemorial* was perfectly legitimate. Poe *could*, of course, as he often did, have coined his own term, such as *immemorable*, from the Latin *immemorabilis*, meaning "not worthy of remembrance," to refer to the year in his life that he'd most like to forget. *Immemorial*, on the other hand, granting that Poe's use of the word goes against the dictionary, immediately puzzles, and should provoke wonder. Keeping in mind that 1847 marked the tragic death of Poe's Virginia, which consequently was the worst year of his life up to that time, that *memorial* is often used to refer to death, and that the Latin *im-* is a variation of *in-*, meaning "not, opposite of," Poe's use of *immemorial* could well have been an inventive way of referring to a year that he wished would not have happened and that he wished would be beyond his memory given that it did. At any rate, surely an artist may be allowed to use an established word in a new and interesting way, breaking with tradition so as to produce a memorable effect, in a word, to be *creative*, without being slapped with a ticket from the dictionary police.

155. *Letters*, 623.

wrote that year, "Deep in Earth" was probably the shortest he ever wrote:

> Deep in earth my love is lying,
> And I must weep alone.[156]

To this we must add that it was only a year after Virginia's death, in a letter to a friend dated January 4, 1848, that Poe finally seemed able to put it into perspective:

> This "evil" was the greatest which can befall a man. Six years ago, a wife, whom I loved as no man ever loved before, ruptured a blood vessel in singing. Her life was despaired of. . . . She recovered partially and I again hoped. . . . Then again—again—again & even once again at varying intervals. Each time I felt all the agonies of her death—and at each accession of the disorder I loved her more dearly & clung to her life with more desperate pertinacity. But I am constitutionally sensitive—nervous in a very unusual degree. I became insane, with long intervals of horrible sanity. . . . I had indeed, nearly abandoned all hope of a permanent cure when I found one in the *death* of my wife. This I can & do endure as becomes a man—it was the horrible never-ending oscillation between hope & despair which I could *not* longer have endured without the total loss of reason. In the death of what was my life, then, I receive a new but—oh God! how melancholy an existence.[157]

Poe finishes the letter by stating, "My ambition is great." He writes about his plan to start his own magazine, *The Stylus*, and to secure prepublication subscriptions. But especially interesting for our purpose is what he wrote at the start of the letter, not only that "My health is better—best. I have never been so well," but "also have been working at my book."[158]

---

156. *P&T*, 87.

157. *Letters*, 641.

158. *Letters*, 639–640.

The death of Virginia, as the death of a loved one does to many, got Poe thinking, like never before, about the meaning of life. What Poe did was take his power of imagination and use it to combine his abiding interest in science and theology with his literary theory on beauty and deformity and plot. To be sure, many of his ideas in *Eureka* had their precursors in his earlier writings. But as we have seen in the second-last section of Chapter 1, those precursors (or at least those that are clearly so) are mainly to be found in the few years leading up to Virginia's death, specifically 1844 and 1845. (During the intermittent year, 1846, Poe's life was marked by quite severe poverty, the advanced illness of Virginia to the point of being a semi-invalid, and a corresponding deterioration in his own health, both physical and mental.) What Virginia's death ultimately catalyzed in Poe, then, what his recovery from her death ultimately meant, Poe supplying the essence and Virginia in a way the matter, (the dead matter, after all,) was the *synthesis* of Poe's ideas about the Universe, of everything he knew and supposed and felt as a ratiocinative artist to be true, one work, born from the womb of the lost Lenore, a unified whole, all the ideas on the topic that had been developing in his mind during the few years before Virginia's death, along with some new ones. It was his ultimate act of both creative and recognitive (to his mind at least) imagination.

Less than three weeks following the letter quoted above, Poe wrote to another friend that, in order to help raise money for his magazine, he planned on giving a public lecture, such that "my subject shall *not be literary* at all. I have chosen a broad text—'The Universe.'"[159] The close to three-hour lecture was given on February 3, 1848, (almost to the day of the one-year anniversary of the death of Virginia,) and published by Putnam as *Eureka* on or around July 11 of the same year.[160] To think that Poe wrote *Eureka* or the lecture on which it is based mainly to raise funds for *The Stylus*, or more widely, as Benjamin Fisher seems to claim, because "Poe wrote what he calculated would remunerate him, thereby leaving us a legacy of works that are far more heterogeneous than

159. *Letters*, 647.

160. *Poe Log*, 714, 720, 742.

they are homogeneous,"[161] overlooks not only the remarkable content of *Eureka*, but also the background and context of its writing. Poe was hoping to make some money, (sure, what's wrong with that? and especially when poor?,) but that does not *explain* why he wrote *Eureka*. To think that it is an American's attempt to one-up and cash-in on the financial success of *Vestiges*, say, is to catapult oneself far beyond the bounds of the rightly lauded (but often violated) principle of charity of interpretation. *Eureka*, instead, is not only *sui generis*; it is Poe's *magnum opus*, possibly his *raison d'être*, (*subconsciente*,) and arguably the *summum bonum* of his life.

A further class of evidence as to Poe's sincerity about *Eureka* comes from what he says about it in his correspondence. For example, in a letter to a friend dated May 19, 1848, slightly less than two months before the publication of *Eureka*, Poe apologizes for not replying promptly, stating as his reason "duties that, just now, *will not* be neglected or even postponed—the proof-reading of a work of scientific detail, in which a trivial error would involve me in very serious embarrassment."[162] It is not entirely clear what Poe means by a "trivial error," (a trivial error in fact or theory?, a spelling error?,) but it is hardly likely that a hoax would occupy him in such a way, or that he would refer to *Eureka* as a work of "scientific detail," unless he wanted to include this "dear friend" in the hoax. The dear friend, by the way, was Mrs. Jane Locke, a poetess who had written a poem about Poe—"An Invocation to Suffering Genius"—and whom Poe thought of marrying, as he thought she was a young widow (he was wrong on both counts, he quickly found out).[163]

A yet further class of evidence comes from what Poe wrote about the nature of genius. In an essay entitled "Fifty Suggestions" (1849), specifically in number 23,[164] Poe argues that the pub-

161. *Fisher*, 100.

162. *Letters*, 665.

163. *Poe Log*, xxxi. There is also the testimony of his aunt, Maria Clemm: "When he was composing 'Eureka,' we used to walk up and down the garden, his arm around me, mine around him, until I was so tired I could not walk. He would stop every few minutes and explain his ideas to me, and ask if I understood him." *Poe Log*, 714.

164. *E&R*, 1301–1302.

lic has the wrong conception of genius, that of the person with "the abnormal predominance of some one faculty over all the others . . . at the expense and to the detriment, of all the others." When you have someone like that, claims Poe, you have what he calls "pseudo-genius," not only because what is taken for genius is at bottom a "mental disease" or an "organic malformation of mind," but because "The works of such genius are never sound in themselves and, in especial, always betray the general mental insanity." The "true genius," on the other hand, says Poe, "is necessarily, if not universal in its manifestations, at least capable of universality," and it is only "*Taste*" which leads it to pursue one subject rather than another, for "With equal zeal, it would succeed equally in all." Poe was well known for having an enormous ego, for "self-inflation," as it is sometimes called today. At any rate, that Poe is referring to himself in the quotation above, and subtly to *Eureka*, is a conclusion difficult to resist when we turn to the beginning of Suggestion 23, published, it must be kept in mind, roughly ten months following the publication of *Eureka*, and therefore written maybe eight months following, in the wake of a reception in which it was clear that the critics and the general readers alike were, on the whole, just not getting it:

> Let a man succeed ever so evidently—ever so demonstrably—in many different displays of *genius*, the envy of criticism will agree with the popular voice in denying him more than *talent* in any. Thus a poet who has achieved a great (by which I mean an effective) poem, should be cautious not to distinguish himself in any other walks of Letters. In especial—let him make no effort in Science—unless anonymously, or with the view of waiting patiently the judgment of posterity. Because universal or even versatile geniuses have rarely or never been known, *therefore*, thinks the world, none such can ever be.

It should be noticed that Poe, on at least one occasion, claimed that with "The Raven" (1845) he had written "the greatest poem that ever was written."[165] And "talent," it should also be noticed, is precisely the term used in a review of *Eureka* to characterize Poe's "fanciful

165. *Ackroyd*, 101.

speculations."[166] Clearly from Suggestion 23, then, it would seem that Poe thought of *Eureka* as a genuine "effort in Science."

Finally, that *Eureka* was not intended by Poe as a hoax is attested by the fact that, as Patrick Quinn puts it, "Poe revised the text of *Eureka* published by Putnam in 1848. . . . [with] more than 300 alterations and additions which he made in four separate copies of the first edition."[167] It is unlikely that someone hoping to perpetrate a gag on friends and public alike would go to such lengths. One would expect, of course, a hoax or two (or more) *within* the pages of *Eureka*,—indeed otherwise it would be surprisingly unPoelike,—but that the whole was intended by Poe as a hoax is, all things considered, improbable.

So why then did Poe perpetrate hoaxes? Calling a side of Poe "Hoaxiepoe," the side that "laughs—immoderately," Daniel Hoffman claims that "Hoaxiepoe and Edgar the Metaphysician are one and the same, a regular practitioner of that aesthetic which keeps the reader off balance and always on the defensive against the superior cunning of the author." When "the Balloon Hoax was the talk of the town," he even goes so far as to say, "I think this day may have been the happiest day of Poe's career."[168]

A far more interesting and charitable discussion is provided by Lynda Walsh in her book devoted to the topic of hoaxes in science, in which she provides a long and useful chapter on Poe. Against those who view Poe's scientific hoaxes as the product of "the insecure genius who uses his hoaxes to humiliate his readership so he can feel superior," Walsh points out that a good hoax "materializes a double readership—the readers who fall for it, and those who catch on and read it as a coconspirator of the hoaxer rather than as his/her victim." Accordingly, she argues that the motive behind Poe's scientific hoaxes, including the hoax elements in *Eureka*, was ultimately to produce the latter of the two audiences, "an audience Poe invoked, created, in fact, with the clues he left in his hoaxes for the acute observer." It is they and them alone who are "those few to whom he dedicated *Eureka*, those 'who love me and whom I love.'"[169]

---

166. *Critical Heritage*, 280.

167. *P&T*, 1372. The text Quinn uses in *P&T* is the "reconstructed text."

168. *Hoffman*, 153, 279, 152.

169. Lynda Walsh (2006). *Sins against Science: The Scientific Media Hoaxes of Poe, Twain, and Others*. Albany: State University of New York Press, 117, 118–119.

A further or additional possibility, in fact the most charitable and holistic given Poe's advocacy for a Smithsonian "College for the advancement of *Science*," as we have seen above in the section devoted to Poe's "A Chapter on Science and Art" (1840), is that, in line with wanting to promote the cause of science in America, Poe with his hoaxes (in addition to having some fun) was attempting to wake up a basically gullible reading public and engender critical reasoning. We can see this especially in Poe's analysis, (which Walsh does indeed take into account,) provided in his 1840 "Note" to "Hans Pfaall" and in his 1846 "Literati" entry on Richard Adams Locke, on why the "Moon-Hoax" (1835) of the latter was so successful and yet "no one should have been deceived": notably, the gross scientific illiteracy of the general public, its tendency to be a sucker for sensational claims, its focus more on style than on content, and its lack of analogical reasoning, of "scrupulous attention to general analogy and to fact."[170] But the purpose of Poe's hoaxes, arguably, was not just to engender critical reasoning in the public toward the quackery and twaddle that was ubiquitous in the numerous magazines and newspapers; it was also to engender critical reasoning toward the theories and claims to knowledge of genuine science itself. And this is precisely what Poe himself did in *Eureka*: in using an imaginary epistle dated 2848, in using but also criticizing the science of his day, and in producing the nine anticipations of twentieth-century science, to which we now at last shall turn.

170. *P&T*, 997; *E&R*, 1217.

CHAPTER 5

The Scientific Anticipations of *Eureka*

. . . the novelty & *moment* of my views. What I have propounded will (in good time) revolutionize the world of Physical & Metaphysical Science. I say this calmly—but I say it.

—Edgar Allan Poe[1]

Entrée

Looking back over more than one hundred and sixty years now, we can safely say that Poe's *Eureka* did not *itself* revolutionize the world of physical science, let alone the world of metaphysics (no one uses the phrase "metaphysical science" anymore, at least not among the scientifically literate). But in *Eureka* Poe anticipated, remarkably, at least nine theories and developments in twentieth-century science, and in their own good time they proved novel and momentous indeed.

I use the word "anticipated" cautiously. In the Introduction to their annotated edition of Poe's *Eureka*, Stuart and Susan Levine state that "There exists a literature by Poe fans about how Poe predicted twentieth-century physics and astronomy. This does not

1. *Letters*, 650.

strike us as a fruitful argument; Poe did not magically predict Einstein or intuit subatomic physics."[2]

One should be able to see the problem here. Words like "fan" and "magically" tend to function much as words like "fruitcake" and "nut." They are negative, or more accurately derogatory, and do not serve to further the discussion.

There is a further problem, widespread in academia outside of the natural sciences, specifically in the humanities, including philosophy, which is the view, strongly expressed, that no one is in any meaningful sense *ahead of his or her time*, that scientists and their theories, in particular, are products of their contexts along with the rest of us. We shall see this in Chapter 6 when we examine modern contextualist history of science. We can also see it in the Introduction by the Levines, who state that *Eureka* "is plainly a product of the science of the day and of a moment in the history of science when the promise of science seemed cosmically bright," that "*Eureka* is a window on the 1840s."[3]

In a very real sense, of course, Poe was a product of his time, as are we all. How could he, and we, be otherwise? But that, in itself, cannot be used to deny that he was *nothing more* than a product of his time. Any good philosopher will tell you that it all depends on what one means by "ahead of his time." In Marshall McLuhan's meaning, "in the electrical age there is no longer *any* sense in talking about the artist's being ahead of his time" (italics mine). This is because for McLuhan "The artist is the man in any field, scientific or humanistic, who grasps the implications of his actions and of new knowledge in his own time. He is the man of integral awareness."[4] Much more recently Jerome McGann, on whether writers can be "in advance of their time," makes a similar point concerning "a commonplace of cultural studies: that every historical moment is pregnant with futurities and that certain writers are their midwives," such that in McGann's view, for example with regard to Poe's "anticipations of a 'Big Bang'

2. *Eureka*, xx.

3. *Eureka*, xxvi–xxvii.

4. Marshall McLuhan (1964). *Understanding Media: The Extensions of Man.* New York: McGraw-Hill, 65.

cosmology in *Eureka*," Poe "draws them out—leaps to them—by his suppositious poetic style."[5]

One can surely make a strong case for Poe as a man of "integral awareness," but I disagree with McLuhan that in the age of information there can no longer be a legitimate meaning of "ahead of his time." Similarly, one can make a case for an historical moment being "pregnant with futurities," but I disagree with McGann's loaded metaphor of Poe as a midwife. Poe did not bring something out of the womb of his environment; instead, he was the parent. But even the parental metaphor breaks down, since Poe *created*, with his *mind*, something genuinely new, something more mature than infant even, rather than just another baby drawn from a preexisting meme pool. What I have in mind is a meaning of "ahead of his time," then, that literally goes beyond that of integral awareness and the metaphor of midwifery, a meaning that is pursued in this chapter and the rest of this book, namely, that of *scientific imagination*. In *that* sense, contra McLuhan, McGann, and many others, one can indeed (or so I shall argue) be ahead of one's time. I do not mean, of course, "educated guesses" or "happy guesses" that turn out to be right, a meaning in philosophy of science that we shall explore in the next chapter. Instead, keeping in mind that a definition is that which so describes its object as to distinguish it from others, by "ahead of one's time" I mean *the educated imagination that takes in information that was available to others at the time but that arranges and adds to it in a strikingly new and superior way, a way that anticipates future understanding of the domain in question, future knowledge.* This also doubles as a definition of *scientific imagination* when the domain is scientific, while the ability to see what those with scientific imagination saw could be called *second-order scientific imagination.* Both definitions, of course, assume that there is such a thing as knowledge and superior understanding, a position we shall examine in the next chapter, especially in the section on Thomas Kuhn.

The view that nobody is ever ahead of their time is often expressed in a form that is little more than a dogma or mindless mantra, and as such is perhaps the greatest barrier to an examination, or rather the reception, of the *poetic dénouement* of this

5. *McGann*, 107.

book, namely, the illumination of scientific imagination. And it highlights a core antinomy, or dissonance, that pulls virtually each and every college and university in opposite directions. One of the things that is remarkable about Poe's *Eureka*, viewed today, is its capacity to bring out this opposition, these opposing forces pulling at the brains of those who read it, including especially academics.

One can even see this opposition, this paradox, within individual authors themselves. Take again Stuart and Susan Levine. In the very same Introduction quoted from above, in the same pages in which they affirm they employed the services of a "scientific consultant," they can also be found stating that Poe in *Eureka* "did think through intelligently the implications of what was known in his day," that "Poe's understanding of space and time is unusually advanced for his day," that "Poe's general scheme may loosely be said to be equivalent to the 'Big Bang' theory," that "his notion of the 'reciprocity' of matter and energy is *roughly* akin to the modern understanding of the relationship represented by Einstein's e = mc2," that "Poe comes close to anticipating astrophysical speculation that the universe is ultimately going to coalesce," that "Poe's idea of seeing electricity as one among several other forces, as yet not well defined, is an interesting guess or projection," that "At best, one can say, 'Poe loosely predicted much of what science believed a few years ago,'" and that "it is fair to say that Poe . . . was thinking in the right directions, certainly speculating creatively."[6]

Well, which is it? Was Poe ahead of his time or not? Or do we have here a case of Schrödinger's cat, both simultaneously alive and dead according to the logic of the consensus interpretation of quantum physics? (I will discuss Schrödinger's cat in the first section below.) I suggest that the Levines are not alone but are rather symptomatic of a modern reading of *Eureka* outside of natural science, that one feels strongly pulled in opposite directions, in the one direction toward the denial that Poe was ahead of his time because no one is, (that is the received view in the humanities,) and in the other direction toward the affirmation that Poe remarkably anticipated many developments in twentieth-century science (that would be the natural response in the natural sciences).

---

6. *Eureka*, xx, xxi, xxii.

At any rate, the point and purpose of this chapter is *not* to claim that Poe "magically predicted" anything in twentieth-century science, but instead that he provides us with a prime example of scientific imagination, which in turn justifies the following chapters. I intend to prove, then, once and for all, in the sense of beyond a reasonable doubt, that *Eureka* vindicates Poe and provides his credentials for taking him seriously on the topic of scientific imagination.

In what follows, the method I employ is to juxtapose the words of twentieth-century scientists in describing a particular theory or development with Poe's own words, to which I add the necessary background. The order of topics follows Poe's order of presentation in *Eureka*, inasmuch as that is possible, since the order is not clear and distinct, as Poe often touches on a subject or idea only to expand on it later in the text, and even so his substantive order of topics is not necessarily the most logical order in terms of scientific reasoning. What we have to keep in mind, after all, is that Poe named the book *Eureka*, and that it was not meant as a systematic treatise (although it contains plenty of argumentation) but mainly as a work of inspiration.[7] This is in spite of his claim, apart from what he calls "the sole absolute *assumption* of my Discourse" that we shall examine in §II below, that the rest of *Eureka* involves "a train of ratiocination as rigorously logical as that which establishes any demonstration in Euclid," which he then claims is "so thoroughly corroborated by induction."[8] (The effervescent hyperbole in Poe's prose should be received, in my view, as part of the fun of reading him, rather than as a reason not to.) Later in *Eureka* Poe seems to acknowledge the lack of systematic presentation, stating that rather than following an arrangement that is "merely

7. According to legend, Archimedes (d. 212 B.C.) shouted "Eureka! Eureka!" ("I've found it! I've found it!") when the answer suddenly came to him—after spending a long time trying to figure out the answer—on how to determine the amount of pure gold in what might be an amalgam. That Poe considered *Eureka* to be *fundamentally* a work of educated inspiration should be clear not only from what he says in the opening epistle against deduction and induction in favor of intuition, but also from what he says in favor of the discoveries of Kepler and Champollion, which fit his scheme nicely and which we shall discuss in Chapter 7.

8. *P&T*, 1301, 1303.

natural" he is following an arrangement that is "almost altogether . . . moral," which explains "If then I seem to step somewhat too discursively from point to point of my topic." His purpose is to produce "that chain of *graduated impression* by which alone the intellect of Man can expect to encompass the grandeurs of which I speak, and, in their majestic totality, to comprehend them."[9] Whatever the order of the presentation of his topics, the result of the series of juxtapositions presented below should be much the same, namely, a crescendo-like unity of effect, culminating in a lucid awakening on the part of the reader, an impression of the greatest force and vivacity, that Poe was not a fraud but the real deal, that he had and exercised true, genuine, scientific imagination, and that the nature of it needs to be investigated.

I. Rejection of Axioms as Intuitively True

In "The Purloined Letter" (1844), Poe has his ace detective Dupin reject the supposition, "of which I am confounded at the universality with which it has been conceived," that "the truths of what is called *pure* algebra, are abstract or general truths."[10] Again he says, "Mathematical axioms are *not* axioms of general truth." Poe has Dupin argue that geometry and mathematics are "merely logic applied to observation upon form and quantity." Genuine axioms exist there, for Poe, but they break down when applied to other topics. One of these is morality. Two motives, says Dupin, when combined, are "not, necessarily, . . . equal to the sum of their values apart." Similarly, in chemistry, he says, "the axiom fails." Poe is clearly referring to the *synergistic effect*, which is the effect of a whole being greater than the sum of its parts. This effect, of course, is well known. In math, 2 + 2 = 4, never 5. But in medicine, for example, an ounce of this can make you feel a little sick, an ounce of that can make you feel a little sick, but when both are taken together they can make you feel more than twice as sick—in fact, you might end up on your sickbed as a nearly liquid mass of loathsome—of detestable putridity.

9. *P&T*, 1330.

10. *P&T*, 692.

By the time Poe came to write *Eureka* he went much further. Now he rejected *all* axioms as necessarily or self-evidently true, including those in geometry and mathematics and even in logic itself. The doctrine is espoused in his imaginary epistle that opens *Eureka*, allegedly written a thousand years ahead of his time (the rejection of axioms is also found later in *Eureka*). As Poe has the writer of the epistle put it in 2848, regarding "axioms, or self-evident truths," it is a "now well understood fact that *no* truths are *self*-evident."[11] The claim is repeated a little further along as "*for no such things as axioms ever existed or can possibly exist at all.*"[12]

What is especially interesting are the reasons given. One is based on a kind of relativity from history, that what was once considered an axiom was later no longer considered as such. The letter writer states, "for, even in their own day, many of their long-admired 'axioms' had been abandoned:—'*ex nihilo nihil fit*,' for example, and a 'thing cannot act where it is not,' and 'there cannot be antipodes,' and 'darkness cannot proceed from light.'" "How absurd, then," says the letter writer, "to persist in relying upon a basis, as immutable, whose mutability had become so repeatedly manifest!"

This argument is borrowed from a classic of Poe's time, John Stuart Mill's *A System of Logic* (1843), a work of philosophy of science in the inductive mode that Poe quotes from and had clearly studied, probably using the 1846 Harper & Brothers New York edition.[13] Mill, Poe notices through his letter writer, rejected the "Ability or inability to conceive" as "a criterion of axiomatic truth," and rightly, says the letter writer, for then "a truth to *David Hume* would very seldom be a truth to *Joe*." But the letter writer then uses this against Mill. "We will select," says the letter writer, "no axiom of an unquestionability so questionable as is to be found in Euclid. We will not talk, for example, about such propositions as that two straight lines cannot enclose a space, or that the whole is greater than any one of its parts. We will afford the logician [Mill] *every* advantage." Notice, at this point, that Poe's letter writer *does* think these axioms are not really such, that

11. *P&T*, 1263.

12. *P&T*, 1266.

13. *Eureka*, 121n13.

they and all supposed axioms *can* be false. Instead, the letter writer goes deeper into the well, to the rock bottom, and says, "We will come at once to a proposition which he [Mill] regards as the acme of the unquestionable—as the quintessence of axiomatic undeniability. Here it is:—'Contradictions cannot *both* be true—that is, cannot cöexist in nature.' Here Mr. Mill means, for instance,—and I give the most forcible instance conceivable—that a tree must be either a tree or *not* a tree—that it cannot be at the same time a tree *and* not a tree." And what, asks Poe's letter writer, is Mill's reason for believing this? "The sole answer is this:—'Because we find it *impossible to conceive* that a tree can be anything else than a tree or not a tree.'" Poe clearly, then, has charged Mill with self-contradiction, that his own rejection of the psychological criterion of axiomatic truth returns to bite his acceptance of the co-called law of contradiction, leaving the matter as Poe's letter writer puts it: "Thus all—absolutely *all* his argumentation is at sea without a rudder."[14]

This is profound, and on a number of levels. For one, it is an indictment of the lack of imagination in many thinkers, limiting in actuality science itself. As Poe's letter writer puts it, "That a tree can be both a tree and not a tree, is an idea which the angels, or the devils, *may* entertain, and which no doubt many an earthly Bedlamite, or Transcendentalist, *does*." Angels or devils, call them what you will, geniuses or madmen, the true giants of scientific discovery and understanding are not hampered in their imaginative abilities (or at least not always) but rise above the common man, and so are as angels. It is the *hoi polloi* that look upon them as deformed, failing all the while to see their own gross deformity. (We shall return to the significance of madmen and imagination in Chapter 7.)

Next, Poe's analysis begs comparison with Erwin Schrödinger's famous thought experiment *against* realism in quantum physics, known as "Schrödinger's cat." As Schrödinger argued back in 1935, if the realist interpretation of quantum physics is true, then if a cat in a sealed chamber would be in a quantum state, (due to a "diabolical device" involving radioactive decay,) it would be in a superposition of states, both dead and alive simultaneously,

14. *P&T*, 1267, 1268.

until someone looked into the chamber to observe it, which would "collapse the wave function," (a wave function is a mathematical description of a probability distribution,) thereby resulting in the cat being either dead or alive with a fifty-fifty chance of one or the other. As Schrödinger put it, "The ψ-function [wave function] of the entire system would express this by having in it the living and the dead cat (pardon the expression) mixed or smeared out in equal parts. It is typical of these cases that an indeterminacy originally restricted to the atomic domain becomes transformed into macroscopic indeterminacy, which can then be *resolved* by direct observation."[15]

In 1935 there was a growing realist interpretation of quantum physics, (that quantum physics is not only *predictive* but actually *descriptive* of reality,) and Schrödinger at the time was trying to oppose it. The collapse of the wave function "prevents us," he says, "from so naively accepting as valid a 'blurred model' for representing reality. In itself it would not embody anything unclear or contradictory." Since 1935, however, despite the best efforts of Einstein and Schrödinger, and some others, the realist interpretation of quantum physics continued to grow into a consensus, which became hardened following the Alain Aspect experiments conducted in 1982. According to the modern consensus, the quantum world, contrary to "common sense," really is as weird as experiments in quantum physics have routinely indicated, not only with the subatomic world as inherently and irreducibly probabilistic or statistical, as in radioactive decay and the quantum jumps (transitions) of electrons, but as inherently indeterministic as well, as in the position and momentum of an electron and the duality of Schrödinger's cat, such that in a quantum system in which an observation has been made, thus collapsing the wave

15. E. Schrödinger (1935). "Die gegenwärtige Situation in der Quantenmechanik." *Naturwissenschaften* 23, 807–812, 823–828, 844–849. John D. Trimmer, trans. (1980). "The Present Situation in Quantum Mechanics: A Translation of Schrödinger's 'Cat Paradox' Paper." *Proceedings of the American Philosophical Society* 124 (5), 328. For a good summary of the cat paradox, see John Gribbin (1999). *Q is for Quantum: Particle Physics From A to Z*. London: Phoenix Giant, 434.

function, "The state of the system is 'sharp' in these observables *only*" (italics mine).[16]

Now recall Poe's statement above, *against* the law of contradiction, according to which "a tree must be either a tree or *not* a tree—that it cannot be at the same time a tree *and* not a tree." The comparison is striking.

Quantum physics, of course, as anyone who has studied it knows,—with its double-slit experiment along with wave–particle duality, the phenomenon of entanglement, and its uncertainty principle combined with an underlying indeterminacy, which includes the principle of quantum transitions as events of pure chance combined with irreducible laws of probability,—*absolutely defies* the axioms of common sense, and yet it is the most powerfully corroborated theory in all of science. There has even been developed a *quantum logic* to go along with it, logic with not two truth values for propositions (classical logic) but three (quantum logic): "true," "false," and "indeterminate."[17] Now add Einstein's theory of relativity with its curved space and so much more. One may also go outside of physics, into biology, for example, with Darwin's rejection of what in his day was called the "axiom" of species immutability and independent creation, to see that Poe's rejection of axioms as necessarily or self-evidently true has been vindicated by modern science over and over again.

But Poe's point is more, which is that the pursuit of knowledge and understanding is hampered in most people because their imaginations are weak and underdeveloped, because they *cannot* imagine what they *ought*, for evidential reasons, to believe, and so they do not believe it. Darwin made a similar point in *On the Origin of Species* (1859), to which we shall turn in the following chapter, in defense of his argument for evolution by natural selection and against critics, which is that "His reason ought to conquer his imagination."[18] Poe, as we shall see in Chapters 6 and 7, takes the point a major step further, not only that the lack of imagination is

16. Fritz Rohrlich (1983). "Facing Quantum Mechanical Reality." *Science* 23, 1253.

17. Hans Reichenbach (1951). *The Rise of Scientific Philosophy.* Berkeley: University of California Press, 189, 227. Susan Haack (1974). *Deviant Logic: Some Philosophical Issues.* Cambridge: Cambridge University Press, ch. 8.

18. Charles Darwin. *On the Origin of Species.* London: John Murray, 188.

a hindrance to science, but also that great achievements in science were also great achievements in imagination.

But is it proper to say that Poe in *Eureka* really *anticipated* the rejection of axioms in twentieth-century physics? Arthur Quinn, in what is still overall the best biography of Poe, put the question to a physicist back in 1940 (or thereabouts). This is what Dr. Paul R. Heyl wrote to Quinn:

> Poe's discussion of axioms is interesting. In 1848 the axioms of geometry were still regarded as self-evident truths. Poe questions this [assumption]. Our whole concept of geometrical assumptions has changed, due to the work of Lobachevsky and of Bolyai, which resulted in non-Euclidean geometry. We now regard the axioms of geometry as mere space definitions. By leaving out the parallel axiom it is possible to develop a geometry quite different from that of Euclid, but just as self-consistent. True, it is inconceivable, but inconceivability no longer troubles the mathematicians. Nothing but inconsistency can do that.
>
> Lobachevsky's first publications antedate 1848, but were not translated from the Russian until much later. Bolyai (a Hungarian) published an abstruse article in Latin in 1831, but it attracted little attention until 1866 when it was translated into French. It is impossible that Poe should have heard of these men and their work.[19]

The nineteenth century saw further developments in non-Euclidean geometry. When János Bolyai first published his work on non-Euclidean geometry, the "Prince of Mathematicians" Carl Friedrich Gauss, a good friend of Bolyai's father, claimed that he had already developed the same ideas but refrained from publishing them. Non-Euclidean geometry was extended later in the century most notably by Gauss' student Georg Friedrich Bernhard Riemann, who famously developed in the 1850s a geometry of curved space, for the sheer mathematical beauty of it. But none of this was applied, or even imagined to apply, to physics. As Steven Weinberg, a Nobel laureate in Physics, puts it, "Gauss and

19. *Quinn*, 544.

Riemann and the other differential geometers of the nineteenth century had no idea that their work would ever have any application to physical theories of gravitation."[20] It was only when Einstein was working on his theory of gravitation that non-Euclidean geometry was combined with physics. As Weinberg notes, Einstein asked his friend Marcel Grossmann if there existed a mathematics to handle curved space in three or four dimensions, who replied that there was and he taught it to Einstein. Weinberg adds, "The mathematics was there for Einstein to make use of."

It therefore needs to be remarked, given that quantum physics and Einstein's theory of relativity both began in the early twentieth century, that in rejecting axiomatic truth in general and Euclidean axioms in particular, Poe in *Eureka* anticipated the application of the new geometry to the physical world. Einstein and the quantum physicists did the real work, of course, but Poe clearly had the requisite foundational vision many decades earlier.

II. Big Bang Cosmogony

The word *cosmogony* refers to the origin of the Universe, *cosmology* to the nature of the Universe. The Big Bang theory, which is not to be confused with the TV show, is both a cosmogony and a cosmology (one often finds the words used indiscriminately). The theory, which is really a *fact* now, (in the sense of *beyond a reasonable doubt*,) rests on the related fact that the Universe is expanding, like a balloon, that every galaxy is moving farther away from every other. The basic idea of the Big Bang is that this process traces back in time to a beginning, to when all the matter and energy of the Universe existed as an extremely small and exceedingly compressed *something*, which then suddenly exploded.[21] The current estimate is that the Big Bang occurred 13.7 (some say 13.8)

20. Steven Weinberg (1992). *Dreams of a Final Theory.* New York: Pantheon Books, 154.

21. The Big Bang theory is not to be confused with the creation of the Universe in the Judeo-Christian-Islamic tradition, which is a very different concept. To think otherwise is to be guilty of gross conceptual sloppiness, more specifically *presentism* (reading present ideas into the past without sufficient evidence).

billion years ago, but that is not an essential part of the theory per se. Neither is what caused the Big Bang, nor what was happening before the Big Bang (a consideration that physicists usually take to be meaningless). Perhaps in the future, if these questions get settled, (if they are not already,) then the so-called Big Bang theory will be more inclusive. Theories, after all, tend to evolve.

Of course, whether Poe "anticipated" Big Bang theory depends on how one defines "Big Bang theory" (and also on how one defines "anticipated.") Here is how Weinberg defines the theory in his classic on the Big Bang, *The First Three Minutes*: "The theory that the expansion of the universe began at a finite time in the past, in a state of enormous density and pressure."[22] I quote this definition to shut up those who would *define away* Poe's anticipation of Big Bang theory.

How did this theory *in modern science* begin? First of all, it is to be noted, as Weinberg points out, "in the 1950s, the study of the early universe was widely regarded as not the sort of thing to which a respectable scientist would devote his time."[23] But there was indeed some important activity going on in this regard. Einstein's formulation of what he called his *general theory of relativity*, his revolutionary theory of gravity, reached its definitive formulation in his paper "The Foundation of the General Theory of Relativity" (1916). A year later, in his "Cosmological Considerations on the General Theory of Relativity" (1917), which is the application of general relativity to the Universe as a whole, Einstein added a totally invented "cosmological constant" to his field equations of gravitation, representing an anti-gravitational force, solely for the purpose of keeping the Universe stable, in a theoretical sense. This was because his original equations entailed that the Universe must be either expanding or contracting. As Steven Hawking puts it, "so strong was the belief in a static universe that it persisted into the early twentieth century. Even Einstein. . . ."[24]

22. Steven Weinberg (1977). *The First Three Minutes: A Modern View of the Origin of the Universe.* New York: Basic Books, 159.

23. Weinberg (note 22), 4.

24. Steven Hawking (1988). *A Brief History of Time: From the Big Bang to Black Holes.* New York: Bantam Books, 40.

A few scientists, however, took Einstein's original equations seriously, notably the Russian mathematician Aleksandr Friedmann. But it was the Belgian mathematician, physicist, and Catholic priest, Georges Lemaître, who is most notable for our tale, for he launched the modern theory of the Big Bang. In 1927, in an obscure Belgian journal, he argued for an expanding Universe by altering Einstein's equations and combining them with the observation that galaxies are surrounded by a red glow, which he took to be a shift toward the red end in the spectrum of light, a kind of Doppler effect, meaning that the galaxies are moving away from us. In 1929, Edwin Hubble, using the new telescope at Mt. Wilson, confirmed that the red glow was indeed a red shift and hence that the Universe is expanding, he noticed that the red shift is greater in more distant galaxies, and he even measured the rate of expansion, (known as the Hubble constant,) a finding close to Lemaître's prediction in 1927, which caused Einstein to deeply regret his cosmological constant. But physicists were still resistant to the idea of the Universe having a beginning. Lemaître's 1927 paper was brought to the attention of the general scientific community in 1931 by his former professor, the astrophysicist Sir Arthur Eddington, who had Lemaître's 1927 paper translated and published in an English journal. The paper, it should be emphasized, does not contain a theory of the beginning, but ends only with "It remains to find the cause of the expansion of the universe."[25] Around the same time, Eddington made it known in the journal *Nature* that although he accepted "mainly through the work of Prof. Lemaître" that the Universe "is expanding rather rapidly," he expressed his feeling that "Philosophically, the notion of a beginning of the present order of Nature is repugnant to me."[26] Lemaître quickly replied with a letter published seven weeks later in *Nature*. There he

25. Abbé G. Lemaître (1931). "A Homogeneous Universe of Constant Mass and Increasing Radius accounting for the Radial Velocity of Extra-galactic Nebulæ." *Monthly Notices of the Royal Astronomical Society* March 1931 (91), 489.

26. Eddington surmised that "the original radius of space was 1200 million light years," that "this equilibrium was unstable," that "An expansion began, slowly at first," and that the present radius is "not less than ten times the original radius." Sir Arthur S. Eddington (1931). "The End of the World: from the Standpoint of Mathematical Physics." *Nature* March 21 (127), 447, 450.

wrote, "If the world has begun with a single quantum, the notions of space and time would altogether fail to have any meaning at the beginning; they would only begin to have a sensible meaning when the original quantum had been divided into a sufficient number of quanta. If this suggestion is correct, the beginning of the world happened a little before the beginning of space and time. I think that such a beginning of the world is far enough from the present order of Nature to be not at all repugnant."[27]

Although Lemaître would later use the term "primordial atom" for what in 1931 he referred to as "a unique atom," "the total mass of the universe," his more common phrases in his 1931 letter are in terms of a quantum,—"a single quantum," "the original quantum," "a unique quantum,"—so that he was clearly referring to what he believed was the original quantum of energy from which the Universe began, "all the energy of the universe," as he put it, "the whole story of the world" following not necessarily like "the disc of a phonograph" but possibly "written step by step." Lemaître's original model in 1927 was in terms of relativity theory, not quantum theory, but he did not present a theory of the origin of the Universe, only of its expansion, whereas by the time of his 1931 letter he realized the necessity of a quantum approach, and suggested as a possibility that the Universe began by the "unique atom" dividing "into smaller and smaller atoms by a kind of super-radio-active process." This approach was further developed by George Gamow and two of his students in the late 1940s, who added that the "primordial atom" must have been in an extremely hot state, not a cold state as Lemaître had supposed, which would allow for nucleosynthesis (the formation of atomic nuclei) as the Big Bang explosion cooled.[28]

---

27. G. Lemaître (1931). "The Beginning of the World from the Point of View of Quantum Theory." *Nature* May 9 (127), 706. For the (almost) complete story, see John Farrell (2005). *The Day Without Yesterday: Lemaître, Einstein, and the Birth of Modern Cosmology.* New York: Basic Books. As Mather and Boslough put it, "A letter Lemaître wrote to *Nature* magazine in 1931 was effectively the charter of what was to become the Big Bang theory." John C. Mather and John Boslough (2008). *The Very First Light.* New York: Basic Books, 39. Mather is an astrophysicist who was awarded the Nobel Prize in Physics in 2006.

28. Farrell (note 27), 136.

During the time of the weaning of the Big Bang theory in physics, a competitor theory emerged, known as the "Steady State" theory, begun in 1948 by Fred Hoyle and others—Hoyle actually coined the term "Big Bang" in a 1950 radio broadcast, as a term of disparagement.[29] According to this theory, the Universe has always been more or less in a steady state, with new matter and eventually new galaxies continually being created to fill the gaps between the galaxies as they move apart from one another, so that the Universe will always and has always been basically the same. This theory was destroyed, and the Big Bang model became the "standard model," with the discovery in 1965 of the cosmic microwave radiation background, the afterglow of the Big Bang, which was, as Weinberg puts it, "one of the most important scientific discoveries of the twentieth century."[30]

While Lemaître coined the term "primordial atom" for the initial state of the Big Bang, the initial state in the standard model is typically described mathematically as a "singularity." Steven Hawking, for example, writes of "the big bang singularity," which he says was "when the universe was infinitely small and infinitely dense," when "the universe is thought to have had zero size, and so to have been infinitely hot."[31] Similarly, Weinberg refers to "a state of infinite density *and* infinite temperature."[32] This language follows from the mathematics of relativity theory, which entails that if the Universe is expanding then it must have begun with a singularity, a point in space-time where, because of the infinities already mentioned, the laws of nature do not apply. From the perspective of quantum theory, however, the initial state leading to the Big Bang, as with Lemaître's "primordial atom," is defined physically, without infinities. As John Gribbin puts it, "the entire physical Universe expanded out from a seed much smaller than a proton."[33] This seed, by the way, was remarkably homogeneous. As the astrophysicist Peter Shaver recently put it in his popular account, all the Universe was "in an extremely compressed state,"

29. Farrell (note 27), 231.

30. Weinberg (note 22), 122.

31. Hawking (note 24), 46, 117.

32. Weinberg (note 22), 140.

33. Gribbin (note 15), 379.

a state "extremely dense, hot and uniform," consisting of "only a few variables and virtually no complexity." Nucleosynthesis "occurred in the first minutes of the universe, and produced the light elements."[34]

When it comes to the origin of our universe, Poe provided a theory surprisingly similar in many respects to modern Big Bang theory. What *led* him there is another matter. It was not, of course, Einstein's equations or quantum theory or evidence of a Doppler shift in receding galaxies. Poe himself says it was "*Intuition*," which he defines for the second time in *Eureka* as "*the conviction arising from those inductions or deductions of which the processes are so shadowy as to escape our consciousness, elude our reason, or defy our capacity of expression*."[35] In fact Poe calls his theory of the origin of our universe "the sole absolute *assumption* of my Discourse." It is that God created, purely by "Volition," what Poe calls "the primordial Particle," a unit of matter "in its utmost conceivable state of—what?—of *Simplicity*," matter "in its absolute extreme of *Simplicity*."[36] Poe in this context uses the term "Imparticularity," but it apparently has basically the same meaning as "unparticled matter," Poe's term in "Mesmeric Revelation" (1844) for matter "without particles—indivisible—*one*," in other words, "infinitely rarified matter." There is no Big Bang theory of the Universe in "Mesmeric Revelation," but there is the idea that unparticled matter "is God," and as such "not only permeates all things but impels all things—and thus *is* all things within itself," and also is what constitutes "what men called mind."[37] Later, in *Eureka*, Poe leaps further, claiming not only that God created "the primordial Particle," but that God *is* the primordial Particle: "the first and most sublime of Acts—that act by which a God, self-existing and alone existing, became all things at once, through dint

34. Peter Shaver (2011). *Cosmic Heritage: Evolution from the Big Bang to Conscious Life*. New York: Springer, 19, 21.

35. *P&T*, 1276–1277.

36. *P&T*, 1277. Poe later in *Eureka* uses a variety of phrases for the same meaning, specifically, "the absolute, irrelative, unconditional Particle," the "Particle Proper," and the "absolute Particle Proper." *P&T*, 1295, 1303, 1304.

37. *P&T*, 720.

of his volition, while all things were thus constituted a portion of God."[38]

Following "the first and most sublime of Acts" came the second, the Big Bang proper, what Poe calls "the ultimate purpose for which we are to suppose the Particle created," which is "the constitution of the Universe from it, the Particle." This, he says, was "effected by *forcing* the originally and therefore normally *One* into the abnormal condition of *Many*." It was what he calls a "diffusion from Unity." "From the one Particle," he says, "as a centre, let us suppose to be radiated spherically—in all directions—to immeasurable but still to definite distances in the previously vacant Space—a certain inexpressibly great yet limited number of unimaginably yet not infinitely minute atoms."[39] This too is God. As Poe puts it near the end of *Eureka*, "the material *and* spiritual God—*now* exists solely in the diffused Matter and Spirit of the Universe; and . . . the regathering of this diffused Matter and Spirit will be but the re-constitution of the *purely Spiritual* and Individual God."[40]

Poe's concept of "radiation" here, by the way, is not at all like the radiation of light, and he is very clear about this. With light, he says, "there is a *continuous* outpouring of *ray-streams*, and *with a force which we have no right to suppose varies at all*." What he has in mind, instead, is "that of a *determinate* radiation—one finally *dis*continued."[41] He later provides some reasons for rejecting the idea of continuous radiation, one of which is that the idea "if not positively disproved, is at least not in any respect warranted by telescopic observation of the stars."[42] This answer leads to another, which is more evidently an anticipation of his solution to Olbers' paradox, which we shall examine in §V below. If one supposes, he says, "Matter to have been diffused by a continuous or infinitely continued force," then one is led to the supposition of an infinite Universe of matter, such that "there is an *infinity* of atoms on all

38. *P&T*, 1314. That Poe uses the phrase "a God" was no accident and will make sense once we turn to his multiverse theory.

39. *P&T*, 1277, 1278.

40. *P&T*, 1357.

41. *P&T*, 1292.

42. *P&T*, 1304.

sides of the atom proposing to move." Because of the gravitational force of all atoms in the Universe affecting all atoms in the Universe, "the atom in question must remain stationary forever." In that case, says Poe, "there could have been no aggregation of Matter—no stars—no worlds—nothing but a perpetually atomic and inconsequential Universe."[43]

Poe goes on in *Eureka* to discuss the nature of particled matter and laws of nature, and he applies Laplace's nebular hypothesis to explain the formation of stars, planets, and moons. Whether Poe's *reductio ad absurdum* of the infinite Universe of matter assumption works, the comparison of Poe's cosmogony with twentieth-century Big Bang theory—minus, of course, the theology—is nothing short of remarkable. Even more remarkable is Poe's anticipation of the idea that the expanding Universe would reach a point of maximum expansion and then enter a contraction phase, ending with the Universe as a singularity once more. In twentieth-century physics, following the Big Bang consensus, the competing hypotheses were between an expanding Universe ending in heat death, maximum entropy, and an expanding Universe with a contraction phase following the current expansion phase, possibly repeating over and over again, an "endless cycle of expansion and contraction," every big bang eventually followed by a "big crunch," for the simple reason, as Weinberg puts it, that "Some cosmologists are philosophically attracted to the oscillating model, especially because, like the steady-state model, it nicely avoids the problem of Genesis."[44] The philosophical sentiments turned scientific over growing indirect evidence for what they called "dark matter,"

43. *P&T*, 1305. It is to be noted that the same point, specifically about infinite matter in an infinite space entailing that each particle of matter must rest "*in equilibrio*," was made by the English classical scholar and theologian Richard Bentley in his letters to Isaac Newton. In his four letters in reply to Bentley, dated between December 1692 and February 1693, Newton rejected the conclusion by denying the concept of a "mathematical center," by denying the hidden assumption "that all infinities are equal," by denying that gravity is "essential and inherent to matter," and by invoking the intervention of "an intelligent Agent." H.S. Thayer, ed. (1953). *Newton's Philosophy of Nature: Selections From His Writings.* New York: Hafner Press, 46–58. For a modern discussion of the problem, see Stephen Hawking (1993). *Black Holes and Baby Universes and Other Essays.* New York: Bantam Books, 87–88.

44. Weinberg (note 22), 153, 154; Hawking (note 24), 173.

non-luminous matter in space that was hypothesized to explain, by virtue of its gravitational force, the way the Universe is expanding. The issue then became one of *quantity*. As the physicist John Gribbin puts it in his entry on dark matter, "many cosmologists believe that the best way to explain how the Universe was born . . . requires that there be enough matter in the Universe to make it gravitationally 'closed,' so that the present expansion will eventually be brought to a halt."[45] The debate got only more complicated with the discovery in 1998, using NASA's Hubble Space Telescope, that the expansion of the Universe is accelerating, (so much for the Hubble constant?,) such that now the debate is over what they call "dark energy," energy that has "a repulsive force across the Universe."[46]

Unlike modern physics, Poe thought the Universe is in a contraction phase, that it is currently returning to the original Unity. He says, for example, "we have reached the conclusion that, on the hypothesis that matter was originally radiated from a centre and is now returning to it, the concentralization, in the return, proceeds *exactly as we know the force of gravitation to proceed*."[47] One can focus on the differences between Poe's theory and modern Big Bang theory, including what Poe says here about the force of concentralization, along with his conjecture that the Big Bang was

45. Gribbin (note 15), 119.

46. Shaver (note 34), 40. The implications of the science for Poe's God, of course, are dire, as it means Poe's God is gradually losing its individuality and will eventually end in heat death. What we have, then, is the ultimate murder mystery in advance. Is the death of God by misadventure, (a miscalculation on God's part, too much dark energy, not enough dark matter,) or does God have a dark side (literally) that is responsible for the slow and agonizing killing of itself? In either case, Nietzsche's madman was not only metaphorically but literally wrong. Thinking of the future as the present, it was not we who killed God. All of this is assuming, of course, that the science—the cosmic CSI in this case—is in. I for one do not think it is, and here is why. Given $E = mc^2$, for all anyone knows it is *possible* that some of the dark energy in the Universe will turn into dark matter, enough even to eventually decrease the rate of the expansion of the Universe, possibly even to the point of halting the expansion altogether and initiating a contraction phase. This is a mere hypothesis, of course, but it does provide a glimmer of Hope in the context of Poe's theology.

47. *P&T*, 1292.

followed by a finite series of progressively smaller big bangs so as to fill the finite Universe of space, what he calls "*a limited sphere of Space*," so as to produce a "generally equable distribution,"[48] from which, *in the contraction phase only*, stars and planets would gradually form from coalescing nebulae due to gravity. One can also quibble about whether Poe's rejection of axioms is consistent with "the sole absolute *assumption* of my Discourse," which he defends as "at least preferable, as a logical basis, to any axiom ever propounded, or to all axioms combined."[49] His topic, after all, is the Universe, which he is trying to make sense of as a whole.

At any rate, whatever the differences between Poe's theory of the origin of the Universe and the modern theory of the Big Bang, those differences should not be used to obscure the fact that the similarities between the two theories are highly remarkable, especially given that Poe was writing in 1848. As the astronomer Alberto Cappi puts it in his critical assessment of "Poe's physical cosmology" from a modern scientific point of view, "the astronomical knowledge of the first half of the 19th century is inserted by Poe in the framework of an evolving universe. This revolutionary and extraordinary synthesis is what gives *Eureka* a modern flavour."[50] He also claims that "the collapsing *Newtonian* universe described by Poe represents a self-consistent prerelativistic model of the Universe," that "Poe seems to have had the mind of a cosmologist," (an example, I might add, of *it takes one to know one*,) that Poe was "the first to conceive a Big Bang," that Poe's early awareness that for a consistent Newtonian cosmology "it was necessary to accept a non-static, finite universe of matter" is "one of the most important aspects of *Eureka*," that "*Eureka* contains . . . the first modern application of the anthropic cosmological principle," (which we have not yet got to,) and that Poe's multiverse theory, (which we also have not yet got to,) which is a "similar concept of parallel universes . . . presently found in quantum mechanics or inflationary cosmology," is "the most imaginative."[51]

48. *P&T*, 1295–1296, 1301, 1304.

49. *P&T*, 1303.

50. Alberto Cappi (1994). "Edgar Allan Poe's Physical Cosmology." *Quarterly Journal of the Royal Astronomical Society* 35, 185.

51. Cappi (note 50), 187, 189, 191, 185, 179, 188.

What is yet further remarkable is Poe's criticism of scientists in his time for failing to entertain the idea of an expanding or contracting Universe, which is similar to Lemaître's criticism of Einstein and his followers, which is that *they should have reasoned from what they had to the Big Bang cosmogony.* Poe finds J.P. Nichol, for a start, on stellar nebulae, stating that "all around them, on every side, there are volumes of stars, *stretching out apparently as if they were rushing towards a great central mass in consequence of the action of some great power.*" He finds Alexander von Humboldt, "whose generalizing powers have never, perhaps, been equalled," admitting that objectively rather than perspectivally "we find *many groups of them [stars] moving in opposite directions,*" and yet Humboldt states that "the data as yet in hand render it not necessary, at least, to conceive that the systems composing the Milky Way, or the clusters, generally, composing the Universe, are revolving about any particular centre unknown." Instead Humboldt says that "It is but Man's longing for a fundamental first Cause, that impels both his intellect and his fancy to the adoption of such an hypothesis." Finally, Poe finds John Hershel stating that "without a rotary motion and a centrifugal force, it is hardly possible not to regard them [stellar nebulae] as in a state of *progressive collapse.*"[52] So then why did not Herschel entertain the possibility that each stellar nebula, and indeed the Universe as a whole, is collapsing toward a center (the latter preceded, of course, by an expansion)? Poe's answer, in the case of Herschel specifically, (but he implies the same for the others,) is that he failed "Simply on account of a prejudice;—merely because the supposition is at war with a preconceived and utterly baseless notion—that of the endlessness—that of the eternal stability of the Universe."[53]

That Poe did not share that prejudice, but went on to anticipate the Big Bang theory of modern astrophysics, was not because of a willingness to be incautious where real scientists were cautious. Poe had good reasons that fit into the picture, notably his solution to Olbers' paradox, which his fellow theorists did not solve, almost certainly again because of what Poe calls their

52. *P&T*, 1346, 1346–1347, 1345.

53. *P&T*, 1348.

preconceived prejudice for an essentially infinite Universe of stars. (We shall turn to Olbers' paradox when it arises in the order of Poe's theorizing in *Eureka*.)

Poe even entertained the possibility of an oscillating Universe, though not for philosophical but for aesthetic reasons. Near the end of *Eureka* he says:

> But are we here to pause? Not so. On the Universal agglomeration and dissolution, we can readily conceive that a new and perhaps totally different series of conditions may ensue—another creation and radiation, returning into itself—another action and reaction of the Divine Will. Guiding our imaginations by that omniprevalent law of laws, the law of periodicity, are we not, indeed, more than justified in entertaining a belief—let us say, rather, in indulging a hope—that the processes we have here ventured to contemplate will be renewed forever, and forever, and forever; a novel Universe swelling into existence, and then subsiding into nothingness, at every throb of the Heart Divine?[54]

It will be noticed that the idea expressed here by Poe is reminiscent of what he elsewhere says is the essence of poetry, a recurring theme in his literary theory examined in Chapter 2, namely, "the *Rhythmical Creation of Beauty*" using words, which we have seen goes back at least to his writings of 1842, and which gives us a strong reason to believe that Poe called *Eureka* a "Prose Poem" mainly because of its subject, the Universe, and the way he conceptualized it. We have also seen in the second-last section of Chapter 1 that in the period of 1844–1845 Poe was expressing the idea of God as "unparticled matter, permeating & impelling, all things" and that "The Universe is a Plot of God." But there was yet no clear inkling of a Big Bang, let alone of a Big Crunch. Those ideas properly began, apparently, with the writing of *Eureka*.

At this point, in order to properly contextualize the matter, let us indulge in an exercise of imagination. Try to imagine how alone Poe must have felt, intellectually and spiritually, for the last sixteen

54. *P&T*, 1356.

months of his life, from the publication of *Eureka* in July 1848, his self-avowed magnum opus, to his tragic death in October 1849. To fully get the point, imagine the spirit of Poe standing behind you and breathing over your shoulder, as he reads with you the following passage written by the physics phenom Stephen Hawking in his second bestseller: "Despite these difficulties with the idea of a static and unchanging universe, no one in the seventeenth, eighteenth, nineteenth, or early twentieth century suggested that the universe might be evolving with time."[55] No one except Poe, of course. And he didn't get his theory published—no mere *suggestion*—in an obscure Belgian journal, but in America by a major publisher, in full view, only to be ignored and forgotten. Hawking continues in the passage to say Newton cannot be blamed for failing to see the light, but Einstein "should have known better," given his formulation of relativity in 1915. I should say, instead, that Einstein and his generation of physicists should have known better along with the two generations of physicists preceding them, but for the negative image of Poe created mainly by his literary executor, a crime against the legacy of Poe which eventually got its poetic justice—the split and splatter of Ludwig Griswold's chest by the razor's edge of the huge pendulum in the 2012 movie *The Raven*.

III. Fine-Tuning of the Laws of Nature

What is the relation of the fundamental laws of nature to one another? Are they independent, or are they locked together in some way? At first glance they would seem *logically* independent. Newton's inverse square law of gravity, for example, would seem to be independent of the speed of light in a vacuum. Conceivably one could alter one of them and not the other, without contradiction. There is thus no *logical* reason for why they should be connected. But maybe they are *physically* connected, such that altering one would automatically alter the other. And then maybe they are *teleogically* connected, in that slight alterations would fail to produce certain ends, such as us. Of course none of this is something

55. Hawking (note 43), 88.

that scientists can experiment with—at least not yet (as far as I can tell). So we are back to our original problem situation.

Up until the second half of the twentieth century, the fundamental laws of nature were conceptualized by scientists as independent, for there appeared no good reason to think otherwise, as independence seemed the simpler, and hence the default, hypothesis. As scientists came to learn more about the nature of and the relations between what came to be known as the "fundamental constants" of the Universe, however, a number of scientists came to believe that the fundamental laws of nature are not independent but interrelated, "fine-tuned" as some of them liked to put it. Today it is known as the *anthropic principle*, or the *fine-tuning argument for God's existence*, which is that the fundamental laws of the Universe, the cosmic constants, seem fine-tuned collectively so as to produce stars, planets, life, and ultimately intelligent life like us. For example, if the force of gravity would be roughly twice of what it is, then stars would shine roughly 100 times brighter than they do and burn out much sooner, such that a star like our Sun would burn out after roughly 100 million years, clearly not enough time for life on Earth to evolve beyond the stage of bacteria. Gravity is but one variable. Consider another. The synthesis of carbon atoms in stars, which is absolutely essential for life as we know it, requires an energy interplay between the strong nuclear force, the force that holds protons and neutrons together in the nucleus of an atom, (a force much stronger than gravity,) and the electromagnetic force, the force of attraction between electrons and atomic nuclei. If the strength of the strong nuclear force would be greater or less than what it is by as little as 1 percent, then carbon atoms would never be able to form, and none of us would be here.[56] And if the strong nuclear force would be different from what it is by a few percent, then not even stars would be able to form.[57] Depending on who one reads, there are as few as twenty of these fundamental constants and as many as fifty.

56. Paul Davies (2006). *The Goldilocks Enigma: Why Is the Universe Just Right for Life?* New York: Houghton Mifflin, 138, 143.

57. Paul Davies (1983). *God and the New Physics.* New York: Touchstone, 187–188.

The term "anthropic principle" was coined by the astrophysicist Brandon Carter in 1973, and since then the literature on this topic has mushroomed, with some, such as the physicists Paul Davies, John Barrow, Frank Tipler, and the geneticist Francis Collins, accepting the existence of God as the best explanation,[58] and others, such as the physicists Steven Weinberg, Lee Smolin, Victor Stenger, and the biologist Richard Dawkins, rejecting it,[59] to name but a few from each side of the debate.

Interestingly, Poe's view of the Universe in *Eureka* is heavily law-based, and he does indeed seem to have anticipated the modern fine-tuning argument. He rejects what he calls "the fashion with astronomical treatises," which is to explain certain phenomena by appeal to "the finger of the Deity itself" when explanations based on laws of nature fail. This was the case, he says, with the synchronous orbit of the Moon, never showing its dark side. It was thought necessary to suppose that God "found it necessary to interpose, specially, among his general laws, a set of subsidiary regulations, for the purpose of forever concealing from mortal eyes the glories, or perhaps the horrors, of the other side of the Moon." And then the correct answer was figured out scientifically by astronomers. The same is going to be the case, says Poe, "of two forces so seemingly independent"—what he calls "the gravitating and tangential"—in the case of the position of planets to suns. He says the planets have "an impetus mathematically adapted to the masses, or attractive capacities, of the suns themselves." This, he says, is an "absolutely accurate adaptation . . . of two forces so seemingly independent." Poe then generalizes to "the conclusion that *each law of Nature is dependent at all points upon other laws*, and that all are but consequences of one primary exercise of Divine Volition."[60]

---

58. Davies (notes 56 and 57). John D. Barrow and Frank J. Tipler (1986). *The Anthropic Cosmological Principle.* Oxford: Oxford University Press. Francis S. Collins (2006). *The Language of God: A Scientist Presents Evidence for Belief.* New York: Free Press.

59. Weinberg (note 21). Lee Smolin (1997). *The Life of the Cosmos.* Oxford: Oxford University Press. Victor Stenger (2003). *Has Science Found God? The Latest Results in the Search for Purpose in the Universe.* Amherst, NY: Prometheus Books. Richard Dawkins (2006). *The God Delusion.* Boston: Houghton Mifflin.

60. *P&T*, 1312, 1313, 1314.

Poe's reasons for arriving at this conclusion, of course, are suspect, having to do with "the very idea of God, omnipotent, omniscient." But, like them or not, in claiming "*the infallibility* of his laws" and that there is between them an "absolutely accurate adaptation,"—which keep in mind is in accordance with the purpose of the Universe in Poe's view, examined in Chapter 3,—it is safe to say that Poe anticipated a major development in twentieth-century cosmology, or at least a genuine debate among scientists, even if the majority of modern physicists do not accept the God hypothesis (and they do not).

IV. Non-Existence of Laws of Nature Before the Big Bang

Did the fundamental laws of nature exist before the explosion of the Big Bang? Or did they come into existence subsequent to it? In other words, did the singularity, or original quantum, from which our universe sprang, already contain the fundamental laws, or did they emerge later, following the explosion? Here's how Stephen Hawking puts it: "Hubble's observations suggested that there was a time, called the big bang, when the universe was infinitesimally small and infinitely dense. Under such conditions all the laws of science, and therefore all ability to predict the future, would break down. If there were events earlier than this time, then they could not affect what happens at the present time."[61] I take that to be a *no*, although I would not bet my life on it. A little more clear on the matter is Andrei Linde, one of the founders of multiverse and inflationary theory, to which we shall turn in §VI below. According to Linde, in reference to standard Big Bang theory on the singularity preceding the Big Bang, "One usually assumes that the current laws of physics did not apply then. They took hold only after the density of the universe dropped below the so-called Planck density, which equals about 10^{94} grams per cubic centimetre."[62]

Poe on laws of nature in *Eureka* is, again, nothing short of fascinating. Let us begin with the law of gravity, which Poe fully

61. Hawking (note 24), 8–9.

62. Andrei Linde (1998). "The Self-Reproducing Inflationary Universe." *Scientific American* 9 (1), 99.

accepts as a "fact," which "Newton deduced . . . from the laws of Kepler"[63] and which Poe takes to have been empirically proven beyond a reasonable doubt by "the Maskelyne experiments."[64] We have seen that in the beginning, according to Poe's cosmogony, there was only God, who by an act of volition turned into "the primordial Particle," and then there was the Big Bang. Gravity, says Poe, in a number of places and in a number of ways, and most clearly and poetically in the following, is to be understood "as the rëaction of an act—the expression of a desire on the part of Matter, while existing in a state of diffusion, to return into the Unity whence it was diffused."[65] This, he says a little earlier, is "the *modus operandi* of the Law of Gravity."[66]

The question then arises, "When did gravity begin?" Poe's answer is that gravity, as an agent or force, accurately described by Newton's inverse square law, did not begin until the Universe began to contract following its maximum point of expansion. He says, for example, that "There could have been no rëaction had the act [the 'exercise of the Divine Volition'] been infinitely continued. So long as the act *lasted*, no rëaction, of course, could commence; in other words, no *gravitation* could take place." Poe then says, "But gravitation *has* taken place; therefore the act of Creation has ceased: and gravitation has long ago taken place; therefore the act of Creation has long ago ceased."[67] Later in *Eureka*, near the end, Poe says, "Going boldly beyond the vulgar thought, we have to conceive, metaphysically, that the gravitating principle appertains to Matter *temporarily*—only while diffused—only while existing as Many instead of as One—appertains to it by virtue of its state of radiation alone—appertains, in a word, altogether to its *condition*, and not in the slightest degree to *itself*."[68] In order to save Poe from contradiction, it would seem that "particulate matter" only came to exist following the point of the maximum expansion of the Universe, when God's act of creative volition ceased.

63. *P&T*, 1269–1270.

64. *P&T*, 1319n.

65. *P&T*, 1296.

66. *P&T*, 1290.

67. *P&T*, 1320.

68. *P&T*, 1348.

The other fundamental force, agent, or law of the Universe Poe calls "Repulsion" or "*Electricity*, with its involute phænomena, heat, light and magnetism."[69] Earlier in *Eureka* Poe makes it clear that "far less shall we be liable to err in attributing to this strictly spiritual principle the more important phænomena of vitality, consciousness, and *Thought*."[70] But let us not bother here with *that*. The force of repulsion would be necessary, says Poe, early in *Eureka*, so as to "at the same time allow the approach, and forbid the junction, of the atoms; suffering them infinitely to approximate, while denying them positive contact; in a word, having the power—*up to a certain epoch*—of preventing their *coalition*, but no ability to interfere with their *coalescence* in any respect *or degree*." This force would come into effect "on withdrawal of the diffusive Volition."[71] Clearly for Poe the expansion phase of the Universe, the diffusion or radiation of unparticled matter into a homogeneous sphere, was not driven by the force of repulsion. Instead it was driven by God's volition, and Poe apparently thought of this phase as occurring instantaneously. In fact to make the point explicit he added, in pencil as a revision to his copy of *Eureka*, "Here describe the whole process as one instantaneous flash."[72] The point of maximal expansion was then followed by an ever so gradual contraction phase, allowing for the formation of stars and planets and the evolution of life. As Poe puts it, "Our solar system . . . is now to be considered as *an example* of the innumerable agglomerations which proceeded to take place throughout the Universe of atoms on withdrawal of the Divine Volition."[73] For this to happen, however, the force of attraction would have to be stronger than the force of repulsion, otherwise the Universe would expand rather than contract, and keep expanding forever. And indeed Poe states that gravity "*must be the strongest of forces*," which he says is "an idea reached *à priori* and abundantly confirmed by induction."[74]

69. *P&T*, 1315, 1314.

70. *P&T*, 1282.

71. *P&T*, 1280.

72. *P&T*, 1400. Poe's note corresponds to the end of the last paragraph in *P&T*, 1295, which is paragraph 96.

73. *P&T*, 1323.

74. *P&T*, 1298.

Early in *Eureka*, commenting on both of these forces, Poe says "*No other principles exist.*"[75] A little later he states that "The Thought of God is to be understood as originating the Diffusion. . . . *Then* comes Rëaction, and through Rëaction, 'Principle,' as we employ the word. It will be advisable, however, to limit the application of this word to the two *immediate* results of the discontinuation of the Divine Volition—that is, to the two agents, *Attraction* and *Repulsion*." From these two, then, you get higher-level laws: "Every other natural Agent depends, either more or less immediately, on these two, and therefore would be more conveniently designated as *sub*-principles."[76]

Even though Poe's presentation of the above matter is somewhat *diffuse*, it should be abundantly obvious that according to Poe none of the laws of the Universe, including its two most fundamental laws, existed as features of "the primordial Particle," the "singularity" or "original quantum" in the language of modern physics. The laws came into existence later, only after the start of the Big Bang. This is truly remarkable.

V. Olbers' Paradox

In 1823 the German astronomer Heinrich Wilhelm Matthäus Olbers had a paper published entitled "On the Transparency of Space," in which he wrote the following: "If there really are suns throughout the whole of infinite space, and if they are placed at equal distances from one another, or grouped into systems like that of the Milky Way, their number must be infinite and the whole vault of heaven must appear as bright as the Sun." He also put it in a simpler form: "if the fixed stars stretch away to unlimited distance, the entire sky must be ablaze with light."[77] Olbers was not the first to express the paradox. In fact in his paper he examines and then rejects the solution given by Edmund Halley, of Halley's

75. *P&T*, 1282.

76. *P&T*, 1301.

77. Olbers' paper is reproduced in translation in Edward Harrison (1987). *Darkness at Night: A Riddle of the Universe.* Cambridge, MA: Harvard University Press, 223–226.

Comet fame, and Newton and others were well aware of the problem.[78] But the name of the paradox became associated with Olbers due to the contingencies of history along with (or because of) two of his contributions to the problem. As the physicist and astronomer Edward Harrison points out in his book on the topic, in Olbers' paper "we encounter for the first time the realization that stars need not be uniformly distributed and may be grouped into milky way systems that nowadays we call galaxies." The second is that "for the first time we encounter the powerful line-of-sight argument."[79] In other words, in an infinitely large universe with infinitely many stars the paradox remains whether the stars are grouped into galaxies. In either case, in whatever direction we look in the night sky our line of sight should end not only with a star or galaxy, but the light from that star or galaxy should be multiplied by light from other stars and galaxies. In the case of a single star, the light from the star is dimmer the further we are from it not because light rays get weaker the further they are from their source but because they are less packed together. In an infinite universe with infinite stars, however, in whichever direction we look in the night sky our eyes should be blasted by light rays. This is because the light coming from the star or galaxy directly in our line of sight is combined with all the light coming indirectly from other stars and galaxies in our sky, with a total effect similar to looking directly at the Sun. But of course this is not at all what we experience.

78. Halley is especially interesting here. In his paper "Of the Infinity of the Sphere of Fix'd Stars," published in 1721 and reproduced in Harrison (note 77), 218–219, in which he deals with the paradox in question, he also argues that in a universe with a finite number of stars "the whole would be surrounded on all sides with an infinite *inane* [emptiness], and the superficial stars would gravitate towards those near the center, and with an accelerated motion run into them, and in process of time coalesce and unite with them into one. . . . But if the whole be Infinite, all the parts of it would be nearly *in aequilibrio*, and consequently each fixt Star, being drawn by contrary Powers, would keep its place; or move, till such time, as, from such an *aeqiulibrium*, it found its resting place." This argument by Halley, or arguments like it, quite possibly played a role, along with arguments by others examined above in §II, in why Poe settled upon a contracting universe.

79. Harrison (note 77), 94.

Olbers' solution was to retain an infinite universe with infinite stars but to reject the assumption that empty space is perfectly transparent. Instead he argued for a degree of absorption of starlight by interstellar matter, such that a "small amount of absorption in an endless star-filled universe is more than sufficient to create the conditions observed on Earth." Olbers, it turns out, was wrong. As Hawking points out, "if that happened [absorption by interstellar matter in an infinite universe with infinite stars] the intervening matter would eventually heat up until it glowed as brightly as the stars."[80]

In *Eureka*, Poe states that "Were the succession of stars endless, then the background of the sky would present us an uniform luminosity, like that displayed by the [Milky Way] Galaxy—since there could be absolutely no point, in all that background, at which would not exist a star. The only mode, therefore, in which, under such a state of affairs, we could comprehend the voids which our telescopes find in innumerable directions, would be by supposing the distance of the invisible background so immense that no ray from it has yet been able to reach us at all."[81] According to Harrison, "The first clear and correct solution to the riddle of darkness, though only qualitatively expressed, came from Edgar Allan Poe," that "only twenty-five years after Olbers wrote his paper on the riddle of darkness, it [*Eureka*] contains the first anticipation of a formally correct solution."[82] This view is seconded by Alberto Cappi, what he calls Poe's "finite age solution."[83] Poe's solution, as Harrison puts it, is that "the light of the 'golden walls' has not yet reached us. When we look far out into space we look far back

80. Hawking (note 24), 6.

81. *P&T*, 1328. Poe's formulation of the paradox is noticeably different from Olbers', and it is unlikely that Poe ever read Olbers' paper. As the Levines point out, the paradox is also discussed in Lecture 1 of the published version of J.P. Nichol's lecture series on the Universe, (the lecture series coincident with Poe's lecture on the Universe,) published in 1848 as *Views of Astronomy. Eureka*, 150n181. Poe apparently had access to that book before he sent *Eureka* to press. *Eureka*, 146n163. Poe also had access to the transcription of the lectures published in the *New York Tribune. Eureka*, 162n229. In any case, Poe's source for his understanding of Olbers' paradox is yet to be determined.

82. Harrison (note 77), 146, 148.

83. Cappi (note 50), 187.

to a time before the birth of the stars." In other words, the stars are "not old enough."[84]

At this point I want to suggest that Poe's solution to Olbers' paradox is the key to understanding the science of Poe's cosmology in *Eureka*. Solve Olbers' paradox by rejecting the assumption of an infinite universe of stars, then you get a finite universe of stars. Accept a finite universe of stars, then the next step is to accept gravitational collapse. Accept gravitational collapse, and the next step is to accept a massive explosion to get the matter out there necessary to make the stars in the first place. Accept a massive explosion, then it's natural to suppose a "primordial Particle" of some sort, one not made of matter at all but that is convertible into matter. And so on. Now add Poe's literary theory and his theological concerns and you have OMG. All of a sudden all the nine topics in this chapter, each of them a speculation forged by Poe in his imaginarium, seem to come together and make sense as a unified whole. But we still have four more of Poe's anticipations of twentieth-century science to go.

VI. Multiverse Theory

In 1979 the physicist Alan Guth developed a version of Big Bang theory that has come to be known as "inflationary cosmology." As Peter Shaver puts it, "He suggested that, at a very early stage, about 10^{-36} seconds after the Big Bang, the universe may have undergone an extremely fast and phenomenal expansion due to a 'phase transition' in response to the rapidly decreasing temperature (a similar phase transition is that from steam to water, or water to ice). The dimensions of the universe may have increased by more than an incredible factor of at least 10^{30} in just 10^{-33} seconds. That's an increase of over a million trillion trillion in less than a billionth of a trillionth of a trillionth of a second."[85] (Sounds to me much like Poe's "one instantaneous flash.") The theory solved a number of problems that had been plaguing Big Bang cosmology, such as the need to explain the homogeneity of the Universe (although it

84. Harrison (note 77), 150, 209.

85. Shaver (note 34), 57–58.

is globular in the sense of stars and galaxies, it is nearly uniform in its distribution of dark matter and in the temperature of the cosmic microwave radiation background). Moreover, the theory made predictions that have since been tested and confirmed, such that the theory is now widely accepted. But most interestingly, the theory raises the possibility that our own universe, with its laws, might be just one little universe produced from "a random fluctuation in a pre-existing 'quantum ocean,'" an ocean that repeatedly produces innumerable universes each from a random fluctuation, each universe starting with its own big bang, each with its own constants and laws, such that, as Shaver puts it, "Our entire universe may be just one speck in a huge sea of frenzied quantum foam of false vacuum, with the seeds of endless universes continually being created randomly by quantum fluctuations." Rather than inflationary theory being part of Big Bang theory, then, the situation by inflationary cosmologists is thought to be the reverse. In short, says Shaver, "Welcome to the Multiverse!"[86]

Almost at the beginning of *Eureka*, Poe distinguishes between two meanings of "Universe" that he intends to employ throughout the work. One is the Universe of space, by which he means "*the utmost conceivable expanse of space, with all things, spiritual and material, that can be imagined to exist within the compass of that expanse*." The other meaning, which Poe says "is *ordinarily* implied," is "the Universe of Stars." "Why this distinction is considered necessary," he says, "will be seen in the sequel."[87] The "sequel" is merely the latter part of *Eureka*. We have seen above in §II that Poe claims that our universe, what we normally call "the" Universe, was created by "a God." Poe's distinction between the Universe of space and the Universe of stars is necessary because he supposes—he shifts between the words "infer," "imagine," and "fancy"—that the Universe of space is populated by many, indeed infinitely many, universes of stars, of which our universe of stars is but one. "The human brain," he says, "has obviously a leaning to the '*Infinite*,'" and so he supposes "an interminable succession of the 'clusters of clusters,' or of 'Universes' more or less similar," a "*limitless* succession of Universes, more or less similar to that

86. Shaver (note 34), 61.

87. *P&T*, 1262.

of which we have cognizance," such that "Each exists, apart and independently, *in the bosom of its proper and particular God.*"[88]

But would this not raise anew the problem of Olbers' paradox, that the night sky should be blazing with light? Poe didn't think so. This is because each universe, begun by its own God, would not necessarily have the same laws of nature. As he puts it in the same passage, each universe "having had no part in our origin, they have no portion in our laws." Poe entertains a number of possibilities here in order to avoid Olbers' paradox, including excessive diffusion for visual perception, lack of light production altogether, and not enough time due to distance. As he puts it rhetorically in a long run-on sentence, "Have we, or have we not, an analogical right to the inference that . . . the rest of which [the other universes] are invisible through distance—through the diffusion of their light being so excessive, ere it reaches us, as not to produce upon our retinæ a light-impression—or from there being no such emanation as light at all, in those unspeakably distant worlds—or, lastly, from the mere interval being so vast, that the electric tidings of their presence in Space, have not yet—through the lapsing myriads of years—been enabled to traverse that interval?" That each universe deserves to be called a universe in its own right, rather than merely a part of one total universe, is because, as Alberto Cappi explains, "these universes do not interact" and "Each universe will have its origin and its laws."[89]

Even the language and imagery that Poe used is remarkably modern. As Poe put it in a letter dated September 20, 1848, with regard to Laplace's nebular hypothesis on the origin of individual solar systems, "The *ground* covered by the great French astronomer compares with that covered by my theory, as a bubble compares with the ocean on which it floats."[90] Poe might well have used the same imagery for his multiverse theory, each universe a bubble, expanding and then contracting, which is the imagery used for modern multiverse theory. As Shaver puts it, "Our entire universe may be just one speck in a huge sea of frenzied quantum foam of false vacuum, with the seeds for endless universes

88. *P&T*, 1329, 1330.

89. Cappi (note 50), 188.

90. *Letters*, 689.

continually being created randomly by quantum fluctuations."[91] Sounds like bubbles to me, and in fact "bubbles" is the common word used in inflationary multiverse theory. Andrei Linde, for example, the theorist who gave us *chaotic inflation* and *eternal inflation*, states that "In essence, one inflationary universe sprouts other inflationary bubbles, which in turn produce other inflationary bubbles. This process, which I have called eternal inflation, keeps going as a chain reaction," to which he adds, "for all practical purposes, one can consider the moment of formation of each inflationary bubble as a new 'big bang.'"[92] Similarly Guth uses "bubbles" in his book on inflationary cosmologies, on "the eternal reproduction of universes," with bubbles "very much like the boiling of water,"[93] as does Weinberg in his review of Guth's book, in which he is attracted to the view that "although our own Big Bang had a definite beginning about ten to fifteen billion years ago, the bubbling up of new big bangs may have been going on forever in a universe that is infinitely old," to which he adds, (although he doesn't like the terminology,) "Loosely put, each bubble of ordinary space could be called a 'universe.'"[94]

But how original was Poe with his own multiverse theory? According to the Levines, "There is a philosophical precedent . . . for Poe's idea of multiple universes and multiple gods in the thinking of the presocratic philosopher Anaximander."[95] This sixth-century B.C. philosopher, however, merely assumed that there was one original stuff, what he called the *aperion*, (meaning "not limited or bounded,") from which, as quoted (or paraphrased) later by Theophrastus, "come into being all the heavens and the worlds in them. And the source of coming-to-be for existing things is that into which destruction, too, happens, 'according to necessity.'"[96] This is hardly a clear "precedent" for Poe's multiverse theory, with

91. Shaver (note 34), 61.

92. Linde (note 62), 103, 104.

93. Alan H. Guth (1997). *The Inflationary Universe: The Quest for a New Theory of Cosmic Origins.* New York: Basic Books, 252, 278.

94. Steven Weinberg (2001). *Facing Up: Science and Its Cultural Adversaries.* Cambridge, MA: Harvard University Press, 172.

95. *Eureka*, xiii.

96. G.S. Kirk, J.E. Raven, and M. Schofield (1983). *The Presocratic Philosophers.* 2nd ed. Cambridge: Cambridge University Press, 118.

its infinity of universes, each universe beginning with a big bang and having its own laws. But alas, if Anaximander's thinking, of which we see through a glass darkly and about which we know next to nothing, is allowed to provide a "precedent" for Poe's multiverse theory, then why is not Poe's thinking in *Eureka*, of which we see through a glass lightly and about which we know much, not allowed to provide an "anticipation" of modern Big Bang theory? The mind reels at the asymmetry in judgment by the Levines here. *Nemo me impune lacessit.*

VII. Space–Time Interdependence

In the Scholium to Definition VIII of the *Philosophiæ Naturalis Principia Mathematica*, first published in 1686, Isaac Newton stated that "I do not define time, space, place, and motion, as being well known to all. Only I must observe that the common people conceive those quantities under no other notions but from the relation they bear to sensible objects." Of time, he wrote: "Absolute, true, and mathematical time, of itself and from its own nature, flows equably without relation to anything external, and by another name is called 'duration'; relative, apparent, and common time is some sensible and external (whether accurate or unequable) measure of duration by the means of motion, which is commonly used instead of true time, such as an hour, a day, a month, a year." And of space, he wrote: "Absolute space, in its own nature, without relation to anything external, remains always similar and immovable. Relative space is some movable dimension or measure of the absolute spaces, which our senses determine by its position to bodies and which is commonly taken for immovable space; such is the dimension of a subterraneous, an aerial, or celestial space, determined by its position in respect of the earth."[97]

The Newtonian revolution, with its framework of absolute space and absolute time, dominated physics and astronomy for over two hundred years, until it was replaced with a new framework, marking a new scientific revolution, namely, the relativity framework of Albert Einstein, introduced to the world with his

97. Thayer (note 43), 17–18.

revolutionary paper of 1905, "On the Electrodynamics of Moving Bodies," in which he introduced what he would later call his *special theory of relativity* (in contradistinction to what he would call his *general theory of relativity*, which generalized his relativity principle to include gravity). Here is how Einstein put it in "The Problem of Space, Ether, and the Field in Physics" (1934), one of his many popular accounts: "With the discovery of the relativity of simultaneity, space and time were merged in a single continuum in a way similar to that in which the three dimensions of space had previously been merged into a single continuum."[98] Accordingly, physicists today talk of "space-time" or "four-dimensional space-time," (both meaning the same,) not "space and time."

In "Marginalia" (November 1844), Poe provides some thoughts on the nature of space and time that are important for understanding what he says in *Eureka*.[99] Beginning with time, he notices that we mark or "appreciate" time by means of events and then proceed to "the erroneous idea that events *are* time," thinking that with more events there is more time. With space, he then says, the situation is analogous, in that we measure space in terms of "the succession of objects" and then proceed to "the false idea that objects *are* space—that the more numerous the objects the greater the space." Accordingly, he says, we can get a vague idea of the distance between the Sun and Uranus because of the idea of the intervening planets, but our mind is "utterly lost" when trying

98. In Sonja Bargmann, trans. (1954). *Ideas and Opinions* by Albert Einstein. New York: Crown Publishers, 281–282. In "On the Electrodynamics of Moving Bodies" (1905) Einstein actually only relativized time. Not that Einstein ever tried to take credit, but it was his former mathematics professor Hermann Minkowski who, in his "Space and Time" lecture (1908) which was based on the work of Einstein and Lorentz before him, relativized space and time into what he called "a kind of union of the two," such that "Three-dimensional geometry becomes a chapter in four-dimensional physics." In W. Perrett and G.B. Jeffery, trans. (1923). *The Principle of Relativity: A Collection of Original Papers on the Special and General Theory of Relativity.* London: Methuen and Company, 75, 80. Minkowski died less than four months later at the age of forty-four, deprived of seeing his lecture in print, let alone of knowing the fate of his massively important concept, which physicists named after him as "Minkowski space-time," and which now is simply "space-time."

99. *E&R*, 1325–1326.

to get an idea of the distance between the Sun and the star Sirius, because we know of no intervening objects between the two solar systems. Poe's point is the simple yet profound one, not original with him of course, that we should expect a fundamental difference between the nature of space and time, on the one hand, and on the other the nature of how our minds work with respect to space and time. His point, then, can be seen to be of a piece with his rejection of axioms in *Eureka.*

When we turn to *Eureka* we naturally expect big things on this topic (given what we have already seen in the present chapter). And, indeed, Poe claims late in *Eureka* that "*Space and Duration are one.*"[100] This, on the surface of it, certainly looks like an anticipation of something like the modern concept of the space-time continuum. According to Stuart and Susan Levine, "The statement seems to be unambiguous evidence that Poe understood space-time in the modern sense. Although most of the more famous ways in which *Eureka* 'anticipates' or 'predicts' twentieth-century science turn out on close examination to be not quite the same as more modern concepts, or shrewd guesses based on not very solid data, this one seems to be a fine extrapolation from firm information that Poe thoroughly understood. It suggests how very good a mind Poe had (Twarog)."[101]

But we have to be more careful here. True enough, the Levines criticize the rest of Poe's paragraph, because "Poe never—despite the dramatic dashes he perhaps added to cover his logical tracks—says exactly why vast cosmic distances are necessary." But we need to look at the context, the bigger picture. The question Poe poses, at the beginning of the paragraph quoted above, concerns the size of our universe, why it is so massive beyond human imagination. Would not a smaller universe have equally served God's purpose? As Poe puts it, "let us take the opportunity of referring to the difficulty . . . *in accounting* for the immeasurable voids alluded to—in comprehending why chasms so totally unoccupied and therefore apparently so needless, have been made to intervene between star

100. *P&T*, 1340.

101. *Eureka*, 157n214. Bruce A. Twarog, a professor of physics and astronomy at the University of Kansas, was one of the scientists that the Levines consulted.

and star—between cluster and cluster—in understanding, to be brief, a sufficient reason for the Titanic scale, in respect of mere *Space,* on which the Universe of Stars is seen to be constructed. A rational cause for the phænomenon, I maintain that Astronomy has palpably failed to assign." Poe is painting in brushstrokes here. But more important, we have seen in §II, following Cappi, that Poe conceptualized the Universe, beginning with a Big Bang, within a Newtonian framework of space and time. So forget trying to see something Einsteinian now on the topic of space and time. What Poe is doing with his "*Space and Duration are one*" is quite different. He does not mean "*one*" in the sense of a continuum. Instead, he means "*one*" in the sense of *co-adaptation*, in line with his critique of the Bridgewater Treatises examined in the previous chapter. In fact, in the very two paragraphs in *Eureka* immediately following "*Space and Duration are one*," Poe gives virtually the same exposition of his concept of "Divine *adaptation*," what he calls "the complete *mutuality* of adaptation," which he then in the next paragraph connects with plots, concluding (or assuming) that "The plots of God are perfect. The Universe is a plot of God." "*Space and Duration are one*," then, for Poe, in the very same sense that we examined in §III above, which is the modern sense of the fine-tuning of the laws of nature, sometimes called the *anthropic principle*. In this sense the vastness of the size of our universe, according to Poe, is not an accident (that would make our universe an imperfect plot) but instead is *mutually adapted* or fine-tuned with time, specifically the age or duration of our universe beginning with the time of maximum expansion and continuing through the contraction phase, all for the purpose of allowing nebular clusters to form in accordance with the respective laws of nature, and consequently clusters of stars (galaxies) with planets orbiting stars followed by life evolving on planets. As Cappi puts it, "It is interesting how Poe justifies the time dimension by using the Anthropic Principle. He has well understood the link between dimension, age, and life, which exists in an evolving universe."[102]

102. Cappi (note 50), 189. It should be added, however, that Poe in *Eureka* did not have much of a theory of the evolution of life per se. We "perceive," he says, "that not merely the *manifestation* of vitality, but its importance, consequence, and elevation of character, keep pace, very closely, with the heterogeneity, or complexity, of the animal structure." Given Poe's theory

Poe anticipated, then, not the modern concept of space-time, but the idea that an evolving finite universe requires us to rethink the relation between space and time. For Newton and other physicists up until 1908, space and time were thought to be independent of each other. For Poe in 1848, space and time, not in themselves but as dimensions of our universe of stars, the actual spatial size of it and the actual duration of it, had to be interdependent. Poe didn't go all the way to the modern view in physics, according to which space and time, because they are connected with matter-energy, do not extend beyond the physical Universe, but what Poe did was nevertheless remarkable.

---

of the finely-tuned condensing Universe, he follows this with "the proposition that *the importance of the development of the terrestrial vitality proceeds as the terrestrial condensation.*" This is in accordance, he says, with the observation of "successive geological revolutions" and "what we know of the succession of animals on Earth." But we get next to nothing by way of specific explanation for the *complexity* of life and its evolutionary progression. What we are given, instead, is the vague statement that "As it [Earth] has proceeded in its condensation, superior and still superior races have appeared." Poe is not even sure if "the successive geological revolutions" merely "attended" or "immediately caused" the succession of plants and animals on Earth. At any rate, he combines the above with the "suggestion" that the successive geological revolutions were in turn caused by "successive planetary discharges from the sun,"—each a nebulous ring from which each planet evolved by a spiralling condensation,—allowing for the possibility that a further solar discharge could result in "a race both materially and spiritually superior to Man." "These thoughts," he adds, "impress me with all the force of truth—but I throw them out, of course, merely in their obvious character of suggestion." *P&T*, 1317, 1323. All of this gives us a further reason, in addition to those examined in the previous chapter, to reject the idea that Poe was influenced (positively or otherwise) by Chambers' *Vestiges of the Natural History of Creation.* And we can certainly dispense with the idea that Poe anticipated Darwin. It is not even clear that Poe entertained the idea of the *evolution* of species. (In this chapter I give him the benefit of the doubt.) Although Poe uses the word "*development*" in reference to the "succession" of one species *after* another, it is not necessarily the same concept as the *evolution* of one species *from* another. To think otherwise is to read present ideas into past texts. Lyell himself, for example, the most famous geologist of his time, discussed in Chapter 3, believed in the *succession* of one species after another, with God creating each succession *de novo*. He did not become an evolutionist until Darwin gradually changed his mind post-1859. In Poe's case, it might be more accurate to call him a *saltationist* than an *evolutionist*.

VIII. Matter–Energy Equivalence

Here is how the physicist Daniel Orange, in a popular book on physics, defines the difference between matter and energy: "*Matter* refers to anything that takes up space, no matter how large or small it is. *Energy* refers to any force that can produce a change in matter."[103] One of the amazing contributions of Einstein, of course, was his argument, which turned out to be one of his many successes, (and also the most dangerous, given that it led to the atomic bomb,) that matter and energy are really the same thing, only in different forms, that "matter" and "energy," in a very real sense, are synonyms, which is encapsulated in arguably the most famous scientific formula of all time: $E = mc^2$. The idea and the mathematical proof, but not the particular formula, (the latter was first made explicit by Einstein in 1912,) was introduced by Einstein in his second paper on relativity, "Does the Inertia of a Body Depend Upon its Energy-Content?" (1905). There he simply states at the end of this short paper that "The mass of a body is a measure of its energy-content."[104] Later, for the benefit of the public, for example in "$E = MC^2$" (1946), Einstein explained "the law of the equivalence of mass and energy" as: "The energy that belongs to the mass *m* is equal to this mass, multiplied by the square of the enormous speed of light—which is to say, a vast amount of energy for every unit of mass."[105]

Einstein speaks of the "equivalence" of mass and energy, but by "equivalence" he does not mean "equal," as with "2 + 2 = 4" or "the equality of man." At this point I must say that trying to get a clear answer from a physicist to a simple question of meaning is sometimes like pulling teeth, the very teeth for which I, in writing the present section of this book, in the disordered chamber of my brain, in the full fury of my *monomania*, have repeatedly longed with a phrenzied desire. Indeed Einstein might as well be called Berenice, for he is no exception to my complaint. What Einstein means by "equivalence," it turns out, after much yanking with my

---

103. Daniel Orange and Gregg Stebbens (1999). *Everything You Need to Know About Physics.* New York: Pocket Books, 5.

104. Perrett and Jeffery (note 98), 71.

105. Bargmann (note 98), 340.

pliers and a close examination of the roots visibly and palpably before me, is numerical identity, one and the same thing in different forms. As Gribbin puts it, "Mass itself is a form of energy, . . . With mass included in the definition of energy, energy can neither be created nor destroyed, but only converted from one form into another. Ultimately, this means that all energy will end up as radiant heat."[106] Or it means that the Universe, if one agrees with Poe, will end up as God returned unto Itself, as pure Unity and ultimate Hot.

A further remarkable feature of *Eureka* is that Poe, while not anticipating, of course, $E = mc^2$,—one would have to be a Bedlamite on laudanum to propose otherwise,—did indeed anticipate the idea that matter and energy are not two different sorts of entities but, behind appearances, one and the same. There is nothing else we can conclude, keeping in mind the modern definitions of "matter" and "energy" with which we began this section and given that Poe states, near the beginning of *Eureka*, that "Attraction and Repulsion are the *sole* properties through which we perceive the Universe—in other words, by which Matter is manifested to Mind—that, for all mere argumentative purposes, we are fully justified in assuming that Matter *exists* only as Attraction and Repulsion—that Attraction and Repulsion *are* matter:—there being no conceivable case in which we may not employ the term 'Matter' and the terms 'Attraction' and 'Repulsion,' taken together, as equivalent, and therefore convertible, expressions in Logic."[107]

There it is, Berenice's "equivalence" and Morella's "converted," Poe's "equivalent" and "convertible." The mind reels in astonishment! For Poe, remarkably, "$m = A + R$." But at this point in *Eureka* Poe is simply laying out his principles, his language or "Logic." It is near the end of *Eureka* that he provides his reasons for the equation. And there we find Poe's concept of "Matter as a Means—not as an End." The means is the gradual contraction of the Universe following the state of maximum diffusion, the gradual coalescence of the diffusion requiring "the existence of parts, particles, or atoms," involving, as physics puts it, "the tendency of 'each atom &c. to every other atom' &c. according to a certain

106. Gribbin (note 15), 151.

107. *P&T*, 1283.

law," the atoms then gradually coalescing over eons of time into stars and planets with life, keeping in mind that light, too, consists of "light-particles."[108]

And then comes Poe's concept of Nothingness. He says, "When, on fulfillment of its purposes, then, Matter shall have returned into its original condition of *One* . . . shall have returned into absolute Unity,—it will then (to speak paradoxically for the moment) be Matter without Attraction and without Repulsion—in other words, Matter without Matter—in other words, again, *Matter no more*. In sinking into Unity, it will sink at once into that Nothingness which, to all finite perception, Unity must be—into that Material Nihility from which alone we can conceive it to have been evoked—to have been *created* by the Volition of God."[109] In a letter dated February 29, 1848, Poe states his "General Proposition" as "Because Nothing was, *therefore* All Things are," and he continues in the letter with "Matter, springing from Unity, sprang from Nothingness;—i.e. was *created*" and that with the return to Unity "then the final globe would be matter without matter—i.e. no matter at all:—it must disappear. Thus Unity is *Nothingness*."[110]

The Levines point out, following earlier critics of Poe, that the attribution of attraction and repulsion to matter, this contribution to molecular theory, was first made by Ruggero Giuseppe Boscovich, a Croation mathematician and astronomer who died in 1787. But along with Cappi, I "do not think it is necessary to see a direct link with Poe's adoption of the two terms."[111] For one, to add substance to Cappi's claim, Boscovich did not make the distinction between particled and "unparticled" matter, which apparently is original with Poe and which anticipates, in a remarkable fashion, Einstein's concept of matter as "frozen energy," the metaphor commonly used today by physicists for laypeople. For another, a direct link between Boscovich and Poe on the nature of matter has yet to be proven, and given the lack of proof, what exactly—aside from satiating a perverse imp of an instinct for calumny—is to be gained by looking back and making the comparison? Nothing.

---

108. *P&T*, 1354, 1355, 1291.

109. *P&T*, 1355.

110. *Letters*, 649. See also *Letters*, 659–660.

111. Cappi (note 50), 181n4.

What we should be doing, instead, is looking forward, A.P. (After Poe). More specifically, what we should be doing, to really appreciate Poe's ideas on matter and the origin of the Universe, is comparing Poe's language with recent physicists, (much more recent than Einstein,) specifically those who might collectively be called the Ligeia of cosmology, namely, the inflationary cosmologists. Here is how Stephen Hawking, for example, puts it in his "no-boundary" contribution to inflationary cosmology:—are you ready for this?—"The inflation . . . produced all the contents of the universe quite literally out of nothing. When the universe was a single point, like the North Pole, it contained nothing. Yet there are now at least ten-to-the-eightieth particles in the part of the universe that we can observe. Where did all these particles come from? The answer is that relativity and quantum mechanics allow matter to be created out of energy . . . . And where did the energy come from to create this matter? The answer is that it was borrowed from the gravitational energy of the universe."[112] And where did the gravitational energy of the Universe come from? At this point Hawking becomes Berenice to my narrator, for he does not mention gravitational energy anywhere else in his book, or in its prequel. But no matter, never mind, (and no energy, never matter). Hawking is not alone in modern physics with his language of nothing. Alan Guth, for example, has a chapter entitled "A Universe Ex Nihilo" in his book devoted to theories of inflationary cosmology, in which he says, "To my knowledge, the first serious suggestion that the creation of the universe from nothing could be described in scientific terms was the 1973 paper by Edward Tyron, 'Is the Universe a Vacuum Fluctuation?' . . . According to quantum theory, the apparently quiescent vacuum is not really empty at all, but on a subatomic level is a perpetual tempest, seething with activity."[113] The quantum vacuum of modern physicists, the Unity/God of Poe, (take your pick,) either way it is much ado about—Nothing.

Come to think of it, the last line from the quotation above reminds me of the second stanza of one of Poe's last poems, "A Dream within a Dream" (1849), which I quoted in Chapter 3 but

112. Hawking (note 55), 97.

113. Guth (note 81), 271–272.

quote here again, and usefully, for the perspective is now modern physics, not theology:

> I stand amid the roar
> Of a surf-tormented shore,
> And I hold within my hand
> Grains of the golden sand—
> How few! yet how they creep
> Through my fingers to the deep,
> While I weep—while I weep!
> O God! can I not grasp
> Them with a tighter clasp?
> O God! can I not save
> *One* from the pitiless wave?
> Is *all* that we see or seem
> But a dream within a dream?[114]

Quoth the Quantum Ocean "Nevermore."

IX. No Material Ether

In his third letter to Richard Bentley, dated 1693 and referenced in §II above, Newton made clear not only that he rejected the idea of gravity as an innate property of matter—the opposite view was ascribed to Newton by the astronomer Roger Cotes in his Preface to the second edition of Newton's *Principia* published in 1713[115] and remains a confusion to this very day—but also that he rejected the idea of gravity as action-at-a-distance. "That gravity should be innate, inherent, and essential to matter," he wrote, "so that one body may act upon another at a distance through a *vacuum*, without the mediation of anything else, by and through which their action and force may be conveyed from one to another, is to me so great an absurdity that I believe no man who has in philosophical matters a competent faculty of thinking can ever fall into it."[116]

114. *P&T*, 97.

115. Thayer (note 43), 118.

116. Thayer (note 43), 54.

Just exactly how gravity operated Newton would not say. Years earlier, however, in a letter to the chemist Robert Boyle, dated 1679, (which is seven years before the publication of *Principia*,) Newton put to pen his "suppositions" about the existence of "an ethereal substance, capable of contraction and dilation, strongly elastic, and, in a word, much like air in all respects, but far more subtle." He supposed that "this ether pervades all gross bodies," he distinguished between "external" and "internal" ether, also between "rarefied" ("finer") and "denser" ("grosser") ether, and he used the concept to help explain not only gravity but many other phenomena as well, including certain properties of light and why "a fly walks on water without wetting her feet, and consequently without touching the water."[117]

This concept of "the ether," (as it is commonly called,) a mysterious material substance permeating bodies and existing between them, indeed permeating the Universe, dominated Newton's thought to the end.[118] It also dominated physics past Poe's time, well up to the Michelson-Morley experiment of 1887, an experiment on light beams that failed to demonstrate the existence of the ether, but demonstrated instead (to everyone's astonishment, including Michelson and Morley) the constancy of the speed of light in empty space, thus bringing the existence of the ether into serious question.[119] In fact the concept of the ether was maintained in physics well into the early 1900s, kept alive by H.A. Lorentz, for example, whom Einstein greatly admired, who in 1904 continued with the distinction between "the ether" and "ponderable bodies," and by J.J. Thomson, another prominent example, who in his 1909 presidential address to the seventy-ninth meeting of the British Association for the Advancement of Science stated that "The ether is not a fantastic creation of the speculative philosopher; it is as essential to us as the air we breathe . . . . The study of this all-pervading substance is perhaps the most fascinating

117. Thayer (note 43), 112–113, 114, 115, 116.

118. A useful discussion on "Newton and the Aether" is to be found in Stephen Toulmin and June Goodfield (1962). *The Architecture of Matter*. New York: Harper & Row, 194–198. My only caveat is that they fail to mention let alone discuss Newton's letter to Boyle.

119. See, e.g., John Losee (2001). *A Historical Introduction to the Philosophy of Science*. 4th ed. Oxford: Oxford University Press, 153–154.

and important duty of the physicist."[120] The existence of the ether was even retained by A.A. Michelson of Michelson-Morley fame, who lamented in 1927, a few years before he died, "But without a medium how can the propagation of light waves be explained?"[121] No matter, the beginning of the end for the concept was Einstein's publication of his special theory of relativity in 1905, in which he argued that "The introduction of a 'luminous ether' will prove to be superfluous inasmuch as the view here developed will not require an 'absolute stationary space' provided with special properties."[122] But it was not until Einstein extended his special theory of relativity to gravity, with his "The Foundation of the General Theory of Relativity" (1916), that the concept of the ether became not only merely dead, but really, most sincerely dead. There Einstein argued that gravity is not a force, but instead the curvature of space-time around a physical body, and he claimed that the approximation to Newton's law of gravity, as well as the explanation of the precession of the perihelion of Mercury, may "be taken as a convincing proof of the correctness of the theory."[123]

In looking, however, for a clear statement from Einstein at this point, on the nature of gravity according to the general theory of relativity, I am experiencing once again the frenzied desire of the narrator of "Berenice," not just a consuming curiosity pervading my soul, but of coveting madly the objects that are the dream of the disordered chamber of my brain. This is because in reading Einstein's 1916 paper he is palpably unclear on the matter (unless, apparently, you understand the mathematics). Neither is he particularly clear in his later writings. So let us pull the teeth in a different way. What shall surely help is to look at different expressions by physicists of Einstein's theory. George Gamow, for example, states that "The great idea, which was included by Einstein in the foundation of his general theory of curved space, consists of the assumption that *the physical space becomes curved*

120. The Lorentz paper is found in Perrett and Jeffery (note 98), 28. The quotation from Thomson is found in Gerald Holton (1978). *The Scientific Imagination: Case Studies.* Cambridge: Cambridge University Press, 48.

121. Quoted in Abraham Pais (1982). *"Subtle is the Lord . . .": The Science and the Life of Albert Einstein.* Oxford: Oxford University Press, 115.

122. Perrett and Jeffery (note 98), 38.

123. Perrett and Jeffry (note 98), 145.

in the neighborhood of large masses; the bigger the mass the larger the curvature."[124] Stephen Hawking puts it this way: "Einstein made the revolutionary suggestion that gravity is not a force like other forces, but is a consequence of the fact that space-time is not flat, as had been previously assumed: it is curved, or 'warped,' by the distribution of mass and energy in it."[125]

Those examples shall do. And what they leave us with is the conclusion that, in accounting for gravity as curved space-time, Einstein left the concept of the ether with absolutely nothing left to do: it had already lost full-time employment over the propagation of light, following the Michelson-Morley experiment, not long after which it became unemployed following Einstein's special theory of relativity, perhaps even comatose, after which it gave up the ghost with Einstein's general theory of relativity, given that there was no longer any action-at-a-distance that required a medium (gravitational effects were fully explained, instead, by the curvature of space-time around bodies). In a sense, then, the concept of the ether simply sank into Einstein's concept of space-time, where it rests in peace. Einstein himself, interestingly, in "The Problem of Space, Ether, and the Field in Physics" (1934), states that in a sense "physical space and the ether are only different terms for the same thing."[126] Similarly, Gamow, likewise focused only on the ether as the medium for the propagation of light, (the same point would apply to the ether as the medium of gravity,) states that "in modern physics the expressions 'light ether' (divested of its alleged mechanical properties) and 'physical space' are considered synonymous."[127] In this way the term "the ether" could have suffered a fate similar to the reconceptualization of "species" following the Darwinian revolution. Such is often the nature of concept change in science, that the extension of a term survives (more or less) a radical change in intension due to a revolution in theory. But the term "the ether" has not in fact survived,

124. George Gamow (1961). *One Two Three . . . Infinity: Facts and Speculations of Science.* New York: Viking Press, 106.

125. Hawking (note 25), 29.

126. Bargmann (note 98), 281.

127. Gamow (note 111), 91.

whereas the term "species" has (to the continued consternation of many in biology and philosophy of biology).

Gamow also states that "the greatest mistake of the physics of the nineteenth century consisted in the assumption that this light ether has properties very similar to those of ordinary physical substances familiar to us."[128] But of course they thought that of the ether *per se*. At this point enter Poe. In *Eureka* Poe notes that astronomers, based on the decaying orbits of comets around the Sun, in particular Encke's comet, concluded that the cause of the decay is the drag of the ether, "an exceedingly rare [i.e., fine] but still material medium pervading space." But, says Poe, "this ether was assumed, most illogically, on the ground that no *other* mode than the one mentioned could be discovered, of accounting for the observed decrease in the orbit of the comet."[129] A combination of causes, says Poe, might equally account for the phenomenon. Astronomers also thought that the decreasing orbit of the Moon was caused by the ether, until Lagrange, as Poe puts it, "came to the rescue" with a simpler theory. Poe himself, while not doing away with the ether entirely, reconceptualizes it, such that "the ether thus conceived is radically distinct from the ether of the astronomers; inasmuch as theirs is *matter* and mine *not*."[130]

If the ether is not matter but is nevertheless real, then what is it? Poe's answer is that it is "a subtle *influence* which we know to be ever in attendance on matter," and he lumps the ether as such with his force of Repulsion, which includes "the various phænomena of electricity, heat, light, magnetism; and more—of vitality, of consciousness, and thought—in a word, of spirituality."[131] The purpose of Repulsion in Poe's scheme, it will be remembered, is to delicately counterbalance Attraction, but not quite equally, only slightly sub-equally, to allow for the gradual contraction of the Universe from the state of maximal diffusion, to allow for the gradual formation of stars and planets and the gradual (or rather saltational) formation of life on planets, including us. And indeed Poe goes on in *Eureka*, a couple of pages further, after reiterating

128. Gamow (note 111), 90.

129. *P&T*, 1350.

130. *P&T*, 1352.

131. *P&T*, 1351, 1352.

his idea of divine adaptation, of the "absolute *reciprocity of adaptation*," to state that "With a perfectly legitimate reciprocity, we are now permitted to look at Matter, as created *solely for the sake of this influence*—solely to serve the objects of this spiritual Ether. Through the aid—by the means—through the agency of Matter, and by dint of its heterogeneity—is this Ether manifested—is *Spirit individualized*. It is merely in the development of this Ether, through heterogeneity, that particular masses of Matter become animate—sensitive—and in the ratio of their heterogeneity;—some reaching a degree of sensitiveness involving what we call *Thought* and thus attaining obviously Conscious Intelligence." Keeping in mind that Poe's ether is not particulate, but an energy or force, he then adds that "When, on fulfillment of its purposes, then, Matter shall have returned into its original condition of *One*—a condition which presupposes the expulsion of the separative Ether, whose province and whose capacity are limited to keeping the atoms apart until that great day when, this Ether being no longer needed, the overwhelming pressure of the finally collective Attraction shall at length just sufficiently predominate and expel it:—when, I say, Matter, finally, expelling the Ether, shall have returned into absolute Unity,—it will then (to speak paradoxically for the moment) be Matter without Attraction and without Repulsion—in other words, Matter without Matter—in other words, again, *Matter no more*."[132]

We are now back to Poe's theology, which is of a piece with his cosmology (and more). But the fact should not be lost that Poe, in trying to work out a consistent and comprehensive understanding of this massive puzzle we call the Universe, rejected the concept, dominant in his time, of the material ether. He did not go in the direction of the modern "mode" of explanation, of course, that of relativity theory. But his vision was nonetheless remarkable, for seeing that within the framework of a finite and evolving universe with a beginning and end—his own concept, which anticipates the modern concept—the concept of the material ether is not required. As Harry Poe aptly puts it, Edgar Poe rejected the existence of the material ether because "It spoiled the plot as an unnecessary

132. *P&T*, 1354, 1355.

addition."[133] Modern physics has rejected the concept of the material ether, too, but as superfluous given Einstein's theory of gravity. Poe took the direction of spiritual ether,—a vague concept in his hands, to be sure,—not, however, for allowing gravity to work at a distance or for explaining decaying orbits or for providing light with a medium, but for the purpose of his theological and poetical view of the Universe, i.e., for the purpose of a sentient universe, i.e., for God experiencing pleasure through seemingly infinite diversity, pleasure heightened by pain.

Poe's theological twist on the ether in particular and on the Universe in general might trouble the reader. But given the genuine debate over the fine-tuning argument *among physicists* in modern cosmology, can anyone be sure, looking a thousand years ahead, say, that some sort of theology will not play a more prominent role in scientific cosmology? Can anyone be so sure today that Poe's direction was altogether *wrong*?

Having examined nine major theories and developments in twentieth-century science that Poe anticipated in *Eureka*, (and having established beyond a reasonable doubt, I believe, that we are justified in using the word "anticipated,") the time has surely now arrived to turn to what Poe had to say about the nature of imagination, especially the educated imagination, in particular scientific imagination—we should want to know his theory. But first we have to provide an intellectual context in order to fully understand his theory, which means we have to look at what philosophers and historians of science have to say about imagination in science, from Poe's time to the present. (In terms of enjoying Poe, this might even allow one to derive positive pleasure from a legitimate source of pain.) We shall then turn to two of the greatest scientists of all time,—*real* scientists, not philosophers or historians, Darwin and Einstein to be specific,—to their use of imagination in their science and what they had to say about the role of imagination in science. In the very last section an argument by analogy shall be

133. *Evermore*, 157.

made for the inclusion of imagination in the process of science, an argument that adds to Poe on the matter. Only *then* shall we turn to Poe's theory of scientific imagination, the penultimate chapter, which shall be modernized in the chapter that follows it.

CHAPTER 6

Imagination in Philosophy and History of Science

> . . . true Science, which makes its most important advances—as all History will show—by seemingly intuitive *leaps.*
>
> —Edgar Allan Poe[1]

Entrée

Modern philosophy of science began with an early twentieth-century movement known as *logical positivism*. There was philosophy of science in Poe's time, to be sure, developed most notably by John Herschel, John Stuart Mill, and William Whewell. And although we shall look at their views briefly below, to help put Poe in context and to understand some of his statements and examples, we have to be cognizant of the fact that it was not what today is called *modern* philosophy of science, for the primary sciences on which modern philosophy of science is based did not yet exist, namely, relativity physics and quantum physics. With the development of these two sciences, philosophy of science came of age. And that meant logical positivism, which did not last, but which spawned numerous *philosophies* of science, and these today collectively constitute modern philosophy of science. Evolutionary biology already existed, of course, beginning most notably

1. *Eureka* (1848). *P&T*, 1264.

with Darwin in 1859, while genetics traces back to the rediscovery of Mendel in 1900, but biology did not come of age until the forging of the Modern Synthesis beginning in the 1930s, and even then philosophy of science remained mainly physics-based, until philosophy of biology finally came of age as a recognized subdiscipline in the 1970s, adding twists and turns to a predominantly law-based field. Prior to this time, philosophers of science paid some attention to biology, including evolutionary biology, but it was marginal, as if only to round out their respective philosophies of science as all-inclusive.

To really appreciate Poe on scientific imagination, which includes unravelling his *theory* of scientific imagination, which is the topic of the next chapter, we need to focus not only on the philosophy of science of Poe's day but especially on modern philosophy of science. The contrast is striking. But more than mere contrast, every section of the present chapter also serves, remarkably, to help us see that Poe actually had a philosophy of science, even a properly corrective one. This chapter, then, should produce a unity of effect similar to the previous chapter, with this and the next chapter producing a unity of effect that shall lead us to the final chapter, devoted to uncovering the nature of scientific imagination itself. It is in that chapter that Poe's self-avowed magnum opus, *Eureka*, receives its ultimate vindication, and hence Poe himself, after all the many years of being denied his due. In the world of law, of course, it is often said that justice delayed is justice denied. In the world of literary legacy, on the other hand, it is a form of poetic justice, as legacy is what Poe lived for. I can imagine Poe would be pleased by it all. But Poe the man is no more, so posterity, in the form of lovers of Poe, will have to be pleased for him in his stead.

What I want to do in this chapter, following a brief excursion into the philosophy of science of Poe's day, is to provide a brief excursion into the main movements in modern philosophy of science that are germane to our topic. There is, to be sure, some work in philosophy of science devoted to the topic of theory formation, and we shall touch on it here and there, also in the last chapter, but on the whole it has played a marginal role, never a central one. Moreover, when theory formation is focused on, it is usually in the sense of trying to find rules or methods, rather than to discover the nature of scientific imagination itself. But no matter. Modern philosophy of science, in the main, when devoted to the nature of science itself, has focused on what to do with scientific theories once they have been made. As Thomas Nickles puts it in his excellent discussion on scientific discovery, "Consequentialism has dominated methodology to the present day."[2]

Specifically, we shall look at the characterizations of science provided by logical positivism, logical empiricism, Karl Popper, Thomas Kuhn, the new experimentalism, the disunity of science movement, inference to the best explanation, the epistemic virtues and values approach, evolutionary epistemology, and finally, contextualist history of science. The latter will serve as a segue to the penultimate sections of this chapter, in which we shall turn to the thoughts on the role of imagination in science by two of the most important scientific minds of our modern age, namely, Charles Darwin and Albert Einstein.

But that is getting ahead of ourselves. Prior to the sections on Darwin and Einstein, what we shall see in this chapter, quite remarkably, is an overall poverty in modern philosophy of science, including recent professional history of science, when it comes to dealing with the role of imagination in the scientific process. It is much like modern philosophers of biology who are interested in the role played by mutations in evolution but not in the causes of the mutations themselves. In the final section of this chapter I shall accordingly draw on my own work on point mutations and make an important analogy. I shall argue that just as it is mistaken to ignore the causes of mutations when trying to understand evolution by natural selection, it is likewise mistaken to ignore the origin and role of imagination when trying to understand the scientific process, what may properly be called *scientific imagination*. Take away mutation and you take away what drives evolution by natural selection. Would you not therefore want to know about the nature of mutation in order to better understand evolution? Similarly, take away imagination and you take away what drives science. Would you not therefore want to know about the nature of scientific imagination in order to better understand science?

2. Thomas Nickles (2000). "Discovery." In W.H. Newton-Smith, ed. (2000). *A Companion to the Philosophy of Science.* Malden, MA: Blackwell, 87.

The latter is Poe, and in using this argument by analogy I shall attempt to help vindicate him, for as we shall see in the next chapter Poe repeatedly inveighed against philosophy of science for failing to capture the heart and soul of what actually drives science, the heart and soul being the *educated imagination*. And what we shall attempt to gain in the chapter following that, the final chapter, is an understanding of the very nature of the imagination that drives science, including many examples of what is properly called "unconscious scientific creativity."

Philosophy of Science in Poe's Time

Philosophy of science in Poe's time was dominated by the inductive approach, which had eclipsed the deductive approach, although Poe heavily criticized both in *Eureka* as "crawling" and "creeping," respectively, in favor of "soaring," as we shall fully see in the next chapter. The inductive approach to the nature of science was represented in Poe's time mainly by Sir John F.W. Herschel, John Stuart Mill, and William Whewell, and it is important to look at some of the highlights of their thinking in order to see how Poe viewed the philosophy of science of his day.

For many decades, Herschel's *Preliminary Discourse on the Study of Natural Philosophy* (1830) served as the standard treatment for nineteenth-century philosophy of science, with subsequent works by others often viewed as fleshing out the ideas. So great was Hershel's influence that Charles Darwin himself, following the publication of his *On the Origin of Species* (1859), sought to know Herschel's response by sending him a presentation copy. Accordingly, Darwin found it "a great blow & discouragement" when he heard that Herschel had dismissed his law of natural selection as "the law of higgledy-piggledy."[3]

According to Herschel, "the whole of natural philosophy [science] consists entirely of a series of inductive generalizations, commencing with the most circumstantially stated particulars, and

3. Frederick Burkhardt and Sydney Smith, eds. (1991). *The Correspondence of Charles Darwin: Volume 7, 1858–1859*. Cambridge: Cambridge University Press, 423.

carried up to universal laws, or axioms, which comprehend in their statements every subordinate degree of generality."[4] These laws of nature are themselves the effects of real forces in nature, which following Newton he calls "veræ causæ."[5] Having established a natural law based on one or more true causes in a particular field, science can then deduce and explain further phenomena, such that "the successful process of scientific enquiry demands continually the alternate use of both the inductive and deductive method."[6] Two of Herschel's many examples of true causes are "the force of magnetism" and "the force of gravity,"[7] the latter "the most universal truth at which human reason has yet arrived."[8] Laws of nature also, according to Herschel, allow for the distinction between scientific and unscientific classification, with the former based on true causes. One of Herschel's examples is, collectively, the laws of crystallography, which are based on "the intimate constitution of the substance" and which determine real "mineral species" from false ones, such that prior to the discovery of those laws, which were accomplished by use of the reflecting goniometer, mineralogy had only "inartificial and unnatural" systems of mineral classification.[9]

And what of the role of imagination in the scientific process? Herschel apparently takes no account of it. The "only ultimate source of our knowledge of nature and its laws," he says, is "experience," which "may be acquired in two ways," the one by "observation," the other by "experiment," and "To these two sources we must look as the fountains of all natural science."[10] Moreover, in an equally telling passage, which immediately follows the one examined above about deduction and induction, he states that "The path by which we rise to knowledge must be made smooth and beaten in its lower steps, and often ascended and descended,

4. John F.W. Herschel (1830). *Preliminary Discourse on the Study of Natural Philosophy.* London: Longman, Rees, Orme, Brown, & Greene, 104.

5. Herschel (note 4), esp. 18, 76, 91, 104–105, 144.

6. Herschel (note 4), 174–175.

7. Herschel (note 4), 150 and 198.

8. Herschel (note 4), 123.

9. Herschel (note 4), 290–296.

10. Herschel (note 4), 76.

before we can scale our way to any eminence, much less climb to the summit."[11]

Not only, as we shall see in this chapter, does Herschel's picture exclude the nature of scientific revolutions, but in the next chapter we shall see Poe in *Eureka* inveigh against the "creeping" and "crawling" approaches of deduction and induction, in favor of "the wings of the eagles," of the educated imagination that "loves nothing so well as to soar in those regions of illimitable intuition which are utterly incognizant of '*path*.'"[12] That Poe saw something and focused on it, with a consuming curiosity that pervaded his soul, in the fury of his *monomania* that is *Eureka*, something most profound about the nature of science, something that others more or less have missed, is the unity of effect intended for this chapter.

Poe, as we have seen in Chapter 4, was familiar with Herschel's work, including the *Preliminary Discourse*, but when it came to his critique of the inductive approach to science he focused his attack on a more recent exponent of that philosophy, namely, John Stuart Mill with his *A System of Logic* (1843). This large book was a classic in its time, which Mill constantly revised through eight editions. As we have seen in §I of the previous chapter, Poe apparently had access to the 1846 American edition published by Harper & Brothers, which was a reprint of Mill's second edition. Mill's *Logic* is famous mainly for Mill's formulation of what have come to be known as "Mill's Methods," the five canons of inductive reasoning, which he calls "The Four Methods of Experimental Inquiry": the method of agreement, the method of difference, the joint method of agreement and difference, the method of residues, and the method of concomitant variations.

The full title of his book is interesting: *A System of Logic, Ratiocinative and Inductive; Being a Connected View of the Principles of Evidence and the Methods of Scientific Investigation*. Poe also used the word "ratiocination" in his writings, particularly in his detective mysteries, and it is useful to compare the meanings. According to Mill, "All valid ratiocination; all reasoning by which, from general propositions previously admitted, other propositions equally or less general are inferred; may be exhibited

11. Herschel (note 4), 175.

12. *P&T*, 1268, 1269.

in some of the above forms. The whole of Euclid, for example, might be thrown without difficulty into a series of syllogisms, regular in mode and figure."[13] This is in §1 of a chapter entitled "Of Ratiocination, or Syllogism," such that the "or" is obviously the "or" of synonymy. Ratiocination for Mill, then, is syllogistic reasoning. And it is important to notice that for Mill induction, properly so called, is "the operation of discovering and proving general propositions,"[14]—not *in itself* the process "from particulars to other particulars,"[15]—while what is called the "Deductive Method" in the investigation of nature "consists of three operations: the first, one of direct induction; the second, of ratiocination; the third, of verification."[16] In Poe's usage, on the other hand, "ratiocination" simply means "profound and searching analysis,"[17] and as we shall see especially in the next chapter his meaning *subsumes* deduction and induction but includes especially the rational imagination, the latter involving processes properly at the unconscious level.

Given that our focus is on imagination, did Mill attribute anything substantial to imagination as part of the scientific process? The answer is that he did, but it is interesting to see how he limits it. Recalling his "Deductive Method," he says "the Hypothetical method suppresses the first of the three steps, the induction to ascertain the law; and contents itself with the other two operations, ratiocination and verification; the law, which is reasoned from, being assumed, instead of proved." And just prior to this he states that "An hypothesis being a mere supposition, there are no other limits to hypotheses than those of the human imagination." Moreover, hypotheses, he affirms, are "absolutely indispensable in

13. John Stuart Mill (1846). *A System of Logic.* New York: Harper & Brothers, 112.

14. Mill (note 13), 172.

15. Mill (note 13), 137.

16. Mill (note 13), 264. Ratiocination for Mill, it should be noted, as we have seen, is syllogistic, which in this case means "to compute from those laws [established in the first step, induction], how the causes will operate in the particular combination known to exist in the case in hand." Verification, the third step, is "comparing this calculated effect with the actual phenomenon." Mill (note 13), 291.

17. *E&R*, 872.

science,"[18] even in the first step of the Deductive Method, genuine induction, since hypotheses "are necessary steps in the progress to something more certain," including "trying one experiment rather than another."[19] Nevertheless Mill put a severe restriction on the use of hypotheses, especially for the Hypothetical Method. Unlike Whewell, for example, Mill would not allow hypotheses incapable of being proven true, of being "brought to the test of observation." Simply accounting for the phenomena in question was not enough, as "a person of fertile imagination might devise a hundred modes of accounting for any given fact, while there are probably a thousand more which are equally possible."[20] Instead, when employing the Hypothetical Method, "men should imagine themselves to be seriously inquiring whether the hypothesis . . . is true; that they should fancy it possible to obtain the assurance that the phenomena are produced in that way and no other."[21] So the Hypothetical Method, when allowed by Mill to be part of the scientific process, must conform to the cannons of inductive method after all, or be banished entirely from the scene.

Mill's logic, along with his empiricist philosophy of scientific method, was largely a response to the philosophy of science of William Whewell. We briefly discussed Whewell in Chapter 4, as the author of one of the Bridgewater Treatises. Whewell was in fact Reverend William Whewell, having a B.D. and later a D.D. A Fellow and then Master of Trinity College, as well as Professor of Moral Philosophy at Cambridge University, Whewell was a historian and philosopher of science known mainly for his *History of the Inductive Sciences* (1837) in three volumes and his *Philosophy of the Inductive Sciences* (1840) in two volumes. Mill and Whewell, these two giants of Victorian philosophy of science, the one a follower of the empiricism of Locke, the other of the mixed empiricism and rationalism of Kant, battled back and forth for years over the true nature of science and of inductive reasoning beginning with the first edition of Mill's *Logic* and onward, in revised editions and in related works, each claiming to be true

18. Mill (note 13), 294.
19. Mill (note 13), 291–292, 290.
20. Mill (note 13), 296.
21. Mill (note 13), 297.

Baconians à la Francis Bacon's *Novum Organum* (1620), the bible of inductive reasoning and of the scientific revolution. It is not known whether Poe ever *directly* read Whewell, (to my knowledge Whewell's two titles listed above were never reprinted by American publishers,) but some of Whewell's ideas are discussed, most of them unfavourably, in Mill's *Logic*, (e.g., in §4 of ch. II of bk. III,) and Poe, being an avid reader, might easily have come across Whewell's ideas elsewhere as well.

At any rate, Whewell is certainly important for understanding philosophy of science in Poe's time. Two of his ideas are especially notable for the purposes of this chapter. One is what Whewell calls "Consilience of Inductions," by which he means "when an Induction, obtained from one class of facts, coincides with an Induction, obtained from another different class. This Consilience is a test of the truth of the Theory in which it occurs."[22] Whewell says such a consilience is "a test," but a consilience of inductions for Whewell is the highest sort of test, for "it gives rise to a constant Convergence of our Theory towards Simplicity and Unity."[23] Whewell goes so far as to claim that "I believe the history of science offers no example in which a theory supported by such consiliences, had been afterwards proved to be false."[24] We shall return to Whewell's concept in the section below on Darwin, as it is one of the keys to understanding *On the Origin of Species*. We shall also return to the concept in the final chapter, as it is a major component of the nature of scientific imagination.

Another important concept of Whewell's is found in a remarkable chapter entitled "Of the Colligation of Facts." For Whewell, "facts are bound together by the aid of suitable conceptions." This binding he calls "Colligation." But where do these suitable conceptions come from? Are they abstracted from the facts themselves? Whewell's answer, most notably and importantly, is *no*. Instead he refers to "that peculiar sagacity which belongs to the genius of a discoverer; and which, while it supplies those distinct

22. William Whewell (1840). *The Philosophy of the Inductive Sciences, Founded Upon Their History*. vol 1. London: John W. Parker, xxxix.

23. Whewell (note 22), vol. 2, 239.

24. William Whewell (1860). *On the Philosophy of Discovery: Chapters Historical and Critical*. London: John W. Parker, 275.

and appropriate conceptions which lead to its success, cannot be limited by rules, or expressed in definitions." He refers to "what are commonly spoken of as felicitous and inexplicable strokes of inventive talent." He says they "are not improperly described as happy *guesses*."[25] What a difference between Whewell and Mill!

But there is more. What about the *process* of guessing? How does that work? Here Whewell gives his most remarkable reflections. He says, "We may, in such cases, conceive the discoverer as inventing and trying many conjectures, till he finds one which answers the purpose of combining the scattered facts into a single rule." Even more strikingly, he claims that "In the minds of most discoverers, this process of invention, trial, and acceptance or rejection of the hypothesis, goes on so rapidly that we cannot trace it in its successive steps." Again, "this often happens so instantaneously that we lose the trace of the progression." Again, "we should find that many more suppositions pass through their minds than those which are expressed in words; many a possible combination of conceptions is formed and soon rejected. There is a constant invention and activity, a perpetual creating and selecting power at work, of which the last results only are exhibited to us. Trains of hypotheses are called up and pass rapidly in review; and the judgment makes its choice from the varied group."[26]

In all the philosophy of science that we shall examine in this chapter, from the Victorian age to the present, Whewell's view is closest to Poe's view, which we shall examine in the following chapter. The similarity on guessing (with some profound differences) is really quite striking. What is equally striking is to compare Whewell and Mill on a prime case from the history of science, one which Poe also focused on, again as we shall see in the following chapter. It is Johannes Kepler's discovery of his first law of planetary motion, which is that planets move in ellipses, not in circles. (That they had to move in circles was the received dogma from the time of the ancient Greeks through to his day.) According to Whewell, "Thanks to his [Kepler's] communicative disposition, we know that he made nineteen hypotheses with regard to the motion of Mars, and calculated the results of each, before he

25. Whewell (note 22), vol. 2, 202, 206.

26. Whewell (note 22), vol. 2, 207, 208.

established the true doctrine, that the planet's path is an ellipse." Equally important, in the chapter following "Of the Colligation of Facts" Whewell states that "in every inference by Induction there is some Conception *superinduced* upon the Facts: and we may henceforth conceive this to be the peculiar import of the term *Induction*."[27] So Kepler guessed nineteen times, each time superinducing a particular conception upon the facts of the orbit of Mars provided him by Tycho Brahe, the twentieth being the happy guess that perfectly fit the facts, that of an ellipse, followed by a further happy guess, that "All the planets do what Mars does."[28]

Mill, on the other hand, plays down the guessing, doesn't like the word "superinduced," and emphasizes the verification stage. "We know from Kepler himself," says Mill, "that before hitting upon the 'conception' of an ellipse, he tried nineteen other imaginary paths, which, finding them inconsistent with the observations, he was obliged to reject." The "successful hypothesis" was a "skillful guess." Kepler "swept all these circles away, and substituted the conception of an exact ellipse." But in doing so "Kepler did not add to, but found in, the motions of the planet, namely, that it occupied in succession the various points in the circumference of a given ellipse."[29] Rather than "superinduce," says Mill, a better word would be "the common word Comparison."[30] And Kepler got it right because "the connecting link . . . *really exists* in the facts themselves," the ellipse "is really *in* the facts," so that "The honor, in Kepler's case, was that of the accurate, patient, and toilsome calculations by which he compared the results that followed from his different guesses, with the observations of Tycho Brahe; but the merit was very small of guessing an ellipse: the only wonder is that men had not guessed it before."[31]

In all of this, if we would get into it more, we would see an explicit debate over the proper meaning of "induction." In the above we can see a difference in emphasis over the role of imagination, which is our focus, but it also speaks to a difference in

27. Whewell (note 22), vol. 2, 207, 215.

28. Whewell (note 22), vol. 2, 216.

29. Mill (note 13), 180, 181.

30. Mill (note 13), 392.

31. Mill (note 13), 183, 390, 391.

the psychologies of the two antagonists. Mill is very left-brained, insistent upon rules and methods, notably with "Mill's Methods" of induction and what he calls the "Deductive Method" of science as a whole, and in the field of ethics with his utilitarian "formula," the "Principle of Utility." Whewell is much more right-brained, meaning more holistic-brained, giving deduction and induction their due but also giving more to imagination and intuition and big picture thinking, with his colligation of facts and his consilience of inductions test.[32]

But even Whewell, in the end, for all of that, places imagination outside the circumference of genuine science, and this is because of his concept of science. He says, for instance, that "When our conceptions are clear and distinct, when our facts are certain and sufficiently numerous, and when the conceptions, being suited to the nature of the facts, are applied to them so as to produce an exact and universal accordance, we attain knowledge of a precise and comprehensive kind, which we may term *Science*." Again he says, "We shall speak of the two processes by which we arrive at science, as *the Explication of Conceptions* and *the Colligation of Facts*." And yet again, this time with regard to the colligation of facts, he states that "The knowledge of such connexions, accumulated and

32. For an excellent discussion on Mill versus Whewell on Kepler, see Laura J. Snyder (2006). *Reforming Philosophy: A Victorian Debate on Science and Society.* Chicago: University of Chicago Press, 101–106. It should be noticed in all of this that Mill's *firsthand* knowledge of natural science was poor. With the exception of collecting botanical specimens, Mill had no experience in observing or conducting scientific research or experiments. He was an outsider looking in, who, I might add, made the common mistake of confusing or inferring scientific processes from published scientific reports. Whewell, on the other hand, was not only a philosopher and historian of science but an actual hands-on scientist. A founding member of the British Association for the Advancement of Science (he coined the term "scientist" in 1833) and later its president, Whewell was Professor of Mineralogy at Cambridge University and contributed works on mineralogy and crystallography, he participated in experiments regarding the mean density of Earth and the effect of gravity on pendulums, (I wonder if that included a pendulum some thirty or forty feet long with a crescent of glittering steel, about a foot in length from horn to horn,) he contributed to tidal research, (for which he won a medal from the Royal Society,) and he regularly communicated with scientists in fields of research other than his own. Snyder (this note), 151–152, 2, 20, 149–150.

systematized, is Science."[33] In all of this, what needs to be noticed is that science is conceptualized not as a process, but as a growing body of knowledge. The two processes labeled in italics above are not *parts* of science, instead it is by them that we "arrive at" science. The processes are distinguished from the product, and only the product gets called "science." Perhaps Whewell was overly influenced by the etymology of the word, which is related to the Italian word *scienza*, meaning "knowledge," both of which come from the Latin word for knowledge, *scientia*. At any rate, this is a good point at which to enter twentieth-century philosophy of science.

Logical Positivism

The philosophy of science that came to be known as *logical positivism* began in 1922, with a group of scientists, mathematicians, and philosophers that gathered informally on a weekly basis for discussions and which quickly came to be known as the Vienna Circle. The founder of the Circle was the physicist and philosopher Moritz Schlick, who hosted the discussions typically every Thursday evening in the chemistry building of the University of Vienna (not the same building where he was later shot dead by his crazed former doctoral student with a double motive). Among the big names who attended with some regularity were the physicist and philosopher Rudolf Carnap, the political scientist and mathematician Otto Neurath, the physicist and philosopher Herbert Feigl, the mathematician Hans Hahn, the mathematician and philosopher Friedrich Waismann, and occasionally the mathematician Kurt Gödel and later the philosopher A.J. Ayer. Beginning as something of a club, logical positivism eventually became something akin to a political party, marked by a manifesto entitled "The Vienna Circle: Its Scientific Outlook" (1929), written by Carnap, Neurath, and Hahn.[34]

33. Whewell (note 22), vol. 2, 169, 171, 202.

34. A.J. Ayer, ed. (1959). *Logical Positivism*. New York: The Free Press, 4.

What united the movement was not only the belief that only science provides the path to knowledge about reality, but also a profound rejection of metaphysics, claims about reality that go beyond the natural sciences, among which they included theological claims. To be abolished were statements such as "the Absolute enters into, but is itself incapable of, evolution and progress," a statement taken from F.H. Bradley's *Appearance and Reality* (1893), along with old favorites such as "God exists" and "God loves us." They did not reject statements such as these simply because they thought they are false, or that they cannot be proven, or that they are a matter of faith or wishful thinking or superstition. Instead, they rejected them because they thought they literally lack *meaning* in the sense of factual content about the world, like gibberish such as "Goo-goo ga-ga." They did recognize, however, that metaphysical statements can make a person feel good or can be expressions of feeling good, such as feeling comfortable in the Universe—so too can "Goo-goo ga-ga," by the way, along with "Everything happens for a reason."

How did they defend this view? First, following David Hume, they divided all meaningful statements into what Hume called (i) "matters of fact," (statements about the world, statements that are true or false depending on their correspondence to reality,) sometimes called *empirical* or *synthetic* statements, such as "The far side of the moon has mountains," and (ii) "relations of ideas," (statements that are matters of definition, and so are true or false solely because of the meanings of the terms, which is what you get, they claimed, in logic, mathematics, and geometry,) sometimes called *analytic* statements, such as "A bachelor is a married male" and "7 + 5 = 12." Like Hume, and against Kant after him, they maintained that all meaningful statements fall into these and only these two categories, with no meaningful statements falling into both or neither.

Second, they developed and maintained what they called the *verifiability theory (criterion, principle) of meaning*, which drew its inspiration from Ludwig Wittgenstein's *Tractatus Logico-Philosophicus* (1921). Here is how Carnap put it: "the meaning of a statement lies in the method of its verification. A statement asserts only so much as is verifiable with respect to it. . . . If something were to lie, in principle, beyond possible experience, it could be neither said nor thought nor asked."[35] Similarly, according to Schlick, "If I am *unable*, in principle, to verify a proposition, that is, if I am absolutely ignorant of how to proceed, of what I must do in order to ascertain its truth or falsity, then obviously I do not know what the proposition actually states, . . . For in so far as I am able to do this I am also able in the same way to state at least in principle the method of verification (even though, often, because of practical difficulties I am unable to carry it out). The statement of the conditions under which a proposition is true is *the same* as the statement of its meaning, and not something different."[36] In all of this one can see the distinction between verifiability in principle and verifiability in practice. The statement, "The far side of the moon has mountains," was not verifiable in practice in the 1930s but was verifiable in principle since one could state the observations which, if obtained, would verify the statement. But Bradley's statement quoted above was taken to be neither, and therefore literally meaningless. Ayer added a further qualification that should be noted, which is between "strong verification" and "weak verification," the former meaning "if, and only if, its truth could be conclusively established in experience," the latter meaning "if it is possible for experience to render it probable" or "that some possible sense-experience should be relevant to the determination of its truth or falsehood." But then the rest is basically the same: "If a putative proposition fails to satisfy this principle, and is not a tautology, then I hold that it is metaphysical, and that, being metaphysical, it is neither true nor false but literally meaningless."[37] The strong–weak distinction was necessary so as to allow for statements of laws of nature, whether true or false, to be meaningful, since statements of laws of nature, such as Newton's theory of

35. Rudolf Carnap (1932). "The Elimination of Metaphysics Through Logical Analysis of Language." *Erkenntnis*, vol. II. Reprinted in Ayer (note 34), 76.

36. Moritz Schlick (1932/1933). "Positivism and Realism." *Erkenntnis*, vol. III. Reprinted in Ayer (note 34), 87.

37. Alfred Jules Ayer (1936). *Language, Truth and Logic.* London: Gollancz, 37, 31.

gravity, are not true by definition, and neither do they refer only to observed bodies, but to all bodies past, present, and future.[38]

Philosophy, it should be noted, does not compete with science in this view, attempting to produce true statements about the world, but consists instead of *logical analysis*.[39] This idea was not new. It was similarly expressed in spirit, though not in detail, long ago by John Locke, often called the "father of British empiricism." According to Locke, philosophy, properly understood, is "to be employed as an Under-Labourer in clearing Ground a little, and removing some of the Rubbish, that lies in the way to Knowledge."[40]

I do not wish here to get sidetracked into a discussion about what caused the demise of logical positivism, such as the nagging problem that the verifiability theory of meaning seems to be neither tautological nor factual and is therefore meaningless, or the problem of what an observational statement is, or the problem of whether logical positivism excludes statements of laws of nature from science and is therefore contrary to actual science, or the dispersal of the logical positivists by WWII, let alone the debate over whether it was Popper's *Logic of Scientific Discovery* (1934/1959) or Quine's "Two Dogmas of Empiricism" (1951) that "killed" logical positivism. I do not even want to get into whether a "long line of scientists . . . had their scientific imagination stimulated by their religious concern."[41] Instead, what matters for the central purpose of this chapter is to notice that the conception of science that the logical positivists had, as logical positivists, was ambiguous and even contradictory, whether on the nature of science in general or on the role of theory formation in particular. This will be seen by

38. A further problem concerns counterfactuals, (contrary-to-fact conditionals,) in that laws of nature, in order to be real, do not require instances of their application in the Universe in order to count as a law, e.g., laws about chemical elements that do not occur naturally. Such laws are, instead, along with laws actually in operation, so-called permanent possibilities of the Universe.

39. Carnap (note 35), 77.

40. John Locke (1690). *An Essay concerning Humane Understanding.* London: Thomas Basset, "The Epistle to the Reader."

41. H.S. Thayer, ed. (1953). *Newton's Philosophy of Nature: Selections From His Writings.* New York: Hafner Press, xi.

reading their literature, but here I will merely illustrate the point with two examples.

Consider first a quotation from the Vienna Circle's founder: "For science, then, our standpoint does not represent something foreign and out of the ordinary, but it has in a certain sense always been more or less taken for granted. It could not be otherwise, because only from this standpoint is a proposition verifiable at all; and since *all the activities of science consist of examining the truth of propositions*, it continuously acknowledges the correctness of our insight and practice" (italics mine).[42] Clearly here the *activity* of scientific imagination (at least in the sense of theory formation) is not considered part of science. Consider next the following by Otto Neurath on the unity of science, a core theme in logical positivism, the idea that science, properly understood, "comprises all scientific *laws*," that these laws "are capable, without exception, of being combined with one another," that "The unified language of unified science, which is derivable by and large from modifications of the language of everyday life, is the language of physics," that "Unified science formulates statements, corrects them, and makes predictions," and that "The physicalist language, *unified language*, is the Alpha and Omega of all science. . . . there is only *Unified Science*, with its laws and predictions."[43] And what of scientific imagination in the "Alpha and Omega of all science"? Darkness there and nothing more.

A year later, however, as a dying ember wrought its ghost upon the floor, Schlick seems to have caught a glimpse of Poe's lost Lenore. Claiming that so-called protocol statements—"the statements [when 'written down or memorized'] concerning what is 'immediately perceived'"—constitute "the *beginning* of science," both temporally and logically, the point from which begins "the construction of science," Schlick then states:

> From them ["protocal statements"] would gradually arise the rest of the statements of science, by means of the process called "induction," which consists in nothing else than

42. Schlick (note 36), 89.

43. Otto Neurath (1931/1932). "Sociology and Physicalism." *Erkenntnis*, vol. II. Reprinted in Ayer (note 33), 286, 287, 293.

> that I am stimulated or induced by the protocol statements to establish tentative generalizations (hypotheses), from which those first statements, but also an endless number of others, follow logically. If now these others express *the same* as is expressed by later observation statements that are obtained under quite definite conditions which are exactly specifiable beforehand, then the hypotheses are considered to be confirmed so long as no observation statements appear that stand in contradiction to the statements derived from the hypotheses and thus to the hypotheses themselves. So long as this does not occur we believe ourselves to have hit correctly upon a law of nature. Induction is thus nothing but methodically conducted guessing, a psychological, biological process whose conduct has certainly nothing to do with "logic." In this way the actual procedure of science is described schematically.[44]

"In his rash exertions to rescue a favorite portion of his hunting stud, he himself perished miserably in the flames."

"I—n—d—e—e—d—!" ejaculated the Baron, as if slowly and deliberately impressed with the truth of some exciting idea.

"Indeed," repeated the vassal.

"Shocking!" said the youth [the Baron], calmly, and turned quietly into the palace.

Poe, I believe, would have been both delighted and appalled by the logical positivists. He would have been delighted by the stern passion and the immense learning of the logical positivists, especially when it came to the physical and mathematical sciences. He would have delighted also in the fact that Schlick in particular uses the term "guessing" in reference to "the actual procedure of science." But he would have been appalled at the obeisance to induction and at not taking guessing beyond the chamber door of consciousness and into the realm of illimitable intuition and the educated imagination. He would have been positively appalled, however, at the anti-metaphysics of the logical positivists, with

44. Moritz Schlick (1934). "The Foundation of Knowledge." *Erkenntnis*, vol. IV. Reprinted in Ayer (note 33), 210, 220.

their "Commit it then to the flames" of David Hume.[45] And in this, Poe would be in good company, with friends of metaphysics such as Karl Popper (more on him below) and modern titles in philosophy of science such as *The Metaphysics of Evolution* and *Metaphysics and the Origin of Species*.[46] And the metaphysics hasn't been confined to the ontology of trifles such as species, classes, individuals, and relations. As we have seen in the previous chapter, metaphysics in the sense of the supernatural is still in the game, most notably in the fine-tuning argument for God's existence debated by scientists and philosophers of science.

Poe on the whole, then, would probably have been more appalled than delighted by logical positivism, and increasingly so by the major movements in philosophy of science that followed, (with some minor exceptions,) which is the main focus of this chapter, the gross failure to include imagination in the scientific process, let alone to give it the main role in what drives science, as emphasized repeatedly by Poe in *Eureka*. In the next chapter we shall reconstruct Poe's theory of scientific imagination and then update it in the final chapter. But for the moment we are finished with logical positivism, except to say that its major oversight is arguably its main legacy, evidenced by the philosophies of science that emerged from its ashes in response to its perceived failures.

Logical Empiricism

Logical empiricism is the name sometimes given to the so-called Berlin school of logical positivism, given that logical positivism, properly speaking, was centered in Vienna prior to WWII. The two schools of thought, along with their names, are usually confused together in the secondary literature, as their advocates shared much in common philosophically and often collaborated together.

---

45. Ayer (note 33), 10.

46. David L. Hull (1989). *The Metaphysics of Evolution.* Albany: State University of New York Press. Michael T. Ghiselin (1997). *Metaphysics and the Origin of Species.* Albany: State University of New York Press.

But they also had differences that justify in keeping them apart.[47] Two of the names principally associated with this philosophy of science, namely, Hans Reichenbach and Carl Hempel, supply the focus of the present section. Without getting into all that distinguished logical empiricism from logical positivism, at the core of logical empiricism is the so-called *hypothetico-deductive method.* Scientific method, according to this view, does not consist of generalizations about nature derived from particular observations, the so-called *inductive method* of science that most of us were taught in grade school science class and that was shared by logical positivism. Instead, scientific method begins with a hypothesis from which observation statements are deduced. If some or all of the observation statements are found to be true, then the hypothesis gains inductive support, possibly even confirmation, while if they are found to be false, then the hypothesis is disconfirmed, possibly even falsified. (Confirmation in this view generates a number of paradoxes, one of which is known as the *raven paradox*, a discussion of which is not germane to our purpose, only the mention of Poe's bird, the omen of death.)

Reichenbach developed the basic structure of logical empiricism in his book, *Experience and Prediction* (1938). What is important for our purposes is not that he claimed in that book to be refuting logical positivism, but that he introduced a profound distinction to better understand the view of science provided by logical empiricism. This is the distinction between what he called the *context of discovery* and the *context of justification*.[48] Here, especially in the former, if anywhere in modern philosophy of science, is where we should expect to find something major about the role of imagination in science, and hence possibly a welcome home for Poe's ideas on science. But do we?

Here is how Reichenbach put the distinction, in his later book on philosophy of science published two years before his death. Looking first at the context of justification, he says "the scientist

47. I follow the useful distinction between logical positivism and logical empiricism outlined by Wesley C. Salmon (2000). "Logical Empiricism." In W.H. Newton-Smith (note 2), 233–242.

48. Hans Reichenbach (1938). *Experience and Prediction: An Analysis of the Foundations and the Structure of Knowledge.* Chicago: University of Chicago Press, 7, 382.

performs an inductive inference, since he wishes to say not only that the facts are derivable from his theory, but also that the facts make his theory probable and recommend it for the prediction of further observational facts. The inductive inference is employed not for finding a theory, but for justifying it in terms of observational data." The matter of justification is a little more complicated, he says, since "A set of observational facts will always fit more than one theory." Nevertheless, "The inductive inference is used to confer upon each of these theories a degree of probability, and the most probable theory is then accepted."[49]

The real interest for us, of course, is the context of discovery. It should be noticed, first, that by his very labels Reichenbach includes the context of discovery in the scientific process: it is one of the contexts of science. But equally important is what he has to say about it, about how a scientist comes up with a scientific theory in the first place. Quite simply, he says, "The scientist who discovers a theory is usually guided to his discovery by guesses; he cannot name a method by means of which he found the theory and can only say that it appeared plausible to him, that he had the right hunch, or that he saw intuitively which assumption would fit the facts." Plausible. Hunch. Intuition. Guesses. Poe would have loved this! Even more, Reichenbach discusses Kepler as one of his prime examples of scientific discovery.[50] Kepler, as we shall see in the next chapter, was Poe's prime example.

But the excitement quickly subsides once we see Reichenbach drop the ball precisely where he should have run with it. He says, "The act of discovery escapes logical analysis; there are no logical rules in terms of which a 'discovery machine' could be constructed that would take over the creative function of the genius. But it is not the logician's task to account for scientific discoveries; all he can do is to analyze the relation between given facts and a theory presented to him with the claim that it explains these facts. In other words, logic is concerned only with the context of justification. And the justification of a theory in terms of observational data is the subject of the theory of induction."[51]

49. Hans Reichenbach (1951). *The Rise of Scientific Philosophy.* Berkeley: University of California Press, 231, 232.

50. Reichenbach (note 49), 100–101.

51. Reichenbach (note 49), 231.

Logic can do this—logic cannot do that—what is wanted are logical rules. The problem here is that *logic* is not the be-all and end-all of a responsible philosophy of science. In terms of the latter, we should want to know, given the etymological meaning of *philosophy* as the love (*philia*) of wisdom (*sophia*) and that *it is not wisdom to ignore evidence*, (copyright © 2013 by David N. Stamos,) I say we should want to know *everything we can possibly know about the context of discovery.* In other words still, we should want to understand *the nature of scientific discoveries*, possibly even to account for them if we can. If the context of discovery is part of the process of science,—and it is,—and the purpose of philosophy of science is to understand science,—and it is,—then we should want to understand that part of the process as best we can. In other words still, we should want to go beyond *logic* and get into *evidence*. It would have been fine if Reichenbach, in 1951, would have said, "We need to understand as best we can how the context of discovery works, but at present we lack sufficient empirical methods, a situation that hopefully will be remedied in the future." But he did not say that. *Poe*—a hundred years earlier—pointed in that direction, and emphatically, as we shall see in the next chapter, while in the final chapter we shall turn to the modern evidence *as best we can* so as to understand the nature of scientific imagination. In other words, we shall pick up the ball precisely where Reichenbach dropped it.

Carl Hempel is not much better than Reichenbach on this matter, by the way. His collection of essays entitled *Aspects of Scientific Explanation and Other Essays in the Philosophy of Science* (1965) is considered both "the zenith" and "the most representative expression of the logical empiricist approach."[52] But it is not to that book that we shall turn, since it confines itself to the nature of scientific explanation and confirmation, but to his short introductory textbook published a year later. Here I quote in full what he has to say about our topic:

> There are, then, no generally applicable "rules of induction," by which hypotheses or theories can be mechanically derived or inferred from empirical data. The transition

52. Salmon (note 47), 237–238.

> from data to theory requires creative imagination. Scientific hypotheses and theories are not *derived* from observed facts, but *invented* in order to account for them. They constitute guesses at the connections that might obtain between the phenomena under study, at uniformities and patterns that might underlie their occurrence. "Happy guesses" of this kind require great ingenuity, especially if they involve a radical departure from current modes of scientific thinking, as did, for example, the theory of relativity and quantum theory. The inventive effort required in scientific research will benefit from a thorough familiarity with current knowledge in the field. A complete novice will hardly make an important scientific discovery, for the ideas that may occur to him are likely to duplicate what has been tried before or to run afoul of well-established facts or theories of which he is not aware.[53]

This is only a little better. In addition to recognizing, as Reichenbach did, the role of genius and that there are no rules to creative imagination, we also find a stress on scientific imagination as an educated imagination. And on the page that follows we find that "the scientist may give free rein to his imagination, and the course of his creative thinking may be influenced even by scientifically questionable notions." Kepler's work on planetary motion is then given as an example. This is all in the direction of Poe. Moreover, following the phrase "happy guesses" we find a footnote pointing out that "This characterization was given already by William Whewell," that A.B. Wolfe in his "Functional Economics" (1924) stressed the requirement of "scientific imagination," and that in *Conjectures and Refutations* (1962) Karl Popper "refers to scientific hypotheses and theories as 'conjectures.'"

The problem is that we should want more, much more. Whewell was not enough, and Popper, as we are about to see, is a dead end.

Poe himself would surely have questioned the essentiality of testability in the context of justification, and what we shall see in

53. Carl G. Hempel (1965). *Philosophy of Natural Science.* Englewood Cliffs, NJ: Prentice-Hall, 15.

the remainder of this chapter is that had he made the criticism he would surely have been right. He would also have criticized the logical empiricists for a very lopsided treatment of the process of science, calling them "lame of a leg," glad that they included the context of discovery, (rather than just the context of justification, let alone only the body of knowledge constituting the product of science,) but critical that their emphasis was on the context of justification. Poe, instead, took imagination to be the very heart and soul of science, or to vary the metaphor, its *wings*. That he was once again right, and remarkably so, is the focus of the rest of this book.

Karl Popper

Karl Popper was a Viennese philosopher of science who had close ties to the Vienna Circle. He was even considered by many to be a logical positivist himself, a charge he repeatedly denied. For one, he never accepted the verifiability criterion of meaning. In fact he was not against metaphysics, arguing that metaphysical ideas are often fruitful ideas that lead to scientific research, such as the atomism of the ancient Greeks. Moreover, Popper was anti-induction. Not only did he argue that David Hume was right, that inductive reasoning can play no role in knowledge, including scientific reasoning, but he even argued that people, not just scientists, do not in fact reason inductively. Instead we reason, he said, deductively, as per the *modus tollens* of deductive logic. Symbolically the law of logic is as follows:

$$p \rightarrow q$$
$$\sim q$$
$$\therefore \sim p$$

This is a valid form of argument. If $p \rightarrow q$ is true and $\sim q$ is true, then $\sim p$ must (logical "must") be true. If it is true that "If it is raining outside then my Jeep is wet," and if it is true that "My Jeep is not wet," then it must be true that "It is not raining outside."

Popper took this law of logic and did something quite profound with it. He said it is the way the pursuit of knowledge works, including scientific knowledge, which he took to be a refined and special case of everyday knowledge. Scientific theories, he said, *genuine* scientific theories, unlike pseudo-scientific theories such as astrology and Freudian psychoanalysis, entail testable predictions, observation statements, and *real* scientists, unlike pseudo-scientists, expose their theories to tests rather than do everything they can to avoid them. In other words, genuine scientific theories are *in principle* falsifiable, which is to say they entail observation statements any of which, if found to be true, would falsify the theory. Darwin exposed his theories to testing, as did Einstein, but astrologers do not. Astrologers never say if this or that were found then it would falsify astrology. When they do an astrological reading for someone and that person's life ends up not fitting the prediction, they simply explain away the failure using more astrology. What scientists do, instead, is try to fix the theory or replace it with a better one, one that explains the phenomena covered by the falsified theory and also the phenomena that falsified it. When a scientific theory is tested against observation and it passes the tests, however, Popper would never say that the theory is proven. He would not even say that the theory gained in probability. That would be inductive reasoning, which he rejected because observations are always limited but scientific theories are about unlimited observations, resulting in an inductive probability of zero given any set of possible observation statements. (For example, no matter how many ravens are observed to be black, the probability of the truth of "All ravens are black" will remain zero, since we cannot possibly observe all ravens past, present, and future.) All Popper would allow is that a scientific theory that passes attempts at falsification receives *corroboration*. In passing attempted falsifications, it would continue to play a productive role in science by explaining observed phenomena and predicting hitherto unobserved phenomena. In this way science progresses, with ever more inclusive theories replacing falsified ones.

Popper's first book on philosophy of science, *Logik der Forschung*, was published in 1935, but it did not receive widespread attention until it was translated into English in 1959 as *The Logic of Scientific Discovery*. Popper then followed this book with others, developing his ideas further and applying them to new domains, all of it as elucidations of what he called "critical rationalism."

What is remarkable about Popper's philosophy of science is that although it never gained much of a following in the world of philosophy of science, it did gain a strong following in the world of science, with many scientists of many different backgrounds endorsing Popper's characterization of science as being basically right. This includes big names, (two of the following are Nobel laureates,) such as Stephen Hawking in physics, Niles Eldredge and Stephen Jay Gould in paleontology, John Maynard Smith and Sir Peter Medawar in biology, and Jacques Monod in genetics.[54]

One problem with Popper's view is that he does not allow for a scientific theory to be proven true, in the sense of *proof beyond a reasonable doubt*. Scientific theories can only be proven false, he says, never true. This has prompted at least one prominent scientist to label Popperian falsificationists as a kind of "truth-heckler."[55] There is something to this assertion, given that scientists often *do* claim that there comes a time when the evidence for a scientific theory is so massive that it is no longer accurate to speak of it as a theory, (let alone as a hypothesis,) but as a *fact*, as with evolution in biology and the Big Bang in cosmology.[56]

But even worse, a linguistic idiosyncrasy of Popper's has virtually gone unnoticed in the secondary literature on Popper—including even by Popperians. In the context of falsifiability as the demarcation of science from pseudoscience, Popper sometimes writes of "scientific laws," sometimes of "scientific theories," and sometimes of scientific "laws or theories,"[57] such that virtually

54. Some of these scientists are discussed in my two Popper papers: (1996). "Popper, Falsifiability, and Evolutionary Biology." *Biology & Philosophy* 11 (2), 161–191. (2007). "Popper, Laws, and the Exclusion of Biology from Genuine Science." *Acta Biotheoretica* 55, 357–375.

55. Richard Dawkins (2003). *A Devil's Chaplain: Reflections on Hope, Lies, Science, and Love.* Boston: Houghton Mifflin, 16.

56. See, e.g., Jerry A. Coyne (2009). *Why Evolution Is True.* New York: Penguin, xvi–xvii. Richard Dawkins (2009). *The Greatest Show on Earth: The Evidence for Evolution.* New York: Free Press, 8. Steven Weinberg (2001). *Facing Up: Science and Its Cultural Adversaries.* Cambridge, MA: Harvard University Press, 174–175. Peter Shaver (2011). *Cosmic Heritage: Evolution from the Big Bang to Conscious Life.* New York: Springer, 64.

57. Karl R. Popper (1959). *The Logic of Scientific Discovery.* London: Routledge, e.g., 36, 59, 45n, 62, 425.

everyone took the "or" to be the "or" of inclusive disjunction, (as in, "Do you have a brother or sister?,") so that a scientific theory could be a statement of a law of nature (which is universal, like Newton's law of gravity) but need not be, in that it could also be a statement about a particular part of the Universe (like a theory about what caused the extinction of the dinosaurs). I seem to be the only one to have fastened upon the pale blue eye of Popper's philosophy, the dull blue eye with a hideous film over it that chills the very marrow in my bones,—the only one to have recognized that when Popper refers to scientific "laws or theories" he is using the "or" of synonymy, as when we say in logic, for example, "False Dichotomy or Bifurcation," which are two different names for the same fallacy.[58] The main implication of this is enormous, for it means that a scientific theory for Popper must not only be falsifiable in order to be scientific but it must also be a statement of a law of nature. And what this means, given the nature of theories in biology alone, is that Popper's philosophy of science excludes all or virtually all of biology![59]

Now it makes sense when Popper says all knowledge, from "the amoeba to Einstein," is "trial and error-elimination,"[60] "learning from our mistakes," "*conjectures and refutations*,"[61] and that scientific knowledge is "*common-sense knowledge writ large*."[62] What fundamentally distinguishes scientific knowledge for Popper is that its theories are inter-subjectively falsifiable statements of laws of nature.

---

58. Popper's linguistic idiosyncrasy also extended to his particular meaning of "testability," as in "*the criterion of the scientific status of a theory is its falsifiability, or refutability, or testability*." Karl R. Popper (1963). *Conjectures and Refutations: The Growth of Scientific Knowledge.* London, Routledge, 37, 39. Here we also find the locution "law or theory" regarding science, e.g., 54, 261, (and sometimes with a comma instead, e.g., 281). One can in fact find many kinds of examples of this linguistic idiosyncrasy in Popper's writings.

59. See my (1996), especially my (2007), (both in note 54,) for the case that Popper is to be understood as using the "or" of synonymy and for its implications.

60. Karl R. Popper (1979). *Objective Knowledge: An Evolutionary Approach.* Rev. ed. Oxford: Oxford University Press, 261, 25.

61. Popper (note 58), ix, 46.

62. Popper (note 57), 22.

But enough of his Evil Eye. What, we must ask, given the purpose of this chapter and given the influence of Sir Karl, did Popper have to say about the origin of scientific theories, about how they are produced? Given our malady, shall we find in Popper surcease of sorrow? Shall we find Popper dreaming dreams no mortal ever dared to dream before? *Id est*, shall we find in Popper the solution to our mystery of mysteries, the unriddling or unravelling of scientific imagination?

In a word, *no*. As Popper puts it, "The initial stage, the act of conceiving or inventing a theory, seems to me neither to call for logical analysis nor to be susceptible of it. The question how it happens that a new idea occurs to a man—whether it is a musical theme, a dramatic conflict, or a scientific theory—may be of great interest to empirical psychology; but it is irrelevant to the logical analysis of scientific knowledge."[63] This reminds me of a criticism made by the evolutionary biologist Ernst Mayr against philosophy of science in general: "What I do not understand is why most philosophers of science believe the problems of the philosophy of science can be solved by logic."[64] In confining the most fundamental focus of philosophy of science to "logical analysis," in refusing to apply a "rational reconstruction" to anything but "the *subsequent tests* whereby the inspiration may be discovered to be a discovery, or become known to be knowledge,"[65] Popper reveals the most serious flaw in his philosophy of science. If (i) the fundamental question, as he puts it, is "the problem of cosmology: *the problem of understanding the world—including ourselves, and our knowledge, as part of the world*,"[66] and if (ii) "*science starts only with problems* . . . and not from observations" and "every worthwhile new theory raises new problems; problems of reconciliation, problems of how to conduct new and previously unthought-of observational tests,"[67] then (iii) it seems we have a

63. Popper (note 57), 31.

64. Ernst Mayr (2004). *What Makes Biology Unique? Considerations on the Autonomy of a Scientific Discipline.* Cambridge: Cambridge University Press, ix.

65. Popper (note 57), 31.

66. Popper (note 57), 15.

67. Popper (note 58), 222.

kind of behaviorism here: philosophy of science confined to what is observable, to problem–theory–tests, excluding that which cannot be observed, (so far,) the generation of the theory. This is not philosophy, the love of wisdom, directed toward science. And this is because science does not take the directions it takes simply because of the testing of theories. As Poe perceived, it is the *generation* of the theories that propels science, *without which there would be nothing to test.*

What philosophers of science should be doing, then, is not logic so much as *ontology*, figuring out the nature of the entity they are studying, and for that, again, *it is not wisdom to ignore evidence*. It should therefore be the ultimate job of philosophers of science to do this, to do *metascience*, to figure out the nature of science—it is not the job of scientists, (although they certainly can contribute to the topic,) whose job is to do science itself.

Popper will allow that "Intuition and imagination . . . are indispensable as the main sources of our theories," although he says their "most important function . . . is to help us in the critical examination of those conjectures which are the means by which we probe into the unknown."[68] Elsewhere he conjectures that "the amoeba's actions" and "Einstein's actions" are not so really different given that "their methods of almost random or cloud-like trial and error movements are fundamentally not very different." The main difference, he says, is in "their attitudes towards error."[69]

Alas, we shall have to look elsewhere if we want to understand the nature of scientific imagination, let alone what promotes it. Popper is a dead end—stone, stone dead.

Thomas Kuhn

Arguably the most widely read and influential book of twentieth-century philosophy of science is Thomas Kuhn's *The Structure of Scientific Revolutions*, first published in 1962 and then again in 1970 with a Postscript designed to reply to misunderstandings.

68. Popper (note 58), 28.

69. Popper (note 60), 247.

Kuhn himself received a Ph.D. in physics before he turned professionally to history and philosophy of science, and part of the value of his book is that it forces philosophers to take a very close look at actual history of science rather than to simply rely on the published products of scientists, let alone on science textbooks (which are idealized for teaching purposes). Even his staunchest critics have to admit that Kuhn's philosophy of science is historically rich to an unprecedented degree. But it is with his conceptual reevaluation of the nature of science that Kuhn has made his greatest impact, one that—without exaggeration—prompts a binary division of the history of philosophy of science into before Kuhn and after Kuhn, or more precisely before *Structure* and after *Structure*.

At the core of Kuhn's view of science is what he calls a "paradigm." In the Postscript he tells us that he uses this term in two different ways, one in the sense of "the entire constellation of beliefs, values, techniques, and so on shared by the members of a given community," the other in the sense of "the concrete puzzle-solutions which, employed as models or examples, can replace explicit rules as a basis for the solution of the remaining puzzles of normal science."[70] According to Margaret Masterman, however, Kuhn uses "paradigm" in *Structure* in at least twenty-one different ways.[71] One of those ways is the sense of a classic work. "Paradigms" such as Newton's *Principia*, says Kuhn, "served for a time implicitly to define the legitimate problems and methods of a research field for succeeding generations of practitioners."[72] In this sense of "paradigm" Newtonian physics defined the way to do science for a scientific community, (or rather a community in an expanded sense, one composed of physicists, mathematicians, and astronomers,) while Newton's explanation of tides using the gravitation of the Sun and Moon served as an example of a paradigm in Kuhn's second sense.

70. Thomas S. Kuhn (1970). *The Structure of Scientific Revolutions.* 2nd ed. Chicago: University of Chicago Press, 175.

71. Margaret Masterman (1970). "The Nature of a Paradigm." In Imre Lakatos and Alan Musgrave, eds. (1970). *Criticism and the Growth of Knowledge.* Cambridge: Cambridge University Press, 59–89.

72. Kuhn (note 70), 10.

So far so good, really. The controversies start with Kuhn's distinction between what he calls "normal science" and "revolutionary science." During normal science what Kuhn calls "puzzle-solving research" is conducted within the framework of the existing paradigm, and it is the abilities of the scientists themselves, working within the paradigm, that are repeatedly tested, not the paradigm itself. During this period, which he says is the norm for the life of most scientists, scientists are not engaged in critical reasoning but rather in "mopping-up operations,"[73] which consist of fact gathering, making predictions, attaining greater precision or articulation, applying the paradigm to new phenomena, developing related technology, and so on. That scientists are normally not critical thinkers is elsewhere stated very explicitly by Kuhn: "critical discourse recurs only at moments of crisis when the bases of the field are again in jeopardy. Only when they must choose between competing theories do scientists behave like philosophers."[74]

A paradigm reaches a crisis point, according to Kuhn, when failures and anomalies have accumulated to the point that it is no longer the scientists working within the paradigm that come into question but the paradigm itself. This period of crisis "generally," he says, precipitates a scientific revolution, during which one or more new paradigms (in the sense of a theoretical framework) are developed and raised against the existing paradigm, which is then followed by a new period of normal science using the winning paradigm.[75]

Now where in all of this does scientific imagination fit in? Clearly the kind that concerns us most, the kind behind a revolutionary new theory, is not going to be found in the heads of scientists during normal science, given Kuhn's view of normal science as "mopping-up operations." We shall have to look, instead, at what Kuhn calls "extraordinary research,"[76] which occurs when

73. Kuhn (note 70), 24.

74. Thomas S. Kuhn (1970). "Logic of Discovery or Psychology of Research?" In Lakatos and Musgrave (note 71), 6–7.

75. Kuhn (note 70), 67.

76. Kuhn (note 70), 91.

scientists in a community come to consider their paradigm as flawed and in need of replacement.

The first thing to notice is that when a new paradigm enters the scene and scientists change from the old paradigm to the new, Kuhn does not see this as a particularly rational activity. The change in a scientific community from one paradigm to another can take decades or even centuries, (at least one hundred years in the case of the Copernican revolution,)[77] but in the head of a single scientist he says the switch "cannot be made a step at a time" but instead is sudden, like a "gestalt switch" in psychology, "all at once . . . or not at all," as with the famous duck/rabbit diagram (think also of a Necker Cube) but confined to a single switch, not back and forth, a profound "conversion experience" with a "transfer of allegiance" involving "faith" that the new paradigm will succeed in solving problems and is on the right track.[78] Between the revolutionary divide, moreover, there is what Kuhn calls "incommensurability," which he characterizes as involving different theoretical kinds, definitions, and problems, misunderstanding between the two sides that he characterizes as "communication breakdown" and "talking through each other," and living in what he calls "different worlds."[79]

But surely there must be involved a profound act of reason, whether conscious or unconscious, for a scientist to change from one paradigm to another. Although Kuhn makes clear that he is not to be understood as claiming that "no arguments are relevant," he also makes clear that during a revolution "The claim to have solved the crisis-provoking problems is . . . rarely sufficient by itself. Nor can it always legitimately be made." Kuhn himself puts more emphasis on factors such as "idiosyncrasies of autobiography and personality," "nationality or the prior reputation of the innovator and his teachers," and "aesthetic considerations." Equally important, even when "at last only a few elderly

---

77. Thomas S. Kuhn (1957). *The Copernican Revolution: Planetary Astronomy in the Development of Western Thought.* Cambridge, MA: Harvard University Press, 1.

78. Kuhn (note 70), 150, 111, 151, 158.

79. Kuhn (note 70), 148, 149, 201, 132, 111.

hold-outs remain," Kuhn claims that "even they, we cannot say, are wrong." This is because "neither proof nor error is at issue."[80]

Kuhn was always very sensitive to the charge that he "glorifies subjectivity and even irrationality" and also that he paints a picture of science as "relativistic."[81] But he has to (if he were still alive) take responsibility for these charges. Beginning with Kuhn himself, who distinguished what he wrote from what his critics said he wrote, it is often said that there are "two Thomas Kuhns," the one who writes of science as relativistic, not progressing in knowledge toward greater understanding of nature, and being more like a religion or political ideology, and the Kuhn who vehemently denied the charge and believed that science was largely a rational activity.[82] I myself think that Kuhn was something of a William Wilson, with the latter Kuhn being his inner double or *doppelgänger*, the one who repeatedly denied the charge of relativism and irrationalism and had a conscience, and the relativist Kuhn being the stronger of the two, repeatedly (though not with "brute ferocity") trying to kill the conscience, and generally having the upper hand, especially when he wrote freehand as opposed to replying to criticisms. I say this because of the recurring dominance in Kuhn's writings of his Darwinian analogy for science evolution. He affirms that "Normal research . . . *is* cumulative," but he denies that "changes of paradigm carry scientists and those who learn from them closer and closer to the truth." In other words, he affirms "scientific development as a succession of tradition-bound periods punctuated by non-cumulative breaks," which means he denies that paradigm change is cumulative. In other words again, he denies that progress in science occurs except "only during periods of normal science."[83] Moreover, he explicitly compares "a succession of theories related by descent" to "an evolutionary tree," such that not only "scientific development is, like biological, a unidirectional and irreversible process," but "the notion of a match between

80. Kuhn (note 70), 152, 154, 153, 156, 159, 151.

81. Kuhn (note 70), 186, 205.

82. Thomas S. Kuhn (1970). "Reflections on my Critics." In Lakatos and Musgrave (note 71), 231. See also, e.g., Alan Sokal and Jean Bricmont (1998). *Fashionable Nonsense: Postmodern Intellectuals' Abuse of Science*. New York: Picador, 75.

83. Kuhn (note 70), 96, 170, 208, 163.

the ontology of a theory and its 'real' counterpart in nature now seems to me illusive in principle," since there is "no theory-independent way to reconstruct phrases like 'really there.'" Scientific theories are therefore neither true nor false but merely useful, like tools, such that "As in manufacture so in science—retooling is an extravagance to be reserved for the occasion that demands it. The significance of crises is the indication they provide that an occasion for retooling has arrived."[84] So just as a biological species is neither true nor false, but its members or parts are only adapted better or worse to a particular environment, so too for Kuhn for a scientific theory or paradigm.

The two Kuhns continued through his later publications. Take, for example, a paper of Kuhn's in which he calls it a "total misunderstanding" to say that his view of science makes theory choice, as his critics put it, "a matter for mob psychology" or that it "cannot be based on good reasons of any kind" or that it constitutes "mere persuasive displays without deliberative substance." I'll never forget my impression the first time I read this paper, which includes the word "objectivity" in its title. Kuhn begins by providing a selection of five "characteristics of a good scientific theory": that it "should be accurate," that it "should be consistent," that it "should have broad scope," that it "should be simple," and that it "should be fruitful." Wonderful, I thought! But no sooner does he bring these up, he then proceeds to tear them down with "difficulties . . . regularly encountered by the men who must use these criteria in choosing." In short, he ends up rejecting the common distinction between the context of discovery and the context of justification since both are riddled with "subjective factors," such that "scientists who share the concerns and sensibilities of the individual who discovers a new theory are ipso facto likely to appear disproportionately frequently among the theory's first supporters." He even believes that some of the "subjective factors" are "present later when the profession agrees." And *that* is all that Kuhn means by "objective"—in the sense, to use one of his recurring analogies, in which the shared language of a community is objective. Scientists, once normal science has resumed, "speak the [new] language like a native." Objectivity is

84. Kuhn (note 70), 205, 206, 76.

really only shared subjectivity. This is why Kuhn states that "No process quite like choice has occurred, but they are practicing the new theory nonetheless."[85]

But wait! Is there no balm in Gilead; is there no physician there? Does not Kuhn have anything *positive* to say about scientific imagination when it comes to the invention of a new paradigmatic theory, something about the creative process that if not true is nevertheless *adaptive*? Notably, Kuhn says, in reference to paradigm anomalies and crises, that

> these are terminated, not by deliberation and interpretation, but by a relatively sudden and unstructured event like the gestalt switch. Scientists often speak of the "scales falling from the eyes" or of the "lightening flash" that "inundates" a previously obscure puzzle, enabling its components to be seen in a new way that for the first time permits its solution. On other occasions the relevant information comes in sleep. No ordinary sense of the term "interpretation" fits these flashes of intuition through which a new paradigm is born. Though such intuitions depend upon the experience, both anomalous and congruent, gained with the old paradigm, they are not logically or piecemeal linked to particular items of that experience as an interpretation would be. Instead, they gather up large portions of that experience and transform them to the rather different bundle of experience that will thereafter be linked piecemeal to the new paradigm but not to the old.[86]

Elsewhere Kuhn states, in explicit response to Popper, including where they agree, that "Where he rejects 'the psychology of knowledge', Sir Karl's explicit concern is only to deny the methodological relevance of an *individual's* source of inspiration or of an individual's sense of certainty. With that much I cannot disagree. It is, however, a long step from the rejection of the psychological

85. Thomas S. Kuhn (1973). "Objectivity, Value Judgment, and Theory Choice." In Thomas S. Kuhn, ed. (1977). *The Essential Tension: Selected Studies in Scientific Tradition and Change.* Chicago: University of Chicago Press, 321, 322, 327, 328, 329, 339.

86. Kuhn (note 70), 122–123.

idiosyncrasies of an individual to the rejection of the common elements induced by nurture and training in the psychological make-up of the licensed membership of a *scientific group*."[87] In his Postscript to *Structure*, written at basically the same time, Kuhn elaborates on this point, and in response to criticism: "Some readers have felt that I was trying to make science rest on unanalyzable individual intuitions rather than on logic and law. But that interpretation goes astray in two essential respects. First, if I am talking at all about intuitions, they are not individual. Rather they are the tested and shared possessions of the members of a successful group, and the novice acquires them through training as a part of his preparation for group-membership. Second, they are not in principle unanalyzable."[88]

But enough. Kuhn's relativism, his entire framework for looking at science, means that there is no such thing as a *genuine* eureka moment, a revolutionary intuition or insight that goes beyond the existing scientific paradigm and provides a more accurate and truthful understanding of the nature of the reality that is the domain of the corresponding science. The only intuitions that matter to Kuhn, and hence that are worthy of study in his view, are the "shared possessions" of a scientific community. *But those are not the intuitions that should be of primary interest if we are realists*. Instead, it is the *successful* intuitions of scientific geniuses that should be our main focus, the intuitions that take an existing science and bring it closer to the truth, whether those intuitions occur inside of a particular scientific community and set the science on a new course, such as those of Darwin and Einstein, or occur outside of the community, like those of Poe. In either case, it seems unlikely that those successful intuitions are simply a matter of blind chance, given their complexity and accuracy, so that realists should want to understand how they work if they want to understand science as a process. But Kuhn's relativism absolutely refuses to admit concepts such as *closer to the truth*, *objectivity*, and *genius*.

Finally, given Kuhn's relativism, there should be little wonder that Kuhn quickly became the darling of the many in the

87. Kuhn (note 74), 22.

88. Kuhn (note 70), 191.

humanities and so-called social sciences, who—filled with their own kind of *ressentiment*, which now masquerades behind the seemingly respectable label "Science Studies," combined with their love of postmodernism—jealously desire to bring the natural sciences down to their level. Consider, for example, the arch-postmodernist Richard Rorty, who uses Kuhn's writings repeatedly throughout his magnum opus[89] to support his view that natural scientists are no more rational than philosophers or literary critics, that "it was a mistake to try to make the natural scientist into a new sort of priest, a link between the human and the non-human," that Kuhn's "pragmatist friends (such as myself) routinely congratulate him on having softened the distinction between science and non-science," that they want "to enlist Kuhn in our campaign to drop the objective-subjective distinction altogether," and that "objectivity" should simply mean "unforced agreement" or "intersubjective agreement."[90] Rorty even goes so far as to claim that "Kuhn's description of how science works was no shock to the scientists whose rationality the philosophers [critics of Kuhn] were concerned to protect."[91]

89. Richard Rorty (1979). *Philosophy and the Mirror of Nature.* Princeton, NJ: Princeton University Press. It is worth noting that Kuhn is discussed on forty-eight pages to Popper's zero.

90. Richard Rorty (1987). "Science as Solidarity." In John S. Nelson et al., eds. (1987). *The Rhetoric of the Human Sciences: Language and Argument in Scholarship and Public Affairs.* Madison: University of Wisconsin Press, 40–41, 42.

91. Rorty (note 89), 333n16. Rorty's statement is wishful thinking at best and outright deception at worst. If by 1979 scientists did not seem shocked by Kuhn it was probably because they had not read him. When they do, I suspect the response is much like that of Steven Weinberg, 1979 Nobel laureate in Physics: "Kuhn then proceeded to the fallacy—much clearer in what he has written recently—that in science we are not in fact moving toward objective truth. I call this a fallacy because it seems to me a simple non sequitur. I do not see why the fact that we are discovering not only the laws of nature in detail, but what kinds of laws are worth discovering, should mean that we are not making objective progress." Weinberg (note 56), 85–86. Or consider the eminent evolutionary biologist John Maynard Smith on Kuhn, recipient of the Darwin Medal from the Royal Society of London: "The fallacy . . . is to suppose that because two scientists are each unable fully to understand what the other is saying, there is no rational way, given time, of settling the issue between them. . . . However committed some people may become to

What in sum, then, shall we say of Kuhn's contribution to our understanding of the nature of science? That there was much of the beautiful, much of the wanton, much of the *bizarre*, something of the terrible, and not a little of that which might have excited disgust.

The New Experimentalism

In support of his view of science and against the traditional, Baconian view, Popper argued that science (and even everyday life) does not proceed from observation to theory but from theory to observation, and also that there is no such thing as theory-neutral observation. This is the view known as the *theory-dependence of observation*, and it permeates his writings. He says, for example, that "My point of view is, briefly, that our ordinary language is full of theories; that observation is always *observation in the light of theories*; that it is only the inductivist prejudice which leads people to think that there could be a phenomenal language, free of theories, and distinguishable from a 'theoretical language.'" Given that Popper seems to be the favorite philosopher of science among actual scientists, it might be illuminating to look at more of his pronouncements on this head, such as: "the real situation is quite different from the one visualized by the naive empiricist, or the believer in inductive logic. He thinks that we begin by collecting and arranging our experiences, and so ascend the ladder of science. . . . it could never add up to a *science*. A science needs points of view, and theoretical problems." Again: "observations . . . are always *interpretations* of the facts observed; . . . they are *interpretations in the light of theories*. This is one of the main reasons why it is always deceptively easy to find *verifications* of a theory,

---

one camp or the other, there is usually a number of scientists well able to understand what is at issue, and to recognize the relevance of experiments supporting one view over the other. The final verdict often combines the new views and the old, but, if so, it is a rational synthesis and not an arbitrary compromise." John Maynard Smith (1980). "Science, Ideology, and Myth." In John Maynard Smith, ed. (1989). *Did Darwin Get It Right? Essays on Games, Sex and Evolution*. London: Chapman & Hall, 47. I could easily multiply examples.

and why we have to adopt a *highly critical* attitude towards our theories . . . ." And again, more interestingly: "Generally, similarity, and with it repetition, always presuppose the adoption of *a point of view*: some similarities or repetitions will strike us if we are interested in one problem, and others if we are interested in another problem. But if similarity and repetition presuppose the adoption of a point of view, or an interest, or an expectation, it is logically necessary that points of view, or interests, or expectations, are logically prior, as well as temporally (or causally or psychologically) prior, to repetition."[92] And yet again: "The problem 'Which comes first, the hypothesis (*H*) or the observation (*O*)?' is soluble; as is the problem, 'Which comes first, the hen (*H*) or the egg (*O*)?'. The reply to the latter is, 'An earlier kind of egg'; to the former, 'An earlier kind of hypothesis'. . . . There is no danger here of an infinite regress. Going back to more and more primitive theories and myths we shall in the end find unconscious, *inborn* expectations."[93]

For Kuhn, "the proponents of competing paradigms practice their trades in different worlds," each paradigm is a gestalt, determining how the scientist sees the world as a whole, "novelty" is seeing that "something has gone wrong," "Anomaly appears only against the background provided by the paradigm," and, as we have just seen in the previous section, there is "no theory-independent way to construct phrases like 'really there'; the notion of a match between the ontology of a theory and its 'real' counterpart in nature now seems to me illusive in principle."[94] In short, "Sir Karl and I . . . both emphasize . . . the intimate and inevitable entanglement of scientific observation with scientific theory."[95]

Popper and Kuhn, these two giants of modern philosophy of science, fundamentally antipodes as we have seen from the preceding sections, share something profoundly in common, which is that in sharing the thesis of the theory-dependence of observation they leave philosophy of science in a parlous state. Popper's version of the thesis would seem to entail, no matter the

92. Popper (note 57), 59n, 106, 107n, 421–422.

93. Popper (note 58), 47.

94. Kuhn (note 70), 150, 65, 206.

95. Kuhn (note 74), 2.

mental acrobatics and contortions that he puts his readers through to avoid it, that it is incoherent to embrace that thesis while also embracing *verisimilitude*, his word for the degree of truth of a theory, in the sense of correspondence to reality. And Kuhn's version of theory-dependence, while rejecting *verisimilitude* to begin with, leaves us with science as noncumulative from revolution to revolution, science without progress except as confined to a paradigm only, with paradigm succession resembling a blind walk in the dark, ultimately going nowhere.

Enter what Robert Ackerman calls "the new experimentalism." According to this view, as he puts it, the "fundamental intuition" of progress along with objectivity in science is "resurrected"—and one might add "rescued"—by the emphasis on experimentation in science, the data of which, he says, "function as solid hinge points for controlling theoretical conjectures." What is especially interesting is his conjecture for "why experiment has been rigorously neglected in history and philosophy of science by comparison to theory." His answer, quite simply, is that up until the twentieth century, although "Almost all physicists were experimentalists (if also theorists)," what happened was that "The emerging age of autonomous theorists [think especially of Einstein] was caught up into emerging positivism," with the latter focusing on "the logical structure of *written* scientific books and papers."[96] His answer, in short, is that historical contingency, not scientific or philosophical necessity, is to blame. That alone is quite interesting.

A major exponent of the new experimentalism is the philosopher Ian Hacking. In his book *Representing and Intervening*, Hacking argues that the new experimentalism lifts science above the problems posed by Popper and Kuhn, most importantly the one we just examined above, the theory-dependence of observation. This is because, as he variously puts it, "experimental science has a life more independent of theorizing than is usually allowed," that "Experimentation has a life of its own, interacting with speculation, calculation, model building, invention and

96. Robert Ackerman (1989). "The New Experimentalism." *British Journal for the Philosophy of Science* 40, 185, 186, 188.

technology in numerous ways."[97] One of his prime examples is microscopy. The level of theory in making and using microscopes to "intervene" in the world is what he calls "low level," nothing much beyond engineering and first-year college physics. But more interestingly, experiments using microscopes can affect "high level" theory, he says, possibly even triggering a scientific revolution. Moreover, while the pure theorist can afford to be an anti-realist about theoretical entities, he or she is like the anti-realist you get in philosophy, based on the "spectator theory of knowledge." The experimenter, on the other hand, is a hands-on thinker, using more engineering than theory, and consequently "must be a realist." This is not only because experiments necessarily involve "non-theory to sort out artifacts from the real thing," but because the experimenter actually gets to "manipulate" entities, thereby measuring and understanding their causal powers.[98] Hence, looking at the history of microscopy alone, we can see not only scientific progress in instrumentation but also in knowledge of the world, in "representing," whether within a scientific paradigm or between them.

A striking corroboration of Hacking's claims, if I may, is one of my favorite examples, which actually extends what he says about electrons in his book. It is the so-called "taming" of the electron and then of the atom in what are known as "traps." For many decades theoretical physicists could posit the reality of electrons as theoretical entities only, while in the same breath deny their reality in nature because of positivist philosophy (to ascribe reality to something unobservable, according to this philosophy of science, is meaningless). Even the realists among them could disagree about the physics of the electron, whether the statistics of its behavior involves genuine chance (the emerging quantum realist interpretation of the subatomic world) or obeys deterministic laws (the traditional interpretation headed by Einstein). In 1973, however, after modifying a device developed in 1936 by Frans Michel Penning for confining electrical currents in radio tubes,

97. Ian Hacking (1983). *Representing and Intervening: Introductory Topics in the Philosophy of Science.* Cambridge: Cambridge University Press, vii, xiii.

98. Hacking (note 97), 274, xiii, 200.

Hans Dehmelt and two assistants managed to capture a single electron in a "Penning trap," which confines the electron in a small vacuum chamber surrounded by a negatively charged magnetic field. In the words of von Baeyer, "the electron circles endlessly within the box, bobbing like a horse on a merry-go-round." Not only did this put a permanent end to the view of Heisenberg and Schrödinger and others, of electrons only as mathematical constructs, not as real entities, but it also allowed for vastly greater knowledge about the properties of electrons, such as the measurement of their magnetism. Then, in 1979, Dehmelt and his team isolated the first atom in a similar trap, a charged barium atom, while in 1986 he published his observations of "quantum jumps" in the same kind of atom. Von Baeyer himself observed a trapped mercury atom, which went dark every time it underwent a quantum jump, more commonly known today as a "quantum transition." Such observations, in a *trap*, of an atom in "free space," free of outside influences except for the ultraviolet light used to illuminate it, demonstrates (beyond a reasonable doubt, I would say) that "quantum jumps are real," as von Baeyer puts it, ending a major doubt about the quantum revolution, which is that quantum jumps are events of pure chance rather than of causal determination.[99]

There is no question that scientific experiments often require a great deal of what may be called scientific imagination. Dehmelt's Nobel Prize in 1989 was well deserved. It reminds one of the imagination behind the experiments on entanglement published by Alain Aspect and his team in 1982, which established that nonlocality is a fact of the subatomic world, (again a major blow to the deterministic worldview of Einstein and others,) that, "as a fundamental truth about the Universe," in the words of Gribbin, "there are correlations which take place instantaneously, regardless of the separation between the objects involved."[100]

Imagination in the progress of experimental knowledge surely counts, then, as scientific imagination. And of course

99. Hans Christian von Baeyer (1992). *Taming the Atom: The Emergence of the Visible Microworld.* New York: Random House, 94–95, 93, 101.

100. John Gribbin (1999). *Q is for Quantum: Particle Physics From A to Z.* London: Phoenix Giant, 29.

experimentalists can have their eureka moments, too. Nevertheless there is something profoundly missing in the view of science emphasized by the new experimentalists, in line as it is with the common idea of a scientist in a lab coat. It is that the imagination focused on is lower-level imagination. The beast we are after, however, is higher-level imagination, exemplified by the likes of Darwin and Einstein, whose imaginations brought together diverse realms of theory and evidence in new ways, ones that revolutionized the way subsequent scientists view and deal with the world. What we are after, then, we lovers of Poe *and* of science, is the nature of scientific imagination in the grand sense.

The new experimentalists, by their very own terms, cannot possibly give us that, given that their focus is on experimentation, engineering, and lower-level theories. But in saying this I am granting their own premises. Alan Chalmers, on the other hand, in his excellent chapter on the new experimentalism, (found in his equally excellent book on the nature of science,) is not so charitable. He does argue that the new experimentalists have delivered some important goods, especially on how experiment can sometimes lead scientists in new directions, directions not determined by prior theory, but he also claims that "it would be a mistake to regard it [the new experimentalism] as the complete answer to our question about the character of science. Experiment is not so independent of theory as the emphasis of the previous sections of this chapter might suggest." Chalmers also points out the obvious, that "theory often does guide experimental work and has pointed the way towards the discovery of novel phenomena."[101] He gives Einstein's theory of general relativity as an example, and we have already seen other examples in this section. Francis Crick, the co-discoverer of the structure of DNA and a Nobel laureate, adds something that is not so obvious, which is that "experimental facts are often misleading or even plain wrong." While no advocate of pure theory, and in fact a harsh critic of "most mathematicians" given that they "dislike reading experimental papers," Crick argues for an interplay between theory and experiment, experimental evidence for providing the needed "constraints . . . through the

101. A.F. Chalmers (1999). *What is This Thing Called Science?* 3rd ed. Indianapolis, IN: Hackett, 206.

jungle of possible theories," for the "jolt" often needed to loosen theorists from their preconceived theories, for providing a "crucial test" of a theory, for discovering phenomena like introns "by accident," and for theory or model building for serving to "unite evidence from several different approaches so that various sorts of tests can be made of it," which is especially powerful, he says, when a theory or model is extended to and supported by different kinds of phenomena, as with the theory of natural selection at the level of cells and molecules.[102] It might nevertheless be said that Crick is understating the importance of experimentation in favor of theory as something higher,[103] but surely most would agree that both are essential parts of science as an actual process.

And what might Poe have said? In *Eureka* he emphasizes (no doubt partly in defense of himself) that "Kepler was essentially a *theorist*."[104] One can say the same about Darwin and Einstein. Einstein dabbled with physical experiments but only did thought experiments in developing his revolutionary theories, and even though Darwin did lots of physical experiments he could easily have written *On the Origin of Species* without them. And of course Kepler, Darwin, and Einstein were unquestionably great scientists. This is not necessarily to disparage experimentalists, but then Poe might also have quoted Dupin in "The Murders in the Rue Morgue" (1841): "They make a vast parade of measures; but, not unfrequently, these are so ill adapted to the objects proposed, as to put us in mind of Monsieur Jourdain's calling for his *robe-de-chambre—pour mieux entendre la musique*."[105] The significance of the French line in this quotation was touched on in Chapter 4, and shall be furthered in Chapters 7 and 8. In short, there is more, much more, to science than measurement, or experiment—it is *the better to hear the music*. And how utterly interesting that Crick himself would say, in the context of upsetting orthodoxy, "Each generation needs a new music."[106]

---

102. Francis Crick (1988). *What Mad Pursuit: A Personal View of Scientific Discovery*. New York: Basic Books, 141, 140, 136, 147, 115.

103. Rosalind Franklin (personal communication). See Crick (note 102), 135.

104. *P&T*, 1270.

105. *P&T*, 412.

106. Crick (note 102), 74.

The Disunity of Science Movement

As John Dupré, a major advocate of this movement, puts it, "it is the denial that science constitutes, or could ever come to constitute, a unified project," partly because of a "metaphysical" thesis, "an assertion of the extreme diversity of the world." Chaos theory, he says in his book, undermines the view of the Universe as a vast machine, while biology destroys any hope for a unified theory of essentialistic natural kinds, leaving us with a philosophy of natural kinds best understood as "promiscuous realism." The search for a unified field theory, moreover, combining relativity physics and quantum physics, "is now seen as a fundamentally misconceived quest." And as for the "sophisticated and shared optimism about the rationality, coherence, and value of science" expressed by many such as Carl Hempel, Dupré accepts that "in substantial part as a result of the influence of Thomas Kuhn, things have fallen apart." Finally, while admitting "prejudices" against astrology and theology and the like, Dupré advocates "epistemological pluralism," one which avoids "the androcentric and ethnocentric biases of much science" but which contends that "there are surely paths to knowledge very different from those currently sanctioned by the leading scientific academies."[107]

From internal evidence afforded by the book itself, I gather that the author suffered from the ills of a mal-arranged marriage—the bitter reflections thus engendered inducing the fable.[108]

Inference to the Best Explanation

"Inference to the best explanation" (IBE) is a common slogan in science and philosophy of science, but the development of IBE as a philosophy of science was accomplished most notably by the philosopher Peter Lipton. Sometimes called *abduction*, following the

107. John Dupré (1993). *The Disorder of Things: Metaphysical Foundations of the Disunity of Science.* Cambridge, MA: Harvard University Press, 1, 3, 5, 7, 9, 11, 10.

108. An elegant expression of a friendly and informed view of science is found in Edward O. Wilson (1998). *Consilience: The Unity of Knowledge.* New York: Alfred A. Knopf.

philosopher Charles Sanders Peirce in the late 1800s, the founder of pragmatism, the basic idea is that scientists infer the hypothesis from the evidence that they think would best explain the evidence in comparison with other explanations. The hypothesis might turn out to be false, but at the time it is the best one they have and so it is the one they use. As Lipton puts it in his book devoted to the topic, when it comes to describing "our inductive practices," IBE "looks to explanation as a key to inference." More specifically, "According to Inference to the Best Explanation, our inferential practices are governed by explanatory considerations. Given our data and our background beliefs, we infer what would, if true, provide the best of the competing explanations we can generate of those data."[109] Elsewhere he states that "Its governing idea is that explanatory considerations are a guide to inference, that scientists infer from the available evidence to the hypothesis which would, if correct, best explain the evidence." Or more succinctly, "Inference to the Best Explanation has it that explanation is prior to inference."[110]

It is worth noting that Lipton's book is exceptional alone for its expression of modesty. Right at the beginning he disavows the "rhetoric of certainty" that is common in philosophical writing, given that any position in philosophy that is "interesting and important . . . is almost always also controversial and dubitable," and accordingly he makes it clear that "Inference to the Best Explanation cannot be the whole story about inference: at most, it can be an illuminating chapter."[111] Keeping that in mind, Lipton's philosophy of IBE "is designed to give a partial account of many inductive inferences, both in science and in ordinary life."[112]

It will be noticed how different this is from earlier accounts that we have examined. Hypotheses or theories according to IBE are not treated in isolation, unlike the standard inductive accounts of scientific method, where the inference is from particulars to

109. Peter Lipton (1991). *Inference to the Best Explanation.* London: Routledge, ix, 58.

110. Peter Lipton (2000). "Inference to the Best Explanation." In Newton-Smith (note 3), 184, 185.

111. Lipton (note 109), 4, 5.

112. Lipton (note 109), 184.

generalizations, and unlike the standard hypothetico-deductive accounts of scientific method, where the hypothesis or theory (or law) is merely invented, (and no matter how,) an observation statement is deduced from the hypothesis or theory, and then the observation statement is subjected to testing, such that the hypothesis or theory receives either confirmation (or corroboration according to Popper) or falsification.

While I cannot possibly do justice to Lipton's treatment of IBE in this section, it is worth looking at some of the basic features of his discussion (especially imagining ourselves reading Lipton through the eyes of Poe).

One of the interesting features of IBE is that scientific explanations need not necessarily refer to laws of nature. As Lipton puts it elsewhere, "one does not need to give a law to give a cause, and one does not need to know a law to have good reason to believe that a cause is a cause."[113] This is in explicit contrast to the deductive-nomological (or covering law) model of scientific explanation, (*nomos* is the ancient Greek word for "law,") according to which a phenomenon is explained if it is deducible from one or more statements of law along with some initial conditions.[114] The implications of what Lipton says here are enormous. In the biological sciences, for example, explanations, especially ultimate explanations, meaning evolutionary explanations, are typically in terms of historical contingencies, almost never (or possibly never) in terms of one or more laws (unless one considers natural selection a law, which is controversial). And for Poe's *Eureka* in particular and for theology in general, (and also for psychology,) what Lipton claims above means that a genuine explanation, possibly even a scientific one, need not be a physical explanation—it can be a personal explanation, too.

A further interesting feature of IBE as understood by Lipton, a feature at its heart and core, is what he calls *contrastive*

---

113. Peter Lipton (1990). "Contrastive Explanation." In Dudley Knowles, ed. (1990). *Explanation and Its Limits*. Cambridge: Cambridge University Press, 248.

114. See Hempel (note 53), 37, 50. So in addition to the context of discovery and the context of justification, the logical empiricists really had a third context for science, the context of explanation, although I don't believe they ever used the label.

explanation. As Lipton puts it, "What gets explained is not simply 'Why this?', but 'Why this *rather than* that?'"[115] To use his simple example, if I am the judge of an essay contest and I have two finalists, Kate and Frank, and I can only give the prize to one and I give it to Kate, it is not enough for me to explain why I gave the prize to Kate, to state all that is great about her essay. I must also list my reasons for why I did not give the prize to Frank.[116] Interestingly, Lipton uses Sherlock Holmes as a further example. When Holmes infers that Moriarty committed the crime, there is no real deduction going on, he says, since the evidence is limited and it is always possible that somebody else committed the crime. "Nevertheless," says Lipton, "Holmes is right to make the inference, since Moriarty's guilt would provide a better explanation of the evidence than would anyone else's."[117]

Science, as one might expect, affords us numerous examples, small and large, of IBE in action. In his book, Lipton focuses on the famous case of Ignaz Semmelweis, a doctor working at the Vienna General Hospital who from 1844 to 1848 grappled with finding out why childbed fever, an often fatal illness, was abnormally high in one of the maternity wards but not in the other. Here, however, I shall focus on Darwin's *On the Origin of Species* (1859), which is an example of IBE writ large. When one reads that book, one finds that Darwin does not make a straightforward case for evolution by natural selection, gathering all the evidence in support of his conclusion. It is true, following Michael Ruse, that we can see Darwin arguing in conformity with Whewell's *consilience of inductions* model, examining every relevant category of evidence, one by one, (more on this later,) and concluding that the only reasonable explanation *for them collectively* is evolution by natural selection, that one could raise cavils against any one of the categories, say the fossil record, and argue that it doesn't prove evolution, but that one could not reasonably do this against all the categories when taken together.[118] There is also, however,

115. Lipton (note 109), 35.

116. Lipton (note 109), 36, 39, 44. See also Lipton (note 113), 250, 253.

117. Lipton (note 110), 185.

118. Michael Ruse (1979). *The Darwinian Revolution: Science Red in Tooth and Claw*. Chicago: University of Chicago Press, 179–180, 188–198.

something else equally profound going on in the *Origin*. Darwin, like a great lawyer, argues throughout his book not only in favor of his own theory but that it is better than the competing theories, the main competitors being creationist biology (whether single-shot or successive creationism) and the minor competitors being non-Darwinian evolutionary theories (principally that of Lamarck, secondarily that of Chambers). Like the essay contest analogy, Darwin argues not only "why this?" but "why this rather than that?"

Another major example is the debate over the mass extinction involving the dinosaurs, a debate that reached a new level in 1980. The insiders, the palaeontologists, had a variety of theories, one of them a nearby supernova, but mostly they appealed to gradual causal processes that are commonly observable on Earth and that resulted in a substantial cooling of the climate. Then a team of four outsiders, led by Luis Alvarez and his son Walter, (the former already a Nobel laureate in Physics,) entered the field and provided a very different theory, one involving "the impact of a large asteroid."[119] The key to the theory is their discovery of high levels of iridium in the 1-cm-thick layer of clay that marks the K/T (C-T) boundary in the rock strata, above which fossil dinosaurs are not found and which had already been dated as forming around 65 million years ago. Iridium is a metal often found in high concentrations in meteorites but only in much lower concentrations in Earth's crust and upper mantle (with of course the exception of the K/T boundary). In their paper, Alvarez and his colleagues argue contrastively, not only in comparing different scenarios and concluding that "the anomalous iridium concentration at the C-T boundary is best interpreted as indicating an abnormal influx of extraterrestrial material," but also in arguing that none of the previously published theories explaining the mass extinction involving the dinosaurs are the best, that "None of the current hypotheses adequately accounts for this [the available] evidence" and that "we have developed a hypothesis that appears to offer a satisfactory explanation for nearly all the available paleontological

119. Luis W. Alvarez et al. (1980). "Extraterrestrial Cause for the Cretaceous-Tertiary Extinction." *Science* 208, 1105.

and physical evidence."[120] For a number of years the palaeontologists resisted, sometimes vociferously, but eventually the field of paleontology converted.[121]

In all of this one can see some of Kuhn's incommensurability thesis, that there is often an inability between scientists to consider and compare competing theories that are *not* within their paradigm, but one can also see how exaggerated the thesis is in the hands of Kuhn (along with his followers). Darwin had no problem explaining and comparing the competing biology paradigms, as did the Alvarezes with the extinction paradigms. Similarly, Poe, as we have seen in the previous chapter, seems to move effortlessly between his own theory of the Universe (which in many respects is remarkably modern) and the paradigm of his time, the former that of a finite and fundamentally changing universe created from the inside by a rapid diffusion from a God particle, the latter that of an infinite and fundamentally unchanging universe created from the outside by God. Darwin and the Alvarezes managed to convert their respective fields in roughly a dozen years, whereas Poe's *Eureka* "fell *dead-born from the press*," to use David Hume's famous description of the publication of his first book, *A Treatise of Human Nature* (1739–40).[122] (Both books are now, of course, not only merely undead, but really, most sincerely undead.) But no matter. Kuhn's thesis of incommensurability should not be accepted as a fundamental truth.

120. Alvarez et al. (note 119), 1102, 1095.

121. Two good books on this topic are: William Glen, ed. (1994). *The Mass Extinction Debates: How Science Works in a Crisis.* Stanford: Stanford University Press. Keith M. Parsons (2001). *Drawing Out Leviathan: Dinosaurs and the Science Wars.* Bloomington: Indiana University Press. It should not go unnoticed that the dinosaur extinction debate has interesting parallels to other debates in science, including the current debate over the main cause of global warming, with the climatologists being the insiders and a substantial number of solar scientists being the outsiders. Given that *politics does not determine good scholarship*, (copyright © 2013 by David N. Stamos,) and that politics (including political correctness) has come to exert a suffocating influence on the debate, for these reasons alone I refuse at this point to say "The science is in."

122. David Hume (1777). "My Own Life." In Norman Kemp Smith, ed. (1947). *David Hume: Dialogues concerning Natural Religion*. Indianapolis: Bobbs-Merrill, 234.

IBE has a lot going for it, part of which is the rejection of Kuhnianism, the rejection of strictly inductive and deductive views of science, and an arguably accurate reflection of the way theoretical scientists often if not usually argue.

Poe would certainly have gravitated toward IBE. Alex Gelfert uses the words "precursor," "affinity," and "anticipation" for the *conceptual* (and not necessarily historical) relation between the philosophy of science expressed in Poe's *Eureka* and Peirce's formulation of abductive inference in the late 1800s.[123] Peirce defined the concept variously throughout his career, stating, for instance, that "*Presumption*, or more precisely *abduction*, furnished the reasoner with the problematic theory which induction verifies. Upon finding himself confronted with a phenomenon unlike what he would have expected under the circumstances he looks over its features and notices some remarkable character or relation among them . . . so that a theory is suggested which would *explain* (that is, render necessary) that which is surprising in the phenomenon." Later he would state that "Abduction is the process of forming an explanatory hypothesis," or that it is "the first starting of a hypothesis and the entertaining of it, whether as a simple interrogation or with any degree of confidence." He also thought of abduction as superior to deduction and induction, in that "It is the only logical operation which introduces a new idea" and "every single item of scientific theory which stands established today has been due to Abduction."[124]

I do not see, however, the supposed anticipation by Poe. It seems to me that Poe had something more different than similar in mind, yes a mental process far above deduction and induction, as we shall see in the next chapter, yes one that possibly provides an explanation, or more accurately a solution, but I doubt that Poe would have called the process a "logical operation." Poe had something inspirational in mind, a mental process of a mysterious nature from which a solution to a vexing problem enters consciousness all of a sudden, aptly called the *eureka effect*, a solution

123. Axel Gelfert (2014). "Observation, Inference, and Imagination: Elements of Edgar Allan Poe's Philosophy of Science." *Science & Education* 23, 596n23, 602, 598, 598n34, 605.

124. All quotations are from Gelfert (note 123), 598, 595, 596, 595, 598.

born from the hidden recesses of the mind (today, the unconscious) in a passionately inspired and educated mind confronted with a pressing problem that requires seeing the big picture, not just details.

I do not even see Peirce as anticipating Lipton, as providing an early version of IBE, given that Lipton sees at the heart of IBE what he calls "contrastive explanation," something apparently missing altogether in Peirce.

Poe, I believe, would have recognized in Lipton's treatment of IBE the importance of contrastive explanation, in making one's case for the original inspiration as being the best of the competitors, as part of the *context of justification*, to use the language of the logical empiricists, specifically Reichenbach and Hempel as we have seen. And this is quite evident in *Eureka*. As we have seen in the previous chapter, Poe not only takes into account the dominant scientific cosmology of his time and argues against it,—the Newtonian view of the Universe, a static universe with infinite space and infinite stars, focusing his critique especially on his contemporaries whom he says should have known better, namely, Herschel, Humboldt, and Nichol,—but in developing his own cosmology he takes into account, as we have seen in Chapters 2 and 3, the aesthetic, theological, and philosophical dimensions that he believes are required to fully make sense of the Universe. Whether Poe got it right is beside the point, which is that if he had the language of IBE at his disposal he probably would have asserted that his view of the Universe is contrastively the best.

Lipton's elucidation and defense of IBE has two features in particular that we need at this point to examine. One, remarkably, is that Lipton ventures into the domain of what we have seen the logical empiricists call the *context of discovery*. One would naturally think that IBE is only a "consequentialist" model, to use Nickles' label examined at the start of this chapter, that it only deals with competing explanations, not with the generation of an explanation. But this is not true, and it is what makes Lipton especially interesting for our purposes. According to Lipton, explicitly rejecting the "happy guesses" approach, (and we must keep in mind the late professor's modesty,) "our explanatory practices guide our inferences." In elaboration, Lipton proposes a model of scientific discovery that involves "two filters, one that selects the plausible candidates, and a second that selects from among them." The first filter actually involves "the generation of candidates," "the process of scientific invention," and is what he calls a "short-list mechanism." But how does that mechanism work? Lipton's answer, in brief, is that the scientist operates at this stage with a "subjunctive question: we ask how good the explanation *would* be if it were true. . . . We construct various causal scenarios and consider what they would explain and how well." The scenarios constructed are "not generated randomly" but consist of "live options," only "serious candidates for an actual explanation," a model which allows for "the feedback between the processes of hypothesis formation and data acquisition that characterizes actual inquiry."[125] Lipton even goes so far as to compare the process to evolution by mutation and natural selection, with "preadaptations" limiting the directions that evolution in a species can take. He calls this process "explanatory selection," with "background beliefs" playing the role of preadaptations in evolution:

> . . . the mechanism of explanatory selection plays a role both in the generation of the short list of plausible causal candidates and in the selection from this list. The background beliefs that help to generate the list are themselves the result of explanatory inferences whose function it was to explain different evidence. We consider only the few potential explanations of what we observe that seem reasonably plausible, and the plausibility judgments may not seem to be based on explanatory considerations; but they are, if the background beliefs that generate them are so based. Those beliefs now serve as heuristics that guide us to new inferences, by restricting the range of actual candidates, much as preadaptations limit the candidate organisms that are generated.[126]

This is a fascinating model, and it marks a major exception to the philosophies of science examined in this chapter (along with that of Whewell). One problem with it, however, is that Lipton is

125. Lipton (note 109), 89, 1, 61, 69, 89, 119, 67, 61, 89.

126. Lipton (note 109), 121.

not clear on whether explanatory selection is exclusively conscious or can be unconscious. He seems concerned with only the former, with his "we are often aware that we are inferring an explanation of the evidence" and his "We are also often aware of making an inferential choice between competing explanations."[127] He does not seem to capture, or does not seem interested in capturing, the phenomenon of eureka moments that are central to Poe's concept of science, reconstructed in the next chapter, of what today is called "unconscious scientific creativity" and of which science provides many actual examples, as we shall see in the second section of the final chapter.

Another concern that I think Poe would have had is with Lipton's emphasis on explanation. As we shall see when we turn to Einstein later in the chapter, not all scientific theories are explanatory theories, and Poe's emphasis on consistency conforms with this point.

But Poe's biggest concern, I believe, would be that Lipton seems much too restrictive on the flight of imagination in science. With his talk that "scientists *infer* from the available evidence *to* the hypothesis which would, if correct, best explain the evidence," (italics mine,) Lipton seems to confine the flight of imagination to an aviary, while with his talk of "restricting the range of actual candidates" in the first of his "two filters" he seems to restrain the flight of imagination on a tether ("jess," in falconry). Poe, as we have seen, (and as we shall see with Einstein,) wants scientific imagination to be able to "soar."

Equally troubling (to my mind at least) is that Lipton's description of hypothesis generation and selection is essentially non-empirical. He does not base it on evidence, or try to gather evidence so that his model is contrastively the best model of the context of discovery. Instead, he infers it from IBE itself, given his view that "a model of induction ought to apply to itself."[128] In other words, *he uses his theory of the context of justification to inform his theory of the context of discovery*. But as the latter is a matter of fact, we should want more. We should want, in fact, "Why this *rather than* that?" And for that we need independent

127. Lipton (note 109), 67.

128. Lipton (note 109), 70.

evidence. Granted, in the second edition of his book Lipton does turn to some evidence for his hypothesis, specifically the studies in cognitive psychology conducted by Kahneman and Tversky published in 1982, which he summarizes as saying, against the Bayesians, "we are sometimes remarkably bad at probabilistic reasoning," and which they summarize as, "man . . . is not a Bayesian at all." What Lipton takes away from these studies, however, is evidence that "causal-explanatory considerations are helping to drive the inferences," that in accordance with "Inference to the Best Explanation our aim in inferring an explanation is not to infer the most probable claim, but rather to infer the most probable of competing explanations. . . . If all we wanted was to maximize probability, we should never venture beyond our data." This clearly applies not only to his model of theory justification but to his model of theory formation as well, given that he uses the former, as we have already seen, to inform the latter, with his language of "an underlying proclivity for explanationist thinking," of what is "embedded somewhere in our epistemic competence," and his claim that "asking what would explain the available evidence is an aid to hypothesis construction. . . . so that will constrain the process of hypothesis generation."[129]

In all of this I applaud Lipton. But as we shall in Chapter 8, we must cast a wider net and go much deeper into the world of empirical evidence if we want to understand scientific imagination. Nevertheless, Lipton has given us a very important perspective on the context of discovery and we shall have to keep it in mind when we turn to Chapter 8, the final chapter in this book, which is devoted to a multilevel empirical investigation into the nature of scientific imagination. As the evolutionary biologist Ernst Mayr once put it, quoted above and I quote it here again, "What I do not understand is why most philosophers of science believe the problems of the philosophy of science can be solved by logic."[130] If scientific imagination is part of the process of science, (and it clearly is in Lipton's philosophy of science, as it is in Poe's,) then

129. Peter Lipton (2004). *Inference to the Best Explanation.* 2nd ed. London: Routledge, 108, 110, 109, 111, 116.

130. Mayr (note 64), ix.

we need to decide how it works by examining as much of the evidence as we can.

The second feature of Lipton's discussion that we must examine concerns what the logical empiricists called the *context of justification*, in Lipton's case the question of what, according to his model of IBE, makes an explanation the "best." A Kuhnian, of course, would likely focus on social, cultural, political, and even personal factors, (i.e., the stuff that science critics love to focus on,) making "best" a matter of community agreement, while a Popperian would focus on falsifiability, as the more a theory prohibits, according to Popper, the more it says about the world, so that the best theory for a Popperian would be the most inclusive one that has also survived the most repeated attempts to falsify it.

Lipton, instead, attempts to answer the question by appealing to two broad considerations, which he calls "the 'likeliest' explanation" and "the 'loveliest' explanation." The former is concerned not only with the "accommodation" of available evidence, (what in philosophy of science is often called "saving the appearances,") making a theory that *fits* the phenomena, (especially mathematically,) but especially with the making of "successful predictions," the latter which provides much stronger support for the truth of the theory. The other consideration, "explanatory loveliness," appeals to "the common observation of scientists that broadly aesthetic considerations of theoretical elegance, simplicity, and unification are a guide to inference."[131] In a later work he expands this list somewhat, referring to "plausible candidates for the explanatory virtues, including scope, precision, mechanism, unification, and simplicity," followed by arguing that "Some of these features have proved surprisingly difficult to analyze."[132] In his book Lipton raises problems as well, arguing, for example, that likeliness and loveliness "sometimes pick out different explanations," that they "sometimes diverge," the reason being that they deal with evidence in different ways. Newtonian mechanics, he says, is as lovely as ever, but it "has become less likely" given more recent competition, while some conspiracy theories are "lovely" but they

131. Lipton (note 109), 61–62, 71, 134, 68.

132. Lipton (note 110), 187.

"may be very unlikely, accepted only by those whose ability to weigh evidence has been tilted by paranoia."[133]

Poe would have loved that Lipton uses the phrase "aesthetic considerations," but he would have replied that a true theory need not make testable predictions, not even to be scientific. Poe's position, and the very fact that Lipton refers to loveliness as involving "explanatory virtues," also that he includes prediction, including falsifiability, among the "theoretical virtues,"[134] brings us to our next major view of the nature of science, known variously as the *epistemic virtues* or the *epistemic values* approach. In the very least, I hope we can agree with Lipton's closing statement in his book, that "Certainly much more needs to be said about what makes one explanation better than another."[135]

Epistemic Virtues and Values

A central problem in philosophy of science is known as the *underdetermination of theory by evidence (or data)*.[136] The idea is simple enough. Whatever amount of evidence we have, it is always finite, but given this finitude, there are always infinitely (or rather indefinitely) many theories that are consistent with the evidence. A common analogy is connecting dots. Whatever set of dots we have, there are always innumerably many ways of connecting them by drawing lines.

Underdetermination creates a problem for lovers of science, who see science as involving the most superior method (or family of methods) for pursuing knowledge and in fact as having the most superior body of knowledge of all, providing us with an age of "spectacular knowledge and insight" developed not from every nation but from "the special genius of ancient Greece."[137] In fact,

133. Lipton (note 109), 61, 62.

134. Lipton (note 109), 143, 144, 186.

135. Lipton (note 109), 189. See also Lipton (note 129), 210.

136. For a determined discussion, see W.H. Newton-Smith (2000). "Underdetermination of Theory by Data." In Newton-Smith (note 2), 532–536.

137. Alan Cromer (1993). *Uncommon Sense: The Heretical Nature of Science.* New York: Oxford University Press, vii.

the underdetermination of theory by evidence, along with the theory-dependence of observation and the hyped incommensurability of scientific paradigms, is one of the favorite weapons (taken, ironically, from philosophy of science itself) of postmodernist critics of science in favor of epistemic relativism and their view that science is just another social construction.[138]

The most powerful answer to this challenge, to my mind, is to bring in a philosophy of science known variously as the *epistemic virtues* or *epistemic values* approach, not only because it works but because the critics will not get it, given that they themselves lack these virtues and values, (and do not want them,) analogous to psychopaths not getting what *remorse* means, because they lack moral virtues and values such as conscience, (and do not want them,) the very stuff of morality.[139] Epistemic virtues can apply to theories and theorists, while only epistemic values can apply to theorists, given that only persons have values.

The philosopher of science Ernan McMullin is a major exponent of this view of science. Although there is no canonical list, McMullin includes: "predictive accuracy," especially in a mature theory, a virtue that includes testability; "internal coherence," since a theory that does not have internal coherence is not likely to be true; "external consistency," consistency with what is already known; "unifying power," in the sense that disparate phenomena and seemingly unrelated areas of inquiry are brought together under a single theory, (what is sometimes called "elegance,"); "fertility," in that a good scientific theory will generate new fields of research; and "simplicity," more commonly known as Ockham's Razor, the idea that one should not multiply explanatory entities beyond what is necessary, since *a priori* the more luxurious a theory the less likely it is to be true. McMullin calls these values *epistemic*, (rather than merely *pragmatic*,) "because they are presumed to promote the truth-like character of science, its character as the most secure knowledge available to us of the world we seek to understand. An epistemic value is one we have reason to believe

138. See Sokal and Bricmont (note 82), x, 51, 70–71. Noretta Koertge, ed. (1998). *A House Built on Sand: Exposing Postmodernist Myths About Science.* New York: Oxford University Press, 40–41, 230–231.

139. See my (2011). "The Philosophical Significance of Psychopaths: Postmodernism, Morality, and God." *Free Inquiry* 31 (5), 46–49.

will, if pursued, help toward the attainment of such knowledge."[140] W.H. Newton-Smith adds what he calls *social epistemic* virtues and values: "the prohibition against suppressing data from one's co-workers"; "the prohibition against the invention of data"; "the requirement that scientific ideas be made public"; the requirement that science be "relatively non-hierarchical"; the existence of a "system for the recognition of experts"; "the way in which the institution encourages innovation and deviance"; and finally, "incorruptibility," meaning "You cannot easily buy them [scientists] off."[141]

In all of this I would add *clarity*, for the simple reason that obscurity (e.g., Heidegger) is arguably not conducive to the pursuit of knowledge in any direct way at all, but only to the *appearance* of depth and knowledge, in other words to *sham intellectuality*, and accordingly should be shunned as a vice. Naturally those who don't value truth, knowledge, and evidence won't care about clarity. But scientists certainly do. For example, consider Steven Weinberg's complaint, in a chapter entitled "Against Philosophy," that "From time to time . . . I have tried to read current work on the philosophy of science. Some of it I found to be written in a jargon so impenetrable that I can only think that it aimed at impressing those who confound obscurity with profundity."[142] Another top example from science is Robert Trivers in his collected volume of theoretical papers. In his introduction to one of his classics he has a section entitled "How to Write a Classic Paper," in which he states, "Write in the language of your discipline but, of course, try to do so simply and clearly," to which he adds, "The virtues of writing simply and clearly should be obvious."[143]

140. Ernan McMullin (1983). "Values in Science." In Peter D. Asquith and Thomas Nickles, eds. (1983). *PSA 1982*. Vol. 2. East Lansing, MI: Philosophy of Science Association, 18. See also Ernan McMullin (2000). "Values in Science." In Newton-Smith (note 2), 550–560.

141. W.H. Newton-Smith (1995). "Popper, Science and Rationality." In Anthony O'Hear, ed. (1995). Karl Popper: *Philosophy and Problems*. Cambridge: Cambridge University Press, 24–27.

142. Steven Weinberg (1992). *Dreams of a Final Theory*. New York: Pantheon Books, 168.

143. Robert Trivers, ed. (2002). *Natural Selection and Social Theory: Selected Papers of Robert Trivers*. New York: Oxford University Press, 13, 15.

It is to be noted that these epistemic virtues and values are a matter of degree depending on the circumstances, the scientists, and the institutions. To use McMullin's example, Einstein in his debate with Bohr over quantum physics emphasized consistency and coherence in his rejection of the realist interpretation of quantum physics, while Bohr emphasized its predictive successes. This points to a feature of this approach to science, which is that it is a *cluster class* approach rather than an *essentialistic* approach—my emphasis on clarity notwithstanding. With the cluster class approach, no one virtue or value must be present for the theory or theorist to be considered scientific. What matters, instead, is more rather than less, an approach that allows for some fuzziness in judgment when judging whether a theory, a person, or an institution is scientific, as well as in arguing for what is contrastively the best of competing theories.

What this specifically means is that a theory or speculation *need not necessarily* be testable, in the sense of making predictions that are in principle verifiable or falsifiable, for the epithet "scientific" to apply. Newton-Smith cites the example of string theory in physics, the highly mathematical theory or family of theories that conceptualizes the most fundamental particles and their relations in terms of extremely tiny one-dimensional strings: "It is not a falsifiable theory. It makes no predictions whatsoever."[144] And yet it counts as a scientific theory, (or family of theories,) given the virtues of internal coherence, consistency with what is known, unifying power, fertility, and simplicity.

What matters, then, in this view of science, is the *overall degree* to which a theory or speculation reflects the list of epistemic virtues and values, whatever that list may be. Modern astrology, for example, fails miserably, while Darwin's biology succeeds massively on every count, and Poe's *Eureka*, I submit, meets the requirement to a sufficient degree. While arguably one cannot say, *today*, what observations would definitively falsify the theological parts of *Eureka*, other parts of it are clearly testable. Poe's claim, for example, that the Universe is in a contraction phase is not only

144. Newton-Smith (note 141), 26. Gribbin concurs. Gribbin (note 100), 472. For a corroborative critical discussion, see Lee Smolin (2006). *The Trouble with Physics: The Rise of String Theory, the Fall of a Science, and What Comes Next.* New York: Houghton Mifflin, esp. xiv and ch. 12.

testable but falsified. But as we have seen in the previous chapter, Poe made other corrections (constituting predictions?) to the science of his day that have largely been borne out, such as his claim that the Universe exploded from a super-concentrated primordial Particle. So one could say *Eureka* has at least a modest degree of predictive accuracy. One could also say it is internally coherent, it was largely externally consistent with the science of his day, and it has some unifying power, while its fertility is doubtful and its simplicity subpar.[145] *C'est la vie*. In terms of epistemic virtues and values, Poe's *Eureka* is still pretty good for someone who was not a professional scientist but fundamentally a poet. I would also add that *Eureka* is remarkably clear once one gets used to Poe's hyperbolic and fanciful style.

And if we look to Poe's own words, in terms of his own view of epistemic virtues and values, what we see is that he embraced pretty clearly some widely recognized ones. As we have seen in

145. Clearly, a supernatural explanation is less simple than a natural explanation if they both include the same natural entities, as the supernatural explanation adds something more. There is substantial debate, however, (as one might naturally expect,) when it comes to the apparent fine-tuning of the cosmic constants, discussed in the previous chapter, and God is used as the explanatory entity. According to the physicist Paul Davies, for example, instead of supposing the generation of multiple universes with random variations in the cosmic constants, with a few universes such as ours allowing for life by coincidence, "It is simpler to postulate one unseen God," meaning "*one* entity of a simple kind" (the latter is from his quotation of Richard Swinburne). Paul Davies (1992). *The Mind of God: The Scientific Basis for a Rational World*. New York: Simon & Schuster, 190. According to the biologist Richard Dawkins, on the other hand, a God that is supposed to do all the things credited to God "*cannot* be simple." Instead, this God would have to be massively complex and hence "highly improbable," such that "His existence is going to need a mammoth explanation in its own right." For Dawkins, natural selection is the much simpler hypothesis, not only for the origin of species but for the origin of universes as well. Richard Dawkins (2006). *The God Delusion*. New York: Houghton Mifflin, 149, 146–147. Conceding Dawkins' point, Davies has since produced a much more nuanced view, one eerily in the direction (without apparently realizing it) of Poe's God in *Eureka*, (even in spite of the "unnerving" problem of dark energy,) in which he identifies a teleological God with the Universe, so that the Universe "explains itself." Paul Davies (2006). *The Goldilocks Enigma: Why Is the Universe Just Right for Life?* New York: Houghton Mifflin, 219–220, 128, 266–269.

Chapter 4 in the section on Poe's criteria of truth, Poe states in *Eureka* that "*a perfect consistency can be nothing but an absolute truth.*" In the same paragraph, focusing now on "our Keplers" and "our Laplaces," he says "their theories are merely corrected—reduced—sifted—cleared, little by little, of their chaff of inconsistency."[146] While this does not exclude testability, Poe certainly does not emphasize it. Nevertheless, Poe's insistence on total consistency can be seen as subsuming testing a theory against evidence. Moreover, as we have seen in Chapter 4, his emphasis on consistency is in line with the correspondence theory of truth—his is not consistency merely for its own sake.

But Poe, as we have also seen in that section of Chapter 4, has a supplementary criterion of truth in *Eureka*, which is the beauty of a theory. In the case of Laplace's nebular hypothesis, it is, he says, "by far too beautiful, indeed, *not* to possess Truth as its essentiality." And then at the end of *Eureka* he combines beauty with symmetry and consistency, claiming that "symmetry and consistency are convertible terms" and that "It is the poetical essence of the Universe—*of the Universe* which, in the supremeness of its symmetry, is but the most sublime of poems."[147]

Poe's emphasis on beauty of theory is certainly a questionable one. In a chapter entitled "The Beauty Myth," while recognizing that unification, as in "bad romance novels," is "The most cherished goal in science" and when accomplished is "the biggest thrill in science," Lee Smolin argues that beautiful theories are not necessarily a good thing. His reason is that a beautiful theory, including one with mathematical beauty, need not "lead anywhere," in the sense of novel predictions and new research, and can indeed be quite false, while a less beautiful theory can lead to wonderful things, including new insights, hypotheses, predictions, "the ring of truth," and ultimately further unifications. One of his examples is Kepler's first theory of the solar system, based on the five Platonic solids: extremely beautiful—Oh! gorgeous yet fantastic beauty!—but a dead end. It was eventually replaced by Kepler with his three laws of planetary motion: not as beautiful with its ellipses—decidedly plain-looking, actually, if not

---

146. *P&T*, 1269.

147. *P&T*, 1312, 1349.

positively ugly—but a theory that had a huge payoff in terms of predictions, successful ones at that, and in terms of future theory, such as Newton's laws.[148]

Other physicists, however, think more highly of beauty. Steven Weinberg, for example, claims that "time and again physicists have been guided by their sense of beauty not only in developing new theories but even in judging the validity of physical theories once they are developed." In his chapter devoted to the topic, however, entitled "Beautiful Theories," Weinberg argues that beauty in scientific theories is not so much a matter of aesthetics (whether an innate sense or beauty for its own sake) but is more like beauty in professional horse training, a developed sense of beautiful horses from the experienced correlation between certain features of horses and their ability to win races. Although not always a guide to a winner, beautiful theories in science, according to Weinberg, involve primarily the features *simplicity*, both in their central ideas and in "principles of symmetry," ("that something looks the same from different points of view,") and *inevitability*, in the sense of structural wholeness, (that the theory works or fails as a whole), such that "our historical experience teaches us that as we look beneath the surface of things, we find more and more beauty" and "the beauty of present theories is an anticipation, a premonition, of the beauty of the final theory."[149]

I think Poe would have related to much of what Weinberg says, with his literary theory and his theology applied to his Big Bang theory of the Universe: in the underlying simplicity of the unity of effect of the Universe, with every part contributing to the whole, in the symmetry of its principles, in the inevitability of the plot of the Universe such that the theory flies or flops as a whole, and in the *dénouement* of the plot, presaged by Beauty, the Universe resolving itself back to its ultimate simplicity, defined by Poe, as we have seen earlier, as both Unity and Nothingness.

In either case, whether the view of Smolin or Weinberg, Poe is in respectable company over a live debate in science and its philosophy.

148. Smolin (note 144), ch. 2.

149. Weinberg (note 142), 90, 136, 165.

But of course the problem of scientific imagination remains. Perhaps the implication of the epistemic virtues and values approach is that genuinely scientific imagination, at both the conscious and unconscious levels, is guided not so much by Lipton's subjunctive question and his two-filters approach, but moreso by epistemic virtues and values, virtues and values that help to demarcate and characterize the scientific imagination as an educated one. Perhaps. And then perhaps we shall find something more or something entirely different in our final chapter.

Evolutionary Epistemology

Evolutionary epistemology means "theory of knowledge from the viewpoint of evolutionary biology." This field has been usefully divided by Michael Bradie into the *evolutionary epistemology of mind*, which attempts to "account for the characteristics of cognitive mechanisms in animals and humans by a straightforward extension of the biological theory of evolution," and the *evolutionary epistemology of theories*, which "attempts to account for the evolution of ideas, scientific theories, and culture in general by using models and metaphors drawn from evolutionary biology."[150]

Although the field is too large to be done justice here, I want to focus on the biggest names in order to get my fundamental point across. Beginning with the evolutionary epistemology of mind, Michael Ruse has done more than anyone to put philosophy of biology on the map, with his many publications and his founding of the journal *Biology & Philosophy*, summarized by his motto of "taking Darwin seriously." In his book of the same name, Ruse argues that "the principles of scientific reasoning or methodology" are not only "rooted in our biology," in the sense that they were evolved in our proto-human ancestors by natural selection, but that they "have their being and only justification in their Darwinian value, that is in their adaptive worth." Science, of course, involves theories and observations, which collectively deal

150. Michael Bradie (1994). "Epistemology from an Evolutionary Point of View." In Elliott Sober, ed. (1994). *Conceptual Issues in Evolutionary Biology.* 2nd ed. Cambridge: MIT Press, 454.

with reality via "prediction, explanation, and testing." At the base of this enterprise are what Ruse calls (following E.O. Wilson) *epigenetic rules*, information processors and filters that are "biologically channelled modes of thinking imposed on us by evolution." Specifically, Ruse includes (i) *inductive reasoning*, which includes reasoning from particular instances to generalizations and laws of nature, (regularities,) analogical reasoning, Whewell's "consilience of inductions" and other forms of causal reasoning, and simplicity, and (ii) *deductive reasoning*, which includes logic (principles such as the law of excluded middle and rules such as *modus tollens*) and mathematics (taken widely to include geometry).[151] We shall see what Poe has to say about inductive and deductive reasoning in the following chapter. I say this here because what is striking about Ruse's long chapter entitled "Darwinian Epistemology" is the total absence of scientific imagination from his list, let alone any attempt to get at what it is.

This is especially striking given Ruse's earlier chapter in the same book, entitled "Evolutionary Epistemology," in which he advocates for the view that science is not a product but a process, with the scientist having "an active role in the course of science," that the scientist should be brought "to the fore, making the human being the first object of our study." And yet even here the role of scientific imagination is played down to a remarkable level, which as we have seen in the present chapter is the standard approach. For example, explicitly recognizing the distinction between the "context of discovery" and the "context of justification," Ruse refers to the chemist Kekulé's discovery of the circular nature of the benzene molecule as stemming from "imagining a snake swallowing its tail." It is true, as Ruse puts it, that if it wasn't for independent chemical analysis, the context of justification, "who would remember Kekulé's name today?" But in totally ignoring, as we shall see in Chapter 8, that Kekulé's fundamental insight came to him visually in a daydream, also for the structure of methane, Ruse cannot truly be said to put the scientist "to the fore." He still seems more hung up on product than on process

151. Michael Ruse (1986). *Taking Darwin Seriously: A Naturalistic Approach to Philosophy*. Oxford: Basil Blackwell, 155, 152, 150, 149, 158, 151, 158–159, 156, 157.

after all. It is the same with his treatment of Darwin, who he says, after converting to evolutionism, "Slowly . . . saw that a form of selection transferred to nature would satisfy his need to explain adaptation." Once again a scientific theory, he says, is "explicitly thought up . . . and designed not merely to beat out the competition, but to do the job properly," the job of "understanding nature." Moreover we are told, contrary to the idea of "chance discoveries" in science such as Alexander Fleming's discovery of penicillin as an antibiotic, ("which may or may not be true,") that not only in the case of Darwin but "in the vast majority of cases new scientific ideas appear to order, because they are needed." The most we get—one that Poe would surely emphasize—is the recognition that in the case of Darwin and elsewhere "there do seem to be the two main uses of analogy, discovery and support [justification]."[152]

Perhaps the most serious problem with Ruse's account is that throughout his entire chapter he makes the context of discovery look primarily conscious. Yet as he surely must have known, Darwin himself, (as we shall see in Chapter 8, just as with Kekulé and many others,) affords us a prime example of "unconscious scientific creativity" with his discovery of what he called the "principle of divergence," which he considered to be just as important as his principle of natural selection.

Ruse, nevertheless, thankfully puts us in the right direction with his emphasis on the scientist as part of science and his recognition that science is not a product but a process. It is a direction we shall follow in the final section of this chapter and throughout the final chapter itself, but with Poe's emphasis on eureka moments as our guide.

Turning now to the evolutionary epistemology of theories, it is the view that knowledge evolves in accordance with the process of natural selection, analogous to the evolution of species. Karl Popper was a major exponent of this epistemology, calling it "*the natural selection of hypotheses*": "our knowledge consists, at every moment, of those hypotheses which have shown their (comparative) fitness by surviving so far in their struggle for existence; a competitive struggle which eliminates those hypotheses which are

152. Ruse (note 151), 66, 33, 36, 56–57, 35.

unfit."[153] Survival of hypotheses, as we have seen in our section above on Popper, means surviving the *modus tollens* of falsification. But the later Popper gives us little more than the earlier Popper on the generation of hypotheses. The most that we get, in a reply paper to Donald Campbell, (to whom we shall turn below,) is what he claims Campbell missed, (but "which he of course would concede,") which is that "The blind trials stand not only under the influence of the exploratory drive or instinct, but also under the experience of error—the experience that this is wrong, that this is *not* the solution."[154]

Far more interesting here is the psychologist Donald Campbell himself, who considered Popper "the modern founder and leading advocate of a natural-selection epistemology." According to Campbell, there are niches in the abstract environment known as *science*, helping to account for a "rushing in and rapid development following new breakthroughs, . . . very like the rapid exploitation of a newly entered ecological niche." But whether gradual or rapid, the evolution of knowledge according to Campbell proceeds by a "blind-variation-and-selective-retention process," a process that involves "(a) Mechanisms for introducing variation; (b) Consistent selection processes; and (c) Mechanisms for preserving and/or propagating the selected variations."[155]

Clearly, as lovers of Poe, (a) is what we should want to know about. And what do we find? We find that the variations on which the natural selection of hypotheses feeds are "blind," but not so much in the sense of "random." Instead, Campbell supposes that there are "certain processes involving systematic sweep scanning," blind in the sense that the "variations are produced without prior knowledge of which ones, if any, will furnish a selectworthy [*sic*] encounter." The variations are truly blind because "they are independent of the environmental conditions of the occasion of their

153. Popper (note 60), 261.

154. Karl R. Popper (1974). "Campbell on the Evolutionary Theory of Knowledge." In Paul Arthur Schilpp, ed. (1974). *The Philosophy of Karl Popper.* La Salle, IL: Open Court, 1059–1065. Reprinted in Gerard Radnitzky and W.W. Bartley, III, eds. (1987). *Evolutionary Epistemology, Rationality, and the Sociology of Knowledge.* La Salle, IL: Open Court, 118.

155. Campbell (1974). "Evolutionary Epistemology." In Schilpp (note 154), 413–464. Reprinted in Radnitzky and Bartley (note 154), 89, 71, 56.

occurrence." They are blind because they are "no more likely to occur at any one point in a series of trials than another." They are blind because no "variation subsequent to an incorrect trial is a 'correction' of the previous trial or makes use of the direction of error of the previous one." And they are blind because "In going beyond what is already known, one cannot but go blindly." All of this is Campbell's view even though he admits "Already-achieved wisdom of a general sort which limits the range of trials (such wisdom due to inheritance and learning)." Presumably he is referring here to the "sweep scanning." What is clear is his emphasis on "the profound indirectness of knowing necessitated by the natural selection paradigm, and the inevitable imperfection and approximate character of both perceptual and scientific knowledge at any stage."[156]

In an earlier paper Campbell applies the same kind of thinking to the topic of "the mystique of the creative genius and the creative act," which for him includes scientific genius. Given "a dozen equally brilliant men," he says, we "deify" the one who got the lucky guess, the one who "just happened to be standing where lightning struck." Our "'awe' and 'wonder,'" instead, he says, "should be directed outward, at the external world just revealed, rather than directed at the antecedents of the discovery. Just as we do not impute special 'foresight' to a successful mutant allele [a mutation in a gene] over an unsuccessful one, so in many cases of discovery, we should *not* expect marvellous consequents to have equally marvellous antecedents. . . . insofar as there has been a genuine gain in knowledge, the difference between a hit and a miss lies in the selective conditions thus newly encountered, not in talent differences in the generation of trials." This is for "equally brilliant men." Campbell does, of course, acknowledge individual differences in thinkers that are relevant to his topic, such as differences in "the accuracy and detail of their representations of the external world," in "degree of information and intelligence," in "the number and range of variations in thought-trials produced," in "age differences," in whether "persons have been uprooted from traditional cultures" or "have been thoroughly exposed to two or more cultures," and in the "number of selective criteria"

156. Campbell (note 154), 56–57, 74.

by which one "weeds out the overwhelming bulk of inadequate trials." Nevertheless, he says, "individual differences . . . do not place the joys of creative innovation beyond the less gifted" and it is "likely that many important contributions will come from the relatively untalented, undiligent, and uneducated."[157] At bottom, Campbell's lucky hits are just that—lucky hits.

The profound indirectness of knowing—necessitated by the natural selection paradigm—no mystique of the creative genius and of the creative act—no awe and wonder directed at the antecedents of discovery—important discoveries from the relatively untalented, undiligent, and uneducated. Perhaps the problem is with the *use* of the natural selection paradigm here, in the sense of putting it where it does not belong. If the paradigm "necessitates" a "profound indirectness," then we should not want to project onto the phenomenon in question, the context of discovery, a paradigm that might not fit at all. To think otherwise is to try to force the proverbial square into the round hole. Perhaps the variations in the discovery process are not, after all, all that blind. Perhaps they are *not* "independent of the environmental conditions of the occasion of their occurrence." Perhaps in "going beyond what is already known" they are not blind but *educated*. And what of the number of them? Ruse is surely right when he rhetorically asks, "Did millions of ideas have to be produced before Watson and Crick hit on the double helix?"[158] And what of the "sweep scanning" processes? Campbell, following Poincaré, says it is "usually unconscious."[159] But what is it that informs these unconscious processes? Is it background knowledge? Is it a set of values? Or is it something else? Once again, Ruse is surely right in stating that "The science is being moulded or directed according to the perceived ends of the scientist, in a way quite unlike that which occurs in organic evolution."[160] But Ruse in turn fails, as we have seen, in seeming to have an aversion to unconscious processes in

157. Donald T. Campbell (1960). "Blind Variation and Selective Retention in Creative Thought as in Other Knowledge Processes." *The Psychological Review* 67, 380–400. Reprinted in Radnitzky and Bartley (note 154), 103–105.

158. Ruse (note 151), 59.

159. Campbell (note 154), 57, 63.

160. Ruse (note 151), 57.

the context of discovery—and ultimately, along with his opponents here, to the existence and role of an unconscious teleological *mind*.

Perhaps there is a middle way between the Scylla of the Poppers and Campbells and the Charybdis of the Ruses. Poe, indeed, as we shall amply see in the following chapter, envisioned that middle way, with his theory of the educated imagination and its "intuitive leaps," what we shall see in Chapter 8 has been called "unconscious scientific creativity." We shall also see in that chapter that the history of science affords us numerous examples of these leaps. What is needed, then, is an empirical approach to the matter, rather than to take a Popperian conjecture based on Darwin's great accomplishment and combine it with some quotations from Mach and Poincaré and others from that era, which is the gist of Campbell's method.[161]

But enough of philosophy of science and its philosophers (whatever their professional pedigree). Soft may the worms about them creep! Let us turn at last to the study of scientists themselves, first by route of professional historians of science, and then by looking at two of the greatest scientists of all time.

Contextualist History of Science

There are different ways of doing history, and they need not converge given the very same topic. One way is *history of ideas*, which is popular among philosophers of science and scientists themselves. This is to trace the evolution of particular ideas, their relations with contemporary and earlier ideas, determining who came up with what and how, what someone's concept of something was, and so on. Among professional historians of science, however, history of ideas has lost out to what is known as *contextualist history of science*. As Bernard Lightman, one of its practitioners, puts it, "The hallmark of contextualist studies is their emphasis on the way scientific ideas are embedded in material culture such that there are no insides or outsides of science." Appearing "full-blown" in the 1980s, and allowing for different kinds of contexts, Lightman

161. Campbell (note 154), 62–66.

adds that contextualist history of science "allowed historians to avoid the false analytical distinction between science and society (or base and superstructure), dissolve the categories external and internal, and begin to transcend the science/society dualism."[162]

One can get no clearer picture of this way of doing history of science than by turning to its two main leaders or models, namely, Adrian Desmond and James Moore, explicitly hailed as paradigms of history of science writing.[163] In their enormously successful *Darwin*, they tell us that "Our *Darwin* sets out to be different—to pose the awkward questions, to probe interests and motivations, to portray the scientific expert as a product of his time; to depict a man grappling with immensities in a society undergoing reform," that "We want to understand how his theories and strategies were embedded in a reforming Whig society."[164] There is a subtle yet discernible philosophy lurking at the foot of each of the seven hundred or so pages of text (exclusive of endnotes, bibliography, and index)—and it is best described as Kuhnian-Rortian. Let me present a précis of my case.[165]

In his biography of T.H. Huxley, Desmond tells us that "At the dawn of the twenty-first century 'reason' seems a precarious, value-laden yardstick, and one which has an infuriating habit of changing allegiance." Huxley, we are then told, was attempting to establish "a rival evolutionary priesthood" and Desmond's biography is "the human story behind these sea-changes."[166] As one reviewer put it, prefacing her review by pointing out that with "this new contextual history, we are in danger of throwing the

162. Bernard Lightman, ed. (1997). *Victorian Science in Context.* Chicago: University of Chicago Press, 7.

163. See, e.g., Seymour H. Mauskopf (1992). "Prize Announcements." *Isis* 83, 279. Michael Shortland and Richard Yeo, eds. (1996). *Telling Lives in Science: Essays on Scientific Biography*. Cambridge: Cambridge University Press, x.

164. Adrian Desmond and James Moore (1992). *Darwin*. New York: Warner Books, xviii, xix.

165. See my (2007). "Darwin and the New Historiography." Chapter 10 in my *Darwin and the Nature of Species*. Albany: State University of New York Press, 207–229.

166. Adrian Desmond (1997). *Huxley: From Devil's Disciple to Evolution's High Priest*. Reading, MA: Addison-Wesley, xv, xvi, xix.

proverbial baby out with the bath water," in the nearly 650 pages of text "Huxley the scientist is often difficult to find," and Huxley the scientist, she concludes, "deserves better."[167]

In their second book on Darwin, Desmond and Moore attempt to show that Darwin's theory of common descent for human races—that the human races evolved from a single ancestral species, his position on the monogenism-polygenism debate that was raging in his day, mankind as one species versus mankind as a genus of more than one species—was motivated primarily not by evidence but by a "*moral* passion," such that their new biography of Darwin "is the untold story of how Darwin's abhorrence of slavery led to our modern understanding of evolution."[168] As one reviewer put it, "the weight of context" in their book "pushes the details of the theory [Darwin's theory of human evolution] far to the sidelines—indeed, so far that evidence for common descent in the form of Darwin's Galápagos mockingbirds is shoved right off the page." Prefacing his review with the claim that "Recovering motives and intentions is a principal job of the historian," the reviewer (prominent in his field) judges that the conclusion of the book is lacking in "explicit evidence" and remains "only a suggestion."[169]

Each hitherto "untold story" by Desmond and Moore is really a fabricated *tale*, (using Poe's sense of the term unravelled in Chapter 2,) a story with a *plot*, the plot having a *dénouement*, the latter *constructed* in accordance with their social constructionist agenda. Thoroughly lacking is Dupin's—and the natural scientist's—"My ultimate object is only the truth."[170]

And it is not just Desmond and Moore. Another high-profile example is James Secord, in his work on the post-Lamarckian and

167. Sherrie Lyons (1999). "In Search of Huxley the Scientist: Review of Adrian Desmond's *Huxley: From Devil's Disciple to Evolution's High Priest*." *Biology & Philosophy* 14 (4), 585, 590.

168. Adrian Desmond and James Moore (2009). *Darwin's Sacred Cause: How a Hatred of Slavery Shaped Darwin's Views on Human Evolution*. New York: Houghton Mifflin Harcourt, xviii, xxi.

169. Robert Richards (2009). "The Descent of Man: Review of Adrian Desmond and James Moore's *Darwin's Sacred Cause*." *American Scientist* 97 (5), 415, 417.

170. *P&T*, 420.

pre-Darwinian evolutionist Robert Chambers (discussed in Chapter 4). One gets a sense of the *ressentiment* that pervades contextualist history of science in particular and the humanities in general with his remark that Chambers—with his ambition to apply evolution not just to species but to "everything from nebulae to the human mind. . . . the universal application of a simple principle of development, to do what Newton had done with gravitation in the *Principia*"—"defied the authority which was being created for the sciences in the nineteenth century," adding that "*it is why these writings deserve to be read today*" (italics mine).[171] In a later work, Secord gives us a different (or further) reason, and it has to do with historical influence, not with truth. Taking the approach of contextualism—according to which "Scientific theories, theological doctrines, and political ideologies are seen as forms of work, set in the context of everyday life" and "the most abstract ideas about nature should be approached first and foremost as material objects of commerce and situated in specific settings for reading"—and recognizing that evolution has gained a "pivotal role in the public arena," marked by Stephen Hawking's *A Brief History of Time* (1988) and Steven Pinker's *The Language Instinct* (1994), Secord claims that this role has "little to do with Darwinian biology or Big Bang astronomy. Instead, the critical period is the first half of the nineteenth century, and the turning point is the response of readers to [Chambers'] *Vestiges*."[172]

One wonders what would have happened if Chambers had counterfactually been stillborn. Would a different "sea change" have occurred than the one that in fact happened, one without evolution enjoying a "pivotal role in the public arena" of today, even with contributions of the likes of Darwin, Hawking, and Pinker being much the same? Does not truth, like Dr. Ian Malcolm put it on life itself, find a way?

So bad has the problem become that the senior historian of science Mary Winsor publicly complained about the direction of her entire field, calling its central problem the "Taboo Problem":

171. James A. Secord (1994). *Vestiges of the Natural History of Creation and Other Evolutionary Writings*. Chicago: University of Chicago Press, xlv.

172. James A. Secord (2000). *Victorian Sensation: The Extraordinary Publication, Reception, and Secret Authorship of Vestiges of the Natural History of Creation.* Chicago: University of Chicago Press, 4, 2.

"We have what amounts to an enormous taboo blacking out the very center of the topic we claim to study." With so much focus on context, she says, "we studiously avoid mention of the progressive direction of scientific change." "Professional historians of science," she says again, "flee the issue, busying ourselves with describing everything else about science except its spectacularly progressive nature," adding that "even our internal histories are disappointing."[173]

But the problem is actually far worse than all of that: it is the rejection of objective truth altogether for natural science and scientists, (presumably not, however, for the contextualist historians themselves,) along with the solution to underdetermination examined earlier in this chapter. In other words, the real problem is not that the baby is hidden or abandoned. Nor is it that the baby is thrown out with the bath water. Instead, the real problem is that, in the fury of their collective monomania, the bathwater is retained and the baby is *drowned* in it!,—and not just once, (as with some alleged baptisms,) but over and over again in a professionalized ritual of attempted murder!,—with graduate students observing and serving as initiates even!

Take Moore, for example. A few years following *Darwin*, he confided that "our historiographic aim was shared: to embed Darwin the man, his practices and theories, in a shifting social order," a historiography that sees "progress" in "the degree to which the pure waters of Darwin's science are muddied by the rich surrounding soil of political economy, natural theology, urban radicalism and provincial Dissent."[174] Moore applies the same *mud* to A.R. Wallace, who came up with much the same theory as Darwin's only twenty years later. Especially revealing is Moore's claim that "Once serendipity is ruled out—the last refuge of the perplexed—and once creative genius ceases to be an explanatory concept (in history as in nature), then contextualizing can begin. Now the task is to explain, not the actual causes, or sufficient conditions, but just the *pre*conditions for Wallace's Malthusian moment,

173. Mary P. Winsor (2001). "The Practitioner of Science: Everyone her Own Historian." *Journal of the History of Biology* 34 (2), 240, 241.

174. James Moore (1996). "Metabiographical Reflections on Charles Darwin." In Shortland and Yeo (note 163), 275, 271.

the conditions of its possibility. All or some may be necessary or sufficient, for the event is historically underdetermined and indeterminate." Moreover, once contextualizing begins,—*really* begins,—"Natural knowledge itself," he says, is "seen as a complex cultural product. . . . Theories and practices, instruments and institutions are not fetishized but 'contextualized,' embedded in their formative social matrix."[175]

One senses with contextualist history of science—so eminently rich in detail, the product of enormous industry—that sometimes a good thing can go too far, much like political correctness and multiculturalism. And thus, joy suddenly faded into horror, and the most beautiful became the most hideous, as Hinnon became Ge-Henna.

The same danger, of course, attends the growing pressure to contextualize Poe, including especially *Eureka*. To fully contextualize Poe would mean either ignoring or rejecting what we have seen in the previous chapter, Poe's remarkable anticipation of at least nine major theories and developments in twentieth-century science. It would also mean either ignoring or rejecting the more fundamental question raised by Poe himself, which is the topic of our next two chapters.

Fortunately, there are still to be found professional representatives of the kind of history highly conducive to our search into scientific imagination and creative genius. Andrew Robinson, formerly at Cambridge University, the son of an Oxford physicist, himself having degrees in chemistry and South Asian studies, is a prime example of this kind of historian: history as a means to a higher kind of understanding, history that brings up rather than brings down (although both are needed). The author of over twenty-five books, Robinson does not see biography as Moore sees its role, as "a powerful contextualizing tool"; he does not "pull focus," to use Moore's cinematic metaphor, in the sense of shifting focus from the individual in the foreground to the environment in the background, blurring the former and sharpening the latter in the process.[176] Instead, Robinson does biography in

175. James Moore (1997). "Wallace's Malthusian Moment: The Common Context Revisited." In Lightman (note 162), 295, 290.

176. Moore (note 175), 291, 296.

order to get the clearest possible closeup on the creative mind. As he puts it in his magnum opus, "How the creative process works in both science and art has long interested me professionally," such that in his book he analyzes "exceptional creativity—'genius'—in both scientists and artists by following the trail that led ten individuals from childhood to the achievement of a creative breakthrough as an adult." Agreeing that "most creative ideas appear to come unexpectedly out of nowhere, I am convinced," he says, "they are not as unpredictable as they seem, and that science can shed light on the act of creation," on "the sources, ingredients, and patterns of genius."[177]

Robinson is an Alvarez to the paleontology of the contextualists, (a true son of a physicist, I might add,) and we shall avail ourselves of some of his discoveries and insights primarily in the section on Champollion in the next chapter and in the section on comparative biography and psychology in the final chapter.

But enough of philosophers and now historians. We have devoted most of a long chapter to them. Let us now turn to two of the greatest scientists of all time, namely, Darwin and Einstein, to see what role imagination played in their science and what they had to say about it.

Charles Darwin

Born the same year as Poe, only twenty-four days his junior, Charles Darwin (1809–1882) revolutionized biology, placing it on a new foundation, one that has remained to this very day. Following Ernst Mayr, "Darwinism," or "Darwin's evolutionary theory," or "Darwin's evolutionary paradigm," consisted not of one theory

177. Andrew Robinson (2010). *Sudden Genius?* Oxford: Oxford University Press, xii, ix, xiv. Contextualist historians need not necessarily, by the way, pull focus in order to remain contextualists. As Bernie Lightman expressed to me over lunch with permission for quoting here, "The challenge to contextualist historians of science is to take into account in a balanced way the social, cultural, and political contexts of science and the creative responses of individual scientists working in that context."

but of "five largely independent theories": (i) evolution as such, that species are not fixed but evolve, to which Darwin converted the scientific world in roughly a dozen years; (ii) branching evolution or common descent, that related species evolved from a common ancestral species, which all trace back to the simplest origin of life; (iii) evolutionary gradualism, that evolution proceeds very slowly and not in jumps; (iv) the multiplication of species or speciation, the modes of the formation of new species; and (v) natural selection as the main driving force of evolution and the only cause of adaptations.[178]

All of these theories are to be found in Darwin's *On the Origin of Species* (1859), one of the most famous books in all of science and the only classic that continues to shake the world. Darwin had other theories, to be sure, such as his theory of inheritance he called *pangenesis*, published in 1868 and which proved a dead end. Nevertheless, it can hardly be doubted—except by the ideologically-driven diehards—that Darwin was well ahead of his time in a number of ways. What Darwin called *natural selection*, for example, as Mayr points out, was the only one of the five theories above that was bitterly resisted by many if not the majority of biologists until the beginning of the Modern Synthesis in the 1930s. To this we must add Darwin's explanation for sexual dimorphism in animals, what he called *sexual selection*, which took well over a hundred years to become appreciated and accepted (albeit modified slightly in its descent) by mainstream biology beginning in the 1970s.[179] There is also what Darwin called the *principle of divergence*, his theory of sympatric speciation, (the formation of new species within the parental range of a species,) which did not

178. Ernst Mayr (1982). *The Growth of Biological Thought: Diversity, Evolution, and Inheritance*. Cambridge, MA: Harvard University Press, 505–512. Ernst Mayr (1988). *Toward a New Philosophy of Biology: Observations of an Evolutionist*. Cambridge, MA: Harvard University Press, 196–214.

179. Malte Anderson (1994). *Sexual Selection*. Princeton, NJ: Princeton University Press, xvi. Richard A. Richards (2013). "Sexual Selection." In Michael Ruse, ed. (2013). *The Cambridge Encyclopedia of Darwin and Evolutionary Thought*. Cambridge: Cambridge University Press, 107–108.

rightfully take its place in biology until the 1990s.[180] And there is yet more, such as Darwin's founding of evolutionary psychology.

All of what Mayr calls Darwin's five theories, including sexual selection and the principle of divergence, were in fact developed years earlier by Darwin before their appearance in the *Origin*, which was an "Abstract" of a much larger book he had been working on. In all of this the imaginarium displayed by Darwin is truly spectacular, one that is rightly honored today mainly by those best adapted to understand and appreciate it—namely, professional biologists and philosophers of biology. But that is surely not the end of it. As discussed earlier, Darwin didn't take one category of evidence, like the fossil record, and argue for evolution. Instead, he argued by means of a consilience of inductions, bringing forth powerful evidence from geology, (with its gradualism and its hundreds of millions of years,) paleontology, (the fossil record,) embryology, (the more similar the species, the more indistinguishable the embryos,) comparative anatomy, (homologies, analogies, vestigial characters, and individual variation within species,) inheritance, (heritable variation in physical traits and in animal instincts,) biogeography, (the geographical distribution of plants and animals,) taxonomy, (the classificatory practices of his fellow naturalists, which recognized groups within groups within groups), and artificial selection (the breeding of plants and animals). One can raise cavils against any one of these categories as providing strong evidence of evolution, but one cannot reasonably do so against the whole, against the big picture. Darwin made this point repeatedly in his correspondence surrounding the publication of the *Origin* and for some time afterward. For example, in a letter to the American botanist Asa Gray, dated November 11, 1859, eleven days before the official publication of his self-described "one long argument," Darwin wrote that "I cannot possibly believe that a false theory would explain so many classes of facts, as I think it certainly does explain.—On these grounds

180. Alexey S. Kondrashov et al. (1998). "On the Sympatric Origin of Species by Means of Natural Selection." In Daniel J. Howard and Stewart H. Berlocher, eds. *Endless Forms: Species and Speciation*. Oxford: Oxford University Press, 97. Jerry A. Coyne and H. Allen Orr (2004). *Speciation*. Sunderland, MA: Sinauer, 126.

I drop my anchor & believe that the difficulties will slowly disappear.—"[181]

One can also see, as discussed earlier, Darwin employ inference to the best explanation in the *Origin*. Against the evolutionism of Jean Baptiste de Lamarck, with the theory of inheritance of acquired characteristics at its core, Darwin points out that Lamarck's theory fails to explain neuter castes of insects, for example, and he is "surprised that no one has advanced this demonstrative case . . . against the well-known doctrine of Lamarck." Against the evolutionism of the anonymous author of *Vestiges*, (Robert Chambers,) Darwin claims that its law of progressive development fails to explain "the coadaptations of organic beings to each other and to their physical conditions of life." In each case Darwin's own theory, he claims, explains what the others fail to explain. But Darwin's contrastive explanations are mainly directed against the divine creationism of his time, which dominated his scientific colleagues, whether single-shot creation or (more commonly) "continued creation," the latter meaning successive creation. For example, creationism does not explain, but Darwin's theory does, he says, the fact that the subordinate taxonomic groupings of his fellow naturalists "cannot be ranked in a single file, but seem rather to be clustered round points, and these round other points, and so on," with "group under group." Another example concerns the distinct and unique species of plants and animals on the Galapagos and Cape de Verde Archipelagos, in that they are similar to those on the South American and African mainlands, respectively. A further example concerns the fact that similar bones exist in "the wing and leg of a bat," even though they are used for "totally different purposes." And then there are the many examples of bad or unintelligent design in nature, such as "the sting of the wasp or of the bee . . . which, when used against many attacking animals, cannot be withdrawn, owing to the backward serratures, and so inevitably causes the death of the insect by tearing out its viscera." Related to the latter are examples that appear downright evil, such

181. Frederick Burkhardt and Sydney Smith, eds. (1991). *The Correspondence of Charles Darwin: Volume 7, 1858–1859*. Cambridge: Cambridge University Press, 369. For more of the same, see my *Darwin* (note 165), 97.

that "to my imagination it is far more satisfactory to look at such instincts as the young cuckoo ejecting its foster-brothers,—ants making slaves,—the larvæ of ichneumonidæ feeding within the live bodies of caterpillars,—not as specially endowed or created instincts, but as small consequences of one general law, leading to the advancement of all organic beings, namely, multiply, vary, let the strongest live and the weakest die."[182]

In short, Darwin's *Origin*, as many have rightly recognized, is a massive and most impressive piece of detective work, drawing on facts and evidence that were available to his fellow naturalists but which they failed to bring together into one big and consistent picture. But what is rarely appreciated is the role played by imagination in the *Origin*. While we are told, on the one hand, that "His reason ought to conquer his imagination," given that most critics (then as today) are unable or incapable of imagining the massively slow process of evolution by natural selection with its "transitional grades," we are also told, for example, that "We may imagine that the early progenitor of the ostrich had habits like those of a bustard, and that as natural selection increased in

182. Charles Darwin (1859). *On the Origin of Species by Means of Natural Selection*. London: John Murray, 242, 4, 95, 129, 413, 397–398, 437, 202, 243–244. Darwin's argument for the *fact* of evolution, by the way, using both consilience and inference to the best explanation, has only been amplified by modern biologists, in accordance with the principle of *beyond a reasonable doubt*. The only additional category of evidence, unavailable to Darwin, of course, is that of DNA. In addition to Coyne (note 58) and Dawkins (note 58), see, for example, Theodosius Dobzhansky (1973). "Nothing in Biology Makes Sense Except in the Light of Evolution." *American Biology Teacher* 35, 125–129. David Quammen (2004). "Was Darwin Wrong? No. The Evidence for Evolution is Overwhelming." *National Geographic* 206 (Nov.), 2–35. The reality of the situation is that debates over whether evolution is just a theory ceased a long time ago *within* biology, although the public continues to get the wrong impression with the public debates over evolution that occur *outside* of science. The situation is akin to the debate over who killed JFK. Conspiracy theorists continue unabatedly to fan the flames of suspicion among the public, all while the conclusion that Oswald did it and acted alone has repeatedly been established beyond a reasonable doubt. See Showtime's *On Trial: Lee Harvey Oswald* (1986), National Geographic's *JFK: The Lost Bullet* (2011), but especially Vincent Bugliosi (2006). *Reclaiming History: The Assassination of President John F. Kennedy*. New York: W.W. Norton.

successive generations the size and weight of its body, its legs were used more, and its wings less, until they became incapable of flight." Another example is the observation of black bears "swimming with widely open mouths, thus catching, like a whale, insects in the water," so that "I can see no difficulty in a race of bears being rendered, by natural selection, more and more aquatic in their structure and habits, with larger and larger mouths, till a creature was produced as monstrous as a whale." Another example involves the taxonomic division of animals and plants not only into higher taxa but into species, with subspecies or well-marked varieties below them, and then lesser subdivisions with individual differences at the bottom, such that "These differences blend into each other in an insensible series; and a series impresses the mind with the idea of an actual passage."[183]

Equally interesting, especially given Poe's emphasis that we shall examine in the next chapter, is Darwin's use of *analogy*,—recall Aristotle on the perception of "the similarity in dissimilars,"—both for illustrative and for argumentative purposes. Beginning with the former, for his branching conception of evolution, for example, Darwin provides the "simile" of the "great Tree of Life," with its "green and budding twigs" representing "existing species," "those produced each former year" representing "extinct species," and the limbs and branches representing "the classification of all extinct and living species in groups subordinate to groups."[184] Another example concerns natural selection. To help us visualize the process, Darwin provides the wedge comparison, (it's never done much for me,) which goes back to his early notebooks but was removed from the second edition of the *Origin*. "The face of Nature," he says, "may be compared to a yielding surface, with ten thousand sharp wedges packed close together and driven inwards by incessant blows, sometimes one wedge being struck, and then another with great force."[185] Presumably the idea is that each wedge at the "yielding surface" represents the population size of a species in a given area, with competition and natural selection driving some species into increase and others into decrease or even

183. Darwin (note 182), 188, 135, 184, 51.

184. Darwin (note 182), 129.

185. Darwin (note 182), 67.

extinction.[186] Yet another example is Darwin's analogy between species evolution and language evolution, such as "Rudimentary organs may be compared with the letters in a word, still retained in the spelling, but become useless in the pronunciation, but which serve as a clue in seeking for its derivation."[187]

The latter example possibly functions as an argumentative analogy as well. But there are clearer examples of that. For one, having read in late 1838 the sixth edition (1826) of Thomas Robert Malthus' *An Essay on the Principle of Population* (1798),—the core principle of which is that food sources for human populations can only increase arithmetically while the populations themselves, if unchecked, increase geometrically, so that there will always be competition and suffering given that the poor are incapable of controlling their reproduction,—Darwin immediately drew the analogy to plant and animal populations in the wild. As he puts it in the Introduction to the *Origin*, "This is the doctrine of Malthus, applied to the whole animal and vegetable kingdoms. As many more individuals of each species are born than can possibly survive; and as, consequently, there is a frequently recurring struggle for existence, it follows that any being, if it vary however slightly in any manner profitable to itself, under the complex and sometimes varying conditions of life, will have a better chance of surviving, and thus be *naturally selected*." To this he adds, "I use the term Struggle for Existence in a large and metaphorical sense, including dependence of one being on another, and including (which is more important) not only the life of the individual, but success in leaving progeny."[188] Using Malthus' *Essay* Darwin argued analogically for the existence of *population pressure* in nature, an important component of his concept of evolution by natural selection. Another important component came from his analogy with plant and animal breeding, the latter which he called *artificial selection*, the fact that breeders select the traits of plants and animals which they themselves fancy, resulting in better and

186. For a slightly different exposition, see M.J.S. Hodge and David Kohn (1985). "The Immediate Origins of Natural Selection." In David Kohn, ed. (1985). *The Darwinian Heritage*. Princeton, NJ: Princeton University Press, 194.

187. Darwin (note 182), 455, (see also 40, 310–311, and 422).

188. Darwin (note 182), 5, 62.

even new varieties and breeds. Of great importance is that breeders, says Darwin, refer to the traits of domesticated species of plants and animals as "something quite plastic." Turning to nature, Darwin then argues that the process is similar, but now the selecting agent is various features of the environment, with adaptation to the changing environment being analogous to adaptation to the changing fancies of the breeders, the result in nature being new varieties and subspecies and from them new species and higher taxa. As Darwin puts it in the *Origin*, "Slow though the process of selection may be, if feeble man can do much by his powers of artificial selection, I can see no limit to the amount of change, to the beauty and infinite complexity of the coadaptations between all organic beings, one with another and with their physical conditions of life, which may be effected in the long course of time by nature's power of selection."[189]

In all of this one would expect Darwin to have espoused a philosophy of science that puts a high premium on imagination. But with that expectation one meets with profound disappointment.

Following the publication of the *Origin*, and even before, just as Darwin anticipated, the problem of his *method* became a major bone of contention. His old geology mentor, Adam Sedgwick, for example, wrote to Darwin (a point he repeated in his book review) that "You have *deserted* . . . the true method of induction—& started up a machinery as wild I think as Bishop Wilkin's locomotive that was to sail with us to the Moon."[190] Darwin was also

---

189. Darwin (note 182), 31, 109.

190. Burkhardt and Smith (note 181), 396. In his book review Sedgwick wrote that "Darwin's theory is not *inductive*,—not based on a series of acknowledged facts pointing to a *general conclusion*,—not a proposition evolved out of facts, logically, and of course including them. To use an old figure, I look on the theory as a vast pyramid resting on its apex, . . . The only facts he pretends to adduce, as true elements of proof, are the *varieties* produced by domestication, or the *human artifice* of cross-breeding." Anonymous (1860). "Objections to Mr. Darwin's Theory of the Origin of Species." *The Spectator*, April 7, 285, 334. Similarly Bishop Samuel Wilberforce, aka Soapy Sam, aided by the esteemed anatomist Richard Owen in the background, claimed in his own book review that "when subjected to the stern Baconian law of the observation of facts, the theory breaks down utterly," that Darwin is not a "loyal disciple" of the "true Baconian philosophy," and that Darwin failed to restrain the flight of his "soaring imagination." Anonymous (1860). "*On*

attacked on the nature of his most fundamental causal law, natural selection. Shortly after the publication of the *Origin* Darwin heard that Sir John Herschel called natural selection "the law of higgledy-piggledy [*sic*]," which Darwin took to be "evidently very contemptuous" and a "great blow & discouragement."[191] Darwin's repeated reply in his correspondence to these and similar criticisms is the one we have seen above, his appeal to a consilience of inductions. At any rate, harsh criticism was precisely what Darwin had intended to avoid, given the way the scientific establishment had blasted the evolutionary theories of Lamarck and the author of *Vestiges*, arguing that each had failed miserably in making a *scientific* case. This largely explains why, as Darwin often remarked, he took twenty years in working on his "one long argument" before presenting it to the world. He was already a highly-respected scientist and wanted to keep it that way.

We can therefore perhaps appreciate Darwin's pleasure when he read an article on the *Origin* in *Macmillan's Magazine* by his statesman friend Henry Fawcett. As Darwin put it in a letter to T.H. Huxley near the end of 1860, "it pleases me that he quotes Mill's Logic & declares that I have philosophised in right spirit, & that all the talk about Baconian induction is cant & rubbish.—"[192] In the middle of the following year Fawcett replied to a letter from Darwin and provided the most striking corroboration for his claim in *Macmillan's*, stating not only that "I was particularly anxious to point out that the Method of Investigation pursued was in every respect, philosophically correct," but that "I was spending an evening last week with my friend M^r^. John Stuart Mill and I am sure you will be pleased to hear from such an authority that he considers that your reasoning throughout is in the most exact accordance with the strict principles of Logic. He also says, the Method of investigation you have followed is the only one proper

---

the Origin of Species." *Quarterly Review*, July & October, 239, 249, 262. Change "Darwin" to "Poe," eliminate the reference to domestication, and the above could read as reviews of *Eureka*.

191. Burkhardt and Smith (note 181), 423.

192. Frederick Burkhardt et al., eds. (1993). *The Correspondence of Charles Darwin: Volume 8, 1860*. Cambridge: Cambridge University Press, 514.

to such a subject."[193] To this letter Darwin replied, "You could not possibly have told me anything which would have given me more satisfaction than what you say about Mr. Mill's opinion. Until your review appeared I began to think that perhaps I did not understand at all how to reason scientifically."[194]

This invites a possible explanation for a comment Darwin made in his autobiography, the latter written in 1876, that "My first note-book was opened in July, 1837. I worked on true Baconian principles, and without any theory collected facts on a wholesale scale."[195] Was Darwin looking back and thinking of himself as a true Baconian given the imprimatur of Mill? The "true Baconian" line is often quoted as contradicting Darwin's actual method. But in fairness to Darwin, the context of the line should make it clear that he was only referring to his investigation into how plant and animal breeding worked. In July 1837 he had recently begun to take evolution seriously and was searching for a mechanism, for which he began by looking at breeding. At that point he was in fact "without any theory" on *that* matter, aside from a theory on how evolution itself might work.

At any rate, Darwin's acceptance of Mill's endorsement might seem rather odd, possibly even disingenuous, given Mill's three-stage Deductive Method of science examined in the first section of this chapter—direct induction, ratiocination, (syllogism,) verification—and Mill's debate with Whewell over imagination in science. Or perhaps Darwin never read (or sufficiently read of) Mill.[196] For the fact is that in Darwin's correspondence surrounding the publication of the *Origin* he makes it all but explicitly evident that his

193. Frederick Burkhardt et al., eds. (1994). *The Correspondence of Charles Darwin: Volume 9, 1861*. Cambridge: Cambridge University Press, 204.

194. Burkhardt et al. (note 193), 212.

195. Nora Barlow, ed. (1958). *The Autobiography of Charles Darwin, 1809–1882*. London: Collins, 119.

196. According to Laura Snyder, "While Whewell appears to have influenced Darwin, Mill's work did not appear to have any impact upon him at all. There are no references to Mill's work in Darwin's notebooks dating before the publication of the *Origin*, so it is not clear that he ever read *System of Logic*. If he had read it, surely Darwin would have seen that his theory could not be verified by the kind of direct observations Mill required." Snyder (note 32), 194.

method did not fit Mill's mold. We find him stating, for example, "why may I not invent hypothesis of natural selection . . . & try whether this hypothesis of natural selection does not explain (as I think it does) a large number of facts . . . . (not that I did *invent* it, for I was led to it by studying domestic varieties)," that "On his [William Hopkins'] standard of proof, *natural* science would never progress; for without the making of theories, I am convinced there would be no observation," or "How odd it is that anyone should not see that all observation must be for or against some view if it is to be of any service!," and "*let theory guide your observations.*"[197]

It is true, of course, as we have seen in the first section of this chapter, that Mill allowed for the first stage of the Deductive Method of science, direct induction, to be bypassed and replaced with the invention of a hypothesis, but he then claimed, as we have also seen in that section, that the scientific theorist must be willing to prove the hypothesis true. And in fact in the fifth edition of *A System of Logic* (1862), Mill explicitly applied the point to Darwin's *Origin*,—but alas, in a *footnote*,—claiming that the main thesis of the *Origin* is an "unimpeachable example of a legitimate hypothesis" and that natural selection is "not only a *vera causa*, but one proved to be capable of producing effects of the same *kind* with those which the hypothesis ascribes to it." But then came the rub. Though Darwin, he says, "cannot be said to have proved the truth of his doctrine, he does seem to have proved that it *may* be true which I take to be as great a triumph as knowledge & ingenuity could possibly achieve on such a question."[198]

Darwin, of course, as we have seen above in his letter to Asa Gray, could not agree to *that*. To his mind he spent most of the twenty years leading up to the *Origin* in working on the context of justification, on proving his theory true *beyond a reasonable doubt*. To provide yet another example, in 1861 he wrote in a letter that "I am actually weary of telling people that I do not pretend to adduce direct evidence of one species changing into another, but that I believe that this view is in the main correct, because so many

197. Burkhardt et al. (note 192), 195, 233; Burkhardt (note 193), 269; Frederick Burkhardt et al. (1999). *The Correspondence of Charles Darwin: Volume 11, 1863*. Cambridge: Cambridge University Press, 483.

198. Burkhardt et al. (note 193), 205n3.

phenomena can be thus grouped together and explained. But it is generally of no use, I cannot make persons see this."[199] In a similar vein he wrote, just a little earlier, that "the descent of species with their modifications through natural selection groups together and fairly well explains many phenomena (as classification, morphology, rudimentary organs, embryology, partially Geograph. Distrib. and partially Geolog. succession), and *therefore* I believe in its truth."[200] Darwin was prescient at the end of the *Origin*, (in more ways than one,) looking to "young and rising naturalists," with their "flexibility of mind," as being the ones best able "to view both sides of the question with impartiality."[201] It was a claim he would repeat throughout his correspondence, as for example in another letter written in 1861, in which he wrote, "I feel this confidence, because I find so many young & middle-aged truly good workers in different branches, either partially or wholly accepting my views, because they find that they can thus group & understand many scattered facts."[202] This is a theme we shall revisit in the final chapter, the relation between age and scientific imagination.

Especially significant in all of this is the fact that Darwin's initial theorizing did not fit Mill's picture at all, and it is important to see why.

In 1831, at the age of twenty-two, Darwin embarked on a five-year circumnavigatory voyage as naturalist aboard the H.M.S. *Beagle*. Getting that opportunity was a stroke of luck, especially given that the voyage "determined my whole career," as Darwin put it in his autobiography.[203] Although he began the voyage as a

199. That was and still is, of course, because of a lack of scientific imagination in such persons, something that can now be detected scientifically by an examination of the forehead, a correlation of growth having recently been established between broadness of forehead and broadness of scientific imagination. I am indebted for this point to Dr. B. Parnassus, Distinguished Research Professor Emeritus of Phenomenology and Phrenology in the philosophy department at York University in Toronto, Canada.

200. Burkhardt et al. (note 193), 96, 50.

201. Darwin (note 182), 482.

202. Burkhardt et al. (note 193), 135. For more of the same, see Stamos (note 165), 244n4.

203. Barlow (note 195), 76.

creationist, two problem situations arose in the voracious mind of this autodidact.

One problem involved his reading of the three volumes of Charles Lyell's newly published *Principles of Geology* (1830, 1832, 1833). In this highly important work, devoted in large part to the history and diversity of plants and animals, Lyell argued not only that the Earth is many hundreds of millions of years old, but he also argued for *uniformitarianism*, (Whewell's label,) that geological change is extremely gradual, not catastrophic, and that scientific explanations in geology are to be based on laws of nature, specifically on causal processes that can be observed in the present as they are also the ones that have shaped the natural world in the past. Darwin received a copy of each volume shortly after its publication. The problem was that in the second volume Lyell argued strenuously against the evolutionism of Lamarck while arguing in favor of successive, continued creation. This dissonance between Lyell's geology and his species biology must have been loud in Darwin's mind, especially given that Darwin's own grandfather was a famous evolutionist.

The second major problem situation was created by anomalous flora and fauna, living and fossil, encountered by Darwin during his voyage, especially from his exploration of the Galapagos Islands in 1835. In the latter case, Darwin was struck by the variation of plant and animal species from island to island, and that human inhabitants of the islands could tell which island a specimen came from just by looking at it. After the *Beagle* voyage Darwin consulted with expert naturalists concerning his specimens, who corrected his many misclassifications, including John Gould, who examined bird specimens Darwin had collected from the Galapagos and concluded that they were endemic species, unique to the islands, but related to South American species. Whether Darwin teetered over to evolutionism while still on his *Beagle* voyage or shortly afterward is a topic of debate, but teeter to the other side he did.[204]

204. Probably definitive here is the work of Sulloway, who argues against what he calls "the myth of the *Beagle* conversion." Frank J. Sulloway (1985). "Darwin's Early Intellectual Development: An Overview of the *Beagle* Voyage (1831–1836)." In Kohn (note 185), 145. Notably, Sulloway concludes that Darwin converted to evolution not because of the force of evidence but

The problem now, from a methodological point of view, is that Darwin's early evolutionary theorizing did not fit the Baconian mold (or much of his later theorizing, for that matter). As the cognitive psychologist Howard Gruber puts it in his study of Darwin's creative processes, "The pandemonium of Darwin's notebooks and his actual way of working, in which many different processes tumble over each other in untidy sequences—theorizing, experimenting, casual observing, cagey questioning, reading, etc.—would never have passed muster in a methodological court of inquiry among Darwin's scientific contemporaries."[205] Equally interesting is that in Darwin's early notebooks we find the record of three of his dreams, that Darwin "delighted in far-ranging speculations and saw himself as creating ideas of the same grandeur and cosmic scale as the 'early astronomers' to whom he likened himself," as Gruber puts it, and that his "images are full of feeling," including "love of beauty."[206]

What I find especially interesting, however, is what Darwin wrote in his autobiography many years later. Specifically, he states that "In October 1838, that is, fifteen months after I had begun my systematic enquiry [recall from above that Darwin started his first notebook in July 1837], I happened to read for amusement Malthus on *Population*, and being well prepared to appreciate the struggle for existence which everywhere goes on from long-continued observation of the habits of animals and plants, it at once struck me that under these circumstances favourable variations would tend to be preserved, and unfavourable ones to be destroyed. The result of this would be the formation of new species."[207] We shall return to the "it at once struck me" in the second section of Chapter 8. It is the "happened to read for amusement" part that concerns me here. Gruber is willing to take seriously Darwin's implication that picking up Malthus was a

because of "the most interesting source of his conversion," namely, Darwin's "gifted individualism," his "genius." Frank J. Sulloway (1982). "Darwin's Conversion: The *Beagle* Voyage and Its Aftermath." *Journal of the History of Biology* 15 (3), 389.

205. Howard E. Gruber (1981). *Darwin on Man: A Psychological Study of Scientific Creativity*. 2nd ed. Chicago: University of Chicago Press, 122.

206. Gruber (note 205), 238, 123, 237.

207. Barlow (note 195), 120.

"coincidence," even though a page later he affirms that "forgotten ideas are not really lost: they are reactivated when useful."[208]

I want to go deeper into the well. Darwin borrowed Malthus' *An Essay on the Principle of Population* from his older brother, Erasmus Alvey Darwin, who had in his library the sixth edition (1826). Why did he do this? Surely it would not afford amusement, like poetry and novels. Instead, the principles of population espoused by Malthus were "in the air" that Darwin breathed, they were part of the *zeitgeist* of his time and place.[209] Consequently, at an unconscious level Darwin must have known that Malthus' principles either were or might be relevant to his search for a causal mechanism for evolution, the very problem that consumed his conscious efforts. In other words, Darwin's turning to Malthus was no accident, no "coincidence," but unconsciously driven. If I may be indulged for a moment, I experienced the same sort of thing while writing this book. For example, sometime after writing the Beauty and Hope section of Chapter 3, I chose, for no apparent reason at the time, to browse Carl Jung's *Memories, Dreams, Reflections* before going to sleep one night, a book I had read many years ago and hadn't touched in almost as long. There I found, to my astonishment while reading in bed, Jung's line about hope that I now make much of in that section. I was elated by the find, and I think this sort of thing happens to virtually everyone, and not a few times, where the unconscious mind, with its deep reservoir of memories, prompts the conscious mind to pursue a line of investigation, as part of its problem-solving activity. (For my own part, this apparently involves no respect for sleep.) I also think that most people do not pay attention to this, so that it typically goes unnoticed. But enough of this for now. It is a major theme we shall explore in the next two chapters.

Perhaps in the end Darwin's view of science as expressed in his autobiography, that it "consists in grouping facts so that general laws or conclusions may be drawn from them,"—a view of science that blandly ignores the context of discovery, including analogies and intuitive flashes of insight,—is the consequence of his having

208. Gruber (note 205), 119, 120.

209. See, e.g., Desmond and Moore (note 164), 264–265. Also Janet Browne (1995). *Charles Darwin: Voyaging*. New York: Alfred A. Knopf, 385–390.

lost in later life his former delight in poetry, such as Byron, Shelley, and Shakespeare, and in music and pictures. He retained only his pleasure in fiction: "novels which are works of imagination, though not of a very high order, have been for years a wonderful relief and pleasure to me, and I often bless all novelists. . . . I like all if moderately good, and if they do not end unhappily."[210]

Given that Poe was not a novelist (with one exception) and that his fiction typically lacks happy endings, I doubt that Darwin, especially in his older age as an agnostic, would have enjoyed Poe's writings, including *Eureka*. But if Poe was right, and the Universe is a plot of God, for the purpose of God's pleasure, then Darwin's blessing of all novelists takes on an unexpected meaning, as the Universe, of which we are all a part, shall end happily after all.

At any rate, let us turn next to Einstein for hope of better insights into the nature of scientific imagination.

Albert Einstein

In 1905, at the age of twenty-six, Albert Einstein (1879–1955) had what is often called his *annus mirabilis*, his miraculous year, a phrase that had often been applied to 1666 for Isaac Newton, the year in which, at the age of twenty-four, he developed the foundation of his revolutionary new physics. It was in 1905 that Einstein had published, in the prestigious journal *Annalen der Physik*, (imagine being the editor,) five papers that would—two of them especially—change the science of his time, and all while he was employed as a third-class technical expert at the Swiss Patent Office. Let us take a relatively brief look at the stuff of Einstein's *annus mirabilis*, in chronological order and with the titles of his papers in English translation.[211]

The first paper, "On a Heuristic Point of View, Concerning the Production and Transformation of Light," completed three days after he turned twenty-six, helped to lay the foundation of

210. Barlow (note 195), 70, 43–44, 138.

211. The five papers are reprinted in translation in John Stachel, ed. (1998). *Einstein's Miraculous Year: Five Papers That Changed the Face of Physics.* Princeton, NJ: Princeton University Press.

quantum physics, second only in importance to the revolutionary papers of Max Planck in 1900. It was for this work mainly, and oddly not for his work on relativity theory, that Einstein was awarded the Nobel Prize in 1922.[212] Here Einstein took the theory of Planck, that electromagnetic energy is not continuous but discrete, consisting of packets Planck called *quanta*, and applied it to light. In doing this, Einstein was going against the contemporary consensus of physicists that extended back to the work in physiology and physics conducted between 1800 and 1803 by the English polymath Thomas Young, which took light to be composed of waves of continuous energy.[213] Einstein, in what he called an "assumption," took light to be composed of discrete units of energy, each of which he called a *light quantum* and which later came to be called a *photon*. Einstein was therefore returning to the view of Newton, that light consists not of waves but of *corpuscles*, although of course in a form unimagined by Newton. Humbly beginning his paper with the claim that the wave theory of light "will probably never be replaced by another theory," Einstein then showed how his quantum theory of light solves a number of "contradictions" that had puzzled physicists using the wave theory, most notably the photoelectric effect, which is the release of electrons when light is shone on a metal but only when the frequency of light is above a certain cutoff frequency. Interestingly, of his *annus mirabilis* papers, this first was the only one Einstein would call "revolutionary."[214]

---

212. Abraham Pais (1982). *"Subtle is the Lord . . .": The Science and the Life of Albert Einstein*. Oxford: Oxford University Press, 502–511.

213. For a wonderful biography of Thomas Young, (with an appropriately long subtitle,) see Andrew Robinson (2006). *The Last Man Who Knew Everything: Thomas Young, the Anonymous Polymath Who Proved Newton Wrong, Explained How We See, Cured the Sick, and Deciphered the Rosetta Stone, Among Other Feats of Genius*. Oxford: Oneworld Publications.

214. Stachel (note 211), 5, 167, also 13–14. See also Albert Einstein (1919). "What Is the Theory of Relativity?" Here Einstein refers to special relativity as "simply a systematic development of the electrodynamics of Maxwell and Lorentz" and to general relativity as a matter of "logical completeness," such that Newton provided "the foundation of our whole modern conceptual structure in the sphere of natural philosophy." In Sonja Bargmann, trans. (1954). *Ideas and Opinions by Albert Einstein*. New York: Crown Publishers, 230, 232.

Einstein's second *annus mirabilis* paper, "A New Determination of Molecular Dimensions," was a slightly revised version of what had just been accepted as his Ph.D. dissertation by the University of Zurich, which he had submitted only a month earlier. In this paper Einstein not only argued for a molecular conception of matter, conceived as "rigid spheres," (remarkably, there were still some continuous substance theorists, who rejected atoms and molecules as illusions and were known as "energeticists,") he also provided a method for determining their sizes. What was particularly novel was that Einstein's method was the first to use fluids with a precision that rivalled methods using gases, thereby providing a convergence of support for the "atomic hypothesis."

Einstein's third *annus mirabilis* paper, "On the Motion of Small Particles Suspended in Liquids at Rest Required by the Molecular-Kinetic Theory of Heat," submitted for publication only eleven days after the previous had been submitted, was Einstein's first paper devoted to a phenomenon known as "Brownian motion," (sometimes also called "the drunkard's walk,") named after the Scottish botanist Robert Brown who in 1827 observed under a microscope the erratic movement of a pollen grain suspended in a liquid. Brown could not explain what he saw, and the phenomenon remained unexplained until Einstein provided his proof using the atomic hypothesis. Each zigzag of the suspended particle, Einstein showed, is not the result of a collision with a single molecule, analogous to what we see today in a pinball machine, but instead is the result of simultaneous pressures from molecules on all sides, a statistical fluctuation in the pressures resulting in the particular motion of the suspended particle. As the physicist John Gribbin puts it, "Einstein calculated the precise way in which the statistics worked, and predicted that the average distance of a particle from any chosen starting point increases as the square root of the time since it was first kicked away from the starting point," to which he adds, "This was a completely new prediction, made by Einstein and confirmed by observations."[215] And as the Oxford mathematician Roger Penrose puts it in his Foreword, "This latter analysis alone would have earned Einstein a place in history."[216]

215. Gribbin (note 100), 77, 78.

216. In Stachel (note 211), viii.

Einstein's fourth *annus mirabilis* paper, "On the Electrodynamics of Moving Bodies," inaugurated what Einstein is most famous for—in a word, *relativity*. Although Einstein did not invent what he called the "principle of relativity," (it goes back at least to Galileo,)—the idea not only that being stationary and being in motion appear relative to different observers, but that the laws of motion will be the same for each,—Einstein was the first to make it a universal principle of motion, (both epistemologically and ontologically,) later calling it the "theory of relativity." Propelled by the desire to eliminate the "difficulties" inherent in combining the classical mechanics of Newton with the more recent electromagnetics of Maxwell, Einstein took the principle of relativity as a "postulate" and proceeded to "introduce another postulate, which is only seemingly incompatible with it," the postulate that light in a vacuum travels at a constant velocity relative to its emitting source, a constancy that is independent of whether the source is in relative motion or rest. With these two postulates Einstein focused on the concept of simultaneity like never before. Gone as a result was the Newtonian concept of absolute time, with its "*absolute* meaning to the concept of simultaneity." Gone was the idea that time moves at a constant rate for all bodies, given that "a balanced-wheel clock located at the Earth's equator must, under otherwise identical conditions, run more slowly by a very small amount than an absolutely identical clock located at one of the Earth's poles." Gone was the idea that an object would maintain its dimensions the more it approached the speed of light, given that at the speed of light "all moving objects—considered from the 'rest system'—shrink into plane structures," and likewise "for bodies at rest in the 'rest' system when considered from a uniformly moving system," with the consequence that "superluminal velocities are not possible." And gone was the concept of the ether, that mysterious physical medium lacking mass and other mechanical properties that was supposed to pervade matter and the spaces between it, given that it is "superfluous, inasmuch as the view to be developed here will not require a 'space at absolute rest' endowed with special properties." All of these basic and previously acceptable ideas in physics received from Einstein the trisyllabic croaking of his stately Raven, with the world of physics echoing the same following an initial period of "icy silence" based on "an attitude of wait and see," as his sister put it, Planck himself being the first to break the ice and to see the light[217]—the light that wrought its ghost upon the floor, the shadow that shall be lifted—nevermore!

Einstein's fifth and final *annus mirabilis* paper, ("final" in that it was his last publication for 1905,) "Does the Inertia of a Body Depend on its Energy Content?," may be called a "pendant" to his previous paper, both because of its brevity (it is less than three pages long) and because of what it adds, which is Einstein's application of special relativity to mass and energy. Here we get Einstein's truly explosive conclusion that "The mass of a body is a measure of its energy content," with the proof for the formula (not explicitly expressed until 1912) $E = mc^2$.

Together with the previous paper, Einstein would come to call the theory his *special theory of relativity*, in order to distinguish it from his application of the theory to gravity, which he called the *general theory of relativity*. Hermann Minkowski in 1908, as we have seen in the previous chapter, took Einstein's theory of special relativity a step further by adding (or pulling out of it) the important concept of four-dimensional space-time. (Surprisingly, it took a while for Einstein to warm up to what his former professor had done with his theory.) A few years later Einstein published "On the Influence of Gravitation on the Propagation of Light" (1911), in which he argued not only that light is bent by gravity but that it should be detectable experimentally during an eclipse of the Sun. (Four years later he predicted that the bend would be twice as much as he originally thought, which was confirmed by Eddington's solar eclipse experiment in 1919, and which in turn quickly made Einstein world famous.) Continuing to work intensely on the problem of gravity, on its nature, the problem left untouched by his 1905 papers on special relativity, Einstein pursued a clue that had suddenly occurred to him in 1907 (more on this below). In short, with the help of his friend and former classmate, the mathematician Marcel Grossmann, who in 1912 found for him and taught to him the non-Euclidean geometry developed in the 1850s by Bernhard Riemann, the key he had been searching for, Einstein took Minkowski's space-time the "decisive step"

217. Pais (note 212), 149–151.

further, completing in 1915 his theory of gravity as curved spacetime, which was published as "The Foundation of the General Theory of Relativity" (1916). In the history of Einstein's life, this would prove to be his crowning achievement—his ultimate claim to gravitas, and the main cause of his graying hair.

Einstein did a lot more, of course, along with his own pangenesis type of mistakes, such as his cosmological constant and his unwavering repudiation of black holes—the former which *he* called the "biggest blunder" of his life, the latter which *we* may call the "biggest blunder" of his life. And yet his mistakes are greatly overshadowed by his triumphs, with the years cited above containing the core of his most remarkable accomplishments in science, triumphs which, it hardly needs to be said, changed the foundation of physics. But how did he do it? And was he really ahead of his time or merely a fortunate conduit, a man of "integral awareness" to use McLuhan's phrase?

Let us begin with the latter question. In NOVA's *Einstein Revealed* (1996), Michio Kaku, the theoretical physicist and renowned popularizer of science, states that Einstein in 1916, with his general theory of relativity, a theory which "is in a class all by itself," was roughly "fifty years ahead of his time." The biographer Andrew Robinson provides an impressive list in order to demonstrate "how far advanced" Einstein was in "many disparate areas," and it needs to be quoted in full:

> In 1905, in his follow-up to special relativity deriving the equation $E = mc^2$, Einstein perceived and in essence predicted nuclear energy 33 years before the discovery of nuclear fission (in 1938). In 1916, in his work on general relativity, he predicted the gravitational red shift, which was not confirmed for almost half a century (in the early 1960s). In 1917, in his paper on the stimulated emission of light, he conceived the possibility of a phenomenon not demonstrated until 37 years later, which then led to the laser (in 1958). In 1925, in his two papers on Bose-Einstein statistics, he envisaged a new state of matter, the Bose-Einstein condensate, which took 70 years to be observed (in 1995) and today offers promise of its first application—as has happened many times with the laser—as a so-called

> "atom interferometer" for the measurement of small changes in the strength and direction of gravity. In 1935, in his "EPR" paper written with Boris Podolsky and Nathan Rosen, he introduced the controversial concept of what became known as quantum entanglement, a subject for experimental research three decades later which has potential applications in quantum computing, quantum cryptography and even, some claim, teleportation. And this leaves aside Einstein's 1918 prediction of gravitational waves. Despite decades of intensive searching, no waves have been detected.[218]

One can now add the experimental discovery of gravitational waves in March 2014 using the BICEP2 radio telescope at the South Pole, but it no longer seems necessary. The question now is, how did Einstein accomplish all of this? Was it by amazing powers of *deductive* reasoning? Einstein's answer is *no*: "the fictitious character of fundamental principles is perfectly evident from the fact that we can point to two essentially different principles, both of which correspond with experience to a large extent." Einstein is clearly referring to what in philosophy of science, as we have seen earlier in this chapter, came to be called the *underdetermination of theory by evidence*. In the same passage Einstein says, "this proves at the same time that every attempt at a logical deduction of the basic concepts and postulates of mechanics from elementary experiences is doomed to failure." He then generalizes the point to "the axiomatic basis of theoretical physics."[219]

Was it then by amazing powers of *inductive* reasoning? Einstein's answer again is *no*: "There is no inductive method which could lead to the fundamental concepts of physics. Failure to understand this fact constituted the basic philosophical error of so many investigators of the nineteenth century. . . . The most

218. Andrew Robinson (2005). *Einstein: A Hundred Years of Relativity*. New York: Harry N. Abrams, 109, 111. Containing numerous photos and illustrations along with articles by others on Einstein, as well as Robinson's considerable biographical skills, this book is a gem for anyone interested in Einstein, both the man and his science.

219. Albert Einstein (1934). "On the Method of Theoretical Physics." In Bargmann (note 214), 273–274.

satisfactory situation is evidently to be found in cases where the new fundamental hypotheses are suggested by the world of experience itself."[220] This last sentence points to the need for something profoundly extra, something repeatedly emphasized by Einstein in the previous passage on deduction, (and elsewhere,) what he calls "free invention" (more on this below).

In a nutshell, Einstein accomplished what he did because of (i) a profound sense of wonder, (ii) the application of enormous energy and industry applied to the problems that concerned him, (something that, like athleticism, seems to peak before middle age), (iii) intensive study into the background knowledge necessary for solving those problems, (iv) a critical flexibility of mind, (which often went against authority,) rather than dogmatism and obeisance to tradition, and (v) an enormously powerful imagination, on the same level with great artists.

In terms of the fifth item, anyone who reads Einstein cannot help but notice that thought experiments and references to imagination permeate his writings. I shall provide here only a few of the more salient examples. In "Autobiographical Notes," to begin with, written at age sixty-seven and amusingly referred to as "something like my own obituary," Einstein recalls "a paradox upon which I had already hit at the age of sixteen." In what has come to be one of his most famous thought experiments, he recalls the thought that "If I pursue a beam of light with the velocity *c* (velocity of light in a vacuum), I should observe such a beam of light as a spatially oscillatory electromagnetic field at rest. However, there seems to be no such thing, whether on the basis of experience or according to Maxwell's equations." To this he adds, "One sees that in this paradox the germ of the special relativity theory is already contained."[221] When we then turn to his first paper on special relativity, the fourth of his *annus mirabilis*

---

220. Albert Einstein (1936). "Physics and Reality." In Bargmann (note 214), 307.

221. Albert Einstein (1949). "Autobiographical Notes." In Paul Arthur Schilpp, ed. (1949). *Albert Einstein: Philosopher-Scientist*. La Salle, IL: Open Court, 53. For more on this, and with an interesting twist, see John D. Norton (2012). "Chasing the Light: Einstein's Most Famous Thought Experiment." In Mélanie Frappier, Letitia Meynell, and James Robert Brown, eds. (2012). *Thought Experiments in Science, Philosophy, and the Arts*. New York: Routledge, 123–140. For my own part, I once tried Einstein's thought experiment

papers, we find Einstein again using "certain (imagined) physical experiments," such as his thought experiment with two clocks and then with two rods and clocks.[222] And then there is the thought experiment illustrating the equivalence of gravitational force and inertial force when the mass of the object in each case is the same, which Einstein first used in a lecture to physicists in 1913 while he was developing his theory of gravity. Drugged and then waking to find themselves trapped in a windowless elevator with all of their instruments, he said, it would be impossible for them to determine whether they were at rest on Earth or in deep space in a state of constant acceleration far away from Earth. In either situation their experience would be the same. (Today, of course, we know what happens to astronauts in a spaceship in a free orbit around the Earth, compared with what happens to them when the spaceship's rockets are firing or when the spaceship is spinning like a top, the latter allowing them to walk and eat just like the rest of us down here.) Einstein would use this thought experiment to argue that there really is no difference between the two forces, that they are really one and the same, which is the "equivalence principle" and which he *derived* in late 1915 from his new theory of gravity.[223]

The elevator thought experiment was repeated a little later in a modified form in Einstein's book (written specially in German) explaining special and general relativity to the general public.[224] Especially interesting in that book, however, for lovers of Poe, is another of Einstein's thought experiments. "I stand at the window of a railway carriage which is travelling uniformly," he writes, "and drop a stone on the embankment, without throwing it. Then, disregarding the influence of the air resistance, I see the stone descend in a straight line. A pedestrian who observes the misdeed from the footpath notices that the stone falls to earth in a parabolic curve." Which observer, Einstein asks, correctly observes the fall of the stone "in reality"? "With the aid of this example," he

while my seclusion was perfect, but I involuntarily closed my eyes in horror. The lids clenched themselves together as if in a spasm.

222. In Stachel (note 211), 126–130.

223. See Robinson (note 218), 73, and Pais (note 210), 195.

224. Albert Einstein (1916). *On the Special and the General Relativity Theory, a Popular Exposition*. Robert W. Lawson, trans. (1920). New York: Henry Holt and Company, 66–69.

answers, "it is clearly seen that there is no such thing as an independently existing trajectory (lit. 'path-curve'), but only a trajectory relative to a particular body of reference." Einstein then adds, two pages later, "Let us imagine a raven flying through the air in such a manner that its motion, as observed from the embankment, is uniform and in a straight line. If we were to observe the flying raven from the moving railway carriage, we should find that the motion of the raven would be one of different velocity and direction, but that it would still be uniform and in a straight line."[225]

Einstein provides the raven example "In order to attain the greatest possible clearness" on "the principle of relativity (in the restricted sense)." What I believe Einstein neglected to mention, however, what would add an even further degree of clarity to the matter, is that his choice of bird was no accident, that his raven is actually one and the same with Poe's raven. Not only is Einstein's choice of bird unlikely otherwise, in need of explanation and having none, but there are further so-called accidents that would otherwise remain unexplained, specifically Einstein's "I stand at the window," which echoes Poe's "I stood" and "my window" in "The Raven." I think the only reasonable explanation is that Einstein's raven is in reality none other than the one that eventually left the bust of Pallas just above Poe's student's chamber door, (perhaps after the student died of loneliness, or perhaps because he refused to feed it more,) which is to say that Einstein knew (perhaps at an unconscious level, or perhaps he gave it as a clue) that his work was connected with Poe's work in a most important way. Absorbed as I am by this delicate thought, I shall nevertheless leave it for others to pursue more fully, perhaps historians, psychologists, and metaphysicians, perhaps even ornithologists, but perhaps *not* philosophers and literary critics—deaf and blind to *te Angel ov te Odd* as they tend to be.

And then there are the numerous statements made by Einstein on the role of imagination in science, so many that they are incredibly easy to find—and uncannily as if they were written by Poe himself, as we shall amply see in the next chapter. For example, in an interview conducted in 1929 by the poet and Germanophile

225. Einstein (note 224), 9, 10, 12.

George Sylvester Viereck, Einstein, when asked how he got his ideas, replied, "I'm enough of an artist to draw freely on my imagination. Imagination is more important than knowledge. Knowledge is limited. Imagination encircles the world."[226] In the 1930s (the exact year eludes me) Einstein invited the poet Saint-John Perse, who was in Washington, to come to Princeton for an interview, (Saint-John Perse was the pseudonym of the diplomat Marie René Auguste Alexis Léger, who would win the Nobel Prize for Literature in 1960). Einstein wanted to know how the mind of a poet worked. Told about the role of intuition and imagination, Einstein replied, "It's the same for a man of science. It is a sudden illumination, almost a rapture. Later, to be sure, intelligence analyzes and experiments confirm or invalidate the intuition. But initially there is a great forward leap of the imagination."[227] In one of his lectures later published as an essay, Einstein writes of the "free inventions" of "the concepts and fundamental principles which underlie . . . a theoretical system of physics." He adds that "the purely fictitious character of the fundamentals of scientific theory was by no means the prevailing one in the eighteenth and nineteenth centuries," that "The natural philosophers of those days were . . . most of them possessed with the idea that the fundamental concepts and postulates of physics were not in the logical sense free inventions of the human mind but could be deduced from experience by 'abstraction.'"[228] In another of his published essays Einstein states that the scientific theorist "should be granted the right to give free reign to his fancy, for there is no other way to

226. Quoted in Walter Isaacson (2007). *Einstein: His Life and Universe*. New York: Simon & Schuster, 387. Einstein, by the way, was a genuine artist, playing the violin at a high level almost all his life and then switching to the piano late in life. See Philip Glass (2005). "Einstein and Music." In Robinson (note 218), 153–155. Quoted as saying, "If I were not a physicist, I would probably be a musician," he also stated that "I often think in music. I live my daydreams in music. I see my life in terms of music. . . I get most joy in life out of my violin." According to Glass, himself a composer and performer, "special relativity was a hypothesis based on an act of imagination: the scientist entering the realm of the dreamer, the poet, the artist. That's what interests me most about Einstein."

227. Quoted in Isaacson (note 226), 549, 640n16.

228. Einstein (note 219), 272, 273.

the goal. His is no idle daydreaming, but a search for the logically simplest possibilities and their consequences."[229]

The last of these examples relates to the topic of what does or should constrain the scientific imagination, the matter of epistemic virtues and values examined earlier in this chapter. That is a matter I shall return to below. The second of these examples speaks to the role of eureka moments in science, momentous moments of unconscious scientific creativity, most notably the one that led to Einstein's special theory of relativity and the one that led to his general theory of relativity, the latter involving what he called "the happiest thought of my life." I'm going to save those two experiences for the second section of the final chapter.

We must now ask, why this *open* emphasis on imagination in science by Einstein? It was surely, as Poe well knew before him, partly because imagination is part of the fact of the matter. But there are also the related matters of context and personality. Unlike Darwin, Einstein apparently couldn't care less if others thought his methods were scientific or not. He knew they were, and that was enough. Connected with this was his profound individualism. For example, as he points out in his "Autobiographical Notes," he reached the age of reason at twelve after reading some science and seeing that "much in the stories of the Bible could not be true." After this "crushing impression," he adds, "Suspicion against every kind of authority grew out of this experience, a skeptical attitude towards the convictions which were alive in any specific social environment—an attitude which has never again left me."[230] It was all part of what Abraham Pais calls Einstein's "apartness," a feature of Einstein that recurred throughout his life in multiple domains, including not only religion but his student days, his two failed marriages, and science.[231]

229. Albert Einstein (1934). "The Problem of Space, Ether, and the Field in Physics." In Bargmann (note 214), 282.

230. Einstein (note 221), 5.

231. Pais (note 212), 39. See also "Autobiographical Notes," in which Einstein defends himself in pursuing the "extra-personal world," the nature of the Universe viewed as a "great, eternal riddle," over "the chains of the 'merely-personal.'" Einstein (note 221), 5. See also "The World as I See It" (1931), in which Einstein describes himself as a "lone traveler." In Bargmann (note

There is also the matter of what Einstein called "philosophical prejudices." As he puts it, again in "Autobiographical Notes," "even scholars of audacious spirit and fine instinct can be obstructed in the interpretation of facts by philosophical prejudices." In the case of Ostwald and Mach in their rejection of the reality of atoms, the philosophical prejudice was "the faith that facts by themselves can and should yield scientific knowledge without free conceptual construction."[232]

And then there was Einstein's childlike sense of passionate wonder, which he never grew out of, (contra Eliot, and fortunately for the rest of us). For example, in a piece written only a few years before his death, he states that "There exists a passion for comprehension, just as there exists a passion for music. That passion is rather common in children, but gets lost in most people later on. Without this passion, there would be neither mathematics nor natural science."[233] Around the same time, Einstein is even reported to have said, by his friend and biographer Carl Seelig, "I have no special talents. I am only passionately curious."[234] That passion is intimately connected with the sense of wonder. In his "Autobiographical Notes," Einstein states that "For me it is not dubious that our thinking goes on for the most part without use of signs (words) and beyond that to a considerable degree unconsciously. For how, otherwise, should it happen that sometimes we 'wonder' quite spontaneously about some experience?" He adds that "This 'wondering' seems to occur when an experience comes into conflict with a world of concepts which is already sufficiently fixed in us."[235] This sense of wonder, he further tells us, looking back as far as he could as an old man, began with him at the age of four or five when his father showed him a compass, with its needle moving without a visible cause, an experience that "made a deep and lasting impression upon me." As for thinking typically not in

214), 9. In all of this, Einstein's choice of words is much like Poe's, I might add.

232. Einstein (note 221), 49.

233. Albert Einstein (1950). "On the Generalized Theory of Gravitation." In Bargmann (note 214), 342.

234. Quoted in Isaacson (note 225), 548, 640n11.

235. Einstein (note 220), 9.

words, in 1916 Einstein remarked to the Gestalt psychologist Max Wertheimer that "I very rarely think in words at all. A thought comes, and I may try to express it in words afterwards."[236] This adds new meaning not only to his thought experiments, but also to his eureka moments (a theme we shall return to in Chapter 8).

What I find especially impressive in Einstein is the way his mind brought conflicting theories and data together, this *synthetic* or *integrative* aspect of his imagination, for which the kind of imagination discussed above, specific acts such as chasing a light beam, played a subservient role. Most notably, with his 1916 theory of general relativity, an "intricate web," as Abraham Pais puts it, with which Einstein established himself as "the founder of general relativistic cosmology, the modern theory of the universe at large,"[237] Einstein brought together into one big and consistent picture Newton's theory of gravity, Maxwell's theory of electromagnetism, his own special theory of relativity, the equivalence principle, Minkowski's space-time continuum, Riemann's non-Euclidean geometry, and much more, all the while perfectly explaining the orbital anomaly known as the perihelion of Mercury (which had gone unexplained since Le Verrier first wrote about it in 1843) and predicting (correctly this time) the bending of light passing by the Sun as well as the gravitational red shift. A few years later, Einstein wrote in the *London Times* that "The chief attraction of the theory lies in its logical completeness. If a single one of the conclusions drawn from it proves wrong, it must be given up; to modify it without destroying the whole structure seems to be impossible."[238]

This brings us back to the role of epistemic virtues and values in scientific theorizing. It will be recalled from earlier in this chapter, and also from Chapter 4, that Poe emphasized "perfect consistency" and "beauty." Einstein, for his own part, time and again emphasized simplicity, scope, and internal coherence, with fruitfulness and testability also being important but not quite so. As Einstein put it in his "Autobiographical Notes," "A theory is the more impressive the greater the simplicity of its premises is,

236. Quoted in Isaacson (note 226), 9.

237. Pais (note 212), 266, 22.

238. Einstein (note 214), 232.

the more different kinds of things it relates, and the more extended is its area of applicability."[239] Or again, thirteen years earlier: "the story [of science] goes on until we have arrived at a system of the greatest conceivable unity, and of the greatest poverty of concepts of the logical foundations, which is still compatible with the observations made by our senses."[240] While testability was important to Einstein, and could mean the ultimate undoing of a theory,—as it had with the death of his "fudge factor," his cosmological constant, which he immediately accepted after seeing the photographs of galaxies taken by Hubble,—Einstein did not let himself become easily perturbed by experimental results that conflicted with and even seemed to refute any of his theories. As Robinson points out, Einstein showed "breathtaking confidence" when Walter Kaufmann claimed in 1906 that his experiments on the acceleration of electrons were incompatible with Einstein's special theory of relativity, and when Dayton Miller in 1921 claimed to have proven that the ether exists and modulates light.[241] It turned out that the experimenters were wrong, not Einstein. The survival of Einstein's main theories in the face of testing has remained to the present day, which includes most notably Gravity Probe B, launched in 2004 to test general relativity and completed in 2005 in Einstein's favor.

And what of inference to the best explanation, the companion theme to epistemic virtues and values? We have seen in the corresponding section of this chapter that Peter Lipton applied IBE in the context of justification to the context of discovery, that he suggested that scientific imagination is driven by the desire to produce not only explanations but the best explanation in contrast with competing explanations. Does Einstein fit this scenario? I suggest not. This is because Einstein distinguished between two kinds of physical theories, which he called "constructive" and "principle." Constructive theories, which include most theories in physics, he says, "attempt to build up a picture of the more complex phenomena out of the materials of a relatively simple formal scheme from which they start out." Einstein's example is

239. Einstein (note 221), 33.

240. Einstein (note 220), 294.

241. Robinson (note 177), 70.

the kinetic theory of gases. He then adds, "When we say that we have succeeded in understanding a group of natural processes, we invariably mean that a constructive theory has been found which covers the processes in question."[242] As John Stachel puts Einstein's meaning, constructive theories "purport to explain certain phenomena on the basis of hypothetical entities, such as atoms in motion, introduced precisely to provide such explanations."[243] Principle theories, on the other hand, of which Einstein includes thermodynamics and his own relativity theory, *do not attempt to explain*. "The elements which form their basis and starting point," he says, "are not hypothetically constructed but empirically discovered ones." Hence their "advantages" are not "completeness, adaptability, and clearness," the advantages of constructive theories, but "logical perfection and security of the foundations." Again as Stachel puts it, principle theories "summarize and generalize without purporting to explain." Stachel points out that this is why Einstein for several years after 1905 "referred to the 'relativity principle' rather than to the 'theory of relativity.'"[244] But even many years later Einstein continued to emphasize the formal aspects of relativity. As Pais points out, "in later times he tended to emphasize the simplicity of the theory rather than its consequences. In 1930 he wrote, 'I do not consider the main significance of the general theory of relativity to be the prediction of some tiny observable effects, but rather the simplicity of its foundations and its consistency.'"[245] That Einstein came to see relativity as something of an explanatory theory does not change the fact that for Einstein, at least in terms of relativity, explanation came second in his scientific theorizing, not first. If Lipton was right about the context of discovery, then, which is highly questionable, he was at most only partly right, not fully right.

In response to a questionnaire submitted to Einstein in 1932, he stated that "The real goal of my research has always been the simplification and unification of the system of theoretical physics."[246]

242. Einstein (note 214), 228.

243. Stachel (note 211), 19.

244. Stachel (note 211), 27n18. See also Pais (note 212), 27–28.

245. Pais (note 212), 273.

246. Quoted in Stachel (note 211), 15.

In the case of Poe, of course, *Eureka* was not confined to physics, such that his *constructive* theory of the Universe with his ultimate hypothetical entity, God, (conceived as a *poet*, no less,) can hardly be called the simplest possible given the evidence, (neither was Einstein's view of the Universe either, by the way). Nevertheless, Poe's priority in that work was clearly unification,—of science, his literary theory, and his theology,—all for the purpose of providing a grand theory of the nature and meaning of the Universe.

In sum, given all of the above, if any scientist was closest to Poe, it was surely Einstein. And this is not simply because both believed, to use Einstein's words, that "The eternal mystery of the world is its comprehensibility," that both experienced "a rapturous amazement at the harmony of natural law, which reveals an intelligence of such superiority," and that both were motivated in understanding the Universe by a conviction Einstein called a "cosmic religious feeling."[247] This is profound enough, and the comparison could well fill a book. Neither is it because both embraced pantheism and rejected the concept of a personal God, the God of Providence, and the belief in an afterlife that usually goes with it. Nor is it because each considered himself a "lone traveler," to use Einstein's phrase, and each proved to be well ahead of his time. Instead, what makes Poe and Einstein closest, in my view, is their joint emphasis on the educated imagination in science,—as something higher and more important than inductive and deductive reasoning, also as something that needs to be constrained by epistemic virtues and values,—and especially on the importance of *ansätze* ("educated guesses," the plural of *ansatz*), ultimately as eureka moments, as "sudden illumination[s]" and "great forward leap[s] of the imagination," followed by "years of hard labour on the calculating treadmill"[248]—echoes of Poe's "double Dupin"

247. Albert Einstein (1936). "Physics and Reality." (1934). "The Religious Spirit of Science." (1930). "Religion and Science." All are reprinted in Bargmann (note 214), 292, 40, 38. For more on Einstein and religion, see Max Jammer (2005). "Einstein on Religion, Judaism and Zionism." In Robinson (note 218), 185–190. See especially Max Jammer (1999). *Einstein and Religion: Physics and Theology*. Princeton, NJ: Princeton University Press.

248. For the attribution of "Ansatz" to Einstein, see Pais (note 212), 202. The treadmill line is also from a physicist, Philip Anderson (2005). "Einstein's Scientific Legacy." In Robinson (note 218), 125.

(discussed in the following chapter). Perhaps Poe, then, was right after all in his view that it is the imagination, the enormously educated and passionately industrious imagination, that drives science. Let us now make an argument by analogy for that view, before we turn to an explication of Poe's theory of scientific imagination in the next chapter.

Mutation and Imagination, an Analogy

Ernst Mayr (1904–2005) was one of the most renowned of twentieth-century evolutionary biologists. A major contributor to the Modern Synthesis, forged during the 1930s, '40s, and '50s, the synthesis of the subdisciplines of biology centered around the core of mutation and natural selection, Mayr was also a major contributor to philosophy and history of biology, with works such as *The Growth of Biological Thought* (1982), *Toward a New Philosophy of Biology* (1988), and *What Evolution Is* (2001). Remarkably, Mayr continued to write and get published right up until shortly before he died, which was a few months shy of turning 101.

What makes Mayr especially unique for our purposes, however, is the evolution of his concept of natural selection. For a long time Mayr referred to natural selection as the second of a "two-step process." In *Growth*, for example, he states that "Natural selection is not deterministic, and therefore not absolutely predictive," but the reason he gave was profoundly dissonant (for those who have the ears to hear): "natural selection is a two-step process. At the first step, the production of genetic variability, accident, indeed, reigns supreme. However, the ordering of genetic variability by selection at the second step is anything but a chance process."[249] Ah, but wait, Professor Mayr. You can't have it both ways. If only the second step is natural selection proper, then not only can you not call natural selection a two-step process, but you also cannot say natural selection is not deterministic. And if you want to say natural selection is not deterministic, as you clearly do, then you have to include the generation of heritable variation (of which mutation is the most fundamental and important) as part

249. Mayr (note 178), 490, 519–520.

of the process of natural selection. Only then can you have natural selection as a genuinely—fundamentally, ontologically, not merely epistemologically—statistical process.

Mayr's concept of natural selection above,—which by the way was and still is the standard, received view in biology and its philosophy, with phrases such as "mutations feed natural selection,"—was maintained by Mayr, as far as I can tell, for at least another ten years. But to Mayr's credit, unlike the rest of evolutionary biologists and philosophers of biology, he eventually saw the light. In an interview of "the grand old man of evolution" conducted in the fall of 1999 by Michael Shermer and Frank Sulloway, (each of them famous in their own right, the latter along with Stephen Jay Gould also a former graduate student of Mayr's at Harvard,) Mayr states that "Darwin's metaphor of selection turned out to be wrong. Natural selection is not a process of selection, it is a process of *elimination*." The metaphor of selection, of course, implies a division between that which is selected and selection itself. For Mayr, "Nothing is being selected. Nature is just eliminating the less fit." He then adds, "The only change I make [to Darwin's basic ideas] is that I consider the production of variation as part of natural selection. They are inseparable. Each is meaningless without the other. Natural selection is a two-step process: (1) variations produced, and (2) variations sorted, with the elimination of the less fit so that you end up with a 'selection' of the best."[250] Behold the difference: natural selection proper is now not the second in a two-step process, but both steps together, a *double* process, the first part consisting ultimately of mutagenesis.

The difference is utterly profound.

In an earlier work, devoted mainly to arguing that point mutations are rooted in quantum chance, I traced the ambiguity in Mayr and others back to Darwin and attempted to defend Mayr's final view. I also provided precedents, theory, and evidence for the view that a single mutation in an organism, what may be called a *key mutation*, or a single sequence of mutations in a number of organisms, known as *mutational order*, can either of them alter

250. Michael Shermer and Frank J. Sulloway (2000). "The Grand Old Man of Evolution: An Interview with Evolutionary Biologist Ernst Mayr." *Skeptic* 8 (1), 81.

the direction of evolution in the population to which they belong, its future evolutionary history, all for the purpose of emphasizing the fundamental contingency of evolution.[251] Although I fully intend to defend in more depth Mayr's reconceptualization of natural selection in the near future, here I shall confine myself to a single point.[252] If (i) we agree that evolution by natural selection is a process, (and professional biologists do, of course,) and if (ii) we want to make true statements about evolution by natural selection, (even if one's life depended on it, one could not find a single postmodernist among professional biologists,) and if (iii) we want to say evolution by natural selection is a genuinely contingent process, (and biologists do,) that the paths of evolution could be otherwise, then we *have* to include mutagenesis as part of the process of natural selection. This is because evolution *by natural selection*, without quantum chance, would have to be a deterministic process, fully predictable *in principle*, à la "Laplace's demon" of classical physics, found in his *A Philosophical Essay on Probabilities* (1814). But biologists don't want to say that, and rightly so. They recognize that evolution by natural selection is a fundamentally contingent process, one that really, truly, ontologically could have been (and could be) otherwise. As Stephen Jay Gould put it without fully recognizing what he was saying, (he was thinking of contingencies such as mutated viruses and asteroid impacts,) "Replay the tape [of evolution] a million times from a Burgess beginning, and I doubt that anything like *Homo sapiens* would ever evolve again. It is, indeed, a wonderful life."[253] Laplace's demon would never say that, but for biologists today to say it, and really mean it, they would have to give up classical physics, and moreover they would have to accept that quantum chance affects biological

251. David N. Stamos (2010). "Quantum Indeterminism, Mutation, Natural Selection, and the Meaning of Life." In Chérif F. Matta, ed. (2010). *Quantum Biochemistry: Electronic Structure and Biological Activity*. Vol. 2. Weinheim, Germany: Wiley-VCH, 837–872.

252. The paper shall be entitled, "The New Higgledy-Piggledy: Natural Selection as Mayr's Two-Step Process."

253. Stephen Jay Gould (1989). *Wonderful Life: The Burgess Shale and the Nature of History*. London: Hutchinson Radius, 289.

processes, most notably and fundamentally at the level of the production of mutations.

I think the reason for resistance to Mayr's view, and why it has been confined basically to Mayr, is mainly twofold: (i) to keep biology professionally distinct from chemistry and physics, (the death-knell of which was first sounded in 1953,) with mutagenesis belonging to the latter, not the former, and (ii) the widespread desire, ever since Darwin, to have natural selection be a *domain-neutral* process, so that one can speak of natural selection in domains other than biology, such as in language evolution, in theory evolution, and even in cosmic evolution. But the desire is misplaced nonetheless, for if Mayr is right then the best one can say, given the many differences in those domains, is that *something like* natural selection occurs in those evolutionary processes, but it is *not literally* natural selection.

Now for the real analogy. As we have seen earlier in this chapter, some have thought that science is a product, rather than the process or processes that produced the product, the product being scientific knowledge. This view is difficult to sustain, however, given simply how we talk. We say that someone is going into science, that science is done by scientists, that science gets too little or too much research money, that science has a history, and so on. Unless one has good reasons to say all of this talk is mistaken, we shall then have to conclude that it is proper to speak of science as a process.

Once we agree with that conclusion, we should then see that science has an analogue to mutation in evolution by natural selection, which is the production of theories. Just as no mutation, no evolution by natural selection, so likewise no theories, no science as a process.

There are, of course, important *dis*analogies between the two domains, such that evolution (confining ourselves now to *biology*, excluding theistic *interpretations* of evolution) is not teleological, it has no goal, even though it is obviously adaptive. Science, of course, is in fact teleological, with actual minds (not mindless mutations) at the helm, their ultimate goal being an understanding of the way things really are, with more accurate and more inclusive theories supplanting older ones. In short, science is adaptive in an *epistemic* sense, evolution is adaptive in a *non-epistemic* sense. In this and many other ways the two domains are dissimilar.

But to focus on the many differences between the analogues is to miss the point of the analogy, "the similarity in dissimilars," to use Aristotle's felicitous phrase. Certainly without mutations evolution eventually grinds to a halt, and without scientific imagination science eventually grinds to a halt. But just as mutation is not merely *sine qua non* for the process of evolution by natural selection, like gravity, but is essentially *sine qua non*, in that it *drives* evolution, so too scientific imagination for the process of science. Each is the first half of a double process, the second of which is a sifting process, which *together* determine the history and direction of the respective processes.

If all of this is so, then just as an understanding of the *origin* of mutations is important if one wants to gain a reasonable understanding of the nature and history of evolution, likewise an understanding of the *origin* of scientific theories is important if one wants to gain a reasonable understanding of the nature and history of science. In other words, a full understanding of science requires an understanding not only of the context of justification but of the context of discovery as well. The sections above on Darwin and Einstein should already have made this evident. When we turn to Chapter 8, the section on eureka moments in the history of modern science shall make it amply evident.

There is still a problem, however. If one accepts, as scientists obviously do, that science, unlike evolution, is an *epistemic* process, then one might still object to the inclusion of the context of discovery in the process of science if that context is not itself epistemic. As Thomas Nickles puts it, with whom we began this chapter, for the context of discovery to be "relevant to epistemology," especially to be "an essential topic for epistemology," "it must be subject to normative rules or standards of an epistemic nature."[254] In other words, the approach of Campbell and Popper, as we have seen in the section above on evolutionary epistemology,—blind trial and error, hopefully followed by a blind, lucky hit,—epistemically decouples the context of discovery from the context of justification. It is the same with the approach of Reichenbach and

254. Nickles (note 2), 91.

Whewell long before him, as we have seen in earlier sections of this chapter, with their characterization of the context of discovery as consisting, ultimately, of "happy guesses."

I'm not entirely certain that the context of discovery has to be epistemic in order to necessarily count as part of the process of science. But, either way, the key is the underdetermination of theories by evidence discussed earlier in this chapter. The evidence involved with an empirical problem (scientific or otherwise) is always finite, but there are always infinitely many ways of connecting the dots, infinitely many ways of solving the problem. Again as Nickles put it, "The problem is to show how to achieve infinite aspiration ('Find the one true theory in an infinite domain of possibilities!') by finite means." The problem is acute for an epistemic enterprise. But let us add to the problem. Given that the human mind has a finite capacity, that the life of each scientist is finite in duration, (in fact it is short, as with the rest of us, whether we die prematurely or maturely,) and that the number of scientists working on a problem is always finite, it is therefore a miracle that there is any science at all. But there *is* science, and remarkably progressive science, with stellar individuals like Darwin and Einstein who've contributed to it. There must therefore *be* an epistemic nature to the creation of scientific theories *by scientists*, an epistemic nature to the context of discovery in the process of science. But what is it? Whatever the character of the process of discovery, says Nickles, "the discovery process filters out these few [candidates] from the limitless set of potential laws, theories, or explanations."[255] But as we have already seen in this chapter, the discovery process for major scientific theories, (if not for many minor theories,) is not going to be, because it can *not* be, deductive or inductive logics or methods, and is unlikely to be algorithmic at all.

Given, moreover, in line with Poe's own emphasis, that at least *much* of scientific theory generation consists of eureka moments, generated by the unconscious, (a point driven home by the catalogue of examples provided in Chapter 8,) the conclusion that forces itself upon us is that the guiding *parameters* of the scientific imagination must themselves be epistemic. And that would lead us

255. Nickles (note 2), 91.

to the epistemic virtues and values examined earlier in this chapter. Poe emphasized consistency and beauty, which was a legitimate emphasis. But, as we have seen, that list is not exhaustive. And what of the *principles* of the scientific imagination, the principles that function as a pull causation on the *practice* of science? *Explanation*, emphasized by the IBE approach, alone will not do. It would not even seem always necessary. *Prediction* often seems just as important, given the major example of quantum physics, which lacks explanation (no one knows why the quantum world behaves the way it does) but is no less scientific for it. And then there is *understanding*, which seems what Poe was ultimately about. It is, of course, the fundamental principle of the human spirit, not just of science. To fully understand is to not only know *that* or to know *how*, but also to know *why*.

Having arrived, at long last, following a massive struggle with philosophy and history of science, at the legitimization of Poe's inclusion of, even emphasis on, imagination in science, it is time now to return specifically to Poe, this time to a detailed examination of his theory of scientific imagination, (this is a Poe book, after all,) before we finish with a modern understanding. It is my hope that, in the end, we shall come to appreciate, like never before, Poe and his book *Eureka*, regardless of how far we are willing to go with him in agreement.

CHAPTER 7

Poe's Theory of Scientific Imagination

. . . they have dared to confine the Soul—the Soul which loves nothing so well as to soar in those regions of illimitable intuition which are utterly incognizant of *'path.'*

—Edgar Allan Poe[1]

Entrée

In the previous chapter we examined the main movements in modern philosophy of science in relation to the topic of scientific imagination. As we have seen, we found not only that they each are seriously wanting, but also that Poe actually had what could be called a philosophy of science, even a properly corrective one, a philosophy of science that places the main emphasis on what came to be called the context of discovery and that views eureka moments as the main driving force of science. In the chapter before that, Chapter 5, we examined Poe's remarkable anticipation of nine major theories and developments in twentieth-century science, what have turned out to be successful ones at that. At some point the reader should want to know how that was possible. Coincidence, or luck, fail as an explanation given what Poe, in a number of places in his writings, appropriately

1. *Eureka* (1848). *P&T*, 1269.

calls "the theory of probabilities,"[2] or alternatively "the Calculus of Probabilities."[3] Just as it is possible, though highly unlikely, that the motive behind the murders in "The Murders in the Rue Morgue" (1841) was money or gold, given that nearly 4,000 francs worth of gold were left at the murder scene, (which had recently been withdrawn from the bank by the mother,) so that, as Dupin puts it, "if we are to suppose gold the motive of this outrage, we must also imagine the perpetrator so vacillating an idiot as to have abandoned his gold and his motive together,"[4] likewise it is possible, though highly unlikely, that Poe scored at least nine lucky guesses in *Eureka* with respect to twentieth-century science. Chance, luck, coincidence, these fail to explain, especially given that (i) Poe kept abreast of the developments of science up to his time and was an autodidact, (ii) Poe seemed to be interested in everything and combined polymathy with monomania, and (iii) Poe had an extraordinary faculty of imagination, evidenced by his poems and tales alone, for which he is loved by millions.

Given all of the above, we should therefore want to know what Poe himself thought of imagination, specifically scientific imagination, what he said about its nature and how it works. The problem is that Poe, quite surprisingly, never wrote an essay on the topic, such that we have to try to piece together his theory of scientific imagination from his various writings.

That is the focus of the present chapter. Our search for insights begins by looking at Poe's use of doubling and of mesmerism in a number of his tales. After a detour through a madhouse, we then look at three of his examples of extraordinary scientific imagination, primarily that of Kepler, as well as three of his examples of failed scientific imagination, which includes Leibniz. We then turn to an analysis of Poe's fictional personification of scientific imagination, the ace detective Auguste Dupin, who is featured in Poe's three detective mysteries, where we employ much of what we learned in Chapter 6. Following that, we look at what Poe has to say about deduction and induction and finish with an elaboration of what Poe calls the "poetic intellect." The unity of effect

2. E.g., "The Murders in the Rue Morgue" (1841). *P&T*, 421.

3. E.g., "The Mystery of Marie Rogêt" (1842–43). *P&T*, 507.

4. *P&T*, 422.

produced by this chapter shall then lead us to the next and final chapter, where we attempt to modernize what Poe had in mind.

Double Consciousness

In Poe's time the distinction that we take for granted today, between the conscious and the unconscious mind, did not yet exist. It is Freud, more than anyone else, beginning with his landmark *The Interpretation of Dreams* (1900), to whom we owe the distinction, combined with the iceberg model, (textbookery be damned,) which was a truly revolutionary concept.[5] There were, of course, gropings toward the concept, to be found in many thinkers from Plato to Nietzsche and Charcot. But we are guilty of *presentism*, of reading present ideas into the past without sufficient evidence, if we read these authors and think they had the concept, too. Even among Freud's contemporaries, (I don't mean his students and immediate followers,) including Janet, there were seemingly none who proposed the existence of a *dynamic* unconscious as the *bulk* of the mind. At most, what precursors and contemporaries of Freud may be said to have had—at *best*, in the spirit of enormous generosity—were *proto-concepts* of the unconscious mind. In the case of Poe, we seem to find two of these, one referring to *doubling* and the other to *mesmerism*, both which Poe seemed to include under the umbrella term *intuition*.

Let us begin with doubling. According to the philosopher of psychology David Livingstone Smith,—an enormously important source for our purposes,—one way psychologists and philosophers in the nineteenth century dealt with what we today (post-Freud) call the *unconscious mind* was what Smith calls *dissociationism*, which he defines as the view that "a single mind can be split or 'doubled.'" As Smith puts it, "Once the mind has been doubled, both portions are fully conscious, but neither has introspective access to the other's mental contents. A person's consciousness is therefore 'unconscious' of the contents of their secondary or

5. See, e.g., Sebastian Gardner (1991). "The Unconscious." In Jerome Neu, ed. (1991). *The Cambridge Companion to Freud*. Cambridge: Cambridge University Press, 136–160.

tertiary consciousnesses in much the same way that the conscious mind of one person is 'unconscious' of the conscious mind of another person." This theory of the mind, says Smith, was "modeled on the clinical phenomenology of multiple personality disorder and was particularly widespread in France, where research into multiple personality was enjoying great popularity."[6]

Research into multiple personality had also been undertaken in the United States, however, in the early 1800s even, notably by Benjamin Rush. Subsequently known as the "father of American psychiatry," Rush had earlier served as Surgeon General for the Revolutionary Army and was a signee of both the American Declaration of Independence and the American Constitution. For years a well-known doctor in Philadelphia, Rush was also a professor at the University of Pennsylvania, the author of *Medical Inquiries and Observations upon the Diseases of the Mind* (1812), and a pioneer in the humane treatment of lunatics in American mental institutions.[7] Around the time that Rush was lecturing to his medical students in Philadelphia on multiple personality,—the

6. David Livingstone Smith (2004). "Freud and Searle on the Ontology of the Unconscious." In Jon Mills, ed. (2004). *Psychoanalysis at the Limit: Epistemology, Mind, and the Question of Science*. Albany: State University of New York Press, 79–80. Another way that scientifically-informed psychologists and philosophers (those who rejected mind-body substance dualism) dealt with the unconscious mind is what Smith calls *dispositionalism*, which is the view that unconscious states and processes are not mental at all (so forget about introspection) but instead are neurophysiological dispositions which under certain conditions cause conscious mental states and processes. This approach does not concern us here, but it is important to note that the kind of evidence that Freud appealed to for his revolutionary new concept of mind, and which Smith himself makes much of, will play an important role in the next and final chapter.

7. Rush also served on the medical staff at the Pennsylvania Hospital from 1783 until his death in 1813, where he was in charge of the treatment of the insane. Accordingly, Rush might in fact have served as the inspiration for Poe's "The System of Doctor Tarr and Professor Fether" (1844), with its "system of soothing." In the very least, the perceptive third eye will have noticed that the words "rushed" and "ushered" are to be found in the tale, (*P&T*, 716, 700,) which should be a surprise were it merely coincidental, given Poe's literary theory. "Usher," of course, is the surname of the mental case in Poe's "The Fall of the House of Usher" (1839), whose namesake in real life was touched upon in Chapter 1.

phrase reluctantly used by Rush was "two minds" for a report on a case described as "two distinct minds,"[8]—twenty-five-year-old Mary Reynolds, who lived some twenty-seven miles outside of Meadville, at the opposite end of Pennsylvania, had begun shifting between two personalities, which continued for a period of eighteen years, from 1811 to 1829, (each personality shift occurred during her sleep, each personality had no memories of the other, and each personality was markedly different in character,) after which time she settled into the second personality.[9]

What would come to be widely regarded as the first well-documented case of multiple personality disorder in the United States, and also one of the most famous, the Reynolds case was first written about by Dr. Samuel Latham Mitchill, a professor at the College of Physicians and Surgeons of New York and a correspondent of Rush's. His article is in the form of a letter addressed to "the Rev. Dr Nott," dated January 16, 1816. Mitchill states that while making an official visit to the Military Academy at West Point in early December 1815 the case of "Miss R——" was related to him by "Major Ellicott, . . . the mathematical chair in that seminary, . . . the subject of which is related to him by blood." (Major Andrew Ellicott was in fact the chair but was related to Mary Reynolds only by marriage, as an uncle.) In his publication Mitchill actually uses the phrase "double consciousness," both in the text and in the title.[10] It is also interesting that he ends with the statement, "My

8. Eric T. Carlson (1981). "The History of Multiple Personality in the United States: I. The Beginnings." *American Journal of Psychiatry* 138 (5), 668.

9. Eric T. Carlson (1984). "The History of Multiple Personality in the United States: Mary Reynolds and Her Subsequent Reputation." *Bulletin of the History of Medicine* 58 (1), 72–82.

10. Samuel Latham Mitchill (1816). "A Double Consciousness, or a Duality of Person in the Same Individual." This was originally published in the *Medical Repository*, the first medical journal in the United States, of which Mitchill was editor at the time and one of the founding editors. There is enormous confusion, however, over the original publication date for Mitchill's letter. Many present-day sources cite it as an 1817 publication, *Medical Repository* 3 (Feb. 1), 185–186. (This is what comes up from the American Periodicals database.) Even the *Christian Examiner* (May 14, 1822), 15, reprints it as "published in the Medical Repository for 1817." Some present-day sources, however, such as Carlson (note 9), cite it exactly as above but as 1816. My bet is that it was published in 1816, around the middle of May, as it seems to have

learned friend, the Reverend Timothy Alden, of Meadville, knows Miss R——, and is engaged in drawing up the history of her curious case, and he has promised, in due time, to send it to me complete." Alden, a graduate of Harvard's School of Divinity, moved to Meadville in May 1815 where he founded Allegheny College that same year. Mary Reynolds at the time lived with her brother, John, who was one of Alden's "nearest neighbours." Alden's first meeting with both Marys was one day and the next, requiring him to be "introduced to her anew" on the second day. Alden originally wrote his article as a letter addressed to "the honorable Samuel L. Mitchill, M.D. etc. of New-York," dated June 21, 1816. But Alden apparently decided not to send the letter. Instead, he published it the next month in the first issue of the magazine he founded.[11] What is interesting for our purposes is that Alden describes the phenomenon as "a person vested with a *two-fold consciousness*, or, more definitely, with *two distinct consciousnesses.*"

Poe moved to Philadelphia in early 1838,[12] where he met Dr. John Kearsley Mitchell [*sic*], a prominent physician and professor who quickly became his personal physician, (and later Virginia's,) no doubt partly because he lived only three blocks south of Poe.[13] His son, Dr. Silas Weir Mitchell, "once met Poe in his

---

been quickly republished, first in the *Reader* (May 29, 1816), 88, and then in *The Country Courier* (June 3, 1816), 3–4. I have been unable to find a pdf of the original *Medical Repository* issue to confirm the exact date and pages for Mitchill's letter.

11. Timothy Alden (1816). "Letter." *Allegheny Magazine* 1 (July 5). Alden's letter was reprinted in a number of publications, including *The National Register* 2 (Sept. 28, 1816), 68–70. Establishing priority is important in the interest of good scholarship, but the dozens of hours I spent on the present case left me pondering, weak and weary, over Poe's immortal reply to Dr. Moran when on his deathbed. According to the doctor in his letter to Maria Clemm, Poe "said the best thing his [Poe's] best friend could do would be to blow out his [Poe's] brains with a pistol." *Poe Log*, 846. The square brackets, of course, are in the interest of clarity.

12. *Poe Log*, 248.

13. This is based on the address of the boardinghouse given for Poe (202 Arch St.) in *Poe Log*, 248, and the address given for Mitchell (288 Walnut St.) in *Letters*, 214. Even when Poe moved to a small house in the autumn of that year, (Sixteenth St. near Locust), he was still close to Mitchell's residence, roughly ten blocks west. *Poe Log*, 255. There is a further connection between Poe and Mitchell, namely, Maelzel's Chess-Player. Created by Wolfgang von

father's office," as Quinn puts it.[14] This younger Mitchell would many years later, in his capacity as a Philadelphia neuropsychiatrist, use a variety of sources to write "the best report" of the Reynolds case.[15] Again using the phrase "double consciousness," and even "two persons in one body," Mitchell right at the start of the paper points out that "My attention was first drawn to it [the case of Mary Reynolds] by a MS. found among the papers of my father, the late Prof. John K. Mitchell [the elder Mitchell died in 1858]."[16] The manuscript he refers to, as he makes clear in the first paragraph, was written in 1835–36 by John V. Reynolds, a nephew of Mary Reynolds, who then gave it (we are not told when) to Archibald Alexander, a professor of theology at Princeton, who then gave it (we are not told when) to Mitchell's father.

The name "Weir," of course, features in one of Poe's best-known poems, "Ulalume—A Ballad" (1847), in which he writes twice of "the misty mid region of Weir" and thrice of "the ghoul-haunted woodland of Weir."[17] Following Quinn, this should not necessarily be understood as a geographical location, but possibly as a reference to "a well known family name in Philadelphia—Dr. S. Weir Mitchell being named after a member of it."[18] Richard Kopley and Kevin Hayes, along with Daniel Hoffman before them and Lewis Leary before Hoffman, and most recently Jerome McGann, who attributes the insight to T.O. Mabbott, "the most learned of all Poe scholars," agree that the reference is not to a geographical location but to Robert Walter Weir, a contemporary artist and art

---

Kempelen in 1769, it was later exhibited by Johann Maelzel, who died at sea on July 21, 1838, during a return boat trip from Havana, where he had given exhibitions of the chess-playing automaton. Poe had already exposed the automaton as a fraud in his "Maelzel's Chess-Player," published in the *Southern Literary Messenger* for April 1836. *E&R*, 1253–1276. Mitchell purchased the automaton from a friend of Maelzel's in early 1840. *Letters*, 214. Was it Poe who convinced Mitchell to purchase "The Turk"?

14. *Quinn*, 268; "CA. 1840?" according to *Poe Log*, 313.

15. Carlson (note 9), 75.

16. S. Weir Mitchell (1888). "Mary Reynolds: A Case of Double Consciousness." *Transactions of the College of Physicians of Philadelphia* (10), 366, 367.

17. *P&T*, 89, 91. The only variation is "This" instead of "the" on the second of these pages.

18. *Quinn*, 534.

teacher in New York, with whom Poe "likely made acquaintance during the 1840s."[19] Given, however, that Silas Weir Mitchell was eighteen when Poe composed "Ulalume" and was a voracious learner, a student at the time at the University of Pennsylvania and about to enter Jefferson Medical College, (he completed his medical degree in 1850 at the age of twenty-one,) I suggest instead that "the misty mid region of Weir" and "the ghoul-haunted woodland of Weir" are references to the interest in physiology and abnormal psychology respectively of the young Mitchell, already manifest in his father and wonderfully predicted and projected for the son by Poe especially with his thrice-repeated wonderful line. And if I am correct, this would make Poe remarkably prescient, given that the son eventually became known as the "father of American neurology"—during his career he produced 225 scientific and medical works, including classics such as *Injuries of Nerves and Their Consequences* (1872) and *Rest in the Treatment of Nervous Disease* (1875)—and given that Weir had a double life writing tales and poems, many of these works dealing with psychiatric themes. My favorite is "Pained Unto Death" (1890), and I would bet that Silas was one of us—a real lover of Poe.

Poe, it should be added, moved from Philadelphia to New York City in April 1844,[20] but he returned for a visit during late July 1847. Given that he was "exceedingly ill" during that visit, he likely would have visited his friend and former physician, Dr. Mitchell.[21] Possibly even this is when the doctor's son met Poe in his father's office, rather than when Poe lived in Philadelphia. Roughly three months later Poe composed "Ulalume" with the Weir lines.[22]

As if all of this is not weird enough already, (pun intended,) it needs to be noticed that "Reynolds" is what Poe repeatedly called out during the evening of Saturday October 6, 1849, having been

---

19. Richard Kopley and Kevin J. Hayes (2002). "Two Verse Masterworks: 'The Raven' and "Ulalume.'" In *Companion*, 199; *Hoffman*, 71; *McGann*, 131.

20. *Poe Log*, 456.

21. During his visit Poe stayed at the Western Hotel located at 288 Market St. *Poe Log*, 703. This is one block south of the boardinghouse where Poe first lived during his residence in Philadelphia and is only two blocks north of where Dr. Mitchell lived.

22. *Poe Log*, 704, 705, 707.

admitted to Washington College Hospital in Baltimore during the afternoon of Wednesday October 3. According to the attending physician, Dr. John Joseph Moran, Poe was in a state of "violent delirium" starting sometime on the Saturday (or possibly the day before, Moran is not clear). The delirium "continued until Saturday evening . . . when he commenced calling for one 'Reynolds', which he did through the night up to *three* on Sunday morning," at which time he calmed down, gently moved his head to the side, uttered "Lord help my poor soul," and sank quietly into death.[23] To this day no one knows why he repeatedly called out (or screamed) "Reynolds."

One theory is that the name comes from the explorer Jeremiah Reynolds, "projector of the voyages to the South Seas, whose very language he [Poe] had used in that tale [*The Narrative of Arthur Gordon Pym* (1837–1838), Poe's one and only novel]. He could easily have known Reynolds."[24] Perhaps indeed Poe, on the edge of death, if I may elaborate upon Quinn's theory, thought of himself as about to embark on his own "expedition of discovery," as he put it in his review of an 1843 pamphlet containing "a synopsis of the Reynolds Expedition of Discovery," a review in which he says, (as if in identification with Reynolds, as seemingly prescient of himself as the future author of *Eureka*), "To the prime mover of this important undertaking . . . He has seen his measures adopted in the teeth of opposition, and his comprehensive views confirmed in spite of cant, prejudice, ignorance and unbelief."[25] Unfortunately the theory does not harmonize at all with what Poe wrote in *Eureka* about death as the loss of personal identity, so unless Poe changed his mind between the writing of *Eureka* and the last evening of his life, or in his madness simply forgot about his view in *Eureka*, (or never really believed it in the first place,) the theory seems implausible.

Another theory is that Poe repeatedly called out "Reynolds" because Henry R. Reynolds was the name of the judge at Ryan's Tavern in Baltimore, which was being used as a polling station for both a municipal and a Congressional election and which was

23. *Poe Log*, 846.

24. *Quinn*, 640.

25. *E&R*, 1248, 1251.

where Poe had been found, either inside or in front, "rather the worse for wear," not simply because he was in tattered and soiled clothes that were not his own and did not fit, but because he was in a delirious and barely conscious state, "in need of immediate assistance," as the person who found him put it, after which he was taken to the hospital. John Walsh asks, in what makes the most sense to him, "Was the judge's name perhaps repeatedly spoken aloud in the crowded room, at last penetrating Poe's stupor and arousing vague associations now untraceable?"[26] Was Poe, then, simply mimicking like a myna bird, or rather like the croaking of his famous raven, or rather still like one of the lunatics in Dr. Alliam's *Maison de Santé*? All I am saying is that it makes very little sense.

A further theory is that the Reynolds thing is a myth, that Poe never called out the name "Reynolds" at all, that Moran was either "incredibly muddle-headed" or a "chronic liar."[27]

I want to take this opportunity to put forward a very different possibility. But first, let us give the good doctor the benefit of the doubt with regard to his original report, given that his report of "Reynolds" was written in a letter to Maria Clemm not long after Poe's death, the letter dated November 15, 1849, that it was based on his memory as well as the official "Records of the case," and that Poe was recognized in the hospital as famous, treated as "an object of unusual regard" by Moran and the other medical personnel.[28] Moran also reports in the same letter that Poe, on his second day at the hospital, suffered hallucinations, what he describes as "vacant converse with spectral and imaginary objects on the walls."[29] Although there was some "tremor of the limbs," his condition was "not violent or active delirium." On the Saturday, however, the day before Poe died, Moran describes him as being in the worst state ever, in "a violent delirium, resisting the efforts of two nurses to keep him in bed."[30]

26. John Evangelist Walsh (2000). *Midnight Dreary: The Mysterious Death of Edgar Allan Poe*. New York: St. Martin's Minotaur, 46, 122, 180.

27. William T. Bandy (1987). "Dr. Moran and the Poe-Reynolds Myth." Lecture given to the Edgar Allan Poe Society of Baltimore and available from their website.

28. Walsh (note 26), 40, 42.

29. *Poe Log*, 845.

30. *Poe Log*, 846.

That condition persisted until the evening, as already quoted, which is when he started calling the name "Reynolds," which he continued to call until he died at 3:00 the next morning. If, as now seems likely, Poe was suffering from the effects of a brain tumor,[31] it is quite possible that he was experiencing, on that evening and night uniquely, what today are called *autoscopic hallucinations*: seeing a solid double of himself, perhaps even one that was interactive. These hallucinations are known to have many causes, including various diseases of and injuries to the brain.[32] To Poe's brain tumor add the possibility of intoxicants or even brain trauma suffered in the days leading to his hospitalization, including the possibility of foul play,—the last five days, prior to his being found in Baltimore, remain to this day a mystery, but when he was found he was initially thought to be in a severely intoxicated state,—and the probability of autoscopy increases, especially if Poe was suffering from increased swelling of the brain during that evening.[33]

In short, my suggestion is that Poe suffered a series of autoscopic hallucinations during the final night of his life, a night preceded by "violent delirium," in which case he might well have screamed out—in the twisted logic of madness, in a partially self-aware fit of his own double consciousness—the last name of the only real case of doubling that he knew of, learned from his physician and friend, the elder Dr. Mitchell, namely, that of Mary —Reynolds!

And maybe it wasn't even a piece of twisted logic at all, but instead an intuitive recognition of "the similarity in dissimilars," which is to say that Poe, in having the autoscopic hallucinations

31. Leon Neyfakh (2007). "Poe's Mysterious Death: The Plot Thickens!" *The New York Observer* (October 16). Of possible relevance is that Poe was diagnosed with "a lesion on one side of the brain" in March 1847 and with "cerebral congestion" in November 1848. *Hutchisson*, 208, 230.

32. Oliver Sacks (2012). *Hallucinations*. New York: Alfred A. Knopf, 262–269.

33. It would be alcohol or laudanum in the case of intoxicants. Walsh (note 26), 43–47. Walsh's own thesis is that intoxication was forced upon Poe in combination with a severe beating, in other words that Poe was the victim of second-degree murder. Walsh (note 26), 116–120. For a more wide-ranging discussion, see "The Mysterious Death of Edgar Allan Poe," provided by the website of The Edgar Allan Poe Society of Baltimore.

before him, made the connection with the case of Mary Reynolds, each doubling different and yet similar, a matter of two persons connected with one brain. If this is the case, then Poe, for his "intuitive perception" of the metaphor, (to return to Aristotle on poetic genius,) should be accorded the honor of *poet to the very end*, and his "Lord help my poor soul" his very last expression of *hope*.

There are some subjects upon which I take pleasure in being minute, but I have digressed in the above (except for Poe lovers). Given the interest in double consciousness in medical and theological circles, (along with the related phenomenon of mesmeric consciousness, our next topic,) which spread into the public and was exploited by the literati,[34] doubling naturally became a major theme in modern literary studies devoted to the Victorian era, with scholars seeing double all over the place. Poe scholars have not been an exception. But let us begin with the most commonly cited example of doubling in Poe, "William Wilson" (1839), written by Poe probably in early 1839, (roughly a year after he moved to Philadelphia,) given that it was submitted for printing in May 1839.[35] In this tale we are told by the narrator that he was followed from childhood by a person of the same name, born on the same day of the same year even, who moreover looked the same, but who differed in having a "moral sense . . . far keener than my own," one even with "apparent omnipresence and omnipotence," but whose voice constitutionally never raised "*above a very low whisper.*" This other William Wilson, with "purposes" or "objects" that totally frustrated the understanding of the narrator with his "natural rights of self-agency," so "perseveringly interfered with my [the narrator's] affairs, and harassed me with his insinuated counsel," that in the end the narrator plunges his sword into him "repeatedly" and with "brute ferocity"—the murder in effect a kind of suicide, or semisuicide if you will, the surviving

34. Or perhaps more accurately, "If the literature of the nineteenth century was in love with duality, so was its psychology, . . . Literature and science collaborated in spreading the gospel which enjoined the plurality of the mind." Karl Miller (1985). *Doubles: Studies in Literary History*. Oxford: Oxford University Press, 329.

35. *Poe Log*, 261.

half (given that the narrator survived to tell the tale) a psychopath "*dead to the World, to Heaven and to Hope!*"[36]

Many see doubling in other of Poe's tales, such as *The Narrative of Arthur Gordon Pym* (1837–38),[37] "The Fall of the House of Usher" (1839),[38] "The Angel of the Odd" (1844) and "The Imp of the Perverse" (1845),[39] "The Cask of Amontillado (1846),[40] and, most important for our purposes, the detective mysteries beginning with "The Murders in the Rue Morgue" (1841), with Poe's ace detective, Auguste Dupin, as the double of the narrator.[41] Since Dupin is key for understanding Poe's concept of the scientific discoverer as the ace detective, we have to be especially careful with this doubling hypothesis. Taking it seriously can lead to the view that Poe thought of scientific imagination as a kind of personal double. But the fact remains that "William Wilson" provides the only clear and unequivocal example of "double consciousness" in Poe's tales, (combined with interactive autoscopic hallucinations at that,) and there the double is the narrator's moral conscience, not a scientific discoverer. Even the shared name, William Wilson, indicates, as Benjamin Fisher puts it, that for Poe "the theme of will is foremost in this tale (Will, I am, will's son)."[42] Moreover, Poe understood scientific imagination as consisting predominantly of eureka moments, as we shall see later in this chapter, not as an alter ego inside the same body, another conscious person, so

36. *P&T*, 345, 355, 343, 349, 354, 349, 356, 357. Poe considered this one of his best tales. *Letters*, 194, 199, 450.

37. E.g., John T. Irwin (1980). *American Hieroglyphics: The Symbol of the Egyptian Hieroglyphics in the American Renaissance*. New Haven, CT: Yale University Press, 119–120.

38. E.g., *Silverstein*, 150–151; *Hutchisson*, 96–97.

39. E.g., *Hoffman*, 207–209.

40. E.g., *Hoffman*, 214, 218–221; Walter Stepp (1976). "The Ironic Double in Poe's 'The Cask of Amontillado.'" *Studies in Short Fiction* 13 (4), 447–453; Chris Beyers (2013). "Byron." In *Context*, 258.

41. E.g., *Hutchisson*, 117.

42. *Fisher*, 67. William Engel adds that "the letter 'W' . . . by a phonetic slip connotes 'double you.'" William E. Engel (2012). *Early Modern Poetics in Melville and Poe: Memory, Melancholy, and the Emblematic Tradition*. Burlington, VT: Ashgate, 111. And that, of course, is doubled, too.

that the doubling hypothesis does not fit at all. Poe instead, as we shall see, was ultimately searching for something *deeper* in the psyche when it came to scientific imagination, not something on the same level as the conscious self, which is what the concept of the double entails. Viewing Poe's Dupin, then, as being a scientific *doppelgänger*, (German for "double-goer,")—we shall solve in detail what Poe meant by "a double Dupin" below,—will therefore lead us in the wrong direction, in fact to a dead end, one that is not only merely dead, but really, most sincerely dead. This in turn, quite interestingly, provides a powerful check against seeing a lot of doubling in Poe's tales, and consequently serves as an example of scientific understanding informing literary interpretation—what is the ultimate woe if not the perpetual horror, the Conqueror Worm, of the humanities.

Mesmeric Consciousness

Mesmerism was the term used in Poe's day for what we today would call the forerunner of *hypnotism*. The term is named after the Austrian physician Franz Anton Mesmer, who first published his theories in 1766. Mesmer used magnets to produce cures in patients, but later found that passing his hands over the bodies of the patients worked just as well. Consequently he concluded that there is a mysterious force or fluid pervading living bodies analogous to the magnetic force found in magnets, which he called *animal magnetism*. Many later researchers came to the conclusion that there was no mysterious force or fluid involved at all, that the power of the mesmerist resided, instead, (when the subject was not faking it,) in the ability to appeal to the imagination of the patient. Concurrent with this debate, the term *mesmerism* widened to include not only heightened states of suggestion while in a mesmeric trance but also sleepwalking, what they called *somnambulism*, along with clairvoyance, precognition, and other kinds of superior knowledge espoused while in a trance. Not surprisingly, many connected these latter mental states to communion with the divine.

An important text on mesmerism in Poe's day was Chauncy Hare Townshend's *Facts in Mesmerism*, published in 1840 in

London. (It was reprinted unabridged in 1841 by a Boston publisher but with fewer pages for more economical typesetting.) Townshend was an ordained minister, a poet, a collector, and a well-known mesmerist, in fact "a member of fashionable intellectual circles and a friend of . . . Charles Dickens, who collaborated with him on mesmeric experiments."[43] In the "Introductory Epistle" Townshend states quite forthrightly that mesmerism is "capable of eliciting the highest state of moral and intellectual advancement, to which a man, in this existence, can probably attain." Especially interesting for our purposes is his chapter entitled "On the Mesmeric Consciousness," in which, using the term "consciousness . . . as a general expression for every act and state of the mind," a designation that admits of kinds, he argues ultimately that mesmeric consciousness exhibits "the two phenomena of intellectual activity and corporeal deadness, heightened, indeed, to an extraordinary pitch." Much of his reasoning is based on his claim that "Consciousness, whether it relate to sensation or to intellectual exertion, acts more forcibly the more it is brought to bear upon a single point," which he supports with the observation that "where one sense is actually wanting (as in the blind or deaf, for instance,) the vivacity of impressions received through the other organs is greatly increased." In the final chapter, "The Mesmeric Medium," Townshend pulls out all the stops, claiming, for instance, that "the student in this science, who at length arrives at the knowledge of an ethereal medium, connected with all animated beings and with thought in particular, beholds

43. Alison Winter (1998). *Mesmerized: Powers of Mind in Victorian Britain*. Chicago: University of Chicago Press, 140–141. Interestingly, Poe claims to have had "two long interviews with Mr D. when here," in his letter dated July 2, 1844. *Letters*, 450. Although the letter was written in New York City, where Poe had moved a few months earlier, Poe's meetings (or meeting) with Dickens occurred in Philadelphia during the latter's American tour while Poe still resided in that city, and occurred on or around March 7, 1842. *Poe Log*, 361–362. It is possible that Dickens got Poe onto mesmerism, including Townshend's book, as Poe's first tale involving mesmerism, "A Tale of the Ragged Mountains," was published in March 1844. *Poe Log*, 455. At any rate, there was mutual respect between Dickens and Poe, perhaps more than usually recognized. Harry Poe, for example, argues rather convincingly that Edgar Poe, not Dickens himself, was the model for David Copperfield, the fictional character in Dickens' famous novel of the same name. *Evermore*, 4–6.

a flood of light illuminating mysteries which have probably long weighed on his heart and brain. He beholds all life and intelligence at once connected and individualized—reciprocally connected in all its parts universally with God."[44]

Bruce Mills suggests that mesmerism "informed Poe's theorizing on the art of the short prose tale. The notion of the single effect articulated in his review of Nathaniel Hawthorne's *Twice-Told Tales* resonates meaningfully with Townshend's study of uniquely receptive states of consciousness. If Poe's work of the 1840s underscores one central pattern, it may be the effort to develop artistic choices consonant with psychological truths arising from an understanding of mesmerism."[45] This is certainly an intriguing possibility for understanding Poe's doctrine of duration for a tale or poem and for "unity of effect," which are central to his literary theory and which we examined in Chapter 2. But in this section I want to argue that mesmerism is a dead end if we want to understand Poe's theory of scientific imagination, even though admittedly it is a better candidate than multiple personality.

There is no question that Poe knew of Townshend's book and thought highly of it. In "Marginalia" (November 1846), Poe reviews a book by William Newnham, *Human Magnetism: Its Claim to Dispassionate Inquiry, Being an Attempt to Show the Utility of its Application for the Relief of Human Suffering* (1845). In that book, using words such as "pretended science," "delusion," and "fraud and jugglery," the author claims that "the existence of counterfeit coin is rather a proof that there is somewhere the genuine standard gold to be imitated." Poe accuses the author of "begging the question," given that just because something can be faked it does not follow "that any undemonstrated original exists," just as one may "feign himself a sphynx or griffin" but that does not demonstrate the existence of those creatures. And for those who think—naively, or more accurately stupidly—that Poe believed in every kind of phenomena that he wrote about in his tales and poems, they need to consider, first, the audience for

44. Chauncy Hare Townshend (1841). *Facts in Mesmerism, or Animal Magnetism: With Reasons for a Dispassionate Inquiry Into It.* Boston: Charles C. Little and James Brown, viii, 287, 298, 299, 502.

45. Bruce Mills (2013). "Mesmerism." In *Context*, 323.

whom Poe wrote, and second, his point that "the belief in ghosts, or in a Deity, or in a future state, or in anything else credible or incredible—that any such belief is universal, demonstrates nothing more than that which needs no demonstration—the human unanimity—the identity of construction in the human brain—an identity of which the inevitable result must be, upon the whole, similar deductions from similar *data*." So for Poe extraordinary claims, no matter how widely believed, require extraordinary evidence—a *consensus gentium* is no criterion of truth. The review is of further interest given that Poe seems quite clearly to express a belief in mesmerism, with the only exception of *prévision*, (what today is called *precognition*). Newnham defends *prévision* in chapter 11 of his book on the basis of its cross-cultural occurrence, the great number of witnesses, and the continued belief in the phenomenon in the face of criticism. Poe states right at the beginning of his review of Newnham's book that "while agreeing with its general conclusions, (except where it discusses *prévision*,) I invariably find fault with the reasoning through which the conclusions are attained." Of equal interest for us is that Poe finishes his review with a defense of Townshend: "Most especially do I disagree with the author of this book in his (implied) disparagement of the work of Chauncey Hare Townshend—a work to be valued properly only in a day to come."[46]

All of this certainly makes matters much more interesting when we turn to Poe's three tales that make fictional use of mesmerism, namely, "A Tale of the Ragged Mountains" (1844), "Mesmeric Revelation" (1844), and "The Facts in the Case of M. Valdemar" (1845).

In the first of these, "A Tale of the Ragged Mountains," the dramatic date of which is 1827, a character named Bedloe receives regular mesmeric treatments for his "acute pains" from "neuralgic attacks," along with the help of morphine, from an old gentlemanly doctor who had become a mesmerist years back, only to have a vision one day while hiking in the mountains outside of Charlottesville in which he relived, from a first-person perspective, a dramatic scene in which he was a soldier fighting against a local uprising in an exotic Eastern city, in which he died shortly

46. *E&R*, 1411, 1410, 1412.

after being struck in the head by an arrow, which was followed by an out-of-body experience, and then a return to his present self in the mountains. Upon returning home to his friends and relating the story, his good but now shocked doctor presents him with a portrait of a young friend of his, a fellow British officer at the time named Mr. Oldeb, which was made in 1780 in Calcutta, shortly before he died from an arrow to the head in an uprising in Benares, a portrait that bore a "miraculous similarity" to Bedloe. This same Bedloe died a week later, the obituary mistakenly spelling his name "Bedlo."[47] A wonderful and entertaining story about reincarnation, without a doubt! But no scientific imagination is there for us to find.

In "The Facts in the Case of M. Valdemar," the long-suffering and dying Mr. Valdemar—"The left lung had been for eighteen months in a semi-osseous or cartilaginous state, . . . The right, in its upper portion, was also partially, if not thoroughly, ossified, . . . and, at one point, permanent adhesion to the ribs had taken place. . . . the patient was suspected of aneurism of the aorta"—agrees to be "mesmerized *in articulo mortis*" (at the point of death) by the narrator, identified only (of course) as "P—," with doctors and nurses in attendance. The transitional state between life and death, remarkably, was "arrested by the mesmeric process" for "*nearly seven months*," without food or drink, during which time "the sleep-waker" [*sic*] rarely spoke, only to confirm that he is dead. After this period, following "the customary passes" aimed at bringing him out of the trance, there followed "a partial descent of the iris," which "was accompanied by the profuse out-flowing of a yellowish ichor (from beneath the lids) of a pungent and highly offensive odor." Then followed the plea of Mr. Valdemar: "For God's sake!—quick!—quick!—put me to sleep—or, quick!—waken me!—quick!—*I say to you that I am dead!*" The "mesmeric passes" were again made to awaken him, at which point "his whole frame at once—within the space of a single minute, or even less, shrunk—crumbled—absolutely *rotted* away beneath my hands," such that "Upon the bed, before the whole company, there lay a nearly liquid mass of loathsome—of detestable putridity."[48] At this

47. *P&T*, 656, 663, 665.

48. *P&T*, 835, 833, 834, 840, 841, 842.

point, unrelated in the tale, the narrator fell to the floor in a faint. The galvanic battery was applied; and he suddenly expired in one of those ecstatic paroxysms which, occasionally, it superinduces. But I have digressed.

When published, the tale caused quite a sensation, many even thinking it was a true, factual account. Originally published in New York in the *American Review* for December 1845, Robert Collyer, a prominent mesmerist in Boston, wrote a letter to Poe, dated December 16, enthusiastically asking for verification of the facts "which I require for publication," adding that he himself had saved a man from premature burial using mesmerism.[49] In London, the *Popular Record of Modern Science* quickly reprinted the tale in its January 10 issue as a factual case with the title, "Mesmerism in America: The Death of M. Valdemar of New York."[50] Another London publisher reprinted the tale in late January as a pamphlet, with the title, *Mesmerism "In Articulo Mortis,"* along with an Advertisement claiming that "credence is given to it in America, where the occurrence took place."[51] Poe was delighted by all of this. One of "the gentlemen" requested by the editor of the *Popular Record* to get the truth of the matter wrote Poe and received a reply, in which Poe, sometime in early 1846, wrote that "I find the Valdemar case universally copied and *received as truth*, even in spite of my disclaimer." He also states that he made a "pronounced effort at verisimilitude for the sake of effect," but in terms of content "I have aimed at merely suggestion and speculation."[52]

Poe's "disclaimer," by the way, was possibly his prefatory note to the republication of the tale in the *Broadway Journal* for December 20, 1845, in which he wrote, in response to controversy over its truth, "We leave [the story] to speak for itself. We may observe, however, that there are a certain class of people who pride themselves upon Doubt, as a profession."[53] Or his "disclaimer" is to be found in the *Broadway Journal* for December 27, 1845,

49. *Poe Log*, 605.

50. *Poe Log*, 617.

51. *Poe Log*, 622.

52. *Letters*, 556. See also *Critical Heritage*, 148–155.

53. Mills (note 45), 327; *Poe Log*, 605.

under "Editorial Miscellany," in which he wrote, in reply to "Dr. Collyer," "We have no doubt that Mr. Collyer is perfectly correct in all that he says—and all that he desires us to say—but the truth is, there is a very small modicum of truth in the case of M. Valdemar."[54] Later, in an entry in *Graham's Magazine* ("Marginalia" March 1848), Poe would neither confirm nor deny the truth of "Valdemar," or of "Mesmeric Revelation," which the London *Popular Record of Modern Science* had also previously printed as true. Instead, he only criticises the magazine for its published reasons for believing "Valdemar" true, and only criticizes the London *Morning Post* for its reasons for being skeptical.[55]

In sum, then, if in terms of content Poe with "Valdemar" aimed only at, as he says, "merely suggestion and speculation" and in so doing included only "a very small modicum of truth," did any of that include his thoughts on the topic of scientific imagination? "Valdemar," quite interestingly, employs the theme of opposites that are only superficially so: wakefulness and sleep, (they're both mental,) "the whiteness of his whiskers, in violent contrast to the blackness of his hair," (they're both hair,) "A male and a female nurse," (they're both human,) life and death, (they're both—soul). But for all the imagination that is to be found in this mesmerizing and horrific tale, (and all the clues that it is a hoax,) there is yet no attribution of scientific imagination to mesmeric consciousness, no scientific insights coming from the "vibratory movement of the tongue" of Mr. Valdemar to contrast with the slow and plodding reasoning of the conscious mind.[56]

"Mesmeric Revelation," however, published the year before, is quite another matter. In this tale the narrator, who "had long been in the habit of mesmerizing the person in question, (Mr. Vankirk,)"—who was also, though not as badly as Mr. Valdemar, "an invalid . . . suffering with acute pain in the region of the heart, and breathed with great difficulty,"—mesmerizes the patient yet again because he, Mr. Vankirk, was never "intellectually convinced" by

---

54. *Poe Log*, 607.

55. *E&R*, 1430–1433. For the reasons given by these two publications respectively, see *Poe Log*, 617, 615. Poe in his private correspondence confirmed unequivocally that "Valdemar" was, "of course," a "hoax," and one that he did not believe *could* be true. *Letters*, 609, 627.

56. *P&T*, 833, 835, 840.

philosophical arguments in favor of the soul's immortality and had "latterly" experienced "a certain deepening of the feeling [of the truth of immortality], until it has come so nearly to resemble the acquiescence of reason, that I find it difficult to distinguish between the two." Vankirk continues:

> I am enabled, too, plainly to trace this effect to the mesmeric influence. I cannot better explain my meaning than by the hypothesis that the mesmeric exaltation enables me to perceive a train of ratiocination which, in my abnormal existence, convinces, but which, in full accordance with the mesmeric phenomena, does not extend, except through its *effect*, into my normal condition. In sleep-waking [*sic*], the reasoning and its conclusion—the cause and its effect—are present together. In my natural state, the cause vanishing, the effect only, and perhaps only partially, remains.[57]

This is of profound importance in trying to understand Poe on scientific imagination. In "mesmeric exaltation" there is a *hypothesized* (clearly Vankirk could not claim to remember it) "train of ratiocination"—as we have seen in the previous chapter, by *ratiocination* Poe meant, not a particular method of reasoning, but merely "profound and searching analysis"[58]—such that, per "hypothesis," "the reasoning and its conclusion—the cause and its effect—are present together," with only the conclusion—"the effect only, and perhaps only partially"—continuing "into my normal condition." With only the conclusion arriving into normal consciousness, minus the train, Poe would seem here to be describing the *eureka effect* which we shall examine below in this chapter and more fully in the next, which he held to be the main driving force of science.

This leads Vankirk "to think that some good results might ensue from a series of well-directed questions propounded to me while mesmerized." The narrator agreeing to the experiment, the questioning begins following mesmerisation, with results, as we have seen in Chapters 1, 3, and 5, that are either found in *Eureka*,

57. *P&T*, 717–718, 718–719.

58. *E&R*, 872.

that prefigure ideas found there, or that are corrected in *Eureka*: such as "the beginning is God"; "The ultimate, or unparticled matter, not only permeates all things but impels all things—and thus *is* all things within itself"; "This matter is God"; "To create individual, thinking beings, it was necessary to incarnate portions of the divine mind"; "Divested of corporate investiture, he [man] were God"; "he can never be thus divested—or at least never *will be*—else we must imagine an action of God returning upon itself—a purposeless and futile action" (Poe clearly had not yet the idea of the Big Crunch); "corresponding with the two conditions of the worm and the butterfly. What we call 'death,' is but the painful metamorphosis"; "pain, which in the inorganic life is impossible, is possible in the organic"; "pleasure, in all cases, is but the contrast of pain." Following these and the rest of the revelations, Vankirk, "with a bright smile irradiating all his features, . . . fell back upon his pillow and expired," his corpse in "less than a minute" having "all the stern rigidity of stone," leaving the narrator wondering, "Had the sleep-waker, indeed, during the latter portion of his discourse, been addressing me from out the region of the shadows?"[59]

This tale, too, (or rather story,) as with "Valdemar" the following year, was reprinted in the London *Popular Record of Modern Science* as a factual case, this time using the title "The Last Conversation of a Somnambule." In the same letter to "a Gentleman," discussed above, dated early 1846, Poe states that "The philosophy detailed . . . is my own—original, I mean, with myself, and had long impressed me. I was anxious to introduce it to the world in a manner that should insure for it attention." And he adds that of

59. *P&T*, 719, 720, 723, 726, 727. Poe uses "sleep-waker," rather than the common term "sleep-walker," both here and at the beginning of the tale, as he would later in "Valdemar." In "Marginalia" (March 1848) he is quite clear that no mistake was made, as "He [Vankirk] is a sleep-waker—*not* a sleep-walker." *E&R*, 1430. Poe likewise uses "sleep-waker" in a letter to James Lowell, dated July 2, 1844. *Letters*, 449. It should be noted that Townshend himself uses "sleepwaker" throughout his book, following Dr. John Elliotson, to whom Townshend's "Introductory Epistle" is addressed, who rejected the term "induced somnambulism" (*somnambulism* means "sleepwalking") as a synonym for "mesmerism" and introduced the replacement term "mesmeric sleepwaking," for the obvious reason that mesmerized patients do not always walk, some even being incapable of walking. Townshend (note 44), ix, 45n1.

its philosophy "I believe actual truth to be involved."[60] Moreover, shortly before he wrote this, in a book review of his anthology of tales entitled *Tales* (1845), written anonymously by Poe himself and published in the *Aristidean* (October 1845), Poe states that

> "Mesmeric Revelations [*sic*]," which comes next, has excited much discussion. A large number of mesmerists, queerly enough, take it all for gospel. Some of the Swedenborgians, at Philadelphia, wrote word to Poe, that at first they doubted, but in the end became convinced, of its truth. This was excruciatingly and unsurpassably funny—in spite of the air of *vraisemblance* that pervades the article itself. It is evidently meant to be nothing more than the vehicle of the author's views concerning the Deity, immateriality, spirit, &c., which he apparently believes to be true, in which belief he is joined by Professor Bush. The matter is most rigorously condensed and simplified. It might easily have been spread over the pages of a large octavo.[61]

Poe would, of course, expand and improve on those ideas a few years later in *Eureka* (in which, ironically, he would intuit a time at which *matter was most rigorously condensed and simplified)*. But is "Mesmeric Revelation" the key to understanding Poe on scientific imagination? The truths expressed by Vankirk were truths believed by Poe at the time. Moreover, the lack of temporality between the conclusions and the reasoning, expressed by

60. *Letters*, 556.

61. *E&R*, 870. "Mesmeric Revelations" (plural) is one of a number of mistakes evidently inserted by Poe to avoid suspicion that he wrote the review. The Bush reference, by the way, is to Reverend George Bush, Professor of Hebrew at New York University. Poe had written him on January 4, 1845, enclosing a copy of "Mesmeric Revelation," and apparently received a favorable reply. *Letters*, 473–474; *Poe Log*, 485–486. Poe would later provide a positive review of Bush in his "The Literati of New York City" (1846). *E&R*, 1121–1122. Bush in turn would favorably review *Eureka* in August 1848 in his Swedenborgian magazine, the *New Church Repository*. *Critical Heritage*, 289–291. Poe's point about the Swedenborgians and "Mesmeric Revelation," by the way, appeared a few months earlier in a condensed form in his "Marginal Notes" (August 1845), to which he wrote at the end: "The *story* is a pure fiction from beginning to end" (italics mine). *E&R*, 1367.

Poe's Vankirk near the beginning of the tale, and summed up by him with the phrase "mesmeric exaltation," seem to bear a strong resemblance to the term "eureka," such that they could possibly substitute for each other. The lack of temporality, moreover, suggests a kind of connection with, or tapping into, divine knowledge, which is also implied in Townshend's book, and which would be perceived, from an exalted theological view of divine omniscience, as nontemporal.

Perhaps it is not surprising, then, that the French poet and main expositor of Poe in France, Poe's younger contemporary Charles Baudelaire, appears to take this as Poe's meaning, that "For him [Poe], Imagination is the queen of the faculties; but by this he understands something greater than that which is understood by the average reader. Imagination is not fantasy; nor is it sensibility, . . . Imagination is an almost divine faculty which perceives immediately and without philosophical methods the inner and secret relations of things, the correspondences and the analogies."[62] Similarly, Harry Poe states that "Poe's theory of knowledge . . . lays in his recognition that Imagination represents the highest order of mental activity, the bridge between the physical and the spiritual, the bridge between the arts and the sciences, the bridge between humans and God. Reason and cognition were merely the means of applying what imagination had discovered."[63]

Perhaps, after all, this is what Poe was getting at. But I think we have to be very careful here. For a start, Poe's God, as we have seen in Chapter 3, is not the God of the Judeo-Christian-Islamic tradition, and it is certainly not the God of the philosophical theologians, such as Augustine, Boethius, and Aquinas, a God existing outside of space and time in an eternal, unchanging state. Poe's God *is* the Universe, or rather *this* universe, in its totality, a universe that is dynamic rather than static. In this pantheistic view there would seem no static source, let alone an eternal one, for the scientific imagination to tap into, let alone mesmeric consciousness.

Second, and just as important, if we really want to understand Poe on scientific imagination, then we have to turn to his real-life

62. Charles Baudelaire (1857). "New Notes on Edgar Poe." Francis Hyslop, trans. In *Recognition*, 52.

63. *Evermore*, 210n53.

examples, such as Kepler, and look at what he has to say about them; we also have to look at Poe's fictional personification of scientific imagination, namely, Auguste Dupin, and what Poe has to say about him in his three tales of ratiocination; and then we have to look at what Poe has to say about scientific imagination in his essays, which certainly must include *Eureka*.

What we shall see is not Vankirk's view. According to that fictional character, mesmeric imagination involves *no temporal process* at all, but instead "the reasoning and its conclusion—the cause and its effect—are present together" with only "the effect" making its way to the conscious mind. If Poe really believed this to be the case for actual scientific imagination, we should expect him to apply this "hypothesis" elsewhere. But he does not. What Poe had in mind, I suggest, as we shall see from the rest of this chapter, is best described as "unconscious scientific creativity," a *process* that will be the focus of the next and final chapter.

But first, we must take a detour from the main-road, along a grass-grown by-path, which, in half an hour, nearly loses itself in a dense forest, clothing the base of a mountain. Through this dank and gloomy wood we must ride.

Lunatics, Lovers, and Poets

Madness is a recurring theme in Poe's writings, and it is surely one of the most interesting. Sometimes a lunatic is just a lunatic, of course, as Poe in his writings is quite aware, for example with the patients in "The System of Doctor Tarr and Professor Fether" (1844), of whom, we are told, "There is no argument which so touches the feeble reason of the madman as the *reductio ad absurdum*"—feeding the person who fancies himself a chicken, for example, nothing but "corn and gravel."[64] What is remarkable, I must add, is how presumably sane people, in our universities especially, are commonly *not* affected by that species of argument—evidence human rights. At any rate, madness in Poe's writings does not usually function alone. Instead, the norm for Poe

64. *P&T*, 702.

is to connect madness with something else, and it is these various connections that make Poe especially interesting for us.

One of the connections is between madness and extraordinary education. In "Berenice" (1835), for example, the young narrator, Egæus, whose line of ancestry "has been called a race of visionaries," whose family mansion was noted for "the library's contents," who was actually born in "the library chamber, . . . —into a palace of imagination—into the wild dominions of monastic thought and erudition," who says he "loitered away my boyhood in books, and dissipated my youth in reverie," whose "passions *always were* of the mind," this same narrator suffers from a "disordered chamber of my brain," resulting in extreme attentiveness, what he calls "*monomania*," such as riveting his focus for hours "on the typography of a book," or for an entire night on "the steady flame of a lamp," and ultimately on the perfect teeth of his cousin and fiancée, Berenice, which leads to an unorthodox dental operation (to put it mildly) while she was prematurely buried.[65] Coined around 1810 by the French psychiatrist Jean-Étienne-Dominique Esquirol, with enormous implications for psychiatric specialization and the insanity defense, *monomania* referred to an *idée fixe* to the point of disease or derangement in an otherwise sane mind. The term became wildly popular outside of French psychiatry soon afterward, being used loosely to refer to an obsessive preoccupation over an idea, such as democracy, to single-mindedness in achieving a goal, such as climbing the social ladder, and even to a recurring theme in a piece of music.[66] Given either the technical or the popular meaning, Poe knew exactly what he was doing when he used the term. And likewise it is surely no accident, given Poe's literary theory, (everything in a plot should serve toward the unity of effect,) not only that Poe's most famous tale about monomania is focused on a scholar, since the career of a scholar, (a seeker of truth such as a philosopher or scientist), if it is to be successful, would absolutely seem to require a *sort* of monomania, but also

65. *P&T*, 225, 226, 229, 230, 227.

66. See the important chapter on monomania in Jan Goldstein (2001). 2nd ed. *Console and Classify: The French Psychiatric Profession in the Nineteenth Century*. Chicago. University of Chicago Press, 152–196. According to Goldstein the term *monomania* reached its heyday in the 1830s and '40s and was virtually dead in psychiatry by 1870.

that the ultimate object of the narrator's monomania is a complete set of teeth, teeth being the representatives of truth, knowledge, and wisdom, housed in this case in the head of a person named "Berenice," which means "bringer of victory" in ancient Greek (*Bereniké*).[67]

Most interestingly, *Eureka* is arguably a work of monomania, and at one point Poe even refers to the "monomaniac grasping at the infinite."[68] With Egæus his scholarly monomania got derailed by a fixation on the wrong kind of teeth, evidenced by his feeling one midnight in the library that he "had newly awakened from a confused and exciting dream," by his garments "muddy and clotted with gore," his hands "indented with the impress of human nails," a "spade" leaning against the wall, and a spilled box from the table revealing "instruments of dental surgery, intermingled with thirty-two small white and ivory-looking substances," along with a "violated grave" and "a disfigured body enshrouded, yet still breathing, still palpitating, *still alive!*"[69] Egæus is what scholars call a "digger." And indeed he dug his way (notably while dreaming) to victory, so to speak,—just unfortunately not the right sort of victory. Poe with his scholarly monomania stayed on track, resulting in the pulling of teeth of a very different sort—truths about the Universe, prematurely buried in *Eureka*, Poe's magnum opus and greatest victory. In either case a mad pursuit!

Unlike "Berenice," sometimes the madness is apparent only, a case of superior intellect misunderstood by others as madness. This is the case in "The Gold-Bug" (1843). Throughout the first part of the tale the narrator repeatedly ascribes madness to William Legrand, the focus of the tale, whom he found "well educated, with unusual powers of mind, but infected with misanthropy, and subject to perverse moods of alternate enthusiasm and melancholy." (Today he might be described as bipolar, which sometimes accompanies genius, a topic we shall return to in this and the next chapter.) Legrand "had with him many books," we are told,

67. It is ancient Macedonian, actually, but close enough—unless, of course, you are Greek or Macedonian.

68. *P&T*, 1342. Both works also deal with the topic of "persons of ardent imagination," "men of ardent imagination." *P&T*, 228, 1269.

69. *P&T*, 232, 233.

"but rarely employed them." The "relatives of Legrand," we are further told, thought of him as "somewhat unsettled in intellect." And during the treasure hunt, which became very difficult and risky, the narrator says of Legrand that "The man is surely mad," he claims "plain evidence of my friend's aberration of mind," and he reaches a point where "By this time what little doubt I might have entertained of my poor friend's insanity, was put finally at rest," such that he was sartain he was faced with "A mind disposed to lunacy" and that it was necessary to "get the lunatic home by force."[70] But all this changes once Legrand actually figures out the location of Captain Kidd's buried treasure during their excursion and they dig it up. The second half of the tale involves Legrand's explanation of how he figured it out, including a fascinating solution to a cryptograph left by Kidd—providing us insights into how Poe himself solved cryptographs (his coinage) sent his way for *Graham's Magazine* during 1841–1842, a process begun when he was at the *Southern Literary Messenger* in 1836 and which became a mini career with his contributions to *Alexander's Weekly Messenger* during 1839–1840.[71] What is especially interesting is that Legrand's solutions sometimes came to him unconsciously,—"there seemed to glimmer, faintly, within the most remote and secret chambers of my intellect, a glow-worm-like conception of that truth which last night's adventure brought to so magnificent a

70. *P&T*, 560–561, 569, 572, 575. It is surely significant, in the very least as an indication of Poe's sense of fun, that he puts "sartain" (instead of "certain") twice into the mouth of Legrand's illiterate manservant Jupiter (not into the mouth of the narrator as I imply). *P&T*, 572, 576. John Sartain, a leading artist and publisher in Philadelphia, was a loyal friend of Poe's, whose acquaintance Poe first made in 1839 when he lived in Philadelphia from early 1838 to April 1844. *Poe Log*, 248, 284, 456. One can only assume that prior to writing "The Gold-Bug" in early 1843 Sartain had saved Poe from a debauch—that he had got "the lunatic home by force"—on one or more occasions. And if he did not, that it was not until the July 1849 incident (see note 73 below) that Sartain would help Poe during a state of temporary madness, then the coincidence with Poe's use of Sartain's name in "The Gold-Bug" must be by far the oddest accident of all, at least in the history of American literature, one that is guaranteed to continually astonish the skeptic and render his vull zubmizzion unto te Angel ov te Odd.

71. *Hutchisson*, 111–113; *Letters*, 294n, 320n, 406n. See also Poe's "A Few Words on Secret Writing" (1841). *E&R*, 1277–1291.

demonstration" and "one morning, it entered into my head, quite suddenly, that . . . ,"[72]— furnishing us with a case, in multiple, of unconscious problem solving that we shall return to below and shall examine much further in the following chapter.

Sometimes Poe makes a connection between madness and accentuated sense perception. In "The Tell-Tale Heart" (1843), for example, the narrator describes himself as "nervous—very, very dreadfully nervous," such that "The disease had sharpened my senses. . . . Above all was the sense of hearing acute." And although the narrator repeatedly fancies that we, (or perhaps his lawyer,) to whom he is narrating his tale, will say he is mad, he does not think himself so, because of the precautions he took in creating fear in the old man, in the manner of killing him, and in the care with which he dismembered the corpse and then concealed it. Nevertheless it was the tell-tale beating of the old man's heart from below the floorboards,—"*much such a sound as a watch makes when enveloped in cotton*,"—with its gradual increase in volume, that forced him over the edge into a confession to the police.[73]

72. *P&T*, 582, 592.

73. *P&T*, 555, 559. Ironically, Poe in a temporary fit of madness involving hallucinations and severe paranoia, (unless someone really was out to kill him,) caused possibly by taking cholera medication (calomel) or by drinking too much alcohol (for him) while passing through Philadelphia in early July 1849 on his way from New York to Richmond, (Poe emphatically denied drinking alcohol except for "only a little Port wine," which was sometimes used as a remedy for cholera,) would claim an "acute" sense of hearing in overhearing some men on the train plotting to kill him, this to his friend John Sartain, who received the terrified and haggard Poe and probably saved him from suicide on that occasion. *Poe Log*, 812–816; *Silverman*, 415–418; Walsh (note 26), 72–78; *Letters*, 820–829. To this it should be added, what Poe well knew, that acute or exaggerated hearing was often associated in the medical literature of his time with insanity in general and "moral insanity" in particular. Benjamin Rush, decades earlier, had divided the mind into nine "faculties" that he grouped into three categories—"intellectual faculties," "passions," and "moral faculties"—and from which he gave to the world the concept of moral insanity, which meant that one could be insane (and therefore not punitively responsible) in the sense of having a diseased moral faculty while one's reason was nevertheless intact. In Poe's time the debate was accordingly over the moral insanity defense in real murder cases, which Poe explored in his fiction with "The Tell-Tale Heart" (1843) and "The Black Cat" (1843). See Paige Matthey Bynum (1989). "'Observe how healthily—how calmly I can tell you the whole story': Moral Insanity and Edgar Allan Poe's 'The Tell-Tale

Sometimes madness is mixed with accentuated mental perception into reality. In "Eleonora" (1841), for example, the narrator begins with some reflections that relate only obliquely to the tale that follows, specifically to his memory of the ethereal voice in the night that absolved him of guilt over his remarriage following the death of Eleonora: "I am come of a race noted for vigor of fancy and ardour of passion. Men have called me mad; but the question is not yet settled, whether madness is or is not the loftiest of intelligence—whether much that is glorious—whether all that is profound—does not spring from disease of thought—from *moods* of mind exalted at the expense of the general intellect. They who dream by day are cognizant of many things which escape those who dream only by night. In their grey visions they obtain glimpses of eternity, and thrill, in awakening, to find that they have been upon the verge of the great secret."[74]

And then sometimes Poe mixes madness with extraordinary education, accentuated sense perception, and also accentuated mental perception into reality. In "The Fall of the House of Usher" (1839), for example, the subject of the tale, Roderick Usher, who came from a "very ancient family" known for its "peculiar sensibility of temperament," and also for its "many works of exalted art," whose life in the family mansion included enormous book learning, including sources as disparate as Machiavelli and Swedenborg, and the playing of a variety of musical instruments, himself suffered from "a mental disorder which oppressed him," an "excessive nervous agitation," a "morbid acuteness of the senses,"

Heart.'" In Frederick Amrine, ed. (1989). *Literature and Science as Modes of Expression*. Dordrecht: Kluwer, 141–152. See also Brian Wall (2013). "Narrative Purpose and Legal Logic in 'The Tell-Tale Heart.'" *The Edgar Allan Poe Review* 14 (2), 129–143. It is also interesting to notice the extent to which Poe's narrator in "The Tell-Tale Heart" "corresponds with current psychoanalytic profiles of the 'paranoid schizophrenic' personality." *Zimmerman*, 18–25, 342–353. By the way, it should be noted that the line "you should have seen *me*" in "The Tell-Tale Heart" was later repeated (sans italics, but with "Look at *me!*" shortly before it) in Poe's comic tale "The Literary Life of Thingum Bob, Esq." (1844). *P&T*, 555, 786. This certainly adds an interesting twist to the view that the latter "might well be a précis of Poe's own experiences," that "Thingum's words . . . might serve for placement of Poe." *Fisher*, 12, 13.

74. *P&T*, 468.

and partly because of it maintained with "pertinacity" not only "the sentience of all vegetable things," but also, "in his disordered fancy," says the narrator, the sentience of "the kingdom of inorganization"[75]—the very doctrine, as we have seen in Chapter 3, maintained by Poe himself in *Eureka*.

Turning to *Eureka*, we find Poe say, in his own person, a little over a quarter of the way into the work, that of Newton's law of gravity "Its *modus operandi* has never, to my knowledge, been approached in the way of an effort at explanation. It is, therefore, with no unwarranted fear of being taken for a madman at the outset, and before I can bring my propositions fairly to the eye of those who alone are competent to decide on them, that I here declare the *modus operandi* of the Law of Gravity to be an exceedingly simple and perfectly explicable thing . . . when we regard it from the proper point of view."[76] The "proper point of view," of course, is the theological one examined in Chapter 5, the argument for the finite and evolving Universe which includes God's fine-tuning of the fundamental laws. Stuart and Susan Levine note that "Such passages offer rationalist readers the option of considering the narrative that follows merely a projection of a disordered mind and therefore credible as psychological fiction: . . . Poe's use of the device here should trouble the reader who want [*sic*] to see *Eureka* as entirely serious."[77]

We must not, of course, make the amateurish mistake of confusing Poe himself with the narrators in his fiction, including the ones who suffer from a mental disorder, given the evidence that "Poe's narrators are distinct from one another and from their creator."[78] But let me trouble the rationalist reader even further. On the very first page of *Eureka*, so as to characterize the work as "a survey of the *Universe*," and a unique one "using the word ['survey'] in its most comprehensive and only legitimate acceptation,"

75. *P&T*, 318, 321, 322, 327.

76. *P&T*, 1290.

77. *Eureka*, 136n78.

78. *Zimmerman*, 25. Using what he calls "notional sets," the evidence Zimmerman brings together in his book is quite impressive. His conclusion, of course, is unaffected by the fact that Poe, in what should be obvious given his biography, put *some* of himself into his narrators—which is part of the fun of reading him.

Poe states that "He who from the top of Ætna casts his eyes leisurely around, is affected by the *extent* and *diversity* of the scene. Only by a rapid whirling on his heel could he hope to comprehend the panorama in the sublimity of its *oneness*. But as, on the summit of Ætna, *no* man has thought of whirling on his heel, so no man has ever taken into his brain the full uniqueness of the prospect; . . ." The point is repeated a page further, that what is needed is "something like a mental gyration on the heel."[79] This is troubling given that in "The System of Doctor Tarr and Professor Fether" (1844) one of the dinner guests, who turns out to be one of the resident lunatics, was known to "turn round upon one heel by the hour, in this manner—so—." He was "the man with the tee-totum predilections."[80]

Such crotchets, I suggest, are not to be taken seriously, for they violate the principle of charity of interpretation, which can easily, and should easily, turn upon its violator with the logical force of so grave a weapon as the *reductio ad absurdum*. Moreover, we have seen at the end of Chapter 4 sufficient reason to reject the view that *Eureka* was intended as a hoax. No rational reader should ignore the evidence taken as a whole.

Accordingly a much different and more fruitful approach to viewing Poe on madmen is called for. But what? The truth is in the well, so to speak, and in this case the truth turns out to be quite deep, not shallow at all. So let us proceed down the well—into the abyss of the mind.[81]

---

79. *P&T*, 1261, 1262.

80. *P&T*, 708, 715. A "tee-totum," by the way, and quite interestingly, is a gambling spinning top. Was Poe gambling with the truth in *Eureka*?

81. The "well of Democritus" is "one of Poe's pet allusions." *Critical Theory*, 14n9. Democritus was an ancient Greek philosopher (c. 460–360 B.C.) who along with Leucippus co-founded atomic theory. They argued that everything we perceive, as well as our own minds and ideas, are composites consisting of tiny atoms and void and that the Universe itself consists of an infinity of atoms and void. By Poe's time "the well of Democritus" was part of a popular proverb about truth, the idea that truth lies deep in a well. Immanuel Kant, for example, in his first *Critique* (1781/87, II.IV.3), writes of "those who are naturalists because they know no better, . . . They follow common sense, without parading their ignorance as a method which is to teach us the wonderful secret, how we are to find the truth which lies at the bottom of the well of Democritus." J.M.D. Meiklejohn, trans. (1872). *Critique of Pure Reason*.

First, it must be noticed that never in his writings does Poe glorify madness per se. His mad narrators are interesting and amusing, no doubt, as with the narrator in "Berenice." But the message is never that it would be better, or we would be better off, if we were insane. There are, after all, "mere madmen," those with "feeble reason."[82]

Second, although Poe does not glorify madness per se, he does glorify it in some cases, and more likely as *unappreciated or unrecognized genius*. In "Marginalia" (June 1849), for example, he writes of "the fate of any individual gifted, or rather accursed, with an intellect *very* far superior to that of his race. . . . since his opinions and speculations would widely differ from those of *all* mankind—that he would be considered a madman, is evident."

London: Bell & Daldy, 517. Poe himself, at the head of "A Descent into the Maelström" (1841), quotes a passage from a yet earlier source, Joseph Glanvill(e) (1636–1680), in which the latter similarly refers to "the well of Democritus." *P&T*, 432. The actual line from Democritus, however, is different: "In reality we know nothing; for the truth is hidden in an abyss." Robin Waterfield, ed. (2000). *The First Philosophers: The Presocratics and the Sophists*. Oxford: Oxford University Press, 176. What is especially interesting is Poe's use of the proverb, which involves both its contemporary and ancient meanings. In "Letter to Mr. — —" (1831), for instance, Poe writes, "As regards the greater truths, men oftener err by seeking them at the bottom than at the top; the depth lies in the huge abysses where wisdom is sought—not in the palpable palaces where she is found. The ancients were not always right in hiding the goddess in a well." *P&T*, 13. It will be recalled that at the beginning of *Eureka* Poe writes of the perception of "the sublimity of its *oneness*" from "the top of Ætna." *P&T*, 1261. In "Ligeia" (1838) the narrator, on the topic of the "expression of the eyes of Ligeia," puzzles about "that something more profound than the well of Democritus—which lay far within the pupils of my beloved." *P&T*, 264. See also, e.g., *E&R*, 1062, 1397. And then in "A Descent into the Maelström" (1841), the narrator states that "The depth in the centre of the Moskoe-ström must be immeasurably great; and no better proof of this fact is necessary than can be obtained from even the side-long glance into the abyss of the whirl which may be had from the highest crag of Helseggen." *P&T*, 436. This should remind us not only of Poe's Ætna reference in *Eureka*, but also of his "glancing at a star" metaphor, which involves "the more important knowledge," as Poe puts it in "The Murders in the Rue Morgue" (1841), that we see a star better with "glances . . . in a side-long way," the point being that "Truth is not always in a well." *P&T*, 412. We shall return to Poe's "glancing at a star" metaphor at the end of this chapter.

82. *P&T*, 1317, 702.

Poe even goes so far as to say, "That individuals *have* so soared above the plane of their race, is scarcely to be questioned; but, in looking back through history for traces of their existence, we should pass over all biographies of 'the good and the great,' while we search carefully the slight records of wretches who died in prison, in Bedlam, or upon the gallows."[83]

Third, Poe indeed refers to madness a number of times in *Eureka*, and it is important to see why. One reason, which we have already seen, is to allay a fear that he suspects on the part of the reader. This is the passage focused on by the Levines. In another passage Poe refers to "mere madmen," what he calls "the professional questioners—the mere madmen who deny the Newtonian law of Gravity."[84] This is clearly a negative meaning.

But there is a further use in *Eureka* that is the one, I suggest, that we should really be focusing on. Recall §I of Chapter 5. Although Poe uses the guise of an imaginary epistle written 1,000 years after the publication of *Eureka*, he makes the profound point there, emphasized by the point being common knowledge a thousand years hence, that no axioms are necessarily or self-evidently true. And he has his writer of the epistle, as we have seen, use language that is remarkably similar to that found in quantum physics. As Poe puts it in the epistle, "That a tree can be both a tree and not a tree, is an idea which the angels, or the devils, *may* entertain, and which no doubt many an earthly Bedlamite, or Transcendalist, *does*."[85] A Bedlamite is an insane person, a lunatic, named after the popular name for a mental hospital that began in London in the late 1300s.[86] The Transcendentalists, which Poe often disparaged but among whom he must in some ways be counted, were "lovers

83. *E&R*, 1459–1460.

84. *P&T*, 1317.

85. *P&T*, 1268. To think the epistle is evidence of *Eureka* as a hoax, by the way, is about as stupid as the narrator of "The System of Doctor Tarr and Professor Fether" (1844), who adds at the end of the tale that "although I have searched every library in Europe for the works of Doctor *Tarr* and Professor *Fether*, I have, up to the present day, utterly failed in my endeavours at procuring an edition." *P&T*, 716. The purpose of the epistle, instead, is surely to indicate, albeit in an exaggerated manner, that Poe thought of his views as being well ahead of their time.

86. OED, "bedlam."

and worshippers of Beauty" who attempted to use intuitive inspiration to go beyond common sense and science and perceive transcendent reality.[87]

Poe's recurrent point about madness, then, is surely, especially in *Eureka*, that madness can loosen the straightjacket and break the shackles of axiomatic and linear thinking—"logical" thinking—and open the door, especially for the educated imagination combined with monomania, to deeper truths about reality, to which axiomatic and linear thinking pose substantial barriers or limits. In short, those who are mad, or thought to be mad, can sometimes "think outside the box," to use the current phrase, and accordingly attain genuine insights. And given the development of relativity physics and quantum physics in the early twentieth century, Poe surely got it right in his rejection of axioms, for the two mentioned pillars of modern physics would not have been possible otherwise.

Poe's overall point, then, given his writings on madness taken collectively, is that a touch or more of madness, real or popularly supposed, might sometimes be necessary for pulling the proverbial teeth of knowledge from the mouth of a universe that continuously presents us with both "unparalleled beauty" and the "most appalling distortion," restoring us "to peace" and giving us "back to reason."[88] Here is how the inspiration for the Hollywood movie *A Beautiful Mind* (2001), the Princeton mathematician John Nash, (now tragically late,) put it in 1994 after decades of suffering from schizophrenia, (during which time he still contributed theoretical work,) in a brief autobiographical essay he wrote following his award of the Nobel Prize in Economic Sciences for his

87. *Eureka*, xvi–xvii, 120n11. I suggest that two of the main differences between the Transcendentalists and Poe, looking at those differences from what seems to me Poe's view, is that Poe's transcendentalism is, to a substantial extent, science based, unlike the Transcendentalism of which Ralph Waldo Emerson was the foremost representative, and also that Poe rejected the idea of a supernatural world *behind and beyond* the physical one, the latter which the Transcendentalist "reckons . . . an appearance." To get a fuller picture, compare Poe's *Eureka* with, e.g., Emerson's lecture "The Transcendentalist" (1842). Joel Porte, ed. (1983). *Ralph Waldo Emerson: Essays & Lectures*. New York: Library of America, 191–209. See also Heidi Silcox (2013). "Transcendentalism." In *Context*, 269–278.

88. *P&T*, 229, 231.

contributions to game theory: "at the present time I seem to be thinking rationally again in the style that is characteristic of scientists. However this is not entirely a matter of joy as if someone returned from physical disability to good physical health. One aspect of this is that rationality of thought imposes a limit on a person's concept of his relation to the cosmos."[89]

To all of this we should note that Poe wrote *Eureka* roughly a year following the death of his beloved Virginia, who died at the end of January 1847, that roughly six weeks later he would describe himself as "overwhelmed by a sorrow so poignant as to deprive me for several weeks of all power of thought or action," and that in his first substantial piece of writing following her death, the poem "Ulalume," written in October of that year, he wrote the line, "It was night, in the lonesome October / Of my most immemorial year."[90] Robert Chambers, also extremely well read, wrote *Vestiges of the Natural History of Creation* (1844) "during a recovery from a mental breakdown produced by the constant strain of writing for the *Journal* [*Chambers's Edinburgh Journal*]."[91] Alfred Russel Wallace came up with the idea of evolution by natural selection independently of Darwin while suffering from a malarial fever in February 1858.[92] The sociobiologist Robert Trivers, who suffers from bipolar disorder, wrote the section on humans in one of his many revolutionary contributions to evolutionary theory, his paper "Reciprocal Altruism" (1971), after a week of severe depression during which he did little more than sleep.[93]

---

89. www.nobelprize.org/nobel_prizes/economic-sciences/laureates/1994/nash-bio.html

90. *Letters*, 624; *Poe Log*, 704–707; *P&T*, 89.

91. James A. Secord, ed. (1994). *Vestiges of the Natural History of Creation and Other Evolutionary Writings*. Chicago: University of Chicago Press, xxiii. Robert was in fact co-editor of this highly successful journal with this brother, William.

92. Martin Fichman (2004). *An Elusive Victorian: The Evolution of Alfred Russel Wallace*. Chicago: University of Chicago Press, 97–98.

93. Robert Trivers, ed. (2002). *Natural Selection and Social Theory: Selected Papers of Robert Trivers*. Oxford: Oxford University Press, 13.

Were these achievements accomplished in spite of the madness or partly because of it? We shall attempt to answer this question in the next chapter.

Kepler, Champollian, and Humboldt

We turn next to scientists Poe exalted as prime examples of scientific imagination. Foremost among them was Johannes Kepler (1571–1630), whose discovery that the orbit of Mars is elliptical rather than circular (generalized into the first of Kepler's three laws of planetary motion) marks a key discovery in the history of science. Poe's veneration for Kepler is evident alone from the imaginary epistle with which Poe begins *Eureka*, dated 2848, a time when science is much better understood. There the author writes of "the *only* true thinkers," "the generally-educated men of ardent imagination," "our Keplers," those responsible for "how rapid our progress" since the time in which *Eureka* was penned.[94] But what was it about Kepler that made Poe raise him to the level of a paradigm? In "A Chapter of Suggestions" (1845) Poe states that "Some of the most profound knowledge—perhaps all *very* profound knowledge—has originated from a highly stimulated imagination. Great intellects *guess* well. The laws of Kepler were, professedly, *guesses*."[95] In *Eureka* Poe states it again, repeatedly. Beginning with the imaginary epistle, he states that "Kepler admitted that these laws he *guessed* . . . . Yes!—these vital laws Kepler *guessed*—that is to say, he *imagined* them. Had he been asked to point out either the *de*ductive or *in*ductive route by which he attained them, his reply might have been—'I know nothing about *routes*—but I *do* know the machinery of the Universe. Here it is, I grasped it with *my soul*—I reached it through mere dint of *intuition*.'" Later in the work Poe refers to "the three immortal laws *guessed* by the imaginative Kepler," adding "the suggestion of these laws . . . and his proving them *à posteriori* to have an actual existence."[96]

94. *P&T*, 1269.

95. *E&R*, 1293.

96. *P&T*, 1270, 1331, 1332.

It will be noticed, first, that by "guessing" Poe does not mean a blind shot in the dark, at least not in the context of scientific imagination. "Guessed" is followed by "imagination," as in visualization. But that term, in turn, is followed by "intuition," which in Poe, as today, refers to operations behind the scenes, to following Freud what is called the *unconscious*. We shall return to this theme in various places in this and the following chapter, for it is key.

Poe, of course, was fully aware of the importance of Kepler's discovery, given his statement in the same passage that "In many works of Astronomy I find it distinctly stated that the laws of Kepler are *the basis* of the great principle, Gravitation."[97] No wonder, then, that Poe makes so much of Kepler. But *intuition*, rather than inductive or deductive reasoning, as the basis of what in Poe's day was considered the greatest scientific achievement thus far, Newton's theory of gravity? What did Poe mean? The answer is given in the imaginary epistle, what apparently is well-known in 2848: "Could not any metaphysician have told him [Kepler] that what he called 'intuition' was but the conviction resulting from *de*ductions or *in*ductions of which the processes were so shadowy as to have escaped his consciousness, eluded his reason, or bidden defiance to his capacity of expression?" Moreover in doing so Kepler "proceeded decorously and legitimately."[98]

In the imaginary epistle Poe states that "Kepler admitted." But where did he get that from? It is possible that Poe actually read Kepler, since he freely paraphrases the final sentences in the Proem to Book V of Kepler's *Harmonice Mundi* (1619).[99] It is unlikely, however, that Poe ever read a work by Kepler, given that, as the historian of science I. Bernard Cohen puts it, "Even today the major works of Kepler are not available in complete English,

97. *P&T*, 1332.

98. *P&T*, 1270.

99. *P&T*, 1399n. The passage exists in an English translation, however, in one of Poe's sources on astronomy, specifically J.P. Nichol's (1842) *The Phenomena and Order of the Solar System*. New York: Dayton & Newman, 80. Nichol, in fact, devotes an entire chapter to Kepler, in which he not only explains Kepler's laws but sings praises about Kepler's life and mind, praises in league with Poe's.

French, or Italian translations."[100] Poe could, of course, have read a copy of the original in Latin.

But enough of that. There is a further problem. At least two of the sources on which Poe heavily relied for his understanding of science claimed that Kepler *deduced* his laws from the data provided him on the orbit of Mars by Tycho Brahe. John Herschel, for example, whose philosophy of science we examined in the previous chapter and whose astronomy we examined in Chapter 4, refers to "the great laws of the planetary motions deduced by Kepler, entirely from a comparison of observations with each other, with no assistance from theory."[101] Similarly John Pringle Nichol, whose work served as a source for Poe as we have seen in Chapter 4, repeatedly states that "Kepler had delivered those three Laws, which, it must be recollected, were pure and immediate deductions from the stores of Uraniburg," in other words, "that Kepler's laws are simple results of accumulated observation on the order of planetary motions—we cannot now be deceived."[102]

But let us be careful here. Whenever dealing with doctors on important matters we need to get second and third opinions or more—especially on matters beyond the grave, namely, *legacy*. According to Stuart and Susan Levine, Poe "is somewhat inaccurate" since "Kepler got his model to work but had no physical understanding of why it worked. Newton would provide the basis for a theoretical account of why Kepler's model worked. So the statement in *Eureka* is not quite right. Kepler worked for years to find the laws; there was no inspired guessing of a theory because he produced no theory, although no science historian would deny that the *model*, his visualization, might well have resulted from the combination of induction, deduction, and subliminal association that Poe describes."[103] For my own part, I do not see why a scientific *model* cannot also be a scientific *theory*, unless one confines the latter to statements of causal laws, as did Karl Popper as we

100. I. Bernard Cohen (1985). *The Birth of a New Physics*. 2nd ed. New York: W.W. Norton, 131.

101. John F.W. Herschel (1830). *Preliminary Discourse on the Study of Natural Philosophy*. London: Longman, Rees, Orme, Brown, & Green, 178.

102. Nichol (note 99), 142, 146. Uraniburg was the Danish astronomical observatory named and operated by Tycho Brahe.

103. *Eureka*, 127n23.

have seen in the previous chapter. Moreover, if Kepler's three laws do not count as theories,—the first is that planets orbit in ellipses, the second and third are about mathematical relations involved,—then so also Newton's law of gravity must be rejected as a scientific theory, given that Newton himself, as we have seen in §IX of Chapter 5, admitted that he did not have a physical understanding for why gravity operated in accordance with his inverse square law, or even whether gravity is a property of matter, all the while adding only the "supposition" that the medium for gravity is the ether.

John Stuart Mill, as we have seen in the previous chapter, played down the role of guessing in Kepler and in science in general, while William Whewell had no problem claiming that Kepler guessed. Whewell stated further that scientific discoveries "are not improperly described as happy *guesses*," that "the genius of a discoverer . . . cannot be limited by rules, or expressed by definitions." Interestingly, Cohen states that "Kepler was a tortured mystic, who stumbled onto his great discoveries in a weird groping that has led one of his biographers to call him a 'sleepwalker.' Trying to prove one thing, he discovered another, and in his calculations he made some major errors that cancelled each other out."[104]

Lastly, consider the judgment of none other than Albert Einstein, who stated that "The orbits were empirically known, but

104. Cohen (note 100), 132. The reference, of course, is to Arthur Koestler (1959). *The Sleepwalkers: A History of Man's Changing Vision of the Universe*. New York: Macmillan. According to the science historian James Voelkel, however, "in order to evade hostile editorial censorship ['from Tycho's heirs over the right to publish anything at all based on the Dane's observations'] and to forestall the censure of the contemporary astronomical community, Kepler purposely *misrepresented* his methodology, depicting himself as having been driven aimlessly to his conclusions." James R. Voelkel (2001). "Commentary on Ernan McMullin, 'The Impact of Newton's *Principia* on the Philosophy of Science.'" *Philosophy of Science* 68 (3), 319–320. But given that Voelkel also claims that "the ellipse cannot be found from the observations simply by induction" (324), this still leaves open the question of how much of Kepler's process of discovery involved eureka moments, Poe's "guesses" or "seemingly intuitive leaps." Voelkel, along with what he calls "Two generations of Kepler scholars," does not seem to me to close the door, he only narrows the opening.

their laws had to be guessed from the empirical data. First he [Kepler] had to make a guess at the mathematical nature of the curve described by the orbit, and then try it out on a vast assemblage of figures. If it did not fit, another hypothesis had to be devised and again tested. After tremendous search, the conjecture that the orbit was an ellipse with the sun at one of its foci was found to fit the facts. Kepler also discovered the law governing the variation in speed during one revolution, . . . Finally he also discovered that the squares of the periods of revolution round the sun vary as the cubes of the major axes of the ellipses." Summing up, he says "Kepler's marvellous achievement is a particularly fine example of the truth that knowledge cannot spring from experience alone but only from the comparison of the inventions of the intellect with observed fact."[105]

We may never know that truth about Kepler, beyond a reasonable doubt, and we may debate whether the word "guessing" should apply, but what we may safely conclude is that Poe, in emphasizing the role of intuitive guessing in the case of Kepler, was in very fine company indeed. And if that very fine company is basically right about Kepler, we may go further and agree with Whewell, that even though the ellipse was (as Mill puts it) *in the facts*, Kepler did not see it in the observed facts (the observed facts were gathered from Earth, after all, not from far above or below the Sun); instead, he *superinduced* the concept of an ellipse on the observed facts, he conjectured and imagined and subsequently found that there was a remarkably close fit between the concept and those facts.

Kepler, of course, broke away from the dogma of planetary circles. Darwin broke away from the dogma of static biological species and that design requires a designer. Poe broke away from the dogma of the static universe. Einstein broke away from the dogma of absolute space and time and that gravity is a force. The educated imagination, in each of these cases, is the key, but an imagination allowed to soar high above the terrestrial paths of rules and methods and received views, not an imagination

105. Albert Einstein (1930). "Johannes Kepler." Reprinted in Sonja Bargmann, trans. (1954). *Ideas and Opinions by Albert Einstein*. New York: Crown Publishers, 265, 266.

confined to them. There is, however, something more. Kepler wrote, "Nothing holds me: I will indulge in my sacred fury: I will triumph over mankind by the honest confession, . . . the book is written, to be read either now or by posterity—I care not which; it may well wait a century for a reader, as God has waited six thousand years for an observer?"[106] Somewhat similarly Egæus states, in his obsession over the teeth of Berenice, "Then came the full fury of my *monomania*, and I struggled in vain against its strange and irresistible influence."[107] Fury. Monomania. Truth. Teeth. Wisdom. —Mad Pursuit!

With the remaining two names that head this section we can be fairly brief. In a series of "three general phases" from 1821 to 1824, the French orientalist and philologist Jean-François Champollion (1790–1832) "would finally crack the code" of Egyptian hieroglyphics ultimately by using the Rosetta Stone, which he had begun to study in 1808. Accidentally discovered in 1799 near Alexandria by a French engineer in Napoleon's army while procuring stones for fortifications, the Rosetta Stone contains a legal decree issued in 196 B.C. by a high council of priests celebrating the first anniversary of the coronation of Ptolemy V, the ruler of Egypt. The decree is written in two languages using three scripts: ancient hieroglyphics, demotic, (an Egyptian script used to replace hieroglyphics starting around 600 B.C.,) and ancient Greek. Champollion's discovery of how Egyptian hieroglyphics works earned him not only the world's first professorship in Egyptology created just for him at the Collège de France in Paris in 1831, but also the popular title of the "founding father of Egyptology." His book on ancient Egyptian grammar was published in 1832, shortly after he died at the age of forty-one from a series of strokes. Champollion's legacy, then, in short, was to get the pharaohs speaking again, this time from their memetic tombs, tombs that had been silent for close to two millennia, thereby shedding light on one of the greatest civilizations of the ancient world, a civilization that

106. Nichol (note 99), 80. Poe in *Eureka* has the writer of the imaginary epistle quote the passage as: "*I care not whether my work be read now or by posterity. I can afford to wait a century for readers when God himself has waited six thousand years for an observer. I triumph. . . . I will indulge my sacred fury.*" *P&T*, 1270.

107. *P&T*, 231.

had remained a mystery ever since the last use of hieroglyphic writing by Egyptian priests some 1,600 years ago.[108]

In the imaginary epistle with which Poe prefaces *Eureka*, he writes, "I have often thought, my friend, that it must have puzzled these dogmaticians of a thousand years ago, to determine, even, by which of their two boasted roads it is that the cryptographist attains the solution of the more complicate cyphers—or by which of them Champollion guided mankind to those important and innumerable truths which, for so many centuries, have lain entombed amid the phonetical hieroglyphics of Egypt."[109] The "two boasted roads" are none but the inductive and deductive methods, to which we shall devote a section on Poe's understanding near the end of this chapter. Poe himself, of course, was also a cryptographist, and we have seen that although he has Legrand in "The Gold-Bug" (1843) explain how he solved the mystery of finding Captain Kidd's treasure, including how he solved the cryptograph on the treasure map, he also ascribed to Legrand intuitive insight, with solutions following prolonged efforts coming to him all of a sudden, sometimes upon first awakening after a night's sleep. But in the specific case of the cryptograph, Legrand in the tale seems more methodical than intuitive. He does say, however, that "there is no alternative but experiment (directed by probabilities) of every tongue known to him who attempts the solution, until the true one be attained." He also says, at a particular stage of the explanation of the solution to the cryptograph, that "Here, then, we have, in the very beginning, the ground-work for something more than a mere guess."[110] So the fundamental point might be that of educated guessing on the part of a brilliant mind, combined with a lot of hard work at the conscious level.

Poe shared in the "Egyptomania" that swept through Europe and America at the time, evidenced alone by his long book review of John L. Stephens' *Incidents of Travel in Egypt, Arabia Petræa, and the Holy Land*" (1837) and by his tale "Some Words with

108. Andrew Robinson (2012). *Cracking the Egyptian Code: The Revolutionary Life of Jean-François Champollion*. Oxford: Oxford University Press, 128–129, 127, 14, 131, 15.

109. *P&T*, 1269. The phrase "phonetical hieroglyphics" was actually used by Champollion. *Eureka*, 126n23.

110. *P&T*, 587, 588.

a Mummy" (1845).[111] However, although Champollion quickly became famous for cracking Egyptian hieroglyphics, it remains uncertain whether Poe actually read anything by Champollion, and what exactly he read as secondary sources.[112] At any rate, the influence on Poe is unmistakable. What is especially interesting, as Andrew Robinson points out in his outstanding biography, is that "With Champollion, there is evidence of an archetypal eureka moment, in Paris in September 1822, when he cried out to his brother: 'Je tiens mon affaire!' ['I've done it!']." He had just that morning deciphered the long cartouche of Ptolemy V on the Rosetta Stone, but before he could explain it to anyone "he collapsed on the floor in a dead faint," not reviving until five days later. It is uncertain whether the story would have been available to Poe in his time, but it also remarkable that Robinson refers not only to Champollion "guessing at the unwritten vowels," but also to "Champollion's single-mindedness" and his "tunnel vision," which "allowed him to begin to perceive the system behind the signs."[113] It is of further interest that "Berenice" is the name of the Ptolemaic queen inscribed on the Rosetta Stone, which Champollion deciphered along with the rest of the Stone.[114] I might therefore suggest that it was *in fact* none other than him—Champollion—who served as the inspiration for Poe's Egæus, with his "all or nothing" monomania and his very own "Berenice"!

But I have digressed. There is certainly more to Poe citing Champollion so favorably in *Eureka* than we have examined thus far. In *Views of the Architecture of the Heavens*, first published in 1837, which helped inform Poe on the astronomy and physics of his time, John Pringle Nichol states in the very last paragraph that "In the vast Heavens, as well as among phenomena around us, all things are in a state of change and Progress; there too—on the sky—in splendid hieroglyphics the truth is inscribed, that the grandest forms of present Being are only Germs swelling

111. *E&R*, 923–941, *P&T*, 805–821. See also Travis Montgomery (2013). "The Near East." In *Context*, 53–62.

112. There were certainly plenty of secondary sources on hieroglyphics in general and Champollion in particular that would have been available to Poe. See, e.g., the discussion in Irwin (note 37), 4–11.

113. Robinson (note 108), 254, 142.

114. Robinson (note 108), 120, 138, 167.

and bursting with life to come!"[115] The point was surely not lost on Poe. Not only is the Universe not a machine, but nature is not simply a book either, one of God's revelations. If it were, its meaning ought simply to be read, using deductive and inductive inferences. Instead, the Universe is a massive hieroglyphic text in need of being deciphered, with meaning existing collectively, not isolated in each individual sign, and with an underlying "system" that can be applied to new signs. Poe fancied himself not only the Kepler of the Universe, then, but also the Champollion of the Universe, too.

Alexander von Humboldt (1769–1859) was a German scientist and explorer, one of the most celebrated in Europe in his time, famous primarily for his unfinished *Kosmos*, published in five volumes from 1845 to 1862, in which he attempted to unify scientific knowledge into one big picture of the Universe. *Kosmos* was republished in England in three different English translations starting in 1845, (which were also published in the United States,) and it is easy to see why Poe would dedicate *Eureka* to Humboldt, "with very profound respect."[116] As the Levines put it, *Kosmos* is "at once a scientific treatise and a highly charged emotional poetic response to the cosmic environment."[117] Humboldt, then, in Poe's eyes, is the way a scientist should be, having an area of specialization, of course, (Humboldt was principally a geologist, although he was published in other areas as well,) but not being against and perhaps even willing to try to grasp the big picture as a whole. Dedicating *Eureka* to Humboldt was therefore a smart move, (which is not to doubt Poe's sincerity,) as *Kosmos*, written by a bona fide scientist, represented a contemporary and legitimate precedent for what Poe was trying to accomplish in *Eureka*. Because of Humboldt, no one could say that what Poe was trying to do in *Eureka* was unscientific or unbecoming to science.

Poe's only criticism of *Kosmos*, as we have seen in Chapter 4, was that Humboldt did not do a good enough job. While he states that *Kosmos* is the "nearest approach" to *Eureka*, he also in practically the same breath states that "I do not know of a treatise

115. J.P. Nichol (1840). *Views of the Architecture of the Heavens: In a Series of Letters to a Lady*. New York: H.A. Chapin, 115.

116. *P&T*, 1258.

117. *Eureka*, xii.

in which a survey of the *Universe*—using the word in its most comprehensive and only legitimate acceptation—is taken at all."[118] While Poe admires Humboldt, stating in *Eureka* that his "generalizing powers have never, perhaps, been equalled," he also criticizes *Kosmos* for treating the Universe "*not* in its individuality but in its generality."[119] As we have seen in Chapter 4, this surely means that *Kosmos* failed in providing what Poe in his literary theory called a "unity of effect." The latter is what Poe attempted in *Eureka*, "a survey of the Universe that the mind may be able really to receive and to perceive an individual impression," a line that is followed with his "top of Ætna" imagery, in which "Only by a rapid whirling on his heel could he hope to comprehend the panorama in the sublimity of its *oneness*."[120] Indeed, with *Eureka*, Poe is trying to solve an ancient dilemma that characterizes much of early Western philosophy, which is whether the Universe is a one or a many, in what sense or senses it is the first of these or the second or both.[121] It is a pity that philosophers, and not just scientists, given that both of their fields trace back to the early Greeks of Ionia, have hitherto failed to appreciate Poe's effort. That Poe scholars have failed is to be expected, albeit unforgivable, given the territoriality that dominates academia.

Poe probably thought of himself as finishing what Humboldt had tried to do with modern science. This would add to the explanation of why he dedicated *Eureka* to Humboldt. But unlike Humboldt with *Kosmos*, Poe in his time failed to get his due for *Eureka*,—and for well over a century since,—partly, no doubt, because he was a poet and not a scientist, but largely because of the *style* of *Eureka*. It does not read like a serious work of science. It not only begins with an imaginary epistle dated 2848, but the entire work is perforated with "whizz!—fizz!," what Poe in one of his tales has his narrator claim is necessary for making a living at magazine writing.[122] But if one can manage to get past the style

118. *P&T*, 1261–1262.

119. *P&T*, 1346, 1262.

120. *P&T*, 1261.

121. See, e.g., Michael C. Stokes (1971). *One and Many in Presocratic Philosophy*. Cambridge, MA: Harvard University Press.

122. "The Literary Life of Thingum Bob, Esq." (1844). *P&T*, 786.

of *Eureka* and look at the *content*, as I hope I have demonstrated in Chapter 5, then what one finds is something of profound interest, and which demands a thorough investigation.

Leibniz, Newton, and Laplace

Having looked at three of Poe's positive examples of scientific imagination, it will also help to look at three of his examples that he did not view so highly, and to see why.

We shall begin with Gottfried Wilhem Leibniz (1646–1716), a German philosopher considered by many to be one of the greatest thinkers of his time. A true polymath, his contributions to various fields are too long to list, but among them he invented the infinitesimal calculus independently of Newton, he invented the binary number system, (which is used in modern computers,) and he contributed to the physical development of mechanical calculators. Especially interesting is that against Newton's view of space and time as absolute he seems to have anticipated Einstein in thinking them relative, stating, for example, in a letter dated February 25, 1716, that "I have said more than once, that I hold space to be something merely relative, as time is; that I hold it to be an order of coexistences, as time is an order of successions. For space denotes, in terms of possibility, an order of things which exist at the same time, considered as existing together."[123] So a universe without things would be a universe without space and time. (The very horror of the thought makes souls shake as the quivering plumes upon the hearse.) Leibniz is probably best known, however, as one of the three great rationalists of the seventeenth century, along with Descartes and Spinoza, and as providing one of the strangest metaphysics of all time.

Leibniz wrote prolifically, but oddly left behind no *magnum opus*, no one work that fully captured his rich and intricately connected philosophy. Instead it was distributed mainly throughout short essays, notes, and letters. We shall therefore confine ourselves to his short essay entitled "The Principles of Philosophy, or, the

123. H.G. Alexander, ed. (1956). *The Leibniz–Clarke Correspondence*. Manchester, UK: Manchester University Press, 25–26.

Monadology" (1714), originally written in French and intended by Leibniz as a condensed version of his mature views. In short, the Universe is composed of what he calls "monads" (from the ancient Greek word *monas*, meaning "a unit," "a one," "alone"). A monad is a "simple substance," such that a monad has "no parts, neither extension, nor shape, nor divisibility." Monads are "the true atoms of nature," out of which "composites" like you and rocks and stars are made. Monads are not physical, however, and may be called "souls," since they have "perception" and "appetition"—the latter "The action of an internal principle which brings about the change or passage from one perception to another." But since not all monads have "sensation" and "memory," the term "soul" is best reserved for the latter kind of monads, a subset of which is "the rational soul, or *mind*." Although the Universe consists of infinitely many monads, no two monads are qualitatively identical, otherwise God—the "supreme substance," "unique, universal, and necessary," the "absolutely perfect" and "absolutely infinite" monad, "the primitive unity or the first simple substance," out of which "all created or derivative monads are products, and are generated, so to speak, by continual fulgurations of the divinity from moment to moment"—would lack a "sufficient reason" for why one particular monad has the position it does rather than another monad. In everything God does, God must have a sufficient reason for why this rather than that, as God is supremely rational. Given God's perfection, and that "there is an infinity of possible universes in God's ideas, and since only one of them can exist," it follows that our actual universe is "the best," one of maximal "fitness," in its entirety "a plenum" (since empty spaces would violate the principle of sufficient reason) with a "harmony pre-established between all substances," such that "each simple substance is a perpetual, living mirror of the universe" and "every body is affected by everything that happens in the universe."[124]

How could one of the greatest mathematicians of his time produce such a wild metaphysics, a universe in which space, time, and matter are total illusions, a metaphysics that makes Poe's *Eureka*

124. Roger Ariew and Daniel Garber, trans. (1989). *G.W. Leibniz: Philosophical Essays*. Indianapolis: Hackett, 213, 214, 215, 217, 218, 219, 220, 221, 223, 220, 221.

seem tame by comparison? Professional philosophers who love metaphysics might defend Leibniz by complaining that my presentation of his philosophy is superficial, that I didn't examine his logic, which is the foundation of his metaphysics. They might say that I didn't go deep enough into Leibniz's law, (the identity of indiscernibles,) that I wholly ignored his explicit adherence to the law of noncontradiction, ("we judge that which involves a contradiction to be false, and that which is opposed or contradictory to the false to be true,") and that I also ignored his implicit adherence, first brought out by Bertrand Russell, to another doctrine in traditional logic, the doctrine that all true statements are of the subject-predicate form, which entails (as Russell showed) that relations do not have a reality of their own but must be reduced to properties of the subject term.[125] *But then if Leibniz was so smart, why did he not come up with nine or so anticipations of major developments and theories in twentieth-century science?* Certainly the relativity of space and time hardly even counts as one, given that his monadology entails that space and time are interior to monads and are therefore illusions, apparent alone from his claim that monads lack "extension" along with his claim that "he who sees all can read in each thing what happens everywhere, and even what has happened or what will happen."[126]

I could do a Bertrand Russell here, and show that Poe, unlike Leibniz, was not bound to traditional logic, (which Russell corrected to allow for relations,) not only that he rejected, as we have seen in Chapter 5, the law of noncontradiction, but also that he was not bound to the subject-predicate form of statement. And as for Leibniz's so-called law, (the identity of indiscernibles,) it is neither a law of logic nor an empirical law but a theological principle, based on Leibniz's principle of sufficient reason *for God*, and Poe did not share the "imaginary axioms" of Leibniz's theology,—or anyone else's for that matter, as we have seen in Chapter 3,—evident alone from his multiverse theory.

Instead we shall confine ourselves to Poe's answer to the above question in italics, which is certainly warranted given Poe's

---

125. See, e.g., Bertrand Russell (1946). *A History of Western Philosophy*. London: Allen & Unwin, 572–573, 575.

126. Ariew and Garber (note 124), 217, 213, 221.

accomplishments examined in Chapter 5. In *Eureka* Poe comments on mathematical physicists, who generally seem content to discover the mathematical principles governing nature, without getting into what the principles are beyond the mathematics, let alone why they are what they are and not something else, what Poe calls the "principle as existing behind the Law." Leibniz, he says, "was a marked exception to the general rule," his "temperament was a singular admixture of mathematical with the physico-metaphysical." But when it came to understanding something so important as gravity, not just the mathematical principle, Leibniz not only failed to find a "*physical*" principle, he failed "amid his old familiar haunts in the kingdom of Metaphysics." And the reason he failed "in search of the treasure . . . was, perhaps, because his fairy guide, Imagination, was not sufficiently well-grown, or well-educated, to direct him aright."[127]

I take Poe to mean that Leibniz's scientific imagination failed, not because he lacked the potential to do something great with it, but because it was not grounded (for whatever reason) in facts, in empirical evidence. Whether Poe's criticism of Leibniz is justified, it must be admitted that it is not for nothing that posterity conferred upon Leibniz the label *rationalist* rather than *empiricist*.[128] And Poe is surely right that a key ingredient in a *successful* scientific imagination is that it must be an *educated* imagination, with knowledge of all the relevant known facts and theories being a necessary, though not a sufficient, condition for success.

Poe makes this point perfectly clear in a letter dated June 26, 1849, in response to an essay sent to him "on the igneous liquidity of the Earth." There he writes, "Before theorizing on such topics, it is always wisest to make ourselves acquainted with the

127. *P&T*, 1289, 1290.

128. At the beginning of his second letter to Samuel Clarke, Leibniz claims that "the principle of contradiction . . . is sufficient to demonstrate every part of arithmetic and geometry," and that with "the principle of a sufficient reason . . . one may demonstrate the being of a God, and all the other parts of metaphysics or natural theology; and even, in some measure, those principles of natural philosophy, that are independent upon mathematics: I mean, the dynamical principles, or the principles of force." Alexander (note 123), 15, 16. The Leibniz–Clarke correspondence was published in 1717 and was widely read thereafter.

actually ascertained facts & established doctrines."[129] The line, by the way, is in the context of some comments by Poe on *Eureka*, and it surely evokes lines from *Eureka* about "the *only* true thinkers," "the generally-educated men of ardent imagination," "our Keplers—our Laplaces."[130] Their guesses were not *mere* guesses, flights of fancy, but products of a well-developed, passionate, and educated imagination.

If asked to name the paradigmatic scientist of all time, most today would probably name Albert Einstein. In Poe's day they would have named Sir Isaac Newton (1642–1727). With his *Philosophiæ Naturalis Principia Mathematica* (1686)—delivered to the Royal Society twenty years after its central law had been formulated when Newton was twenty-four—Newton revolutionized physics, uniting celestial and terrestrial mechanics with a fundamental set of laws, specifically the three "axioms or laws" of motion (inertia, change in motion, and action-reaction) and the law of universal gravitation, a mathematical formula which states that the gravitational force of an object is directly proportional to its quantity of matter and decreases with the inverse square of its distance between its center and the center of any other object. With these laws Newton explained a staggering array of phenomena, including Kepler's elliptical orbits, the orbits of Earth's moon and the moons of Jupiter, the orbits of comets, the flattening of the Earth at its poles and its bulging at the equator, and the ocean tides. With his *Opticks* (1704) Newton did even more, proving with prisms that white light consists of all the colors of the rainbow and arguing that light is made of corpuscles rather than waves.

Not only did Newton revolutionize physics, but with his many successes along with his "rules of reasoning in philosophy" one may safely say that he revolutionized the scientific world. The point was made famous by the English poet Alexander Pope, with his proposed (but subsequently rejected) epitaph for Newton's grave at Westminster Abbey:

> Nature and Nature's Laws lay hid in Night:
> God said, Let Newton be! and All was Light.

129. *Letters*, 813.

130. *P&T*, 1269.

Newton's most important discovery, of course, was the inverse square law of gravity. But how did he discover it? In the "General Scholium" that Newton added to the second edition of *Principia* in 1713, he says it was "deduced from the phenomena," that it was "afterward rendered general by induction," that this is how "the impenetrability, the mobility, and the impulsive force of bodies, and the laws of motion and of gravitation, were discovered," and that any other method has "no place in experimental philosophy."[131] This is the context of his famous *hypotheses non fingo* ("I frame no hypotheses") line. He is also very clear in this passage that although he thought he had proved the existence of the force of gravity, he was providing only its mathematical nature, not "the cause of those properties of gravity." He was not even assuming that gravity is innate to matter, given what he believed to be the absence of evidence. There was nothing new here. It was implied in the first edition of *Principia* and was explicitly stated in his second letter to Richard Bentley (1693).[132]

One cannot, of course, rely on the *Principia* to provide an accurate account of the actual process of discovery that went on in Newton's head. What scientific publications typically give us, if they get into it at all, is a "rational reconstruction" of events.[133] Suffice it to say that many throughout the years took Newton at

131. H.S. Thayer, ed. (1953). *Newton's Philosophy of Nature: Selections From His Writings*. New York: Hafner Press, 45.

132. Thayer (note 131), 53.

133. There is a modern debate over whether Newton can legitimately use the word "deduce." While it is generally agreed that he derived his inverse square law mainly from Kepler's second and third laws, which are mathematical, he could not have strictly deduced the inverse square law using the observational data of Mars upon which Kepler superinduced an ellipse, for the simple reason that that data would not and could not make a perfect ellipse, given gravitational perturbations caused by the orbits of other planets and by the slight movement of the Sun, which in turn was a gravitational perturbation caused by the orbits of the planets. For an excellent discussion, also on what Newton meant by "hypothesis," see Ernan McMullin (2001). "The Impact of Newton's *Principia* on the Philosophy of Science." *Philosophy of Science* 68 (3), 279–310. George Smith (2001). "Comments on Ernan McMullin's 'The Impact of Newton's *Principia* on the Philosophy of Science.'" *Philosophy of Science* 68 (3), 327–338. Ernan McMullin (2001). "Response." *Philosophy of Science* 68 (3), 339–345.

his word, that the process of discovery he followed was deduction from phenomena followed by inductive generalization. This is true of Poe's main sources on modern astronomy and physics. John Pringle Nichol, for example, among his raptures over Newton's greatness, which includes that Newton "was with one of the Almighty's Ideas," states that Kepler's "success . . . . not only prepared the way, but even necessitated the grand final generalization of Newton," that "the principle of gravitation . . . contains nothing which was not in Kepler's three discoveries; *i.e.*, it is a mere exposition of the existence of an order through the motions of the planetary system," and he refers to Newton as a "mathematician" and "this great Geometer," as providing, in relation to Kepler, "Profounder mathematics, and the introduction of important novel conceptions in that science, [which] were requisite for the resolution."[134]

Whether Poe actually read Newton, the comments of Nichol help to shed light on Poe's assessment of Newton as a scientist. In *Eureka* he states that "Newton deduced it [the law of gravity] from the laws of Kepler," while a little later in the work he states that "The great mind of Newton, while boldly grasping the Law itself, shrank from the principle of the Law." He then adds that "The more fluent and comprehensive at least, if not the more patient and profound, sagacity of Laplace, had not the courage to attack it. But hesitation on the part of these two astronomers it is, perhaps, not so very difficult to understand. They, as well as all the first class of mathematicians, were mathematicians *solely*—their intellect, at least, had a firmly-pronounced mathematico-physical tone. What lay not distinctly within the domain of Physics, or of Mathematics, seemed to them either Non-Entity or Shadow." He then adds, "Either Newton or Laplace, seeking a principle and discovering none *physical*, would have rested contentedly in the conclusion that there was absolutely none."[135] In other words, Newton, along with Laplace, (whom we shall return to below,) were lacking in scientific imagination precisely because they had purposely limited it to the mathematical, refusing to venture into thought about the *what* and *why* of the mathematical relations

134. Nichol (note 99), 154, 74, 145, 142, 154, 143.

135. *P&T*, 1269–1270, 1290.

found in nature. Poe himself had no such reservations. As he puts it in the very next paragraph, "with no unwarranted fear of being taken for a madman at the outset, . . . I here declare the *modus operandi* of the Law of Gravity to be an exceedingly simple thing—that is to say, when we make our advances towards it in just gradations and in the true direction—when we regard it from the proper point of view." Then follows the rest of *Eureka*, with its surprisingly prescient science that we examined in Chapter 5. Based on the hindsight provided by that chapter, we need to reflect anew on George Putnam's recollection of Poe's visit and book proposal to the publisher, that with a "glittering eye" and "seeming to be in a tremor of excitement" Poe declared that "Newton's discovery of gravitation was a mere incident compared to the discoveries revealed in this book."[136]

Whether Poe got the *modus operandi* of gravity correct is beside the point here, which is his criticism of the scientific imagination that is afraid to fly. Following immediately the passage quoted from above, Poe turns to Leibniz, whose views we have already examined. Leibniz's power of scientific imagination was lacking, says Poe, but "it is almost impossible to fancy, of Leibnitz, that, having exhausted in his search the physical dominions, he would have not have stepped at once, boldly and hopefully, amid his old familiar haunts in the kingdom of Metaphysics."[137] In other words, Newton's scientific imagination was lacking something that Leibniz's scientific imagination was not, the boldness to go beyond the known physical facts into the nature of things, while Leibniz was lacking something that Newton was not, an educated and developed imagination grounded in the known facts. The complete scientific imagination requires both. Kepler had both, in Poe's view. He was an expert mathematician with a focus on physical facts with no *hypotheses non fingo* mentality to prevent the flight of his scientific imagination. And if one thinks that Poe is wrong here, in his criticism of Newton, one has to keep in mind three things: first, Newton, as examined above, refused to investigate the nature of gravity but was instead satisfied with discovering and applying the mathematical principle of gravity;

136. *Poe Log*, 731.

137. *P&T*, 1290.

second, Newton did not live up to Poe's criterion of truth, examined in Chapter 4, which is to try to achieve complete consistency with everything known and which requires an all-encompassing, multidisciplinary, and courageous mind; and third, what fundamentally drives science according to Poe is not methodical paths, which was Newton's way, whether deductive or inductive, (more on this below,) but the flight of eureka moments, which is the focus of our next and final chapter. Poe's Dupin, (and by implication Poe himself,) had it all. But we still need more on Laplace before we turn to Dupin.

Pierre-Simon Laplace (1749–1827), during his prime, was the foremost scientist in France and is still considered one of the greatest scientists of all time, having made important contributions to a number of fields, most notably mathematical astronomy, the theory of probability, and the theory of heat. Two features of his science are of special importance to us in this section. One is his philosophy of science, which combines determinism with a fundamental reliance on mathematics in scientific theories and explanations. It is this philosophy that is behind his reply to Napoleon Bonaparte, who once asked Laplace, "Where does God fit into your system?" Laplace famously replied, "Sir, I have no need of that hypothesis." This philosophy of methodological naturalism connects with the second feature of Laplace's science that concerns us, which is his *nebular hypothesis* for the origin and nature of the solar system.[138] First presented at the end of his *Exposition du Systeme du Monde* (1796), which is a non-mathematical introduction to astronomy, the detailed mathematical version of the hypothesis was presented in his *Traité de Mécanique Céleste*, the first two volumes of which were published in 1799 and the last in 1825.

---

138. The label "nebular hypothesis" was not actually used by Laplace, but instead was coined by William Whewell in his Bridgewater Treatise of 1833, in which he claimed that the hypothesis still demanded "an intelligent Author, an origin proceeding from free volition not from material necessity." Others resisted the nebular hypothesis itself, such as Wilhelm Olbers in 1812, who used the anomaly of the moons of Uranus (more on this below) to deny its inductive appeal. For this and a lot more, including especially the role of the nebular hypothesis in concepts of progress in Victorian science, see Simon Schaffer (1989). "The Nebular Hypothesis and the Science of Progress." In James R. Moore, ed. (1989). *History, Humanity and Evolution: Essays for John C. Greene*. Cambridge: Cambridge University Press, 135.

The contrast with Newton is profound. For all that Newton accomplished for having "cultivated mathematics as far as it relates to [natural] philosophy," as Newton himself puts it in the Preface to the first edition of *Principia* (1686), Newton retained a role for God in astronomy. Given that all the planets orbit the Sun in the same direction, and that the orbits of comets go through the planetary orbits, "The most beautiful system of the sun, planets, and comets could only proceed from the counsel and dominion of an intelligent and powerful being," as he puts it in the General Scholium to the second edition of *Principia* (1713). The same thinking was universalized by Newton for every solar system in the Universe. Newton is even more explicit in his four letters to Richard Bentley (1692/3), which were published in London in 1756. There he writes that the distribution of matter into stars and planets fails to be explicable "by mere natural causes," that he is "forced to ascribe it to the counsel and contrivance of a voluntary Agent," also for "Why there is one body in our system qualified to give light and heat to all the rest," also for why "there is no natural cause which could determine all the planets, both primary and secondary [moons], to move the same way and in the same plane, without any considerable variation," and also for why "this frame of things could not always subsist without a divine power to conserve it."[139]

With the nebular hypothesis Laplace did away with God. Very simply put, the hypothesis—which in a much modified form is the dominant model in astronomy today[140]—is that massive amounts of matter (particles, gas) floating around in space gradually come together, due to the mutual gravitational attraction of the particles, into a *nebula*, (Latin for "cloud"). As the giant nebula (in the case of our solar system, some 10,000 times the size of the Sun) gradually collapses or coalesces, it begins to slowly rotate, producing a centripetal force. As the speed of rotation increases, the nebula slowly shrinks and becomes flattened. Eventually gravity

139. Thayer (note 131), 42, 47, 48, 56.

140. See Peter Shaver (2011). *Cosmic Heritage: Evolution from the Big Bang to Conscious Life*. New York: Springer, 8–9, 82–85. According to Carl Sagan, "It was not until more than a century later . . . that the nebular hypothesis was confirmed." Carl Sagan (1997). *Billions and Billions: Thoughts on Life and Death at the Brink of the Millennium*. New York: Random House, 55.

is no longer able to contain the centripetal force, and the nebula throws off an outer ring, which in turn spirals and condenses into a planet, with its shrinking throwing off a spiraling ring which in turn condenses into a moon. As this very gradual process repeats, more planets with moons are formed, with the remaining gas ball in the center forming a sun, the entire system spinning in the same direction as the spin of the sun. The hypothesis had precedents in the speculations of Emmanuel Swedenborg in 1734, Thomas Wright in 1750, and Immanuel Kant in 1755. Empirically it was supported by the telescopic observation of Saturn's rings, (first discovered by Galileo in 1610,) by the observations of the Sun and planets, but especially by the telescopic observations of various kinds of nebulae in the night sky by William Herschel in the late 1700s, which included starless nebulae, individual stars each surrounded by a vast nebula, small nebulae of stars, and large nebulae like the Milky Way (spiral galaxies had not yet been observed, with Lord Rosse first observing them in 1845). In fact with a little imagination one could see solar systems and galaxies in various stages of formation. Although Laplace did not invent the nebular hypothesis, he was thought by many to have perfected it, combining the mathematics of Newtonian gravity, the observations of William Herschel, and his own development of probability theory.

As John Pringle Nichol, one of Poe's sources, puts it in *Views of the Architecture of the Heavens*, "the Cosmogony [of Laplace] has thus every mark of truth: . . . Laplace's bold and brilliant *induction* (may I not now so name it?) includes and resolves all! The theory is so beautiful and perfect, . . ." The perfection largely consists of the fact that all the planets orbit around the Sun in basically the same plane, in line almost perfectly with the equator of the Sun, and in the same direction as the direction of the rotation of the Sun on its axis, which is the same as the direction of the rotation of each planet on its own axis, and that the moons do the same for their respective planets.[141] Likewise in *Eureka*, Poe states that "From whatever point we regard it ['the view of the Nebular Theory as its author himself conceived it'], we shall find it *beautifully true*. It is by far too beautiful, indeed, *not* to possess

141. Nichol (note 115), 101.

Truth as its essentiality—and here I am very profoundly serious in what I say."[142]

But there was a serious anomaly to the system. Nichol states that the "Nebular Cosmogony" is "checked by one emphatic indication," meaning one anomaly, the "retrograde motions" of "the only two satellites of this remote body [Uranus, the farthest known planet at the time] which have been seen since the time of Sir William Herschel," to which he adds, "It is scarcely to be doubted that their motions are retrograde."[143] "Whatever this influence is," Nichol continues, (the correct causal explanation to his mind did not yet exist,) "it cannot invalidate the theory of Laplace" since "The laws of nature never destroy, but only modify each other."[144] Years later, in his seven lectures given in New York

---

142. *P&T*, 1312.

143. William Herschel discovered Uranus in 1781 and the two moons in 1787. By 1797 he thought he had "proved" the existence of four more, but they were later found to be spurious. His supposed proof of the extra moons and his "announcement" of the retrograde motion of the first two moons was published in William Herschel (1798). "On the Discovery of Four Additional Satellites of the Georgium Sidus. The Retrograde Motion of Its Old Satellites Announced; And the Cause of Their Disappearance at Certain Distances from the Planet Explained." *Philosophical Transactions of the Royal Society of London* 88, 47–79. Herschel, by the way, refers to Uranus as "Georgium Sidus," meaning "George's Star," as he wanted to name Uranus in honor of King George III. But the astronomical community wanted all the planets to be named after Greek and Roman gods, so the name Uranus was eventually adopted, after the son and husband of Gaia (mother Earth) in Greek mythology who in turn was the father of Saturn.

144. Nichol (note 115), 101, 102. See also Nichol (note 99), 85. I shall try to explain the retrograde anomaly clearly, given present-day knowledge. Viewing our solar system from way above the north pole of Earth, the Sun rotates on its axis counterclockwise. Accordingly, all the planets orbit the Sun in the same direction as the rotation of the Sun (counterclockwise). The same direction is known as *prograde*, the opposite direction is known as *retrograde*. All the planets in our solar system also rotate on their axes prograde, counterclockwise like the Sun, with the exception of Venus and Pluto, which are retrograde because they are upside down. Uranus is the only planet in our solar system that is on its side, with its north pole facing the Sun for half of its orbit and its south pole facing the Sun for the other half. Uranus rotates on its axis prograde, in the sense that it would rotate counterclockwise were its north pole upright like that of Earth. The moons

during January and February 1848 (unfortunately eclipsing Poe's lecture on the Universe) and immediately published serially in the *Tribune*, Nichol states that "we know that the planet [Uranus] has three Satellites" and that they constitute the "one exception" to the nebular hypothesis, in that the moons "are not moving round Uranus in the plane to which the other planets are confined. They sweep out from it, nor do they move in the same direction as the other planets, but in the opposite. They move backward." Nichol then asks rhetorically, "Are we to accept this single exception as overthrowing the general order? or as arising from some external cause with which we are not acquainted?"[145]

Poe in *Eureka* likewise turns to the same problem discussed by Nichol. Mentioning that Uranus has "three moons," (Poe might have got this from Nichol's fifth lecture published in the *Tribune* or from a recent source common to both, but in either case they were both mistaken, as more moons of Uranus were not discovered until 1851,) he states that "In the revolution of the satellites of Uranus, there does appear something seemingly inconsistent with the assumptions of Laplace; but that *one* inconsistency can invalidate a theory constructed from a million of intricate consistencies, is a fancy fit only for the fantastic. In prophesying, confidently, that the apparent anomaly to which I refer, will, sooner or later, be found one of the strongest corroborations of the general

---

of Uranus known to Nichol and Poe orbit Uranus *prograde*. But since Uranus is on its side, and takes 84 years to orbit the Sun, the orbits of those moons will appear (a) prograde (counterclockwise) from Earth when the north pole of Uranus is facing the Sun, and (b) retrograde (clockwise) from Earth when the north pole of Uranus is facing away from the Sun. One can therefore predict (or rather retrodict) that in the years of the anomaly as observed by Herschel, Uranus was in that part of its orbit in which its north pole was (b) facing away from the Sun, so that its two moons would appear from Earth to be orbiting Uranus clockwise. And this was in fact the case at the time! Consequently Nichol years later, simply following Herschel years earlier, refers to the orbits of those moons as *retrograde*. (The research for this footnote along with the necessary visualization was conducted while I had the cholera, or spasms quite as bad, but I challenge anyone to find as clear an explanation of the retrograde anomaly in any other source.)

145. Nichol (note 115), 27.

hypothesis, I pretend to no especial spirit of divination. It is a matter which the only difficulty seems *not* to foresee."[146]

---

146. *P&T*, 1309, 1312. Unlike Nichol, Poe did indicate an explanation for the anomaly. In a note to the above he states that "I am prepared to show that the anomalous revolution of the satellites of Uranus is a simply [*sic*] perspective anomaly arising from the *bouleversement* of the axis of the planet." (The word "inclination" in the 1848 edition was replaced by Poe with "*bouleversement*" in an 1849 annotation to the text. *Eureka*, 60n. The word *bouleversement* is French for "violent disturbance" or "reversal"—*Merriam-Webster Online*.) The dominant theory today is that genuinely retrograde moons are captured satellites, not moons formed by the nebular development of the planet. The moons of Uranus known to Herschel, Nichol, and Poe, however, are not in fact retrograde moons but prograde, as per note 144 above. According to the Levines, who take *bouleversement* to mean "somersault," "Poe has the right answer," "Uranus *is* tilted," "The note is essentially correct." *Eureka*, 142n149. This is truly remarkable! (Both for Poe getting it right and for the Levines admitting it.) But Poe has not in fact *explained* the unusual tilt of Uranus in terms of the nebular hypothesis, let alone how it would corroborate it. In fact he only says that he "is prepared to show." Did Poe then have in mind something like the modern pole shift hypothesis for planets, maybe even the modern theory of geomagnetic reversal, or was it some sort of collision theory (which is the modern view)? Unfortunately Poe did not tell us, and dead men tell no tales. What makes the matter only more interesting, (and possibly disturbing—see Chapter 4, the section on *Vestiges*,) is that Robert Chambers, in *Vestiges* (1844), states in a footnote that "The orbital revolutions of the satellites of Uranus have not yet been clearly scanned. It has been thought that their path is retrograde compared with the rest. Perhaps this may be owing to a *bouleversement* of the primary, for the inclination of its equator to the ecliptic is admittedly high; but the subject is altogether so obscure, that nothing can be founded on it." Secord (note 91), 9. In *Explanations* (1845), a sequel published separately in 1846 by Wiley and Putnam in New York and subsequently under the same covers with *Vestiges*, Chambers states that in *Vestiges* "I have thrown out the hint of a possible *bouleversement* of the whole of that planet's system." The suggestion "has been scoffed at," he says, "but it is only the supposition of a greater degree of obliquity in the inclination of the axis of the planet to the plane of its orbit than what we find in several others." Secord (note 91), E22. Do we have here a case of plagiarism on the part of Poe, or of independent convergence on the same solution, (and even French word,) or did they both purloin from a common (and unmentioned) source? I shall leave this mystery to others, noting that it has been overlooked in the works dealing with a possible *Vestiges–Eureka* connection that I have examined, namely, those by Scholnick and the editors of *Letters*. I shall here merely add, in defense of Poe, that his use of the word *bouleversement* is not uncommon in his writings,—e.g., in "Hans Pfaall" (1835),

So what did Poe have against Laplace? We have already seen, in the discussion on Newton above, that Poe thought of Laplace as primarily a mathematician and secondarily a physicist, and overall as deficient, for not caring to determine "the *modus operandi* of the Law of Gravity," for avoiding "the kingdom of Metaphysics."[147] Later in *Eureka* Poe refers to "the Nebular Cosmogony of Laplace" as "the most magnificent of theories," but rejects the word "Cosmogony" (which, recall, was used by Nichol) because "'Cosmogony' is far too comprehensive a term for what he really discusses—which is the constitution of our solar system alone—of one among the myriad of similar systems which make up the Universe of Stars."[148] In other words, Laplace's nebular hypothesis is at best only about the origin of solar systems, not about the origin of the Cosmos (Universe). And clearly Poe thought that Laplace did not think big enough, that he should have turned to the origin of the Universe as a whole and should have seen it as finite and in the process of condensation. In this light we should remember Poe's comment in one of his letters examined in Chapter 5, that "The *ground* covered by the great French astronomer compares with that covered by my theory, as a bubble compares with the ocean on which it floats."[149] Likewise Poe remarks that Laplace (and Comte after him) did not show "that Matter at any period actually existed as described, in a state of nebular diffusion, but that, admitting it so to have existed throughout the space and much beyond the space now occupied by our solar system, *and to have commenced a movement towards a centre*—it must gradually have assumed the various forms and motions which are now seen, in that system, to obtain."[150] In other words, Poe himself with his theory of the Universe supplies the rationale for the assumption

"Thingum Bob" (1844), "Rationale of Verse" (1848), and "Marginalia" (June 1849); *P&T*, 990, 785, *E&R*, 46, 1458,—such that the word may have had a fairly common currency among the Anglo-American literati of the time. At any rate, whatever the truth behind the mystery, I only ask that Poe be accorded the benefit of the doubt until if and when reasonable doubt is no more.

147. *P&T*, 1290.

148. *P&T*, 1306.

149. *Letters*, 689.

150. *P&T*, 1317.

ascribed to Laplace, the nebular diffusion, while the line in italics emphasizes Poe's earlier point that Laplace failed to see the bigger picture, both the cause of the nebular diffusion in the first place and that his "Nebular Theory" implies a movement of the Universe towards a center. This perhaps explains Poe's final comments about Laplace in *Eureka*. Laplace *assumed*, following others of his time, including the "ancient imagination" of the Epicureans, says Poe, that atoms exist "in unlimited succession, extended throughout the Universal space." (He then repeats his point, examined in Chapter 5, that "under such circumstances, there could have occurred no movement at all.") Moreover Laplace *assumed*, following the astronomers with their telescopes, that the so-called nebulae observed through telescopes were "in reality, what their designation implies," and so "he depended less upon what he saw than upon what he heard." And "very properly" it is, says Poe of "this great man," that he had "an inferior faith in his own merely *perceptive* powers." What Laplace had, then, was "an almost miraculous mathematical instinct," an instinct that "led him, blindfolded, through a labyrinth of Error, into one of the most luminous and stupendous temples of Truth."[151]

Poe did not only "stand on the shoulders of giants," then, as Newton famously said of himself in a 1676 letter to Robert Hooke about seeing further than others,—he also liked to give them a good backhand to the side of the head while he was at it, too.

Poe's "Double Dupin"

It is common to compare scientific work with detective work. John Stuart Mill, for example, compares solving "the inquiries of geology" with "judicial inquiries," both of which "aim at discovering a past event by inference from those of its effects which still subsist," such as "whether a man was murdered or died a natural death," to which he adds that the "celebrated speculation of Laplace, now very generally received as probable by astronomers, concerning the origin of the earth and planets, participates essentially in the strictly inductive character of modern geological

151. *P&T*, 1322.

theory."[152] Einstein provided an interesting twist from the viewpoint of physics, stating that "For the detective the crime is given, the problem formulated: who killed Cock Robin? The scientist must, at least in part, commit his own crime, as well as carry out the investigation."[153] A little more recently, André Lwoff, at the time the head of the Department of Microbial Physiology at the Institut Pasteur in Paris, in his review of James Watson's *The Double Helix*, (a book we shall return to in the next chapter,) claims that "The book is the history of a scientific endeavour, a true detective story that leaves the reader breathless from beginning to end."[154] More recently still, the logician Irving Copi, in an effort to "describe the general pattern of scientific research," states that "A perennial favorite in this connection is the detective, whose problem is not quite the same as that of the pure scientist, but whose approach and technique illustrate the method of science very clearly." He then proceeds to employ Sir Arthur Conan Doyle's "immortal creation, Sherlock Holmes," as "our hero in the following account."[155] The most perceptive comparison, in my opinion, is by the philosopher of science Joseph Agassi, who states

152. John Stuart Mill (1846). *A System of Logic*. New York: Harper & Brothers, 298.

153. Albert Einstein (1938) "The Evolution of Physics." Quoted in Andrew Robinson (2005). *Einstein: A Hundred Years of Relativity*. New York: Harry N. Abrams, 32.

154. André Lwoff (1968). "Truth, Truth, What Is Truth (About How the Structure of DNA was Discovered)?" *Scientific American* 219, 133–138. Reprinted in Gunther S. Stent, ed. (1980). *The Double Helix: Text, Commentary, Reviews, Original Papers*. New York: W.W. Norton, 224–234.

155. Irving M. Copi and Keith Burgess-Jackson (1996). *Informal Logic*. 3rd ed. Upper Saddle River, NJ: Prentice Hall, 249. Doyle's Holmes, by the way, has very little to do with its precursor, Poe's own creation, as we shall see below, and accordingly with the scientific mind. Confining ourselves to the first chapter of Doyle's second Sherlock Holmes novel, (or story, given that it was first published complete in *Lippincott's Monthly Magazine*,) *The Sign of the Four* (1890), in which Holmes explains his detective method, we find that it is an "exact science" involving "analytical reasoning from effects to causes," that it is a process of "unravelling," and that it consists of three parts: "knowledge," meaning background knowledge, "the power of observation," which includes "an extraordinary genius for minutiae," and "deduction," such as "Eliminate all other factors, and the one which remains must be the truth." As for guessing, "I never guess," says Holmes, "It is a

that whether we are dealing with philosophy and history of science, (Popper and Kuhn,) science fiction, (Mr. Spock and Captain Kirk,) or detective fiction, (the police force and the amateur sleuth, or Watson and Holmes,)—one can think of more pairs of examples, (such as Jonathan Quayle Higgins III and Thomas Sullivan Magnum, or Dana Scully and Fox Mulder)—"the tension is between the mechanical or drab view of science and the romantic or mysterious one. And the closer the two get, the greater the tension." Agassi is also surely right about all three domains when he says, "to the very end, our ambivalence about the nature of reason—is it an automatic algorism or is it obscure intuition—is preserved."[156]

With exceptions such as Agassi, the utter irony of it all is that most people outside of Poe studies are completely unaware that Poe literally invented the detective mystery, the *genre*, supplying all the key elements that we have come to know so well.[157] He did this with his three tales featuring his amateur detective "the Chevalier C. Auguste Dupin,"[158] namely, "The Murders in the Rue Morgue" (1841), "The Mystery of Marie Rogêt" (1842–43), and "The Purloined Letter" (1844). Poe, moreover, knew very well what he had accomplished. In his anonymous review of his collection of tales, entitled *Tales* (1845), (the anthology includes all three of the Dupin pieces,) Poe writes of them specifically that "They are all of the same class—a class peculiar to Mr. Poe. They

shocking habit,—destructive to the logical faculty." Now *that*, for reasons we shall see, is what I call *pure* fiction!

156. Joseph Agassi (1982). "The Detective Novel and Scientific Method." *Poetics Today* 3 (1), 99–108. But what if in the future "our ambivalence about the nature of reason" is no longer preserved but resolved by science to the satisfaction of the public? Would it then follow, as Agassi puts it, that "the detective novel is dead"? Can the progress of science possibly kill Dupin, Holmes, and Colombo? If the answer is yes, then *for God's sake please don't anyone read the next and final chapter*, for while writing it I kept thinking to myself, "It is like confessing a murder."

157. J. Brander Matthews (1907). "Poe and the Detective Story." *Scribner's Magazine* 42, 287–293. Reprinted in *Recognition*, 82–94. Peter Thoms (2002). "Poe's Dupin and the Power of Detection." In *Companion*, 133–147.

158. *P&T*, 507.

are inductive—tales of ratiocination—of profound and searching analysis."[159]

Poe's Dupin is not just the ace detective featured in his new genre, however, he is also Poe's fictionalized personification of scientific imagination, or rather of the scientific mind in full. Dupin, then, is a composite. This should be clear on a number of levels.

Let us begin with the name, "Dupin." According to Arthur Quinn, the name is "borrowed from the heroine, Marie Dupin, of a story, 'Marie Laurent,' the first of a series of 'Unpublished Passages in the Life of Vidocq, the French Minister of Police.' These appeared in *Burton's* from September to December, 1838, and are signed 'J.M.B.' Poe must have read these because he refers to Vidocq in 'The Murders in the Rue Morgue' as 'a good guesser,' but one who 'impaired his vision by holding the object too close.'"[160] This is certainly plausible. More recently, Richard Kopley, following T.O. Mabbott, tells us that "In September 1840, Poe received a letter concerning the terms for a possible position as a French teacher for 'M[onsieur] C. Auguste Dubouchet, a gentleman of your acquaintance.' Poe borrowed the initial and middle name for his detective, and coupled them, most probably, with the surname of the jurist André-Marie-Jean-Jacques Dupin, about whom Poe would have read in R.M. Walsh's . . . *Sketches of Conspicuous Living Characters of France* (which Poe reviewed in the same issue of *Graham's Magazine* in which 'Rue Morgue' first appeared). Notably, Dupin's writings are characterized in Walsh's volume as offering an 'abundance of facts and logical deduction' and revealing their author to be 'a perfect living encyclopedia.'" Moreover, "in Walsh's *Sketches*, André Dupin . . . is said to have indulged in punning," so that he is "Poe's model for his detective."[161] This is also plausible, except that it much better fits Doyle's Holmes than Poe's Dupin, given that the method of the

159. *E&R*, 872. That Poe here uses the word "inductive" was evidently to avoid suspicion that he was the author of the review. The same mistake is made by Poe's narrator of "Marie Rogêt," who characterizes Dupin's reasoning as a series of "inductions." *P&T*, 508. Poe's real characterization of Dupin's reasoning is very different, as we shall see below.

160. *Quinn*, 310–311.

161. Richard Kopley (2008). *Edgar Allan Poe and the Dupin Mysteries*. New York: Palgrave Macmillan, 28, 43, 10.

former is deductive while the method of the latter, as we shall see below, is not deductive at all (or even inductive). What should be further noticed, in spite of possible historical sources, or perhaps in addition to them, is that the name "Dupin" appears to be a contraction, possibly of "deduction-plus-induction," (perhaps implying a transcendence of the two methods, especially in light of Poe's definition of "intuition" discussed below,) with the first name, "Auguste," emphasizing the impressive nature of his genius, and "the Chevalier" (beginning with the second Dupin tale, and meaning a knight of the French Legion of Honour) adding distinction to the whole. But I am willing to dismiss the contraction claim as a drunk 'n stupid crotchet, albeit as a remarkable one (if such is possible without being an oxymoron).

Much more striking is that in *Eureka* Poe states that "Now, I have elsewhere observed that it is by just such difficulties as the one now in question—such peculiarities—such roughness—such protuberances above the plane of the ordinary—that Reason feels her way, if at all, in her search for the True."[162] Attached to the end of the word "elsewhere" is the sign for a footnote, to which Poe at the foot of the page has "*Murders in the Rue Morgue*." The precise purpose of this reference is something we shall examine below. But it is enough to see that Dupin is relevant to understanding Poe on scientific imagination. That he is central to that understanding becomes apparent once we look at the key features of the character and how he operates in the tales.

Most of what we need to know is found in "The Murders in the Rue Morgue" (1841), in which we are first introduced to Dupin. We are told that books "were his sole luxuries" and the narrator is "astonished . . . at the vast extent of his reading." The narrator also remarks on "the vivid freshness of his imagination." Dupin, he says, had a "peculiar analytic ability" and took "an eager delight in its exercise." We are also told that Dupin had "perhaps . . . a diseased intelligence."[163] Also notable are Dupin's

162. *P&T*, 1293.

163. *P&T*, 400, 401, 402. The point about a "diseased intelligence" needs to be taken with a grain of salt, just as the narrator in "The Gold-Bug" (1943), as we have seen in the section on madness above, thought that Legrand suffered from an "aberration of mind" and was "disposed to lunacy," meanwhile the impression was largely or entirely due to the narrator's own

powers of observation. He is quoted as saying, "observation has become with me, of late, a species of necessity." We are also told that the narrator, during his stay in the city, rented for them a "tottering" mansion, (shades of "Usher"?,) that they "should have been regarded as madmen," that their "seclusion was perfect," that they "admitted no visitors," and that "We existed within ourselves alone."[164]

The latter brings us to the bugbear of doubling, which as we have seen in the first section of this chapter has led many Poe scholars down a blind alley without knowing it. It is time now to deal with it fully. Early in "Rue Morgue" the narrator states that "I often dwelt meditatively upon the old philosophy of the Bi-Part [*sic*] Soul, and amused myself with the fancy of a double Dupin—the creative and the resolvent."[165] The variety of possibilities put forward by Poe scholars for what exactly Poe meant by "a double Dupin" is staggering. Given what in the previous paragraph I have quoted from "Rue Morgue," John Irwin states that those lines give "the impression that the real split and doubling in Dupin may be less between the creative and the resolvent than between the genius and the madman."[166] Elsewhere Irwin sees "Dupin's own doubleness" as comprising both "animal instinct and bodily knowledge" on the one hand (Dupin puts his mind into that of an orangutan, after all) and "the ideal, the abstract, the logical" on the other.[167] Stanton Garner, noting that "From time to time Dupin's mind actually recedes from this world as much as does the soul of the dead M. Valdemar in the later tale," argues that the "bipartite Dupin" is really for Poe "a Janus who is both of this world and beyond it."[168] Gregory Hayes notices

intellectual shortcomings when in the company of a considerably superior intelligence.

164. *P&T*, 403, 401.

165. *P&T*, 402.

166. Irwin (note 37), 119.

167. John T. Irwin (1994). *The Mystery to a Solution: Poe, Borges, and the Analytic Detective Story*. Baltimore: Johns Hopkins University Press, 199–200, 413.

168. Stanton Garner (1990). "Poe, Thoreau, and Poe's 'Double Dupin.'" In Benjamin Franklin Fisher IV, ed. (1990). *Poe and His Times: The Artist and His Milieu*. Baltimore: The Edgar Allan Poe Society, 138.

not only that Dupin quotes Virgil and Ovid, who are "canonical poets of the Augustan golden age," and that his name, C. Auguste Dupin, "evokes that same Augustan period," mimicking "C. Julius Caesar" or "M. Tullius Cicero," but he claims that Dupin, "whose very name echoes 'OeDIPUs," solves a mystery "that brings about the downfall of his other self, the poet-mathematician [Minister] 'D.'"[169] Richard Kopley, looking at "early versions of the tale," states that "Poe's tale ultimately celebrates the double process of creation and resolution, of imagining and unimagining, of writing and the 'kindred art' of reading. And as the tale honors this double process, it honors, too, the brotherhood of writer and reader." Furthermore, Kopley connects this with *Eureka*, in that "Rue Morgue" "appears to anticipate—perhaps prepare the way for—the irradiation and return of Matter and Spirit in *Eureka*. Indeed, Poe's enigmatic cosmological prose-poem may be interpretable, at least in part, as an allegory of writing and reading."[170] Or maybe the "double Dupin" refers just to Poe himself, as another of his cryptonyms, in this case what Poe scholars call a "double '*d*'s" reference to himself as Eddy, a hidden signature using his common name, a dynamic name server if you will, the usual suspects of which include "curling eddies of smoke," "riddled" and "unriddled," "shudder" (in "Usher,") "deepened, so deepened," "imbedded" ("I'm Eddy,") "folded" and "refolded."[171] And then perhaps Dupin and the narrator are simply doubles of each other, alter egos ("We existed within ourselves alone").[172] But of course, as a determination of non significance, even as a did not show, the simplest explanation is not always the best.

I suggest, instead, as an equal if not stronger possibility, (that is my attempt to unadopt "the rhetoric of certainty," to use Lipton's felicitous phrase,) that by "the old philosophy" Poe had in mind Aristotle, not his distinction between the rational and non-rational parts of the human soul, but his division of the rational part into "master" and "servant," the former "scientific," the

169. Gregory Hayes (2013). "Ancient Classics." In *Context*, 226, 227.

170. Kopley (note 161), 41.

171. See Kopley (note 161), 94n23, 10, 12, 18.

172. *Hutchisson*, 117.

latter "calculative."[173] That would nicely fit the "double Dupin," not genius and madman, or rational mind and animal mind, or mesmeric consciousness and normal consciousness, or good genius and evil genius, or writer and reader, or detective and narrator, but indeed as Poe himself puts it, "the creative and the resolvent" (which could still also refer to Poe himself). Poe uses basically the same language in the preface that precedes the tale, in which he distinguishes between what are called the "analytical" faculties, which he says have "the whole air of intuition" and which themselves are "little susceptible of analysis," and what he calls the "faculty of re-solution," which involves "retrograde operations" and which he says "is possibly much invigorated by mathematical study." Poe adds that the latter is referred to "as if *par excellence*, analysis" but that this is a mistake since "to calculate is not in itself to analyse."[174] Aristotle also, quite interestingly, used the word "calculative" (Barnes translation) in describing the "servant" part of the rational soul, and it should be noticed further that the "mistake" Poe refers to in calling calculating "analysis" was later committed in Poe's anonymous review of his own *Tales* (1845), in which the author of the review characterizes "the *rationale* of the process" used by Dupin as "imagining" followed by "analysis."[175]

Given what we have seen in the previous chapter, the lights should go on at this point. Agassi was in the right direction with his insight about our "ambivalence" about reason and that it is reflected in detective fiction, that it seems both an "obscure intuition" and an "automatic algorism." But he did not apply this

173. *Nicomachean Ethics* 1138b8, 1139a1–15. Jonathan Barnes, ed. (1984). *The Complete Works of Aristotle*. Princeton, NJ: Princeton University Press. See also *Politics* 1333a24–26, where the rational part of the soul is likewise divided into two parts, called "speculative" and "practical." For more on Aristotle's theory and its antecedents, see William W. Fortenbaugh (1983). "On the Antecedents of Aristotle's Bipartite Psychology." In John P. Anton and Anthony Preus, eds. (1983). *Essays on Ancient Greek Philosophy: Volume Two*. Albany: State University of New York Press, 303–320.

174. *P&T*, 397.

175. *E&R*, 872. Once again, that Poe here uses a descriptive word of importance, this time "analysis," in contradiction to his meaning elsewhere, was evidently to avoid suspicion that he was the author of the review. For more examples of Poe's duplicity concerning this review, (which after all is harmless and rather amusing,) see notes 61 and 159 above.

insight to Poe's Dupin. Had he done so, he might have seen that what Poe means by "the creative" in the "double Dupin" (also "intuition," "analysis") is what in modern philosophy of science (as we have seen in the previous chapter) is known as the *context of discovery*, and what Poe means by "the resolvent" (also "retrograde actions," "to calculate") is what is known as the *context of justification*.

In his preface to "Rue Morgue" consisting of "observations at random," Poe elaborates on the distinction between "intuition" and "re-solution" by comparing checkers ("draughts") with chess. With the latter, which consists of variations in pieces with their "*bizarre* motions," Poe claims that "to have a retentive memory, and to proceed by 'the book,' are points commonly regarded as the sum total of good playing." But with checkers, the situation is very different. Since the pieces are all the same and the moves admit of little variation, the checkers champion, unlike the chess champion, involves "some *recherché* movement, the result of some strong exertion of the will." More specifically, "the analyst throws himself into the spirit of his opponent, identifies himself therewith, and not unfrequently sees thus, at a glance, the sole methods (sometimes indeed absurdly simple ones) by which he may seduce into error or hurry into miscalculation." In this latter respect Poe likens excellence at checkers to excellence at whist, (a card game, the forerunner of bridge,) wherein "mind struggles with mind," a struggle that involves "intuitive perception"—noting facial expressions, the way cards are held in the hand, body language, etc.—and success "includes a comprehension of *all* the sources whence legitimate advantage may be derived," sources that are "not only manifold but multiform," such that the successful mental processes "lie frequently among recesses of thought altogether inaccessible to the ordinary understanding." The latter ability involves what Poe calls "analytical power," which in turn involves a "*truly* imaginative" mind, while success at chess involves merely "fancy," a "constructive or combining power," and "simple ingenuity."[176]

Looking at Poe's Dupin tales, we should now be able to see the exact two processes at work, the "double Dupin," Dupin in the context of discovery followed by Dupin in the context of

176. *P&T*, 397, 398, 400, 399.

justification. Starting with the former, Dupin does not examine all the facts and then reason deductively or inductively to the solution of the mystery. Instead, after examining all the facts, the solution comes to him intuitively. Take the horrific murder of Madame L'Espanaye and her daughter, Mademoiselle Camille L'Espanaye, in their fourth-floor apartment in a house on the Rue Morgue, reported in the papers in the following days, including interviews of those who lived nearby and also of the main suspect, the unsuspecting Adolphe Le Bon. Using all of this information, combined with actually getting to inspect the murder scene with the dead bodies still in their places, where "Dupin scrutinized every thing—not excepting the bodies of the victims," Dupin then proceeds to solve the murders in a series of what he calls "guesses," including his guess that the murders were conducted by an "Ourang-Outang" that had escaped from the custody of a French sailor belonging to a Maltese vessel that had recently returned from Borneo. Of these guesses Dupin says, "I have no right to call them more—since the shades of reflection upon which they are based are scarcely of sufficient depth to be appreciable by my own intellect, and since I could not pretend to make them intelligible to the understanding of another."[177]

This is the context of discovery as Poe understood it, a matter of intuition and educated imagination, a leap, the result of Dupin's mind bringing together into one picture of enormous consistency all the facts that were available to him, including his recall of a passage from Cuvier in which the latter describes the anatomy of orangutans. That the process is basically what we today call *unconscious* is further made evident in "Rue Morgue" when the narrator reflects on what he calls "the meaning of Dupin." In the midst of Dupin's account of the steps leading to his orangutan theory, a discourse Dupin is about to characterize as a series of educated guesses as we have seen above, the narrator inserts his reflection that "At these words a vague and half-formed conception of the

177. *P&T*, 425. It is interesting that earlier in the tale Dupin, having solved the problem of escape through the window, says, "The riddle, so far, was now unriddled." *P&T*, 419. Now he uses the word "guesses." Given Poe's literary theory, we should expect Poe's choice of words to be careful, not careless, and at times to have a meaning far deeper than its surface, which is especially the case with "guesses."

meaning of Dupin flitted over my mind. I seemed to be upon the verge of comprehension, without power to comprehend—as men, at times, find themselves upon the brink of remembrance, without being able, in the end, to remember."[178] The verge, the brink, of conception and comprehension, denied the narrator's mind by reason of the fact that his unconscious mind was not up to the task, either because of lack of education or power of imagination or both, thereby not yet sending to his conscious mind the right answer,—the exact opposite of Dupin,—is compared to something everyone can relate to, that of a memory, such as a name, at the tip of one's tongue, but held back by something in the mind. It is all so Freudian, but I shall leave that to the next chapter.

This remarkable passage on "the meaning of Dupin," by the way, was co-opted by Poe from a passage written by him in an earlier work, "Ligeia" (1838). Words cannot express the sudden thrill that came over me the moment the connection came to my mind, followed by wonder at the meaning of it. In "Ligeia" the narrator, the lover of Ligeia, himself an opium eater at the time of narration, (indeed the word "opium" occurs at least five times in various places in the tale,) states that "There is no point, among the many incomprehensible anomalies of the science of the mind, more thrillingly exciting than the fact—never, I believe, noticed in the schools—that, in our endeavors to recall to memory something long forgotten, we often find ourselves *upon the very verge* of remembrance, without being able, in the end, to remember."[179]

Basically the same "meaning of Dupin," the same insights into the context of discovery in scientific imagination, was expressed by Poe two years later (and possibly better) in "The Gold-Bug" (1843) with Legrand, as we have already seen in the section above on madness. Poe has Legrand say, for example, in response to a remarkable coincidence about what turned out to be the treasure map, that "Here was indeed a mystery which I felt it impossible to explain; but, even at that early moment, there seemed to glimmer, faintly, within the most remote and secret chambers of my intellect, a glow-worm-like conception of that truth which last night's adventure brought to so magnificent a demonstration."

178. *P&T*, 421.

179. *P&T*, 264–265.

The intuition was followed the next day by "a more methodical investigation" of the coincidence, which eventually led to the "demonstration," the actual finding of the treasure. Even more significant is the way Legrand, after deciphering the cryptograph on the parchment, solved an important part of the riddle indicating the location of Captain Kidd's treasure. After a "few days" of failing to find the solution, "one morning, it entered into my head, quite suddenly, that this 'Bishop's Hostel' might have some reference to an old family, of the name Bessop, which, time out of mind, had held possession of an ancient manor-house, about four miles to the northward of the Island."[180] This solution of the riddle proved to be the correct one, leading ultimately to the discovery of the treasure.

Following the context of discovery comes the context of justification. In the case of Legrand, the justification was the discovery of "a treasure of incalculable value." In the case of Dupin in "Rue Morgue," it was the experiment that confirmed his educated guesses, the truth of the murders. In short, Dupin placed an advertisement in *Le Monde*, "a paper devoted to the shipping interest, and much sought by sailors." The ad announces that an orangutan has been captured, that it is suspected that the rightful owner is a sailor belonging to a Maltese ship, and that the animal can be retrieved at a particular location contingent upon satisfactorily describing the animal and paying "a few charges arising from its capture and keeping." The trap is set, involving pistols just in case. A sailor arrives and offers to pay a reward. Dupin replies to the sailor, while locking the door and putting his pistol on the table in plain view, "My reward shall be this. You shall give me all the information in your power about these murders in the Rue Morgue." Long tale short, the sailor "fell back into his seat, trembling violently, and with the countenance of death itself," revealing all after he had "recovered his presence of mind."[181]

180. *P&T*, 582, 592.

181. *P&T*, 425, 427, 428. Poe is not entirely clear as to why the sailor, who was indeed the owner, answered the ad, given that the narrator at the end of the tale, upon relating the details of the sailor's confession, adds that after leaving the murder scene the animal "was subsequently caught by the owner himself, who obtained for it a very large sum at the *Jardin des Plantes*." (Prior to the *Le Monde* experiment, Dupin states to the narrator that the

Turning to *Eureka*, we see again the exact same double process, between "the creative and the resolvent," and once again it is important to look at Poe's language.

Beginning with the creative, fairly early in the book Poe defines "intuition" as "*the conviction arising from those inductions or deductions of which the processes are so shadowy as to escape our consciousness, elude or reason, or defy our capacity of expression.*" Immediately following, Poe provides what he calls "the sole absolute *assumption* of my Discourse." It is "an intuition altogether irresistible, although inexpressible, [that] forces me to the conclusion that what God originally created—that that Matter which, by dint of his Volition, he first made from his Spirit, or from Nihility, *could* have been nothing but Matter in its utmost conceivable state of—what?—of *Simplicity*."[182] This is, of course, as we have seen in Chapter 5, his own version—the original, first version!—of Big Bang theory. A little further along Poe repeats himself, now using the word "leap": "By the difficulty—the 'peculiarity'—now presented, I leap at once to *the* secret—a secret which I might never have attained *but* for the peculiarity and the inferences which, *in its mere character of peculiarity*, it affords me." He then elaborates on the leap, stating that "I say to myself—'Unity, as I have explained it, is a truth—I feel it. Diffusion is a truth—I see it. Radiation, by which alone these two truths are reconciled, as a consequent truth—I perceive it."[183] And so on he goes.

That Poe's fundamental intuition of *Eureka* is described by him as a "leap" connects, of course, with what we have seen above in the section on Kepler—that Kepler "*guessed*"—that Kepler "*imagined*"—that Kepler reached his "vital laws," he imagines Kepler saying, "through mere dint of *intuition*," which he has Kepler define as "but the conviction resulting from *de*ductions or *in*ductions of which the processes were so shadowy as to have escaped his consciousness, eluded his reason, or bidden defiance

"Ourang-Outang . . . is still at large." *P&T*, 425. But he could not have known that.) If the capture occurred *after* the owner's confession to Dupin, then fine, but if it occurred *before*, then what evidently compelled the owner to answer the ad was that cousin of the Imp of the Perverse known as Greed.

182. *P&T*, 1276–1277.

183. *P&T*, 1293.

to his capacity of expression."[184] Poe's language about leaps, moreover, connects with what he says earlier in the imaginary epistle, about "the progress of true Science, which makes its most important advances—as all History will show—by seemingly intuitive *leaps*."[185]

This is Poe talking about what today is referred to as the *context of discovery*, which we have seen in the previous chapter is the part of science usually excluded from the scientific process by philosophers and historians of science. But what of the *context of justification*? Poe is the opposite of the modern scene, not because he denies the context of justification, but because he seems to give it less attention, saying of "The Keplers" that "their theories are merely corrected—reduced—sifted—cleared, little by little, of their chaff of inconsistency, until at length there stands apparent an unencumbered *Consistency*—a consistency which the most stolid admit—because it *is* a consistency—to be an absolute and an unquestionable *Truth*."[186]

And what about the chaff of inconsistency that must have accompanied *Eureka* before the published version? In Chapter 4, in the section on whether *Eureka* is a hoax, we have seen in Poe's correspondence evidence of correction, in which Poe wrote in apology for not replying promptly, slightly less than two months before the publication of *Eureka*, that he was delayed by "duties that, just now, *will not* be neglected or even postponed—the proof-reading of a work of scientific detail, in which a trivial error would involve me in very serious embarrassment."[187] Given the possibility of "very serious embarrassment," it is unlikely that by "proof-reading" Poe was referring merely to matters of style, such as spelling and grammar. More likely he was referring mainly to matters of theory, leaving him satisfied that the rest of *Eureka* follows from his "sole absolute *assumption*" with "a train of ratiocination as rigorously logical as that which establishes any demonstration in Euclid," that it follows a "deductive process"

184. *P&T*, 1270.

185. *P&T*, 1264.

186. *P&T*, 1269.

187. *Letters*, 665.

which is "thoroughly corroborated by induction."[188] The hyperbole aside, Poe clearly thought that working out the consistency of a scientific theory (both internally and externally) is much like working out the consistency of a plot in a poem or tale, with the two becoming one when it came to the Universe, the "plot of God."

Poe's reference to "proof-reading" and "scientific detail" might only refer to working out the internal consistency of the theory. But what of working out the external consistency? That in the technical sense would be a matter of the context of justification, of testing the theory against facts and establishing it as the most probable. We can see at least two examples of this in *Eureka*.

The first example has to do with a test against the facts that actually makes Poe revise his theory as initially proposed in an earlier portion of *Eureka*. It involves the context of the footnote to "Rue Morgue" in which Poe says, "it is just by such difficulties as the one now in question—such peculiarities—such roughness—such protuberances above the plane of the ordinary—that Reason feels her way, if at all, in her search for the True."[189] Poe is dealing with an observation that creates a problem for his Big Bang theory, an observation that would seem unimportant, that is "seemingly irrelevant," to the casual observer. It is that the Universe of stars does not present itself, especially to astronomers with their modern telescopes, as roughly equal in diffusion. Instead, as was already established in Poe's time, some galaxies are farther away than others from our own galaxy, the Milky Way, and in our own galaxy some stars are farther away than others from our own star, the Sun. This is the "difficulty," says Poe, which follows from his supposition of "equability of distribution effected through *radiation from a centre*." Given that Poe has explicitly rejected a continuous outpouring of atoms from the center, the "unconditional Particle," like light from a light source, with its "*continuous* outpouring of *ray-streams*" or "light-particles,"[190]

188. *P&T*, 1301, 1303.

189. *P&T*, 1293. The exact point made by Dupin in "Rue Morgue" is that "it is by these deviations from the plane of the ordinary, that reason feels its way, if at all, in its search for the true." *P&T*, 414.

190. *P&T*, 1294, 1291.

(apparently because it does not fit his theology,) it follows that in Poe's view one should have a universe of stars that is something like a "hollow sphere of glass."[191]

In solution of this difficulty, Poe takes a *leap*, a "leap at once to *the* secret," which is to suppose a series of diffusions of atoms from the center, each diffusion (except for the first) an "inferior exercise of the same force" as the one that preceded it, "and so on, until these concentric strata, growing gradually less and less, come down at length to the central point; and the diffusive matter, simultaneously with the diffusive force, is exhausted."[192] Poe would later add to one of his copies of the book a note in pencil following the word "exhausted," which is, "Here describe the whole process as one instantaneous flash."[193] The solution allows Poe to say "whether the ascertained *facts* of Astronomy confirm the general arrangement which I have thus, deductively, assigned to the Heavens. Thoroughly, they *do*. Telescopic observation, guided by the laws of perspective, enables us to understand that the perceptible Universe exists as *a roughly spherical cluster of clusters, irregularly disposed*," with each cluster a "nebulæ," one of which is "the Galaxy, or Milky Way."[194]

The second example of Poe testing his theory against the facts involves what Poe calls "the world-renowned instrument of Lord Rosse."[195] By this he means "Leviathan," (what did you think he means?,) the 58-foot telescope built by William Parsons, the third Earl of Rosse, in 1845 (which remained the largest telescope in the world until 1908). This and other giant telescopes built by Parsons created problems for the nebular hypothesis. Parsons was the first

191. *P&T*, 1295.

192. *P&T*, 1293, 1295. Is this "leap" the matter of "scientific detail" that Poe wrote about in the letter referenced above, the correction he was working on, "in which a trivial error would involve me in very serious embarrassment"? We shall probably never know.

193. *P&T*, 1400.

194. *P&T*, 1325. Current astrophysical theory entertains "three main possible geometries of the large-scale universe": "curved like a saddle," "curved like the surface of a sphere," and "flat." It all depends on the relation of the total density of the Universe to what is called the critical density of the Universe. Shaver (note 140), 35.

195. *P&T*, 1319.

to observe spiral galaxies, which seemed to support the nebular hypothesis, but he also observed that what were thought to be gaseous nebulae by earlier astronomers, nebulae supporting the nebular hypothesis such as the Orion nebula, were really clouds of stars, not gaseous nebulae in stages of formation into solar systems. Parsons himself concluded that all nebulae observed by telescopes were in fact stellar nebulae only.

Poe's solution in *Eureka* is twofold. First, he does not charge the Earl and his followers with hasty generalization. In fact, he claims that were some nebulae really to be seen as gaseous nebulae by telescopes rather than as stellar nebulae, then he (Poe) "should consider the Nebular Cosmogony—*not*, indeed, as corroborated by demonstration—but as thereby irretrievably overthrown." His reason is that the discovery of such gaseous nebulae would mean, in his scheme of the Universe, that we would "have to suppose that, in the particular instances of these 'nebulæ,' an unerring God found it necessary to introduce certain supplementary regulations—certain improvements of the general law—certain retouching and emendations, in a word, which had the effect of deferring the completion of these individual stars for centuries beyond the æra during which all the other stellar bodies had time, not only to be fully constituted, but to grow hoary with unspeakable old age."[196] In other words, if the nebular hypothesis actually denotes not only prior but *current* solar systems in production, then the latter could serve no purpose in the Universe as a plot of God, making the plot imperfect by having not only one but a set of unnecessary excrescences.

Second, in conjunction with this, Poe brings in the perspective of time. Even if telescopes should observe a genuine gaseous nebula, which he expects should be the case, he claims it is a mistake to "declare it a real and not merely a perspective nebulosity." The light from nebulae to telescopes takes time, so that "since the light by which we recognize the nebulæ now, must be merely that which left their surfaces a vast number of years ago, the processes at present observed, or supposed to be observed, are, in fact, *not* processes now actually going on, but the phantoms of processes completed long in the Past—just as I maintain all these

196. *P&T*, 1321.

mass-constitutive processes *must* have been."[197] (In other words, given that light has a finite speed, telescopes can never be used to see the present, only the past.) Hence, a few pages earlier, Poe claims that a *demonstration* of the nebular hypothesis would not be a *proof*, since "to show that, certain data being given, a certain existing result might, or even *must*, have ensued, will fail to prove that this result *did* ensue, *from the data*, until such time as it shall be also shown that there are, *and can be*, no other data from which the result in question might *equally* have ensued."[198] To use language from modern philosophy of science examined in the previous chapter, Poe is referring to the *underdetermination of theory by data*, that more than one theory is always consistent with the data (which is always finite). Isolated, the nebular hypothesis fails, as it can never be proven, it can only be demonstrated consistent with the data. Poe's solution is to make the nebular hypothesis part of a bigger picture, his all-encompassing Big Bang theory of the Universe. What Poe is doing, then, is *inference to the best explanation*, (widely conceived,) to again use language from twentieth-century philosophy of science. The nebular hypothesis is true *only if* it is taken as referring to past processes, not to present ones. Poe is trying to take everything in and make sense of the Universe as a whole, appealing to absolute consistency with everything that is known as the prime *epistemic value* (to use language yet again from the modern scene) in determining what makes an explanation of the Universe the best. The result is his theory of the Universe presented in *Eureka*, which he believes is the most consistent, true, and beautiful in comparison with any other actual or possible theory.

Returning to Dupin, the second tale in the Dupin trilogy, "The Mystery of Marie Rogêt" (1842–43), is very different from the first of the trilogy. In this tale Poe himself attempts to solve a famous murder case by making his tale parallel the real case, the murder being that of Mary Cecilia Rogers, who was found dead in July 1841 in the vicinity of New York City. The biggest difference from "Rue Morgue," however, is that although Dupin tells us that the Rogét case "is a far more intricate case than that

197. *P&T*, 1321.

198. *P&T*, 1318.

of Rue Morgue," he also tells us that "There is nothing peculiarly *outré* about it," that "This is an *ordinary*, although an atrocious instance of a crime." And yet the tale is a very long one. Why? Because almost all of it is Dupin showing the bad reasoning of the newspapers and magazines that tried to solve the case, "variable fancies," all of the writers mere "myrmidons" of the Prefect of Police, thinking that "because it was not impossible that either of these numerous modes and motives *could* have been the actual one, they [each] have taken for granted that one of them *must*."[199] Consequently the tale is enormously unlike "Rue Morgue," lacking both the context of discovery on Dupin's part and the context of justification. But the tale is nevertheless important for understanding Poe's theory of scientific imagination.[200]

Two points need to be brought out here. The first has to do with coincidences and probabilities, such that "There are few persons, even among the calmest thinkers, who have not occasionally been startled into a vague yet thrilling half-credence in the supernatural, by *coincidences* of so seemingly marvellous character that, as *mere* coincidences, the intellect has been unable to receive them." (Think of coincidental phone calls, and the fanciful flight of weak intellects to the conclusion of some sort of

199. *P&T*, 519, 520.

200. Poe doesn't actually have Dupin solve or "unravel" the murder, but only indicate the right direction toward the solution. As he wrote in a letter dated June 4, 1842, "I believe not only that I have demonstrated the fallacy of the general idea—that the girl was a victim of a gang of ruffians—but have *indicated the assassin* in a manner which will give renewed impetus to investigation. My main object, nevertheless, . . . is an analysis of the true principles which should direct inquiry in similar cases." *Letters*, 338. In another letter written on the same day, this one to Joseph Snodgrass, ironically the same doctor that would find him delirious in Baltimore on October 3, 1849, and send him to the hospital where he would soon after die, Poe repeats the same message, adding that "The press has been entirely on the wrong scent" and that "Dupin *reasons* the matter through." *Letters*, 340. (In both letters, Poe says Dupin "unravelled the mystery of Marie's assassination.") If anything, Dupin's reasoning in this tale can only count as the context of justification *in the sense of* rejecting proposed theories, but it is surely incomplete if it is also thought to include the context of discovery, since again the mystery is not in fact "unravelled" in the tale, as no specific theory is actually proposed, not even as a sketch.

psychic connection.) What is needed to "stifle" this bad reasoning, says Poe's narrator at the beginning of the tale, is "the doctrine of chance, or, as it is technically termed, the Calculus of Probabilities."[201] This is what, at least in large part, separates genuine scientific imagination from mere fanciful thinking, the latter usually confused for intelligent thinking. (Keep a log of all the times you've thought of someone who has your number but they didn't call; do this for a week if you can, and you'll see what Poe means.) A just reasoner deals with probabilities, not necessarily certainties. (The odds are that it's going to happen a few times in your life, more than a few if you carry a cell phone, that the person you've been thinking about phones you shortly afterward.) Solving murder mysteries or scientific mysteries is not different. One has to calculate upon probabilities, keep an open mind, and always remember that the extremely rare *can* and *does* happen, (of course one always needs strong evidence for causal reasoning, not merely *post* or *cum hoc ergo propter hoc*,) as with the extinction of the dinosaurs by meteor impact, but also that the probabilities for a truth claim can become so vast, and especially when vast and varied, that one can safely and rightly speak of knowledge *beyond a reasonable doubt*. (Imaginary doubt, of course, is not reasonable doubt, and Descartes' principle of knowledge—*beyond all possible doubt*—places the bar too high, making Descartes himself the father, in effect, of the *reductio ad absurdum* specifically known as *postmodernism*.)

Connected with this point is another important point made by (or rather through) Dupin in "Marie Rogêt," a point about taking everything in when trying to solve a mystery, not just what is obvious, but also what might seem trivial or unimportant, perhaps even merely coincidental. The passage needs to be quoted in full:

> In that which I now propose, we will discard the interior points of this tragedy, and concentrate our attention upon its outskirts. Not the least usual error, in investigations such as this, is the limiting of inquiry to the immediate, with total disregard of the collateral or circumstantial events. It is the mal-practice of the courts to confine evidence and

201. *P&T*, 506–507.

> discussion to the bounds of the apparent relevancy. Yet experience has shown, and a true philosophy will always show, that a vast, perhaps the larger portion of truth, arises from the seemingly irrelevant. It is through the spirit of this principle, if not precisely through its letter, that modern science has resolved to *calculate upon the unforeseen*. . . . The history of human knowledge has so uninterruptedly shown that to collateral, or incidental, or accidental events we are indebted for the most numerous and valuable discoveries, that it has at length become necessary, in any prospective view of improvement, to make not only large, but the largest allowances for inventions that shall arise by chance, and quite out of the range of ordinary expectations. It is no longer philosophical to base, upon what has been, a vision of what is to be. *Accident* is admitted as a portion of the substructure. We make chance a matter of absolute calculation. We subject the unlooked for and unimagined, to the mathematical *formulae* of the schools.
>
> I repeat that it is no more than fact, that the *larger* portion of all truth has sprung from the collateral; . . .[202]

It would be interesting to test Poe's claim here with a casebook on the history of scientific discoveries, if only such a book would exist.[203] The point we need to take, at any rate, is that scientific imagination is not to be confused with wild imagination or fancy. Instead, it is the product of an educated mind *taking everything in*. What would seem to be a necessary precondition for that, at least when it comes to big topics, the biggest being the meaning of the Universe, is the attitude of being *interested in everything*, something I brought out about Poe in Chapter 4, which is something that Poe in turn brings out about Dupin in all three of the detective tales.

Turning again to *Eureka*, we can see much of "Marie Rogét" there, too, Poe attempting to take everything in, even the

202. *P&T*, 534.

203. And it does! See Royston M. Roberts (1989). *Serendipity: Accidental Discoveries in Science*. New York: John Wiley & Sons.

seemingly irrelevant so as to produce his theory. We have seen this in virtually every chapter leading right up to the present point. We can also see Poe in *Eureka*, much like Dupin in "Marie Rogêt," criticizing others for being "entirely on the wrong scent." For example, in Chapter 5 we have seen that in *Eureka* Poe criticizes Nichol, Humboldt, and Herschel for not entertaining the possibility that the Universe is not static, even though in their writings they evinced evidence that should have got them thinking that way. That they went in the wrong direction, says Poe, is "Simply on account of a prejudice;—merely because the supposition [of a finite and dynamic universe] is at war with a preconceived and utterly baseless notion—that of the endlessness—that of the eternal stability of the Universe."[204] And in the previous section we have seen Poe likewise criticize Leibniz, Newton, and Laplace for their wrong directions.

With "The Purloined Letter" (1844), the last in Poe's Dupin trilogy, Poe expressed mixed opinions, first that it "is, perhaps, the best of my tales of ratiocination," this was just before it was published,[205] and then a little later, in October of 1845, in his anonymous review of his anthology published earlier that year, that "We like it less than the others, of the same class. It has not their continuous and absorbing interest."[206] At any rate, the tale is quite different from the other two. First, it is known who committed the crime: a government minister, who was seen taking a letter from the "royal *boudoir*." Second, the motive is also known: blackmail in the form of political power, given that the letter (apparently a salacious love letter) was delivered to the owner of the royal *boudoir*, an unnamed female member of the aristocracy. The mystery, then, is to find the letter, thereby saving the woman from scandal and ruining the Minister in the bargain.

Two points are of interest to us in this tale for understanding Dupin, each of them closely related to the other. The first is a kind of reasoning used by an eight-year-old schoolboy, who "had some principle of guessing" that his friends "termed 'lucky.'" In "Rue Morgue," as we have seen, the analogy is to checkers and whist,

204. *P&T*, 1348.

205. *Letters*, 450.

206. *E&R*, 872.

where success depends on developing "intuitive perception" and identifying with the opponent, such that "mind struggles with mind." In "The Purloined Letter" the example now is the schoolboy game of "even and odd," in which each boy hides marbles in his hand and the other guesses even or odd, a correct guess garnering a marble from the opponent, an incorrect guess the opposite. Brilliance here in the eight-year-old meant "an identification of the reasoner's intellect with that of his opponent," which included an analogical kind of reasoning in which he mirrored as accurately as he could the facial expression of his opponent and observed "what thoughts or sentiments arise in my mind or heart, as if to match or correspond with the expression."[207] Of course it is the very same kind of reasoning that Dupin uses in the tale to solve the mystery. It is also the kind of reasoning that Poe must have thought he was using in producing *Eureka*, as we have seen in earlier chapters, Poe solving the mystery of the Universe by analogically identifying with the mind of God as Poet.

Of equal interest is that Poe in "The Purloined Letter" identifies Dupin, for the very first time in the trilogy, as something of a poet. And he does this in a humorous way. The Prefect tells Dupin that the Minister, although "Not *altogether* a fool," is nevertheless "a poet, which I take to be only one remove from a fool." Poe has Dupin reply while taking a puff from his pipe, in what is clearly an understatement, that "I have been guilty of certain dogrel myself." Later in the tale Dupin tells us that the Minister is a "poet *and* mathematician," that because of both "he would reason well," much better than if he were only a mathematician, that mathematical reasoning has mistakenly been taken for "*the* reason *par excellence*," and that "The French are the originators of this particular deception." Part of the reason for the mistake, according to Dupin, is "the great error" that supposes that mathematical axioms are axioms "of general truth." They are not, says Dupin, given that in areas outside of mathematics and algebra "it is usually *un*true that the aggregated parts are equal to the whole,"

207. *P&T*, 689, 690. The idea here—not to be confused with physiognomy—bears some comparison with Darwin's *The Expression of the Emotions in Man and Animals* (1872), in which he argues that similar facial expressions cut across human races and which he puts into an evolutionary perspective. But I am not going to pursue the comparison.

such as motives in morality, or chemical combinations in chemistry.[208] And of course the Prefect turns out to be the real fool in the tale, given that his methods utterly failed, while Dupin figured out the cunning deception of the Minister quite readily.

Poetical thinking is also made important in the tale because of the use of metaphor in poetry. As Dupin puts it, "The material world . . . abounds with very strict analogies to the immaterial; and thus some color of truth has been given to the rhetorical dogma, that metaphor, or simile, may be made to strengthen an argument, as well as to embellish a description." We shall explore this connection further in the last section of this chapter, but it is interesting to notice that Dupin draws a comparison between "*vis inertiæ*" in physics and in "metaphysics," such that, as with larger physical masses, "intellects of the vaster capacity, while more forcible, more constant, and more eventful in their movements than those of inferior grade, are yet less readily moved."[209]

Of further interest is what Poe says about the methods of the Parisian police, including especially the Prefect. In "Rue Morgue" we are told that they are "cunning, but no more," that they use merely "the method of the moment," that their results are "brought about by simple diligence and activity," that they hold the object of their inquiry "too close" and thereby often fail to see the big picture, and that the Prefect of Police, while "well disposed" to Dupin, is "somewhat too cunning to be profound." In short, Poe's point goes back to his chess analogy, in that the police proceed by "the book" and therefore lack (or at least fail to exercise) "the higher powers of the reflective intellect."[210] In "Marie Rogêt" we get little more, mainly that the Prefect failed to look for any "prominences above the plane of the ordinary," which is why his investigation went in the wrong direction.[211]

But it is in "The Purloined Letter" that Poe goes all-out. There we are told that the "measures" of the Prefect "were good in their kind, and well executed," but that he "perpetually errs by being too deep or too shallow, for the matter in hand." We are told that

208. *P&T*, 684, 691, 692.

209. *P&T*, 694.

210. *P&T*, 411–412, 398, 397.

211. *P&T*, 520.

he and his police "consider only their *own* ideas of ingenuity; and, in searching for anything hidden, advert only to the modes in which *they* would have hidden it [the letter]." The claim is clearly that they are very limited in their imagination, but Poe also connects the claim with the previous point, which is that they are bound to a method. Dupin says, "What is all this boring, and probing, and sounding, and scrutinizing with the microscope, and dividing the surface of the building into registered square inches—what is it all but an exaggeration *of the application* of the one principle or set of principles of search, which are based upon the one set of notions regarding human ingenuity, to which the Prefect, in the long routine of his duty, has been accustomed?" This is where the lack of and disdain for poetry comes in. The Prefect is lacking in the "mode of reasoning" that the eight-year-old excelled in, a kind of reasoning by intuition and analogy. The Prefect's reasoning, instead, is "mathematical," such that if he would have been both a "poet *and* mathematician" the Minister "would have been at the mercy of the Prefect."[212]

Poe's criticism through Dupin of police method is reminiscent, quite interestingly, of Karl Popper's criticism of Thomas Kuhn's "normal science." As we have seen in the previous chapter, Kuhn argued that most scientists spend their lives doing normal science—"puzzle-solving research" within a "paradigm"—and that only during a period of crisis or scientific revolution does a scientist become a thinker engaged in "critical discourse." And even then, as we have seen in that chapter, Kuhn's emphasis is on what he calls "subjective factors." Popper, in reply, wrote that "The 'normal' scientist, as described by Kuhn, has been badly taught. He has been taught in a dogmatic spirit: he is a victim of indoctrination. He has learned a technique which can be applied without asking for the reason why (especially in quantum mechanics). As a consequence, he has become what may be called an *applied scientist*, in contradistinction to what I should call a *pure scientist*."[213]

Now if only Popper would have read Poe—and *really* read him. He might have seen that the *pure scientist* is not a deductive

---

212. *P&T*, 689, 690, 691.

213. Karl Popper (1970). "Normal Science and its Dangers." In Imre Lakatos and Alan Musgrave, eds. (1970). *Criticism and the Growth of Knowledge*. Cambridge: Cambridge University Press, 53.

machine, trying to expose his or her theories to falsification, but instead is driven by imagination.

But we are not quite finished with Dupin. In "Rue Morgue" Poe has Dupin criticize Eugène François Vidocq (1775–1857), an ex-convict who founded in 1812 and headed until 1827 the *La Sûreté Nationale*, (the French National Police, the forerunner of Scotland Yard and the FBI,) who then in 1833 founded and ran the first private detective agency. Dupin says that "Vidocq, for example, was a good guesser, and a persevering man," but he lacked "educated thought" and "impaired his vision by holding the object too close. He might see, perhaps, one or two points with unusual clearness, but in so doing he, necessarily, lost sight of the matter as a whole."[214]

Whether what Poe through Dupin says about Vidocq is fair and accurate does not concern us here. What matters, instead, is that in this very short critique Poe nails a number of key features of the scientific imagination in its full nature. One is that it is "educated," not uneducated and therefore wild. Another is that it takes in the big picture, not leaving anything out. This connects with Poe's criterion of truth, examined in Chapter 4, as total and complete consistency, and also with what I have emphasized in a number of places and have ascribed to Poe, that the bigger the topic the more one has to have the attitude of being interested in everything. As Poe has Dupin put it in the same passage, which surely connects with the Mt. Ætna reference at the beginning of *Eureka*, "depth lies in the valleys where we seek her," but for "more important knowledge," he says, it is "upon the mountain-tops where she is found." And this, in turn, connects with what Poe's narrator at the beginning of "Rue Morgue" calls the "analytical power," which he says is not to be confused with "simple ingenuity" or the "constructive or combining power," given that the latter "has been so frequently seen in those whose intellect bordered otherwise upon idiocy."[215]

It is also surely significant that Poe has Dupin be an amateur detective, not a professional police detective. Just like Poe himself, the amateur unriddler and unraveller of the mystery of the

214. *P&T*, 412.

215. *P&T*, 412, 399.

Universe, uncovering its *modus operandi* even, the person who wants to understand the Universe as a whole needs to rise above a specialty and have an all-encompassing synthetic mind. The scientific specialist sees things "too close," albeit with "unusual clearness," but "By undue profundity we perplex and enfeeble thought." The scientific specialist, then, is not well suited to the task of taking everything in, and probably will ridicule any attempt to do so.

With what we have so far, we have come much closer to our goal, but we are not quite there if we are to fully understand Poe on scientific imagination. We still need to look more closely at why he emphasized the poetic intellect in "The Purloined Letter" and elsewhere. And before we do that we need to look more closely at Poe's criticisms of deductive and inductive reasoning.

Against Deduction and Induction

Aristotle is often called the "father of logic," as he is the first person in history to study and write about logic systematically. Aristotle devoted a number of works to logic, covering topics such as terms, propositions, deductive and inductive reasoning, and fallacies. Collectively his works on logic are known as the *Organon*, meaning, in ancient Greek, "organ" or "instrument," in this case for knowledge. At the heart of Aristotle's logic is the *syllogism*, a valid deductive argument having a major premise, a minor premise, and a conclusion, such as, to use a common example, (not Aristotle's,) "All men are mortal, Socrates is a man, therefore Socrates is mortal." The knowledge that is potential in the premises is made actual in the conclusion. There is, of course, much to Aristotle's logic.[216] But it all culminates in his belief that deduction is the highest form of reasoning and that all deductive reasoning boils down to the syllogism.

216. For an excellent introduction by a top Aristotle scholar, see Sir David Ross (1949). *Aristotle*. 5th ed. London: Methuen & Co., 21–64. Also excellent, this time by a top logician, is Bertrand Russell (note 125), 206–212. A more recent and certainly authoritative introduction to Aristotle's logic is Robin Smith (1995). "Logic." In Jonathan Barnes, ed. (1995). *The Cambridge Companion to Aristotle*. Cambridge: Cambridge University Press, 27–65.

When Aquinas in the thirteenth century combined the authority of the Bible with the writings of Aristotle, the latter became authoritative and you got the heyday of what is known as *scholasticism*, a thick, pedantic, and rigid way of reasoning about the Universe and everything in it that lasted until the scientific revolution of the seventeenth century, made possible primarily by Gutenberg, Columbus, and Copernicus. The new way of reasoning about reality, (or rather *renewed*, as it flourished in Alexandria until the hegemony of Christianity,) known as *experimental science*, needed its own bible, or authoritative set of guiding principles, and that was supplied by Sir Francis Bacon with his book *Novum Organum*, meaning "new instrument," written in Latin and published in 1620.[217] The "new instrument" is actually articulated in two books of aphorisms, indicating that there is more to be done and encouraging collaboration. Gone is appeal to tradition and other "Idols," including syllogistic reasoning altogether, in favor of "true and complete induction," (II, §21,) a system which is bottom-up, based entirely on observation and experimentation.

In Poe's day, Bacon was the equivalent of a saint in the world of science, and we have seen in the first section of the previous chapter that Mill attempted to modernize Bacon with what he (Mill) called the "Deductive Method" of science, which was really an attempt to combine the best of Aristotle and Bacon. The Deductive Method consists of three parts: (i) "direct induction," which is "the operation of discovering and proving general propositions," which is to be conducted using what today are called "Mill's Methods" of induction, meaning "The Four Methods of Experimental Inquiry"; (ii) "Ratiocination, or Syllogism," meaning the same thing, which is "to compute from those laws," i.e., to deduce from the general propositions established in the first step, one or more phenomena in the actual world; and (iii) "verification," which is "comparing this calculated effect with the actual phenomenon." (Mill, it will be remembered, allowed for the "Hypothetical method," which skips the first step and is allowable only if the theorist is willing to submit the hypothesis at some point to the canons of inductive proof.)

217. Lisa Jardine and Michael Silverthorne, eds. (2000). *Francis Bacon: The New Organon*. Cambridge: Cambridge University Press.

As Laura Snyder points out in her important book, Mill's *Logic* "became a major success, one that far outstripped the success of Whewell's *Philosophy*." Going through eight editions in Mill's lifetime, it was read widely at Cambridge University and was the standard textbook on logic at Oxford University until the end of the nineteenth century. But even more important, Mill's *concept* of induction won the day over Whewell's much wider concept of induction expressed in his "Of the Colligation of Facts," which we examined in the opening section of the previous chapter, such that Mill's concept of induction, as Snyder points out, "became the standard view of induction, remaining so today." What is especially revealing is Snyder's observation that "In his *Autobiography*, [published posthumously in 1873 but first drafted in 1853–54,] Mill clearly stated his motivation for writing *System of Logic*: to expel the intuitionist philosophy from its 'stronghold' in physical science and mathematics. He saw this as being the crucial precondition for reforming moral and political philosophy." As a utilitarian, Mill worked to increase individual liberty and to achieve equality for women. The appeal to intuition, he feared, allowed conservatives to maintain the truth of their views. Hence his utilitarian calculus or "formula," the "Greatest Happiness Principle" or "Principle of Utility," was of a piece with his "Deductive Method" of science. As Snyder puts it, it all "stems from Mill's denial of any contribution of mind in induction; he thought that leaving room for any such contribution was to open the door to intuitionism."[218]

This is gold, a golden key even, giving us access to vaults and catacombs beneath *Eureka* hitherto unimagined except by Poe himself.

Let us begin by focusing on the imaginary epistle that prefaces *Eureka*, apparently written in 2848. The letter writer either misspells or misconstrues names with views that are rejected as old-fashioned and long ago disproved. Aristotle is the "Turkish philosopher called Aries and surnamed Tottle," who was celebrated as "the founder, or at all events as the principal propagator, of what was termed the *de*ductive or *à priori* philosophy,"

---

218. Laura J. Snyder (2006). *Reforming Philosophy: A Victorian Debate on Science and Society*. Chicago: University of Chicago Press, 99, 100, 95, 100.

which started with "axioms, or self-evident truths." Aries Tottle had some famous disciples, among them "one Tuclid, a geometrician, and one Kant, a Dutchman, the originator of that species of Transcendentalism which, with the change merely of a C for a K, now bears his peculiar name." Aries Tottle "flourished supreme," we are further told, "until the advent of one Hog, surnamed 'the Ettrick shepherd,' who preached an entirely inductive system, which he called the *à posteriori* or *in*ductive." "Baconian," adds the writer, "was an adjective invented as equivalent to Hog-ian," as it was considered "more dignified and euphonius." Aries Tottle, however, was not done. He "recovered ground, and was permitted to divide the empire of Philosophy with his more modern rival," the controversy ending by "the promulgation of a Median law, to the effect that the Aristotelian and Baconian roads are, and of right ought to be, the sole possible avenues to knowledge"—"that there exist *but two practicable roads to Truth*." Mill, referred to as "Miller, or Mill," who "rode a mill-horse whom he called Jeremy Bentham," wrote "the cleverest ancient work on its topic, which is 'Logic,'" and was "much esteemed in his day." Devoting two pages to Mill, the writer exclaims that "absolutely *all* his argumentation is at sea without a rudder," basically for his claim that "A tree must be either a tree or *not* a tree . . . Because we find it *impossible to conceive* that a tree can be any thing else than a tree or not a tree."[219]

219. *P&T*, 1263–1264, 1266, 1268. Given Poe's literary theory, it is to be expected that there are no accidents or real errors in Poe's imaginary epistle of 2848, and this would apply to Poe's "double error," confusing first the name of Bacon with Hog and then claiming that his surname was "Ettrick shepherd," which was in fact the nickname of James Hogg, a British poet and novelist that would be known to Poe's readers. Whatever Poe's purpose for the double error with its humor and clues, it certainly provides a connection to John Stuart Mill, (son of *James* Mill,) who in his *A System of Logic* promoted the "Median law" uniting the approaches of Aries Tottle and Hog. The key word is "Hog." In *Utilitarianism* (1861) Mill deals with the "German, French, and English assailants" of utilitarianism, the long-held criticism that it is "a doctrine worthy only of swine" (2, ¶3). Mill's reply is to add quality to quantity of pleasures, involving a hierarchy, which he capped with the now famous line, "It is better to be a human being dissatisfied than a pig satisfied; better to be Socrates dissatisfied than a fool satisfied" (¶6). Utilitarianism isn't exactly "consistent with Poe's general philosophical scheme," (*Eureka*, 138n111,) given what Poe says about pain

It is Poe's rejection of the "two . . . paths" of deduction and induction that is of especial interest here. He writes of "their pompous and infatuate proscription of all *other* roads to Truth than the two narrow and crooked paths—the one of creeping and the other of crawling—to which, in their ignorant perversity, they have dared to confine the Soul." The deductive road is rejected primarily because of "the now well understood fact that *no* truths are *self*-evident," that "*no such things as axioms ever existed or can possibly exist*," the inductive road is rejected primarily because of its "repression of imagination," and both are rejected collectively because "their restrictions so absurd" serve to "retard the progress of true Science," exemplified in the likes of Kepler and Champollion. There is also a clear condemnation of the professionalization of science, of the scientific monopoly on knowledge, of the "bigots and tyrants" who went "from the sculleries into the parlors of Science—from its pantries into its pulpits." The writer says, "'the means!' they vociferated—'let us look at the means!'—and if, on scrutiny of the means, it was found to come neither under the category Hog, nor under the category Aries (which means ram), why then the savans went no further, but, calling the thinker 'a fool' and branding him a 'theorist,' would never, thenceforward, have any thing to do either with *him* or with his truths." Obsessed with the word "facts," moreover, [and allowing only generalizations strictly based on facts, I should add,] "All attempts at generalization were met at once by the words 'theoretical,' 'theory,' 'theorist'—all *thought*, to be brief, was very properly resented as a personal affront to themselves."[220]

---

that we examined in Chapter 3. Moreover, any "formula" (Mill) or "calculus" (Bentham) for happiness was anathema to Poe, which is in keeping with his rejection of the inductive and deductive "paths" in science. It is therefore surprising that Poe's letter writer does not botch the name of Jeremy Bentham, the famous moral and legal reformer and official father of utilitarian*ism*, (he coined both the term, "utilitarianism," and the phrase, "the greatest happiness of the greatest number,") to whose "doctrine worthy only of swine" Mill was the main successor. For a different (and less twisted) approach to Poe's "double error," and much more on the background to the imaginary epistle, see *Eureka*, 119–128.

220. *P&T*, 1268–1269, 1263, 1266, 1265, 1264–1265, 1266.

Contrasted to the "ground-moles," the "microscopical savans," the "diggers and peddlers of minute *facts*," are "our Keplers—our Laplaces," who "speculate" and "theorize"—Kepler, especially, "was essentially a *theorist*." They are "the *only* true thinkers," otherwise known as "the generally-educated men of ardent imagination." They suffer not "mental slavery," there is no "repression of imagination," but instead they have "the Soul which loves nothing so well as to soar in those regions of illimitable intuition which are utterly incognizant of '*path*.'" No one in 2848 would think that they "must clip the wings of the eagles," for it is only by soaring that one can truly perceive "the great thoroughfare—the majestic highway of the *Consistent*."[221]

What makes Poe's extraordinary attack especially striking is the similarity in language to one of the most respected sources on science and philosophy of science in Poe's time. It will be remembered from the first section of Chapter 6 that John Herschel, who we have seen in Chapter 4 was a source read by Poe, claimed unequivocally that "it is very important to observe, that the successful process of scientific enquiry demands continually the alternate use of both the *inductive* and *deductive* method. The path by which we rise to knowledge must be made smooth and beaten in its lower steps, and often ascended and descended, before we can scale our way to any eminence, much less climb to the summit."[222] Not only the words "inductive" and "deductive," but the word "path" especially, give credence to Poe's focus.

In order to put Poe into the most balanced perspective, however, we have to look beyond the "whizz!—fizz!" of the imaginary epistle prefacing *Eureka* and turn to a letter dated September 20, 1848, that Poe wrote to Charles Hoffman, the editor of a weekly that published a scathing review of *Eureka*.[223] There Poe states that "What I *really* say is this:—That there is no absolute *certainty*

221. *P&T*, 1269, 1265, 1269.

222. John F.W. Herschel (1830). *Preliminary Discourse on the Study of Natural Philosophy*. London: Longman, Rees, Orme, Brown, & Greene, 174–175.

223. *Critical Heritage*, 281–285. The reviewer was John Henry Hopkins, Jr., a theology student known to Poe who in his review claimed that "The Theological portion is intolerable" and suspected that the entire work is "a scientific hoax of the higher order."

either in the Aristotelian or Baconian process—that, for this reason, neither Philosophy is so profound as it fancies itself—and that neither has a right to sneer at that seemingly imaginative process called Intuition (by which the great Kepler attained his laws;) . . . of which the processes are so shadowy as to escape our consciousness, elude our reason or defy our capacity of expression."[224]

That is certainly a watered-down expression of the way Poe puts it in the imaginary epistle, but either way Poe can actually be supported by a modicum of logic and observation. With deductive arguments, it should be obvious, no new term or concept is found in the conclusion that is not already there among the premises. The conclusion of a valid deductive argument merely makes explicit what is implicit in the premises. But no new term or concept is found in the conclusion of an *inductive* argument, either, that was not already there among the premises, even though the conclusion goes beyond the premises and accordingly provides only a strong probability at best that the conclusion is true. This applies not only to Mill's concept of induction, but also to more inclusive ones, which include argument by analogy and (although it is possibly an exception) inference to the best explanation. In fact the problem is even greater. As pointed out by Carl Hempel, the philosopher of science we visited in the previous chapter under Logical Empiricism, one cannot just go around collecting facts. One first has to have a problem along with a hypothesis of a solution, "for what particular sorts of data it is reasonable to collect is not determined by the problem under study, but by a tentative answer to it." But even more, not only is there no "general and mechanical induction procedure . . . available at present," but "Nor can the discovery of such a procedure ever be expected." The basic reason is that an inductive generalization "contains no novel terms," but "scientific hypotheses and theories are usually couched in terms that do not occur at all in the description of the empirical findings on which they rest, and which they serve to explain." Consequently Hempel, as we have seen in the previous chapter, states that "The transition from data to theory requires creative imagination. Scientific hypotheses and theories are not *derived* from observed facts, but *invented* in order to account

224. *Letters*, 688–689.

for them." "They constitute guesses," he says, of the kind that "require great ingenuity, especially if they involve a radical departure from current modes of scientific thinking, as did, for example, the theory of relativity and quantum theory."[225]

Whewell, of course, living well before relativity and quantum theory, clearly saw this point, with his "Colligation of Facts" and the necessity of "guesses." Poe saw it, too, but he added something more, most notably what he called the "poetic intellect."

The Poetic Intellect

It will be recalled from Chapter 2, devoted to Poe's literary theory, that Poe was impressed that Aristotle in *Poetics* "declared poetry the most philosophical of all writing." In Chapter 4, in the section on "Pinakidia," we saw Poe emphasize that many of the ancient natural philosophers were also poets. And in this chapter we have seen Poe's Dupin emphasize the importance of being a poet as opposed to a mere mathematician. Clearly, then, this part of Poe's theory of scientific imagination has also to be investigated, and I have left it for last.

One part of Poe's theory is that imagination does not create any part of that which is imagined. Instead, imagination consists of creating a whole from parts given in experience. Originality resides in the whole, then, not in the parts. In his review of Thomas Moore's *Alciphron, a Poem* (1840), published in 1840, in reply to Coleridge's claim that "the fancy combines, the imagination creates," Poe claims that "The fancy as nearly creates as the imagination; and neither creates in any respect. All novel conceptions are merely unusual combinations. The mind of man can *imagine*

225. Carl G. Hempel (1965). *Philosophy of Natural Science*. Englewood Cliffs, NJ: Prentice-Hall, 12, 14, 15. As we have seen in the previous chapter, Einstein said much the same in "Physics and Reality" (1936), that "There is no inductive method which could lead to the fundamental concepts of physics. Failure to understand this fact constituted the basic philosophical error of so many investigators of the nineteenth century. . . . The most satisfactory situation is evidently to be found in cases where the new fundamental hypotheses are suggested by the world of experience itself." In Bargmann (note 105), 307. Poe would have agreed with Einstein, but in his own unique way, as we shall see below.

nothing which has not really existed." Poe then gives the example of "a *griffin*," which "does not exist. Not the griffin certainly, but its component parts. It is a mere compendium of known limbs and features—of known qualities. Thus with all which seems to be *new*—which appears to be a *creation* of intellect. It is resoluble into the old."[226] Similarly, a few years earlier, in his review of Joseph Rodman Drake's *The Culprit Fay, and Other Poems* (1835), published in 1836, Poe states in a footnote that "What the Deity imagines, *is*, but *was not* before. What man imagines, *is*, but *was* also."[227]

Many of Poe's tales can readily be seen as griffins, in which Poe puts together various elements in a unique and original way, such as "Some Words with a Mummy" (1845) and "The Facts in the Case of M. Valdemar" (1845). But the *doctrine* of imagination, famously residing in John Locke's *Essay concerning Humane Understanding* (1690), in his distinction between simple and complex ideas, (found, for instance, in his chapter "Of Real and Fantastical Ideas,") can actually be found playing a role in Poe's tales. Specifically in "The Domain of Arnheim" (1846), a rewrite of "The Landscape Garden" (1842), Poe has the narrator proclaim, following the philosophy of the genius and fabulously wealthy landscape gardener Ellison, that landscape gardening is "the truest and most natural, if not the most extensive province," of "the poetic sentiment," and this because it is "the fairest field for the display of imagination in the endless combining of forms of novel beauty; the elements to enter into combination being, by a vast superiority, the most glorious which the earth could afford." The landscape gardener does not create the "elements" of the composition, it will be noticed, only the "combination," (much like someone, I might add, who creates a mosaic). Since Ellison's "bias" was toward "the creation of novel moods of purely *physical* loveliness," Ellison,

226. *E&R*, 334. The sheer irony of what is one of Poe's favorite examples is that it is now thought that the mythical griffin—with its head, neck, wings, front legs, and talons of an eagle and its torso, rear legs, and tail of a lion—is historically based on the discovery of a dinosaur fossil. See Adrienne Mayor (2011). *The First Fossil Hunters: Dinosaurs, Mammoths, and Myth in Greek and Roman Times*. 2nd ed. Princeton, NJ: Princeton University Press, 15–53.

227. *E&R*, 511.

says the narrator, "became neither musician nor poet—if we use this latter term in its every-day acceptation." Nevertheless in "the widest and noblest sense he was a poet. He comprehended . . . the true character, the august aims, the supreme majesty and dignity of the poetic sentiment. . . . the creation of novel forms of beauty."[228]

With the "poetic sentiment," illustrated by Poe's Ellison, along with what we have learned from Poe's Dupin and Legrand, we have almost reached our goal. What remains is what Poe calls the "poetic intellect."

In "The Colloquy of Monos and Una" (1841), a dialogue between two lovers not only after their deaths but after the destruction of the world, (due to the linear thinking of people like the "utilitarians," who failed to see "each advance in practical science as a retro-gradation in the true utility,") Monos speaks to Una of what then seems "so utterly obvious" to "our disenfranchised reason," which are "principles which should have taught our race to submit to the guidance of the natural laws, rather than attempt their control." For what then follows we need to keep in mind that the context concerns the natural laws. Monos says, "Occasionally the poetic intellect—that intellect which we now feel to have been the most exalted of all—since those truths which to us were of the most enduring importance could only be reached by that *analogy* which speaks in proof-tones to the imagination alone, and to the unaided reason bears no weight—occasionally did this poetic intellect proceed a step farther in the evolving of the vague idea of the philosophic, and find in the mystic parable that tells of the tree of knowledge, and of its forbidden fruit, death-producing, a distinct intimation that knowledge was not meet for man in the infant condition of his soul."[229]

If we put aside the prophecy here, we can see that the "poetic intellect," according to Poe, is something more than the "poetic sentiment" of Ellison and others, that of poetic minds the poetic intellect is "the most exalted of all." Unlike the poetic mind per se, which deals with beauty, the poetic intellect reaches "truths"

228. *P&T*, 859, 858. For evidence that Poe read (or at least read of) and admired John Locke, see, e.g. "Lines on Joe Locke" (1830), *P&T*, 62, "Morella" (1835), *P&T*, 235, and Poe's essay on Richard Adams Locke in his "The Literati of New York City" (1846), *E&R*, 1222.

229. *P&T*, 450.

that are "the most enduring" by means of "that *analogy* which speaks in proof-tones to the imagination alone." In a footnote that follows two pages later, Monos claims that "reason . . . deals only with the true," but that the soul with a "musical education," which includes "not only the harmonies of time and of tune, but the poetic diction, sentiment and creation, each in its widest sense," is what makes the soul "*beautiful-minded*."[230] The poetic intellect, then, is a kind of synthesis of the purely rational and the purely aesthetic minds, combined with a powerful imagination. Only then can one speak of "proof-tones."

But what does that mean? Seeing truths by analogy, big truths, ultimate truths, is something that takes us back to Aristotle's *Poetics*, a major influence on Poe, as we have seen in Chapter 2. Poe, it will be recalled again, was impressed that Aristotle "declared poetry the most philosophical of all writing." We have seen that Aristotle's immediate reason was that poetry deals with universals about human behavior. But in that chapter I claimed that Poe was probably more impressed by a reason given by Aristotle further along in *Poetics*, that of all the necessary features of being a good poet "the greatest thing by far is to be a master of metaphor. It is the one thing that cannot be learnt from others; and it is also a sign of genius, since a good metaphor implies an intuitive perception of the similarity in dissimilars."

This, I suggest, informs Poe's meaning of the "poetic intellect." A good metaphor is not just an imaginative use of words, a figure of speech with an implied comparison, such as "all the world's a stage." Instead it is a comparison that points to something analogical in reality, like the "selfish gene" metaphor of Richard Dawkins, that genes behave *as if* they were selfish, *in the sense that* all they want to do is make copies of themselves. With that perspective, which is the core of modern evolutionary biology and evolutionary psychology, the lights go on—and on and on, with one profound insight following another.[231]

230. *P&T*, 452.

231. As Dawkins puts it, "If you push novelty of language and metaphor far enough, you can end up with a new way of seeing. And a new way of seeing . . . can in its own right make an original contribution to science. Einstein himself was no mean popularizer, and I've often suspected that his vivid metaphors did more than just help the rest of us. Didn't they also fuel his

But Poe had something even more in mind. Seeing "the similarity in dissimilars" requires a kind of *elliptical* or *non-linear* thinking, looking not directly at things, but indirectly, which can allow one to actually see more than those who only look directly. This fits with Poe's theme of eureka moments traced in the present chapter, what in "Mellonta Tauta" (1849), Poe's morning star to *Eureka*'s sun, he calls scientific progress by "intuitive bounds."[232] Poe expressed both ideas together in "A Chapter of Suggestions" (1845), in which he states that "The intuitive and seemingly casual perception by which we often attain knowledge, when reason herself falters and abandons the effort, appears to resemble the sudden glancing at a star, by which we see it more clearly than by a direct gaze; or the half-closing the eyes in looking at a plot of grass, the more fully to appreciate the intensity of its green."[233] Glancing at a star is in fact one of Poe's recurring metaphors, and we find an even fuller expression of it in "The Murders in the Rue Morgue" (1841). On the topic of "the more important knowledge," Dupin states that "To look at a star by glances—to view it in a side-long way, by turning toward it the exterior portions of the *retina* (more susceptible of feeble impressions of light than the interior), is to behold the star distinctly—is to have the best appreciation of its lustre—a lustre which grows dim just in proportion as we turn our vision *finally* upon it. A greater number of rays actually fall upon the eye in the latter case, but, in the former, there is the more refined capacity for comprehension. By undue profundity we perplex and enfeeble thought; and it is possible to make even Venus

creative genius?" See also Dawkins' reply to the philosopher Mary Midgley, who narrow-mindedly complained in her book review that "Genes cannot be selfish or unselfish." Richard Dawkins (1989). *The Selfish Gene*. 2nd ed. Oxford: Oxford University Press, ix, 278.

232. *P&T*, 875. Poe actually has Pundita claim that "the progress of all true knowledge . . . makes its advances almost invariably by intuitive bounds." "Mellonta Tauta," by the way, is from a line in Sophocles' *Antigone*, which means "These things are in the future." *P&T*, 449. Hopefully *now* we can fully appreciate Poe's meaning, given especially what we shall see in Chapter 8.

233. *E&R*, 1294.

herself vanish from the firmament by a scrutiny too sustained, too concentrated, or too direct."[234]

Now add to this kind of thinking the attitude of being interested in everything. Next add Poe's criteria of truth, examined in Chapter 4, a total and complete consistency with all that is known and knowable combined with the experience of beauty. Mix this combination with Poe's claims about intuition and imagination examined in this chapter. And *voilà*: the result is Poe's concept of the poetic intellect, through which eyes he viewed his magnum opus, *Eureka*.

To sum up this long rigmarole, I have shown that scientific imagination for Poe, in the grand sense of the term, is not a kind of consciousness, including neither double consciousness nor mesmeric consciousness, but instead is a primarily unconscious process, a mental process that occurs behind the scenes of consciousness, a process driven by a well-developed, passionate, and educated imagination, one not bound by rules and methods and that holistically takes everything in, which attempts to formulate solutions to problems perplexing the somewhat monomaniac conscious mind by making the solutions beautifully consistent with everything known to the unconscious mind, which upon completion then informs the conscious mind in the form of a eureka moment, and that this is the engine that drives science.

It is time now, at long last, to attempt a modern update of Poe's theory, what I take to be the ultimate vindication of Poe.

234. *P&T*, 412. Poe's earliest clear use of the "glancing at a star" metaphor is in "Letter to Mr. — —" (1831). *P&T*, 14.

CHAPTER 8

Epilogue

> His reasons, however, have not yet been matured into expression. It remains for a more profound analysis than the world has yet seen, fully to investigate and express them.
>
> —Edgar Allan Poe[1]

Entrée

How shall we provide a modern update of Poe's theory of scientific imagination, examined in the previous chapter and (in a way) the chapter before it? This is the problem of *methodology*. As we have seen in Chapter 6, consequentialism has dominated modern philosophy of science, with the context of justification receiving the lion's share of eminence and attention, while the context of discovery, like Poe's Hop-Frog,—the tyrant king's jester, a dwarf and cripple, viewed as less than a man,—is either ignored, abused, or thrown some crumbs, if not laughed at altogether.

In spite of what we have seen in Chapter 6, however, there have been some book-length attempts at rectifying the situation, and it is important to see, briefly, why they fail. In his book entitled *Patterns of Discovery*, devoted to the topic of "theory-finding," the philosopher of science Norwood Hanson says "my concern is not with the testing of hypotheses, but with their discovery."

1. "The Domain of Arnheim" (1846). *P&T*, 860.

Providing a detailed analysis of elementary particle physics as "a typical research science," as well as a detailed analysis of Kepler's discovery of the elliptical orbit of Mars, Hanson takes the position that "If establishing an hypothesis through its predictions has a logic, so has the conceiving of a hypothesis." And the relevant logic, for Hanson, which has "a perfectly definite logical form," is *abduction* (also *retroduction*) as developed by C.S. Peirce in the late 1800s. Scientists don't reason inductively from data to theories, he says; neither do they come up with guesses, which they are doubtful of in the first place, and then test them. "Kepler never modified a projected explanation capriciously; he always had a sound reason for every modification he made." At no stage was a hypothesis thought up in "Kekulé [dreamlike] fashion." At every stage, instead, Kepler—"who typifies all reasoning in physical science"—was applying abduction and "always had a sound reason for every modification he made." And once he finally arrived at the explanation that worked, that of an elliptical orbit for Mars, he knew he had the right answer because "the ellipse covered all observed positions," so that "his work as a creative thinker was virtually finished. Any mathematician could then deduce further consequences not included in Tycho's lists. It required no genius to take Kepler's idea and try it for other planets."[2]

The problem by now should be clear enough. Hanson's method is *logic* biased—even worse, it is biased toward *conscious* processes through and through. (It also suffers from the problems with IBE as discussed in Chapter 6.) One may agree with Hanson's statement that "the initial suggestion of an hypothesis is very often a reasonable affair." But the statement that immediately follows it shows his hand: "It is not so often affected by intuition, insight, hunches, or other imponderables as biographers or scientists suggest."[3] Poe rejected that viewpoint, as did Einstein within a century later. Without rejecting the role of conscious, rational thinking in the formation of scientific theorizing and imagining, what is needed for our topic is a methodology that goes much

2. Norwood Russell Hanson (1958). *Patterns of Discovery: An Inquiry Into the Conceptual Foundations of Science*. Cambridge: Cambridge University Press, 3, 119, 71, 86, 84, 89, 84.

3. Hanson (note 2), 71.

deeper, one that takes into account, and indeed tries to explain, the role of the largest part of human mental functioning—the unconscious mind.

Equally defective in methodology is the approach of the research physicist and science historian Gerald Holton. In his book, *The Scientific Imagination*, which is devoted to figuring out "how the scientific imagination functions," meaning "the imagination of scientists engaged in the act of doing science," which is important "to ensure that the conditions under which scientific originality can flourish," Holton's methodology is that of examining a "series of case studies." More specifically, his concern "is to find the extent to which, on certain crucial occasions, the imagination of a scientist may be guided by fidelity to one or more *themata*," which are "preconceptions . . . largely independent of the empirical and analytical content" of the scientist's work itself. Likening his work in his book to "a folklorist or anthropologist who listens to the epic stories for their underlying thematic structure and recurrence," Holton claims that "Thematic analysis allows discernment of some constancies or continuities in the development of science, . . . relatively stable structures that extend across supposed revolutions and among apparently incommensurable rival theories." Finding in the physical sciences "less than 100" themata, and that "The appearance of a new thema is rare," Holton concludes that "two themata predominate," namely, "simplicity and necessity."[4]

In his introductory chapter Holton says he is interested in "the context of discovery, trying to understand the 'nascent moment,'" but our hopes are dashed when in the very first case study he says, "In this study I treat the period after the earliest phase of discovery, when the stirrings of a new conception are difficult to document, but before the new work has been absorbed into the mainstream of science through the mechanisms of justification."[5]

4. Gerald Holton (1978). *The Scientific Imagination: Case Studies*. Cambridge: Cambridge University Press, vii, x–xi, viii, 99, 9, ix, 10, xii.

5. Holton (note 4), 4, 26. The case study is Robert Millikan's discovery of the smallest electrical charge, the charge of an electron, largely for which he was awarded the Nobel Prize. Another of Holton's case studies is the "nascent moment" of Enrico Fermi's important discovery, which I will save for the following section.

The fundamental problem here is that he confines his approach to case studies, and more specifically to the "nascent moment" as he defines it. Such an approach (albeit highly valuable) is much too narrow. Confining himself to "after the earliest phase of discovery" is like agreeing that mutations are important for evolution but not caring to inquire into the nature of their origin.

Given the above in addition to what we have seen in Chapter 6, it should now be sufficiently clear that if we want to understand the nature of scientific imagination,—really, truly understand it,—then it is not to the field devoted to the study of science that we should turn, to philosophy of science, or to the field devoted to its history, professional history of science, for they have proven themselves over and over again to be either ill-equipped or ill-willed for the task. Their nature is their fate, as the ancient Greeks would say. So let us leave them to themselves and part in peace.

But that is hardly the end of it. In much more recent years, in what is sometimes called "creativity science," science itself has taken up the task, has turned itself upon itself in order to know its own nature. The paradox is profound but enlightening. In using mainly computer models, brain scans, and neuroscience, much has been learned about how scientific imagination actually works. And yet here again we are beset with a familiar problem, which is that scientists are specialists and tend to stick to their speciality, so that when it comes to providing research on scientific imagination they each give us only a part of the picture (albeit important parts). The knowledge we want, instead, lies not only in the depth of the valleys but also upon the mountaintops, in the sublimity of its *oneness*.

What is needed for a full understanding of scientific imagination, then, so ubiquitous and yet intangible, is the kind of attitude and approach that the extraordinarily brilliant Francis Crick, Nobel laureate in Medicine or Physiology for his co-discovery of the structure of DNA, expressed in his memoir. Following his continued work in molecular biology, Crick became interested in the problem of consciousness. He rejected the approach of cognitive science, (we shall examine his reasons in the section on cognitive science below,) not so much because he found it wrongheaded, but because, although he himself preferred the level of neurons, (a reductionist to the end,) he had to admit that "I think one should approach these problems at *all* levels, as was done in molecular biology."[6]

That is what is needed here, a rejection of the *territoriality* that unfortunately pervades academia, combined with acceptance of being *interested in everything*. Following Crick's lead, we shall proceed by a series of levels beginning from the phenomenological and leading down into the furthest catacombs of the human mind and even beyond, going back up the river of time. More specifically, we shall begin with an examination of a substantial number of eureka moments in the history of modern science. We shall then turn to comparative biography and psychology, to look for commonalities. Following that, we shall explore a level deeper, known as cognitive science, and then a level deeper still, known as neuroscience, focusing on the bi-part brain. We shall then finish with what biologists call *ultimate* as opposed to *proximate* causation, meaning an evolutionary explanation.

In all of this, we shall not pretend to completeness, to bringing it all in—that would be utter madness, as it would require volumes. Instead, we shall admit our bias from the beginning, that we shall be following a specific trail of clues that are in line with Poe's own thinking examined in the previous chapter,—this is a *Poe* book, after all, and this chapter an *epilogue* at that,—and that we hope to provide, but for its relative brevity, the ultimate vindication of Poe, more in *kind* than in detail. What this means is that we shall leave "a more profound analysis" to those far more capable than both of ourselves combined, your most obedient, humble master and servant, unsettled in intellect, suffering from the disordered chamber of his brain, evidenced alone by his teetotum predilections.

Unconscious Scientific Creativity

Is the unconscious mind real? Given that the heyday of Freudian psychology is long past, with Freudian explanations such as psychoanalytic dream analysis and the Oedipus complex in enormous

6. Francis Crick (1988). *What Mad Pursuit: A Personal View of Scientific Discovery*. New York: Basic Books, 150.

disrepute today, one might be inclined to answer *no*. But one would be extremely mistaken, for the Freudian belief in an active, dynamic, causal unconscious is alive and well in modern psychology, cognitive science, and neuropsychology. And yet it has a history of detractors, right up to the present, those who want to deny its reality in one way or another.

The existentialist Jean-Paul Sartre is a prominent example of the denial of the unconscious mind. In the first of his three sections on "bad faith" in *Being and Nothingness*, which he defines as "it is from myself that I am hiding the truth," a hiding that "can even be the normal aspect of life for a very great number of people," Sartre rejects the concept of the unconscious mind, in particular the Freudian concept, in favor of his own view, which is that "Bad faith . . . implies in essence the unity of a *single* consciousness," and "not at two different moments" but at one, "the unity of one and the same psychic mechanism and consequently a double activity in the heart of unity," such that "The effort [of Freud et al.] to establish a veritable duality and even a trinity (*Es*, *Ich*, *Ueberich* expressing themselves through the censor) has resulted in a mere verbal terminology."[7] This view is fully in accord with Sartre's position on free will versus determinism, as expressed in his lecture-turned-essay "Existentialism and Humanism," which is what he calls, following the priority he gives to phenomenological analysis, the "absoluteness of the act of choice," that "there is no determinism—man is free, man *is* freedom." Of a piece with Sartre's axiom of free will is his philosophy of responsibility, which is not only that "Man is nothing else but what he purposes, . . . nothing else but the sum of his actions,"—in other words, that we are what we do,—but that "man is responsible for what he is," totally responsible for all of his choices, his personality, and so forth, with nothing and no one to blame but himself. To bring all of the above together, including bad faith, Sartre's view, in short, is that "Since we have defined the situation of man as one of free choice, without excuse and without help, any man who takes refuge behind the

7. Jean-Paul Sartre (1943). *L'Être et le Néant*. Paris: Gallimard. Hazel E. Barnes, trans. (1956). *Being and Nothingness*. New York: Philosophical Library, 49, 50, 53.

excuse of his passions, or by inventing some deterministic doctrine, is a self-deceiver."[8]

The philosopher of mind John Searle is a more recent example of the rejection of the unconscious mind. Searle takes consciousness to be "caused by the behavior of lower-level elements," and so is "a higher-level feature of the entire system," meaning the brain. Two key features of consciousness are "intentionality and subjectivity," where "intentionality" refers to making "reference to something beyond itself" and "subjectivity" refers to what should be stupidly obvious if you are reading this. On the other hand, in what he calls "a kind of dispositional analysis of unconscious mental phenomena," Searle argues that unconscious mental states are only "neurophysiological states and processes," that they deserve to be called "mental" only if they are "candidates for consciousness, because they are capable of causing conscious states." In this view of the unconscious, then, confined as it is to "an occurrent neurophysiology capable of causing conscious states and events," not only are there no subjective unconscious states and processes, given that "Conscious mental states and processes have a special feature not possessed by other natural phenomena, namely, subjectivity," but "there are no deep unconscious intentional states" either, no "deep unconscious mental intentional phenomena that are not only unconscious but that are in principle inaccessible to consciousness," no "*intrinsic* intentionality," only "*as-if* intentionality" at best, which we often ascribe to things like thermostats, carburetors, and computers, but it is not real intentionality, it is "not a kind of intentionality" at all, only "metaphorical ascriptions," which as such are "not to be taken literally."[9]

Out of this "Darkness there and nothing more" enter the philosopher of psychology David Livingstone Smith. In a chapter devoted to refuting Searle's rejection of the unconscious mind, which could serve to refute Sartre as well, Smith points out that Searle's view is really only a "contemporary representative" of a

8. Jean-Paul Sartre (1947). *L'Existentialisme Est un Humanisme*. Paris: Nagel. Philip Mairet, trans. (1948). *Existentialism and Humanism*. London: Methuen. Reprinted in Walter Kaufmann, ed. (1975). *Existentialism From Dostoevsky to Sartre*. New York: New American Library, 362, 352, 358, 349, 365.

9. John R. Searle (1992). *The Rediscovery of the Mind*. Cambridge: MIT Press, 167, 169, 84, 160, 159, 161–162, 169, 93, 162, 173, 78–79, 82, 172.

view that traces back to the latter part of the 1800s, which he calls *dispositionalism*, the view that "what seem to be unconscious mental events are actually nonmental neurophysiological dispositions which, under the right circumstances, realize their causal powers to produce conscious mental phenomena." Although the dispositionalists did indeed use the word "unconscious," Smith notes that we should not be misled and read into their use the modern meaning. The latter, instead, comes from Freud, who rejected dispositionalism and competing theories of mind "in favor of a theory of radically unconscious yet occurrent mental states." Smith claims, moreover, that dispositionalism was "decisively refuted by Sigmund Freud early in the twentieth century," and part of Smith's argument involves a most interesting quotation from one of Freud's writings published in 1915:

> The data of consciousness have a very large number of gaps in them; both in healthy and in sick people psychical acts occur which can be explained only by presupposing other acts, of which, nevertheless, consciousness affords no evidence. . . . Our most personal daily experience acquaints us with ideas that come into our head, we do not know from where, and with intellectual conclusions arrived at we do not know how.[10]

In line with this quotation is the one that heads Smith's chapter, from a passage written in 1897 by the interdisciplinary "founder of experimental psychology" Wilhelm Wundt (1832–1920), who, in spite of the quotation, was something of a dispositionalist (Smith,

10. David Livingstone Smith (2004). "Freud and Searle on the Ontology of the Unconscious." In Jon Mills, ed. (2004). *Psychoanalysis at the Limit: Epistemology, Mind, and the Question of Science*. Albany: State University of New York Press, 80, 79, 80, 73, 81. Freud, it should be noticed, called the unconscious "mental" because "it is easy to describe the unconscious and to follow its developments if it is approached from the direction of its relation to the conscious, with which it has so much in common. On the other hand, there still seems no possibility of approaching it from the direction of physical events." Sigmund Freud (1913). "The Claims of Psycho-Analysis to Scientific Interest." *The Standard Edition of the Complete Psychological Works of Sigmund Freud*. Vol. XIII, 179.

personal communication): "This unconscious mind is for us like an unknown being who creates and produces for us and finally throws the ripe fruits into our laps."[11]

There is yet more, a consideration of intense interest. In addition to taking on the fine details of Searle's argument, Smith's most powerful category of evidence in favor of Freud's view focuses on the last part of Freud's statement above, presaged by Wundt. As Smith puts it, "Consider the experience of laboring fruitlessly over an intractable intellectual problem, giving up all hope of solving it, and after an interval of time having the solution drop into one's mind . . . . There are many examples of unconscious problem solving in the history of science."[12]

Mysteriously, I am reminded of Poe's "Morella" (1835), in which Morella, whose "erudition was profound," retained her "identity which is termed personal" through the birth of her child,—"in dying she had given birth,"—such that we find "food for consuming thought and horror—for a worm that *would* not die."[13] Yes, *it's alive, IT'S ALIVE!* Oh, in the name of God, Poe's most profound thought, (that he indeed set down in words, with even more distinctness than that which he conceived it,) his Conqueror Worm of true Science, *lives* not just through Dupin and Legrand, with their "Bi-Part Soul," their "guesses," their "glow-worm-like" conceptions, and their "it entered into my head, quite suddenly," or through the author of the imaginary epistle in *Eureka*, with his "generally-educated men of ardent imagination," his "I grasped it with *my soul*," and his definition of "seemingly intuitive *leaps*" as "the conviction resulting from . . . processes . . . so shadowy as to have escaped his consciousness,"[14] but also *through* the likes of Freud and Smith and others, *through* their psychology and philosophy of eureka moments, *reincarnated* in them with no apparent memory of Poe, (the few who love him, without even knowing it, and whom he loves,) *responding* through their own writings with the very voice of Poe, who is saying, in effect, "I am here!"

11. Smith (note 10), 73.

12. Smith (note 10), 81.

13. *P&T*, 234, 235, 237, 238.

14. *P&T*, 402, 425, 582, 592, 1269, 1264, 1270.

Filled with the spirit of Poe, or at least directed by that spirit while he wrote, (standing behind and looking over his right shoulder, to be specific,) Smith in his chapter discusses the case of the nineteenth-century German mathematician Carl Friedrich Gauss, but his most extended discussion is devoted to the case of the French mathematician and theoretical physicist Henri Poincaré, to whom we shall return in the following section. In another work, *Why We Lie*, a large part of Smith's focus is on self-deception, and for that he again makes a case for the unconscious mind using eureka moments, what he now calls "unconscious scientific creativity." It is here that he pulls out all the stops, so to speak. In an appendix to his chapter 4 he briefly discusses or mentions the cases of a number of poets and composers who had all or parts of their compositions come to them all of a sudden, either while they were awake focusing on something else or while they were daydreaming or waking from a dream, namely, Coleridge, Boux, Blake, Houseman, Mozart, Wagner, Beethoven, Tartini, and McCartney. Among scientists and their discoveries, Smith briefly discusses or mentions the cases of Mendeleev, Calvin, Szilárd, Tesla, Loewi, Parkinson, Howe, Kekulé, and Profet, while in chapter 4 itself he does the same for Edison, Einstein, Darwin, and Wiles and provides an extended discussion on the case of Poincaré as his "paradigm case." In each case, he points out, "the conscious mind is a relatively passive recipient of insight," to which he adds that it is common for scientists to describe the experience in words suggestive of an "epiphany."[15]

In an earlier work, the Jungian psychologist Carl Meier, in a chapter entitled "Creative Effects of the Unconscious," similarly provides and discusses a list of artists, namely, Mozart, Goethe, Schiller, Paul, Hoffmann, Lamb, Stevenson, and Nietzsche, and also scientists, namely, Kekulé, Champollion, Mendeleev, and Mayer, each of them examples of a major *einfall*, the German word for "bright idea," which means "literally *infall* and presumably refers to something which falls into us from above, in the form of a finished product or even a foreign body." In the case of the

15. David Livingstone Smith (2004). *Why We Lie: The Evolutionary Roots of Deception and the Unconscious Mind*. New York: St. Martin's Press, 81, 83–84.

scientists in his list, each provides an example of what he calls a "'bright idea' in the genesis of a great discovery."[16]

Smith's "unconscious scientific creativity" and Meier's "bright idea," of course, are somewhat wider concepts than that of a eureka moment, strictly conceived, since the latter involves a profound emotive awakening, while the former do not necessarily require that experience at all. Nevertheless, they are all of a piece, with eureka moments constituting the highest and most perfect examples of the former. Keeping that in mind, in the following I draw on the lists of Smith and Meier and on some other sources, to which I add a few examples I myself have found, all for the purpose of a unity of effect that vindicates Poe as examined in the previous chapter, and that fully prepares us for the sections that follow.

As we have seen in the previous chapter, Jean-François Champollion (1790–1832) is generally credited with deciphering the Rosetta Stone and thereby unravelling Egyptian hieroglyphics for the world. We have also seen that he was one of Poe's favorite examples of scientific imagination, and there does seem to be strong evidence of a major eureka moment at the foundation of his breakthrough. It was in a rented house in Paris, on September 14, 1822, close to midday, that Champollion had "the decisive insight," as Meier puts it, the beginning of his cracking the code of Egyptian hieroglyphics, which Meier says "came to him . . . at the end of a severe crisis, during which he had to fight a battle inside himself against the claims of other hypotheses." As a biographer put it in 1906, "the *illumination . . . had come to him so suddenly*" that he rushed across the street to his brother's office to tell him the news, exclaiming to him "Je tiens mon affaire! ['I've got it!']," after which "he suddenly collapsed and fell like a dead man to the ground." Champollion remained in a state of lethargy for five full days, rising up again "like a flash of lightning" to continue

16. C.A. Meier (1984). *The Unconscious in Its Empirical Manifestations: With Special Reference to the Association Experiment of C.G. Jung*. Boston: Sigo Press, 9, 18.

his work, but he was still too exhausted.[17] Andrew Robinson, whose biography of Champollion proved very useful in the previous chapter, points out that while the steps in Champollion's decipherment "are hard to discern," given that his correspondence during this period was minimal and he reported only his results in his lectures and publications, nevertheless Robinson tells us that in the rented house in Paris "Champollion was able to devote his every waking minute—and probably his dreams as well—to ancient Egypt," that on the morning of September 14, 1822, he received "a hieroglyphic inscription . . . that launched the second phase of his decipherment," (the discovery that there was "a major phonetic component in the hieroglyphic script,") and that his "Je tiens mon affaire!" uttered later in the day was "his own version of Archimedes's cry 'Eureka!'"[18]

It should hardly need to be said (but unfortunately it needs to be said) that Charles Darwin (1809–1882) was one of the greatest scientists in modern history. What Champollion accomplished for the most fundamental insight in Egyptology, Darwin accomplished for the most fundamental insight in biology, anthropology, and psychology, (and arguably much more,) namely, evolution by natural selection. We are therefore extremely fortunate to have Darwin's autobiography, which furnishes us with two substantial examples of eureka moments.

The first example concerns his core principle of natural selection. As already discussed in Chapter 6, in 1831 Darwin began his voyage on the *Beagle* as a creationist but encountered two major problem situations during that five-year voyage. One concerned his reading of Lyell's *Principles of Geology*, in which Lyell applied natural law gradualism to geology but argued against that approach for biology. The other problem situation concerned Darwin's own observations of anomalous flora and fauna, living and fossil, encountered during his *Beagle* voyage, especially from his exploration of the Galapagos Islands in 1835. Following his return to England, many of Darwin's specimens were classified

17. Meier (note 16), 23, 24.

18. Andrew Robinson (2012). *Cracking the Egyptian Code: The Revolutionary Life of Jean-François Champollion*. Oxford: Oxford University Press, 127–128, 140, 129, 142.

by expert naturalists as belonging not to subspecies or varieties of species already known, but to full-fledged species in their own right. Unique to where Darwin found them, they were also closely related to nearby species, which is the case with species Darwin discovered on the Galapagos Islands and species on the coast of South America. Teetering toward evolutionism, Darwin was now hard-pressed to come up with a scientifically respectable explanatory theory, and that meant a causal law for speciation, the origin of species. Again as we have seen in Chapter 6, Darwin's early notebooks reveal what Gruber calls "pandemonium" in Darwin's early theorizing. The key for the present chapter is what Darwin wrote many years later in his autobiography (1876):

> In October 1838, that is, fifteen months after I had begun my systematic enquiry [Darwin started his first notebook in July 1837], I happened to read for amusement Malthus on *Population*, and being well prepared to appreciate the struggle for existence which everywhere goes on from long-continued observation of the habits of animals and plants, it at once struck me that under these circumstances favourable variations would tend to be preserved, and unfavourable ones to be destroyed. The result of this would be the formation of new species.[19]

The "at once struck me" is the language of a eureka moment. Moreover, it fits perfectly Aristotle's line in *Poetics*, examined in Chapter 2, that "the greatest thing by far is to be a master of metaphor. It is the one thing that cannot be learnt from others; and it is also a sign of genius, since a good metaphor implies an intuitive perception of the similarity in dissimilars." Working enormously hard on a problem that had vexed him for a number of years, one involving enormous amounts of data, Darwin all of a sudden, seemingly intuitively, perceived the deep causal similarity in what were thought at the time to be two highly dissimilar worlds, namely, that species are fundamentally "plastic" and the process of selection that modifies species in domestic plant and

---

19. Nora Barlow, ed. (1958). *The Autobiography of Charles Darwin, 1809–1882*. London: Collins, 120.

animal breeding is basically the same process that modifies species in the wild, the main differences being the time-span and the causal agents involved. For roughly the next twenty years Darwin added an enormous amount of evidence and argument in working out the details of his revolutionary new theory, evolution by natural selection, which he presented in his "Abstract," *On the Origin of Species* (1859).

Darwin provides us with another major example of a eureka moment, even more clearly than the previous one. Natural selection explained the nonbranching evolution of a single species over time, (the environment gradually changed,) as well as the branching evolution of a single species into multiple species involving new geographical locations, new ranges, (like islands,) but it failed to explain the branching evolution of a single species into multiple species *within its own range*. Given the evidence that this kind of evolution also occurred, not just the other two, Darwin needed an additional theory to explain it. The solution was what he called the "principle of Divergence," which he developed in the period from 1854 to 1858 and which in a letter dated June 8, 1858, he called "with 'Natural Selection' . . . the key-stone of my Book."[20] According to Darwin's principle of divergence, now widely accepted in biology, within the range of a species (if the range is relatively wide) a species will typically be exposed to a variety of adaptive conditions, what Darwin called "diversified places in the economy of nature," ("empty niches" to use a more recent and now dated phrase,) which the species will tend to fill (barring extinction) in an adaptive manner, so that it will gradually evolve in a branchlike fashion into closely-related but distinct species. What is remarkable for our purposes is how Darwin first came up with this theory. As he put it in his autobiography, "I can remember the very spot in the road, whilst in my carriage, when to my joy the solution occurred to me; and this was long after I had come to Down [Darwin's home from 1842 to his death almost forty years later]."[21]

20. Frederick Burkhardt and Sydney Smith, eds. (1991). *The Correspondence of Charles Darwin: Volume 7, 1858–1859*. Cambridge: Cambridge University Press, 102.

21. Barlow (note 19), 120–121.

The suddenness combined with the feeling of joy and remembering the exact spot are all key features of eureka moments. But Darwin fails to tell us the form of the theory in which it first came to him. Was it primarily verbal, the way he later explains the theory, or was it primarily visual, something like adjacent circles circumscribed by a larger circle? We are not told, but one thing is for sure. As David Kohn puts it, "He conceived of it as a principle of unification. This explains why he called it 'a keystone of my book.' Unification was not a detached value; it was an internal guide that penetrated and directed the development of his research." Also interesting is Kohn's point that Darwin's principle of divergence had a "much longer gestation" than either his conversion to evolution or his first explanatory theory, natural selection.[22] Whether Kohn means by this the context of the unconscious process leading to the discovery of the principle or the subsequent "calculation, revision, and exposition" of the principle, the latter being Darwin's context of justification,[23] is not clear but is beside the point. What is remarkable is the amount of data and theory leading up to the eureka moment along with Darwin's observation that even by the summer of 1844, at which time he had written a long essay on his theory of evolution, (to be published in the event of his premature death,) he had "overlooked one problem of great importance," such that "it is astonishing to me . . . how I could have overlooked it and its solution."[24] Much depends, of course, on what Darwin means here by "I," a riddle I shall attempt to unriddle at the end of the neuroscience section of this chapter.

The so-called co-discoverer of evolution by natural selection, Alfred Russel Wallace (1823–1913), provides us with another case in point. Wallace had developed a theory of evolution that looked very similar to Darwin's, although the units of selection were described by him as "varieties" rather than Darwin's individuals with heritable variations in a population or species, which makes it not entirely similar to Darwin's theory. It also lacked much of

22. David Kohn (1985). "Darwin's Principle of Divergence as Internal Dialogue." In David Kohn, ed. (1985). *The Darwinian Heritage*. Princeton, NJ: Princeton University Press, 247.

23. Kohn (note 22), 250.

24. Barlow (note 19), 120.

the kinds of evidence that Darwin brought forth. At any rate, in 1858 Wallace sent Darwin a copy of his essay on his theory hoping for critical feedback from the already famous naturalist and from the even more famous Charles Lyell before he submitted it for publication, unaware that Darwin himself was an evolutionist, already for the past twenty years, let alone that he was in the midst of writing a very large and detailed book for the purpose of making a truly scientific case for evolution and presenting it to the world. The letter, quite understandably, shocked Darwin to the core, causing him to think that he was now "forestalled" and that his originality "will be smashed."[25] Following, at the suggestion of Lyell, a joint presentation of Darwin's and Wallace's work to the Linnean Society less than a month later, and given that his "big species book" was still far from finished, Darwin followed the suggestion of his friend J.D. Hooker and began writing an "Abstract" of his theory, which he completed in slightly less than eight months, the now classic 490-page *On the Origin of Species*. So how did Wallace arrive at his own version of the theory?

Wallace made his living by hunting and selling specimens, which involved a lot of travel to exotic locations. But in his heart he was a theorist. Converting to evolutionism in 1845 shortly after reading Chambers' *Vestiges*, Wallace, like Darwin, was not satisfied with the evolutionary theories of Lamarck or Chambers. He read much, studied much, and combined with his own observational data he felt in his bones that the "law of nature" that explained the origin of species had not yet been found. In 1855 he wrote a paper proposing a "law" on the appearance of new species in geological time, which was published later the same year. Lyell and another naturalist before him had called Darwin's attention to it, Darwin read it or had already read it, and many modern scholars argue that Darwin should have seen the latent evolutionism in that paper.[26] It was not until February 1858, however, on the island

25. Burkhardt and Smith (note 20), 107.

26. Those scholars are wrong and need to look at the matter from Darwin's eyes, not modern eyes. Darwin dismissed the paper as having "nothing very new" almost certainly because internal evidence, the very content of the paper, strongly suggested to Darwin that Wallace's theory was just another version of successive creationism. See my (2007). *Darwin and the Nature of Species*. Albany: State University of New York Press, 185, 243n17.

of Gilolo in the Malay Archipelago, that the idea of evolution by natural selection (the concept, not the phrase) came to Wallace. In his autobiography, written almost fifty years later, in 1905, he puts it thus:

> At the time in question I was suffering from a sharp attack of intermittent fever, and every day during the cold and succeeding hot fits had to lie down for several hours, during which time I had nothing to do but to think over any subjects then particularly interesting to me. One day something brought to my recollection Malthus's "Principles of Population," which I had read about twelve years before. . . . It then occurred to me that these causes or their equivalents are continually acting in the case of animals also; . . . Vaguely thinking over the enormous and constant destruction which this implied, it occurred to me to ask the question, Why do some die and some live? And the answer was clearly, that on the whole the best fitted live. . . . Then it suddenly flashed upon me that this self-acting process would necessarily *improve the race*, because in every generation the inferior would inevitably be killed off and the superior would remain—that is, *the fittest would survive*. . . . The more I thought over it the more I became convinced that I had at length found the long-sought-for law of nature that solved the problem of the origin of species.[27]

That evening, feeling somewhat better, Wallace wrote up "some notes for a paper on the subject," which he then, over the succeeding two evenings, turned into the paper "On the Tendency of Varieties to Depart Indefinitely from the Original Type," which he then, within a day or two, promptly mailed to Darwin from the neighboring island of Ternate. The words in his autobiography that are most important for our purposes are "had to lie down for several hours, during which time I had nothing to do but to think," "something brought," and "it suddenly flashed upon me," words referring to one of the most striking cases of unconscious

27. Burkhardt and Smith (note 20), 512.

scientific creativity in modern science, given that Wallace was stricken with what was probably a malarial fever.[28]

Julius Robert von Mayer (1814–1878) was a medical doctor who became interested in physics and is responsible for the earliest published expression of the first law of thermodynamics, the conservation of energy, the principle that energy is neither created nor destroyed. The question of priority was hotly debated in the years following the establishment of the principle in science, in favor of one or another physicist properly so called, most notably James Joule or Hermann von Helmholtz. The question of priority doesn't really matter for our purposes, although it should be noted that the prominent English physicist, John Tyndall, gave full credit to Mayer in a monograph, *Heat Considered as a Mode of Motion*, published in 1863. What matters, mainly, is how Mayer arrived at his discovery.

Shortly after completing medical school in Germany, Mayer was bit by the travel bug and secured the position of doctor aboard a Dutch ship headed for the East Indies. The trip lasted from February 1840 to February 1841. Struck by the bright color of venous blood of a crewman while they were off the coast of Java, an observation that began around June 20, this initial problem situation started the domino effect that led to Mayer's eventual eureka moment, (collectively considered,) which he wrote about in a letter to a friend, dated June 16, 1844:

> This theory was by no means hatched out at my writing desk. I had occupied myself, eagerly and continuously, on my journey to the East Indies, with the physiology of the blood, and my observation of the changed physical condition of our crew in the tropics, and of the process of their acclimatization, provided me with a great deal of food for thought; the types of illness from which they suffered, and particularly the state of their blood, drew my attention over and over again to the generation of animal heat in the body as a result of the process of respiration.

28. In accordance with the principle of charity, "There is no reason not to accept Wallace's account in *My Life* as his genuine recollection of the events . . . ." Martin Fichman (2004). *An Elusive Victorian: The Evolution of Alfred Russel Wallace*. Chicago: University of Chicago Press, 98.

> But we cannot gain an insight into physiological matters if we know nothing of the corresponding physical processes— . . . So I concentrated on physics and pursued this subject with such enthusiasm that many people might be inclined to laugh at me, since I asked very few questions about the strange part of the world in which I found myself, but preferred to remain on board, where I could work uninterruptedly, and where I quite often felt like a man inspired—in fact I cannot remember having experienced anything similar, either before or since that time. I had a number of *flashes of insight*—this happened on the roadstead at Surabaya—which I followed up assiduously, without delay, . . . the calm testing of what I experienced at that period has convinced me that it was truth—not merely something which was felt subjectively, but something which can also be objectively substantiated— . . . What is certain is that the day will come when these truths will be the common property of science; by whom, and when, this development will be brought about, who can tell?[29]

Upon arriving back in Germany, Mayer immediately set to work on writing a scientific paper on his discovery, which he considered the sole result of his journey. The paper was submitted to a physics journal in 1841 but was rejected for publication, mainly because it contained a number of errors and lacked the formal math and technical language of classical physics. With the help of his friend Carl Baur, a professor of mathematics, Mayer refined his ideas in a second paper, which indeed got published, in May 1842, with the fundamental equation of work and heat, and it even contained the words, (in German,) "Energy is neither created nor destroyed." After a few years of experts hashing over the matter, however, Mayer was overlooked in the story of the discovery of the first law of thermodynamics, and was even slandered. Angry, Mayer attempted suicide in 1850 by jumping out of a third-floor

29. Meier (note 16), 31. See also Caneva, who interestingly characterizes "the entire course of Mayer's theorizing" as "a search for valid analogies." Kenneth L. Caneva (1993). *Robert Mayer and the Conservation of Energy*. Princeton, NJ: Princeton University Press, 46.

window, which fortunately he survived, but he was in and out of asylums for a number of years afterward.

What is especially interesting about this case is that Mayer suffered a lifetime of symptoms of what Meier, in his clinical experience, calls "manic depressive insanity," what today is called *bipolar disorder*, and he suggests that it actually may have contributed, in a positive way, to Mayer's flashes of insight. Meier notes that "on two occasions during the voyage to the East Indies Mayer fell a victim to inexplicable bouts of depression. We should express this in our terminology by saying that during these periods of relative introversion, libido (i.e., psychic energy) flowed backwards, in such a way that the subject's interests now concentrated on certain contents belonging to his inner relativity and turned away from the outside world; this process is also known as regression." This was not necessarily the context of discovery, however, but apparently it prepared the way. As Meier continues, "Mayer's depressive fits died away and gave place—particularly on his arrival to Java—to a manic mood disorder. Mayer was now possessed by 'the most blissful feelings'—and suddenly we find him working like mad at scientific problems."[30] The words "like mad" would have brought a grin to Poe's face, given all that Poe wrote on madness, much of which we examined in the previous chapter, including the "monomania" of Egæus and the "perverse moods of alternate enthusiasm and melancholy" of Legrand.

Friedrich August Kekulé (1829–1896) was an engineering student who turned to chemistry, where he quickly became one of the great pioneers of what today is known as *structural chemistry*, or more specifically *molecular geometry*, chemistry focused on the arrangement of atoms in molecules. Kekulé provides us with two major examples of eureka moments, and from a time when many chemists and physicists still did not believe in the physical reality of atoms. Both of his examples come from the 1850s. The first one concerns the atomic structure of methane. It had already been surmised, based solely on its weight and chemical properties, that methane consists of one carbon atom and four hydrogen atoms. But in what arrangement? Different arrangements would result in different chemical properties, so one had to get it right.

30. Meier (note 16), 28, 32.

Kekulé was already familiar with the chemical properties of marsh gas, but the configurations of the five atoms repeatedly gave him the wrong results. Then one night, while sitting on a bench atop a horse-drawn omnibus on his way home, Kekulé had a waking dream, a vision, which he later related to his first research assistant, the great-grandfather of the professor of physics who I am here using as my source. As von Baeyer puts it, Kekulé saw atoms that "sprouted arms, whirled about, and seemed to reach out to each other." When the dream ended he immediately knew he had the atomic structure of methane: "a carbon atom, from which four arms grow in the form of a cross, with a hydrogen atom at the end of each," to use von Baeyer's words again.[31] The answer was basically correct, lacking only the three-dimensional structure established by more recent chemistry.

The second example involves the atomic structure of benzene. This was a far more difficult molecule to figure out, but the dream that came to Kekulé this time might have been influenced by a memory that went back to his days as an architecture student, the memory of a gold ring that belonged to his murdered neighbor, about which his testimony helped convict the person responsible. The ring depicted two intertwined snakes biting their own tails, a symbol (with either one or two snakes) known as the *ouroboros*, (from the ancient Greek meaning "tail-devouring snake,") made much of by alchemists (and later by Jung) and which goes back (at least) to the ancient Egyptians. "One night," wrote Kekulé, "I turned my chair to the fire and sank into a doze." Atoms once again danced before his eyes, this time in combinations twisting in a snakelike fashion. All of a sudden, "one of the serpents caught its own tail and the ring thus formed whirled exasperatingly before my eyes. I awoke as by lightning, and spent the rest of the night working out the logical consequences of the hypothesis." His hypothesis was a hexagonal ring of six carbon atoms, each with a hydrogen atom attached to it. This was the first ring-shaped molecule hypothesized in chemistry, which began a whole new branch of structural chemistry, and once again Kekulé got it basically right, what came to be an established fact, even long

31. Hans Christian von Baeyer (1992). *Taming the Atom: The Emergence of the Visible Microworld*. Mineola, NY: Dover, 60.

before the structure of benzene was confirmed by a microscopic image (really a computer reconstruction) in 1988. As von Baeyer puts it, "the image inadvertently demonstrated the power of the human mind, unaided by modern machinery, to synthesize vast amounts of complex chemical information and to render the result in visual terms."[32] The point about visual terms needs to be noted for later, when we turn to neuroscience, but for the present we should notice that it was Kekulé's *unconscious* mind that did the most impressive work—the "creative" rather than the "resolvent," to use Poe's terms for the "double Dupin."[33] The creative work certainly does not appear to be a "happy guess," to use the phrase we have seen in Chapter 6. Instead, Kekulé's unconscious mind appears to have made use of the image of the neighbor's ring, which he had seen roughly fifteen years earlier, the ring embodying the essential insight to the solution of the problem as worked out by his unconscious mind, in a manner not unlike Darwin's sudden impulse to pick up and read "for amusement" Malthus' *Essay*.

Dmitri Ivanovich Mendeleev (1834–1907) was a chemist best known for devising a table of chemical elements, the sixty-one elements known at that time, into what is now called the Periodic Table, which he published in his *Principles of Chemistry* (1869), the definitive chemistry textbook of its time. Although Mendeleev's Table has undergone modification since then, it has basically withstood the test of time, as the modern Periodic Table continues to be connected with his name given that he "essentially laid it out in its current form."[34] There were others, of course, working on a table of chemical elements around the same time. The problem was to produce a table that was not artificial but natural, one that "carved nature at its joints," to use a phrase that goes back to Plato on the topic of natural kinds, which in the case of the chemical elements meant a system that followed valid chemical principles and that allowed for the prediction of new elements, elements

32. von Baeyer (note 31), 61, 61–62, 62–63.

33. See Meier (note 16), 23, where, in a case of coincidence, cryptomnesia, or maybe even synchronicity, the words "creative" and "resolved" are used in relation to the Kekulé case.

34. P.W. Atkins (1995). *The Periodic Kingdom: A Journey into the Land of the Chemical Elements*. New York: BasicBooks, 52.

not yet known to exist. Many chemical elements, of course, have been known since antiquity, elements that had been isolated by using fire, such as copper, iron, and gold. Since the early 1800s, many more elements were discovered and isolated by *electrolysis*, running an electric current through a compound. Today chemical elements are determined by their *atomic number*, which is the number of protons in the nucleus of the atom. Gold, for example, is number 79, one up from platinum and one down from mercury. But in Mendeleev's time neither the concept of atomic number nor the technology to determine atomic numbers existed. So how could a table of chemical elements be made? The answer is that the chemical element cartographers looked at qualitative information, such as comparing weights of large quantities of an element and by observing *valence*, meaning which chemical elements could combine with which to form compounds.

The main source for Mendeleev's discovery is from the correspondence and conversations Mendeleev had years later with a much younger friend, the chemist Paul Walden. According to Walden, in preparation for *Principles of Chemistry* Mendeleev wrote down the chemical elements along with their atomic weights and observable relationships onto cards, each element getting a separate card. He then arranged the cards in various orders and groups, trying to find the most natural arrangement, based primarily but not exclusively on atomic weight. On March 1, 1869, Mendeleev made his remarkable discovery. As Walden puts it in a 1930 publication, "The process of selecting the most appropriate arrangement involved trying out a large number of alternatives; afterwards, Mendeleev described how one evening he experimented once again with different methods of grouping his element cards, and then later, when he was asleep and unconsciously still working on his problem, he had suddenly caught sight of the right system, which he wrote down as he woke up. The exhaustive testing of this system from every possible point of view in chemistry and in physics was a task which absorbed him till the day of his death."[35] What is equally remarkable is that Mendeleev avoided the temptation to completeness by putting gaps in his Table, empty squares for where he thought there should be elements not yet

35. Quoted in Meier (note 16), 26.

discovered. In fact he not only corrected some of the published atomic weights but he predicted the existence and properties of no less than three elements, which were later discovered in reality by others, namely, gallium in 1875, scandium in 1879, and germanium in 1886.

Jules Henri Poincaré (1854–1912) was a mathematician, theoretical physicist, engineer, and philosopher of science best known for laying the foundation for chaos theory and for coming close to anticipating Einstein on special relativity and on mass–energy equivalence. But perhaps he should best be known for his contribution to the topic of eureka moments. In a paper entitled "Mathematical Creation," first published in French in 1908, Poincaré recalls a series of his own eureka moments and provides reflections on the matter. His main example concerns hypothetical mathematical equations in geometry which he calls "Fuchsian functions." He tells us that for fifteen days he strove to prove that they could not exist. Then, he says, "One evening, contrary to my custom, I drank black coffee and could not sleep. Ideas arose in crowds; I felt them collide until pairs interlocked, so to speak, making a stable combination. By the next morning I had established the existence of a class of Fuchsian functions," which he then wrote out in a few hours. Continuing to work on the topic, he then went on a "geological excursion." At one point he "entered an omnibus to go some place or other. At the moment when I put my foot on the step the idea came to me without anything in my former thoughts seeming to have paved the way for it, that the transformations I had used to define the Fuchsian functions were identical with those of non-Euclidean geometry." Although at the time Poincaré did not have the opportunity to "verify the idea," he nevertheless "felt a perfect certainty," with the verification following his return home. After this, dealing with what he thought was an unrelated set of mathematic questions, Poincaré was once again experiencing a period of failure in spite of his best efforts. "Disgusted with my failure," he says, "I went to spend a few days at the seaside, and thought of something else. One morning, walking on the bluff, the idea came to me, with just the same characteristics of brevity, suddenness, and immediate certainty, that the arithmetic transformations of indeterminate ternary quadratic forms were identical with those of non-Euclidean geometry." Arriving home he "deduced the consequences." There was still one problem, however, the solution of which evaded him. Now engaged in military service, he was unable to provide his conscious attention to the problem. "One day," however, "going along the street, the solution of the difficulty which had stopped me, suddenly appeared to me." When home from military service he then wrote out the solution "at a single stroke and without difficulty." And all of this he calls "this single example."[36]

Poincaré tells us that he personally has many more examples, and that in a study on mathematicians published just before he wrote his paper "the majority of witnesses confirm my conclusions." We should also take a brief look at Poincaré's own hypothesis on how mathematical creativity works. First, "this appearance of sudden illumination," he says, is "a manifest sign of long, unconscious prior work." Second, the sudden illumination, which is accompanied by "the feeling of absolute certitude," is "on the one hand preceded and on the other hand followed by a period of conscious work." The prior efforts, he says, are "not as sterile as one thinks; they have set agoing the unconscious machine, and without them it would not have moved and would have produced nothing." Third, "the subliminal self is in no way inferior to the conscious self." It is "not purely automatic," he says, "it is capable of discernment; it has tact, delicacy; it knows how to choose, to divine." In fact, he says, "It knows better how to divine than the conscious self, since it succeeds where that has failed." The question now becomes, the question *par excellence*, "What is the cause that, among the thousand products of our unconscious activity, some are called to pass the threshold, while others remain below?" Fourth is his answer, which in short is "the feeling of mathematical beauty, of the harmony of numbers and forms, of geometric elegance. This is a true esthetic feeling that all real mathematicians know." "The useful combinations," he continues, "are precisely the most beautiful, I mean those best able to charm this special sensibility that all mathematicians know," and "it is this special esthetic sensibility, which plays the role of delicate sieve of which I spoke, and that sufficiently explains why the one lacking it will

36. Henri Poincaré (1910). "Mathematical Creation." George Bruce Halsted, trans. *The Monist* 20 (3), 326, 327, 328.

never be a real creator." He then writes of "the working methods of the two egos."[37] At this point the true lover of Poe will recall, quite naturally, not only Poe's "double Dupin," but also what we have seen Poe say against the methods of deduction and induction in favor of soaring, along with the epistemic virtues of beauty and consistency.[38]

Nikola Tesla (1856–1943), the archetype of the mad scientist, was an electrical and mechanical engineer and inventor known for his invention in 1882 of the AC (alternating current) induction motor, (an electric motor that uses a rotating magnet,) his invention of the AC generator, patented in 1891, which provides our light and electricity, (and which was strongly opposed by his

37. Poincaré (note 36), 321, 328, 329, 330, 331, 332, 335. In case one thinks Poincaré's experience is extremely rare in mathematics, there is also the testimony of the mathematicians collected in the journal *L'enseignment Mathématique*, mentioned by Poincaré at the start of his paper. Another example is Srinivasa Ramanujan (1887–1920), one of India's greatest mathematical geniuses, who claimed to receive his mathematical insights from the Hindu goddess Namagiri during his dreams, with scrolls unrolling before his eyes. Smith (note 15), 82. A much more recent case is that of Andrew John Wiles (1953–), who finally solved Fermat's Last Theorem in 1994. This is what he said of the experience: "suddenly, totally unexpectedly, I had this incredible revelation." Quoted in Smith (note 15), 84. Conjectured in 1637 by Pierre de Fermat, the solution to the theorem discovered by Wiles came after 358 years of failed efforts by mathematicians and in turn stimulated the proof of the modularity theorem.

38. Poincaré's paper, by the way, seems a direct response to Poe in "Rue Morgue" and "Purloined Letter." Not only does Poincaré's paper seem an attempt to support Poe on eureka moments and intuition, but also, seemingly referring to Poe's preface in "Rue Morgue" all but in name, Poincaré uses the examples of "the whist-player" and "the chess-player" to argue that mathematical reasoning is much more than merely having a good memory and applying rules, that in discovering a "mathematical law" the mathematician, like the discoverer of a law in physics, discovers an "unsuspected kinship between other facts, long known, but wrongly believed to be strangers to one another." It will be recalled from the previous chapter that both Dupin and Minister D— are *both* mathematicians and poets. It will also be recalled that Aristotle in *Poetics* states that "to be a master of metaphor" is "a sign of genius, since a good metaphor implies an intuitive perception of the similarity in dissimilars." What a beautiful paper it would be, "Poe et Poincaré," a comparison of the two on our topic, especially if one could find that Poincaré had in fact read Poe. But life is too short to accomplish all that one wants, so one must leave certain goals to others, hoping merely to have pointed the way.

former boss, Thomas Edison, who was heavily invested in DC,) his invention in 1897 of wireless energy transmission and communication, (the latter in his vision making the "whole earth . . . like a brain,") his accidental discovery of X-ray photography, (using Mark Twain as the subject,) his invention of a particle beam weapon, (verified in 1984,) and much else. In his autobiography, entitled *My Inventions* (1919), of particular interest is Tesla's discussion of his invention of the AC induction motor. Following a "complete breakdown of the nerves," a period in which he had fallen into a deep depression, (to which he was prone,) this time apparently from difficulty at trying to make a living, Tesla's health was gradually restored with the help of a friend, who got him into physical exercise. "Back in the deep recesses of the brain was the solution," he writes, about trying to invent a new motor, "but I could not yet give it outward expression." As part of his rehabilitation, he and his friend would go for a walk every evening. "At that age," he says, (he was twenty-six,) "I knew entire books by heart, word for word." And then "One afternoon," during one of these walks and while "the sun was just setting," he was "reminded of the glorious passage" from Goethe's *Faust*. The passage, quite interestingly, includes the lines, prophetic in Tesla's case, "done is the day of toil," "new fields of life exploring," "A glorious dream!," and "wings that lift the mind." "As I uttered these inspired words," he then states, "the idea came to me like a flash of lightning and in an instant the truth was revealed. . . . The images were wonderfully sharp and clear and had the solidity of metal and stone, . . . I cannot begin to describe my emotions." As Bernard Carlson notes, it apparently took Tesla some years to work out all the details, involving experiments he subsequently undertook in 1883 and 1887; nevertheless he does not doubt that Tesla's "major breakthrough" in 1882 "consisted of three related insights."[39]

Albert Einstein (1879–1955), of course, was possibly the greatest scientist of all time. At the end of Chapter 6 we discussed his main theories, his use of imagination in science, and what he had

39. W. Bernard Carlson (2013). *Tesla: Inventor of the Electrical Age*. Princeton, NJ: Princeton University Press, 51, 52. The quotations I use from Tesla's *My Inventions* are included in this book on the pages cited.

to say about the role of imagination in science, which includes the words "imagination is more important than knowledge," "a sudden illumination, almost a rapture," and "a great forward leap of the imagination." It will also be recalled that the special theory of relativity was apparently thirty-three years ahead of its time and the general theory of relativity roughly fifty years ahead of its time. One can quibble about the numbers, but the point remains that Einstein was remarkably ahead of his time. What needs to be noticed at this point is that each of those theories was produced by, or in some way directed by, a eureka moment. With regard to special relativity, Einstein claimed that "After seven years of reflection in vain [1898–1905], the solution came to me suddenly with the thought that our concepts and laws of space and time can only claim validity insofar as they stand in a clear relation to our experiences; and that experience could very well lead to the alteration of these concepts and laws. By a revision of the concept of simultaneity into a more malleable form, I thus arrived at the special theory of relativity."[40]

Even more striking is how Einstein arrived at the general theory of relativity, completed in 1915 and published in 1916. Referring to a day probably in November 1907, Einstein wrote, "I was sitting in a chair in the patent office at Bern when all of a sudden a thought occurred to me: 'If a person falls freely he will not feel his own weight.' I was startled. This simple thought made a deep impression on me. It impelled me toward a theory of gravitation." He would later refer to this sudden thought as "the happiest thought of my life."[41]

Our next example is put together here as a pair of examples, because they both have to do with the development of the atomic bomb. The first in the pair comes from the physicist and inventor Leó Szilárd (1898–1964), a student of Max Planck and Einstein, the former the founder of quantum physics. Throughout the early 1930s the physicist Ernest Rutherford repeatedly referred to the prediction of harnessing atomic energy as "talking moonshine."

40. Quoted in John Stachel, ed. (1998). *Einstein's Miraculous Year: Five Papers That Changed the Face of Physics*. Princeton, NJ: Princeton University Press, 112.

41. Quoted in Abraham Pais (1982). *"Subtle is the Lord . . .": The Science and the Life of Albert Einstein*. Oxford: Oxford University Press, 179, 178.

In 1933 Szilárd read an article in the *New York Times* summarizing Rutherford's view, and it immensely annoyed him. Neutrons had only recently been discovered, and they got Szilárd thinking. Then, in September 1933, while he was standing at a corner in London waiting for the light to change, something interesting happened. "As the light changed to green," he says, "it suddenly occurred to me that if we find an element . . . which would emit two neutrons when it absorbed *one* neutron [this] could sustain a nuclear chain reaction."[42] The following year he filed a patent for the first neutron-induced atomic chain reaction. In 1938 he obtained a position at Columbia University in Manhattan, where shortly afterward he was joined by the physicist Enrico Fermi. They concluded that uranium would be the best element for sustaining a nuclear chain reaction and they conducted the first experiments which led to the Manhattan Project, the development of the first atomic bomb.

Meanwhile we need to backtrack a little and focus now on Enrico Fermi (1901–1954), a student of Max Born, one of the pioneers of quantum physics. In Italy in October of 1934 he and his junior collaborators had been working on "neutron-induced radioactivity." The idea was to see what would happen when an atomic nucleus of a radioactive element was blasted with a beam of neutrons. At first, the results "made no sense," as he put it. What was needed was an intervening "filter" to slow down the approach of the neutrons. Then came the discovery, and here is how Fermi described it:

> One day, as I came to the laboratory, it occurred to me that I should examine the effect of placing a piece of lead before the incident neutrons. And instead of my usual custom, I took great pains to have the piece of lead precisely machined. I was clearly dissatisfied with something: I tried every "excuse" to postpone putting the piece of lead in its place. When finally, with some reluctance, I was going to put it in its place, I said to myself: "No, I don't want this piece of lead here; what I want is a piece of paraffin." It was just like that: with no advanced warning, no

42. Quoted in Smith (note 15), 201.

> conscious, prior, reasoning. I immediately took some odd piece of paraffin . . . and placed it where the piece of lead was to have been.[43]

The result was "miraculous," as one as his collaborators put it, with Fermi figuring out what had happened in the next few hours.

And the rest is history, as the saying goes,—with the experiment (I would say the eureka moment) constituting "the effective beginning of the nuclear age," as Holton puts it,—with Fermi receiving the Nobel Prize in 1938 for this work,—with Einstein sending a letter to President Roosevelt in August 1939 informing him of "Some recent work by E. Fermi and L. Szilard" using "the element uranium," of "a new and important source of energy in the immediate future," and of "extremely powerful bombs of a new type,"—with Szilárd and Fermi going to Chicago as part of the Manhattan Project where they personally helped build the world's first nuclear reactor known as Chicago Pile-1, in which the first manmade nuclear chain reaction was initiated in December 1942 under the west stands of an abandoned football field at the university,—with Szilárd and Fermi filing a U.S. patent for the "neutronic reactor" in December 1944,—and so on until the consummation of Fate, the epoch of fiery overthrow,—in defiance of principles which should have taught our race to submit to the guidance of the natural laws, rather than attempt their control,—a combustion irresistible, all-devouring, omni-prevalent, immediate;—the Earth burst at once into a species of intense flame, for whose surpassing brilliancy and all-fervid heat even the angels in the high Heaven of pure knowledge have no name. Thus ended all, with Peace On Earth and Purity Of Essence.

As Royston Roberts points out, "Fortunately it [the making of the atomic bomb] was done in the United States before it could be done in Germany."[44] What is particular striking about the Szilárd and Fermi example is that it does seem analogous to what I was talking about at the end of Chapter 6, that either a single mutation

---

43. Quoted in Holton (note 4), 155–156.

44. Royston M. Roberts (1989). *Serendipity: Accidental Discoveries in Science*. New York: Wiley, 149.

in a population, known as a *key mutation*, or a particular sequence of mutations in a population, known as *mutational order*, can alter the course of evolutionary history in that population. In the case of the development of the first atomic bomb, had the key insight and related insights of the critical mass of a uranium-235 atomic bomb first occurred to one or more of the Nazi physicists, in particular to Werner Heisenberg, who continuously thought it much too high to be practical during the war, the mushrooms might have popped up in very different places, rather than in Japan, altering the future of both political and scientific history. Instead, the key insight first occurred to the physicist Otto Frisch in February 1940, after he had left Germany for England once Hitler came to power, and then again in collaboration with Rudolph Peierls in England, who together got the critical mass much too low but at least thought it was practical, thus jump-starting the British atomic bomb project, which led to the Manhattan Project, with its *dénouement* in Hiroshima and Nagasaki.[45]

45. Heisenberg, awarded the Nobel Prize in Physics in 1933 for his contribution to quantum physics, was the leading scientist of the Nazi atomic bomb project, which extended from 1939 to 1945. The "true reason" for why Nazi scientists did not develop an atomic bomb, argues Paul Rose, was not because they rejected "Jewish physics," (Heisenberg, for example, actually defended it,) or because they lacked the manpower and resources, (so-called "external circumstances,") but because of a set of false assumptions, most notably Heisenberg's continued belief (common in 1939) that the critical mass required for a uranium-235 atomic bomb was several metric tons, more like 10, (not 56 kilograms, the right answer,) an amount that looked like it would take several decades to procure, along with the arrogant belief that German science was superior to that of other countries. To the very day of August 6, 1945, the day the bomb was dropped on Hiroshima, Nazi physicists, including Heisenberg, one of the ten scientists interned at Farm Hall in England in a state of uncertainty, (pickled egg on their faces superimposed with disbelief of the news,) still did not know how to make the bomb. Heisenberg, however, pretty much figured it out, quite remarkably, between the fourth and before the eighth day following. Consequently, "Had Heisenberg been more clearheaded and more dynamic, Speer [Albert Speer, Hitler's Minister of Armaments] would most likely have been very willing to allocate much greater resources to the enterprise and—who knows?—with German technical ingenuity, some breakthrough might well have been achieved with isotope separation or plutonium production." Paul Lawrence Rose (1998). *Heisenberg and the Nazi Atomic*

Melvin Calvin (1911–1997) was a biochemist best known for his discovery of the chemical pathways of photosynthesis, for which he received the Nobel Prize in Chemistry. Troubled by his preconception of how photosynthesis worked, he wrote,

> One day I was sitting in the car while my wife was on an errand. For some months I had some basic information from the laboratory that was incompatible with everything that, until then, I knew about the cycle. I was waiting, sitting at the wheel of the car, probably parked in the red zone, when the recognition of the missing compound occurred. It occurred just like that—quite suddenly—and suddenly, also, in a matter of seconds, the cyclic character of the path of carbon became apparent to me. But the original recognition . . . occurred within a matter of thirty seconds. So, there is such a thing as inspiration, I suppose, but one has to be ready for it.[46]

Freeman Dyson (1923–) is a theoretical physicist best known for demonstrating the equivalence of two formulations of *quantum electrodynamics*, the application of both Einstein's special theory of relativity and quantum physics to show how light and electrons interact. The problem of developing a workable theory of quantum electrodynamics was a hot problem after WWII, occupying the minds of a great many physicists, with no one developing a theory that the physics community could use to make calculations. One of the main contenders was by Richard Feynman, a former teacher of Dyson's, while another was by Julian Schwinger. The problem was that Feynman's formulations involved great leaps of intuition and diagrams that no one but Feynman could make sense of, while Schwinger's formulations were highly methodical but he was not satisfied that they worked. Dyson was in the "happy position of being familiar with both of them," both of the theories, but he would soon be in the even happier position of unifying them.

Bomb Project: A Study in German Culture. Berkeley: University of California Press, 77, 46.

46. Quoted in Smith (note 19), 201.

In an interview recorded years later, he describes how he arrived at the solution:

> It was the summer of 1948, so I was then twenty-four. . . . I spent six months working very hard to understand both of them clearly, and that meant simply hard, hard work of calculating. . . . At the end of six months, I went off on a vacation. . . . After two weeks in California, where I wasn't doing any work, I was just sight-seeing, I got on the bus to come back to Princeton, and suddenly in the middle of the night when we were going through Kansas, the whole sort of suddenly became crystal clear, and so that was sort of the big revelation for me, it was the Eureka experience or whatever you call it. Suddenly the whole picture became clear, and Schwinger fit into it beautifully and Feynman fit into it beautifully and the result was a theory that actually was useful. That was the big creative moment of my life. Then I had to spend another six months working out the details and writing it all up and so forth.[47]

Dyson's solution was published in two papers the following year and to the satisfaction of the physics community, so that they now had a theory of quantum electrodynamics that they could actually use and that removed the "logjam" to further developments. But only Feynman and Schwinger went on to receive Noble Prizes for their work—one of life's many injustices, *or possibly*, since it raises the problem of authorship, of personal identity, a problem that all of the eureka moments in this section raise, and which we shall attempt to solve in the neuroscience section below.

Our next example is actually a set that comes from two more Nobel Prize laureates, namely, Francis Crick (1916–2004) and James Watson (1928–), the co-discoverers of the structure of DNA, the hereditary molecule, the discovery in 1953 of "the secret of life" as the former put it. In Watson's book on the discovery, one can find him express, at least twice, what seem to be eureka

47. Quoted in Mihaly Csikszentmihalyi (1996). *Creativity: Flow and the Psychology of Discovery and Invention*. New York, HarperCollins, 82.

moments, with his "Suddenly I realized" and "Suddenly I became aware that."[48] Crick is much better, however. In his own book on scientific discovery, he tells us that Linus Pauling "has described how he first hit on the α helix while confined to bed with a cold during his stay in Oxford in 1948 as a visiting professor."[49] Pauling (1901–1994), of course, went on to be the recipient of two Nobel Prizes and is often called the greatest chemist of the twentieth century. Crick, in turn, describes what happened to himself on Good Friday, 1960. He and his colleagues had been working hard on trying to solve the nature of protein synthesis. The right answer, the hypothesis of messenger RNA, came to him as a eureka moment. Here is how he puts it:

> It is difficult to convey two things. One is the sudden flash of enlightenment when the idea was first glimpsed. It was so memorable that I can recall just where Sydney, François, and I were sitting in the room when it happened. The other is the way it cleared away so many difficulties. . . . I woke up that morning with only a set of confused ideas about the overall control of protein synthesis. When I went to bed all our difficulties had resolved and the shining answers stood clearly before us. Of course, it would take months and years of work to establish these new ideas, but we no longer felt lost in the jungle. We could survey the open plain and clearly see the mountains in the distance.

A little later he repeats himself, saying, "One immediately sees how many previously puzzling facts are neatly explained by the new hypothesis. One could kick oneself for not having the idea earlier, it now seems so obvious. Yet before, everything was in a fog."[50]

Our final example is also a set, this time from Robert Trivers (1943–), one of the main contributors to the revolutionary field of sociobiology, the application of evolutionary theory to social

48. James D. Watson (1968). *The Double Helix: A Personal Account of the Discovery of the Structure of DNA*. New York: Atheneum Press. Reprinted in Gunther S. Stent, ed. (1980). *The Double Helix: Text, Commentary, Reviews, Original Papers*. New York: W.W. Norton, 106, 114.

49. Crick (note 6), 58.

50. Crick (note 6), 120, 141.

behavior.[51] What is especially interesting about both of his examples is that, although they belong to a major contribution to a nascent field that was revolutionary at the time, they illustrate the fact that major examples of unconscious scientific creativity do not have to take a long time, but can occur in a matter of weeks or even days. Trivers was at Harvard finishing his Ph.D. and was working on the problem of applying the mathematics of natural selection, developed by R.A. Fisher and then later by W.D. Hamilton, to sex ratios in social insects (ants, bees, wasps, termites). One evening, he says, he received unpublished field data for two ant species from a biologist he had corresponded with, the one species the slave maker of the other. The tables of sex ratios for the two species didn't fit what he expected in theory. He then thought to himself, "I wonder why that might be?" "Then I went off to bed," he writes. "I remember waking early the next morning, six or six-thirty—a full two hours ahead of schedule—and almost literally jumping out of bed, saying 'But that's exactly what you would expect!' I then spent a few moments clearly formulating the logic. . . ."[52] Rather than falsify his pre-existing theory, the data required a revision of the theory, which was later confirmed by another set of data. This is often the way science actually works, but publications usually do not show this, let alone the contribution of the unconscious mind, unless one turns to autobiographical accounts.

Trivers' paper was not quite finished, however. He was then confronted with "an interesting new fact," this one about the sex ratios within ant nests. "Why should this be?," he asked.

> Why should the mother dominate one variable and the daughters the other? I thought about this problem for a

51. Using the word "genius" to describe Trivers, a word he tells us he rarely uses let alone lightly, Barash is not exaggerating when he says, "In a handful of truly extraordinary, paradigm-shifting papers, Trivers [in the 1970s] laid out many of the paths that the rest of us have subsequently followed, and which for some involved morphing into 'evolutionary psychology.'" David P. Barash. (2011). "All the Better to Fool You With, My Dear: A Review of Robert Trivers, *The Folly of Fools: The Logic of Deceit and Self-Deception in Human Life*." *Evolutionary Psychology* 9 (4), 527, 526.

52. Robert Trivers, ed. (2002). *Natural Selection and Social Theory: Selected Papers of Robert Trivers*. Oxford: Oxford University Press, 159.

> few weeks without any progress, when I fell asleep one night and had a dream about Ernst Mayr. [An ornithologist by specialty, Mayr was one of the leading figures in twentieth-century evolutionary biology and fortunately for Trivers was a professor at Harvard at the time.] Professor Mayr and I were in an ant nest, reduced to the size of ants, . . . Ernst Mayr was pointing at the queen and saying, "Bob, it's the chance of the queen dying, it's the chance of the queen dying!" I woke up around six-thirty in the morning, mumbling the words, "It's the chance of the queen dying." Soon enough I thought to myself, Well, Ernst Mayr has never been wrong in real life, as why assume that he would be any less accurate in my dreams?

Trivers, however, still did not have the solution. But Mayr in his dream gave him "the key parameter" to the solution, as he puts it, which "occurred" to him one day after he spent "a few weeks trying various alternatives but eliminating each."[53] The paper was eventually published as "Haplodiploidy and the Evolution of the Social Insects" (1976), and his observation is surely relevant that "Intellectually we were revolutionizing the subject of animal behavior, and I still naively imagined that our proper influence within the social sciences would soon be widely felt."[54] (To this day, of course, the so-called social sciences are still indoctrinating their students with fierce resistance to sociobiology and evolutionary psychology.)

Trivers' dream about Mayr is surely one of the most captivating cases of unconscious scientific creativity ever recorded, like something from a movie. At any rate, we would do well at this point to recall once again Poe's seemingly fantastic claim in the imaginary epistle of *Eureka*, about "the progress of true Science," which is that it "makes its most important advances—as all History will show—by seemingly intuitive *leaps*."[55] These words speak now as proof-tones of prophecy.

---

53. Trivers (note 52), 160.
54. Trivers (note 52), 164.
55. *P&T*, 1264.

And on and on the History of true Science shows, with example after example, each scientist "dreaming dreams no mortal ever dared to dream before," whether awake or asleep, each of them having "unthought-like thoughts that are the souls of thought," to use Poe's words that are particularly apt for this occasion.[56] Naturally, each of the discoveries discussed above was preceded by a problem situation to which enormous background knowledge and intense mental effort were applied in trying to find the solution, typically attended with failure, with the "sudden flash of enlightenment," as Crick puts it, followed by a period of "hard work," the latter "a happy state of affairs."[57] This connects with Crick's fog-and-lost-in-the-jungle metaphor, which is an especially insightful one, as it provides an important clue to the

56. *P&T*, 82, 88. Speaking of souls, if one wants the most extreme example of a eureka moment one can do no better than the events recalled by Carl Jung preceding the writing of his *Septem Sermones ad Mortuos* (1916), (*Seven Sermons to the Dead*,) events that were precipitated by his break with Freud (and much else) and that involved the "haunting" of his house with his family in it. It began, he tells us, "with a restlessness" and "the strange feeling that the air was filled with ghostly entities," then his eldest daughter saw "a white figure passing through the room" in the middle of the night, the second daughter experienced her blanket "snatched away" not once but twice, his son had an "anxiety dream" involving the devil, which he made a picture of the following morning, (he "ordinarily never drew,") in the brightness of the afternoon the front doorbell began "ringing frantically" and yet no one saw anyone outside, Jung himself "not only heard it but saw it moving," (it was the old pull-handle type,) the house felt "crammed with spirits" and with air "so thick it was scarcely possible to breathe," and then "it began to flow out of me, and in the course of three evenings the thing was written." C.G. Jung (1963). *Memories, Dreams, Reflections*. Aniela Jaffé, ed. Richard and Clara Winston, trans. London: Routledge & Kegan Paul, 182, 183. See also Sonu Shamdasani, ed. (2009). *Liber Novus, the Red Book of C.G. Jung: A Reader's Edition*. New York: W.W. Norton, 40–42. Given that a mandala, according to Jung, symbolizes "the wholeness of the personality," *Septem Sermones* is a *literary* mandala, the first of his "My mandalas were cryptograms." Jung (this note), 187. As such, it marks the beginning of Jung's journey to psychic wholeness and his psychology of individuation (more on this in the neuroscience section below).

57. Crick (note 6), 121.

ultimate meaning of scientific imagination, or so I shall argue in the penultimate section of this chapter.

It is also important to notice Crick's acknowledgment that "Of course, messenger RNA was bound to be discovered eventually, but there is little doubt in my mind that this revelation speeded up the process considerably."[58] His point takes us back to the analogical argument I made in the last section of Chapter 6, that eureka moments play a key role in the engine that drives science, helping to shape the paths and timelines it takes, much as, to quote the molecular biologist Gunther Stent, "It could now be seen that gene mutation, being the prime source of biological novelty, is the motor that drives evolution."[59] Take away mutation in evolutionary history, and evolution would grind to a halt, confined to genetic recombination and genetic drift. Similarly, take away the creative unconscious in the history of science, with the "soar" of its eureka moments, and science, too, argued Poe way back in 1848, (lacking only the modern concept of the unconscious mind,) would grind to a halt, confined to "the two narrow and crooked paths—the one of creeping and the other of crawling," namely, the methods of deduction and induction.[60] If this taking away would have happened in A.D. 1200, say, then science, I submit, would still be in the Dark Ages to this very day.

But it is not just scientists, or artists, or Bedlamites, who achieve these moments of seemingly divine inspiration. We all have these experiences, albeit on a much smaller scale, if only we would pay attention when they happen. No one who trustingly consults and thoroughly questions his own soul, will be disposed to deny the entire radicalness of the propensity in question. If I may provide a personal example, when I was an undergraduate I took a course in computer programming. I'll never forget the night before a major assignment was due. For at least two weeks the program I devised would not work. Every alteration I made failed. Then, the night before it was due, I tossed and turned all night as if in a nightmare, with images of computer symbols flying

58. Crick (note 6), 121.

59. Gunther S. Stent (1980). "The DNA Double Helix and the Rise of Molecular Biology." In Stent (note 48), xii.

60. *P&T*, 1268.

around in my dream. The second I awoke the next morning I had a flash of insight, how to fix the program to make it work. I wrote it down and went to campus and ran it, and it worked! What I had that morning was a eureka moment. And I'll bet we've all had them. For example, before writing this chapter I discussed my experience with a Ph.D. student, Brandon Fenton, a former teaching assistant of mine. He said he had a similar experience during a summer job putting up drywall. There was a particular application that nobody could figure out, so they broke for lunch. At the end of lunch the solution came to him all of a sudden. Back at the site they tried what he had in mind, and it worked.

Certainly in learning about eureka moments we would be learning something important about ourselves as humans. But the focus of this chapter, of course, is on the giant and massively complicated examples found in science, everyday eureka moments writ large, one might say, the extremes of the phenomenon, each occurring in a mind with enormous background knowledge combined with an intense conscious preoccupation at solving a specific problem that seemed to fight back with equal resistance, almost like an irresistible force meeting an immovable object. If we really want to understand science as a process, then, if we are not just paying lip service, we should want—really, truly, *want*—to understand these moments as best we can. Poe understood this. He may have overemphasized the role of eureka moments in science, (and then he may have been spot on,) claiming that they are what drive science, but in the very least they are surely an essential part of science as a process. So let us descend now further into the abyss in search of the greatest treasure of all—the treasure of self-knowledge.

Comparative Biography and Psychology

When we look at the case list of eureka moments presented above, which could be increased a thousandfold, it should become obvious that eureka moments are not to be explained exclusively by an appeal to culture or the environment. Otherwise many scientists trying to solve a particular problem should be struck with eureka moments in close succession and sometimes even simultaneously, like popcorn in a Jiffy Pop. But that is not what we find. The *zeitgeist* or "spirit of the times" approach to explaining genius just won't do. It doesn't explain the fact that Einstein's discoveries were well ahead of their time, or Darwin's, or Poe's anticipation of the modern Big Bang theory for that matter. The spirit of *our* time, of course, with its political correctness and multiculturalism, wants to level genius to the rest of us, so that no one is ahead of their time and every child passes grade school and gets a trophy for playing a sport. But clearly if we *seriously* want to understand scientific genius, then we have to go beyond environmental explanations and ignore our modern thought police. And to do this we have to look into the lives of geniuses themselves, even into their brains if we can.

Let us begin with comparative biography in our search for insights. Of course it would be madness to try to be comprehensive here, and so I shall rely on the comprehensiveness of others, in fact on one other, the biographer Andrew Robinson, author of over twenty-five books, many of them biographies of creative geniuses. We have already used some of his insights on Champollion. Here we shall use his book *Sudden Genius?* Never interested in biography merely for the sake of getting the history right, Robinson pursues the topic for the purpose of understanding "exceptional creativity—'genius'—in both scientists and artists," and his thesis is that comparative biography, combined with modern science, "can reveal much about the sources, ingredients, and patterns of genius." At the core of his research is the "eureka experience," which he characterizes as "a pivotal revelatory episode."[61] Robinson, it turns out, is a lover of Poe without knowing it, (he has not read Poe, personal communication,) and in the following we shall attempt to relate his insights to Poe wherever we can.

In addition to ten chapters devoted to the biographies of five great scientists and five great artists,—da Vinci, Wren, Mozart, Champollion, Darwin, Curie, Einstein, Woolf, Cartier-Bresson, and Ray,—Robinson devotes a number of chapters to various questions concerning creative genius. One chapter is devoted to the topic of inheritance, in which he draws from biographical data and research by psychologists. His conclusion is that there

61. Andrew Robinson (2010). *Sudden Genius? The Gradual Path to Creative Breakthroughs*. Oxford: Oxford University Press, xi, xiv, xvii, 316.

is yet no substantive evidence to warrant the view that genius is mainly inherited, although there is much evidence to support the view that the brain is "plastic," that it can gradually "rewire" itself to a remarkable degree as a result of repetitious behavior or following disease or trauma. That chapter is followed by a chapter devoted to intelligence, in which Robinson concludes that "almost all geniuses . . . have an IQ well above average (100)—but . . . to have an IQ well above average is no guarantee of genius." Much the same is true of creativity tests, that they score poorly in predicting creativity in the real world, which is to be expected "Given the artificial conditions of tests, compared to the complexity of the real world in which creativity must flourish."[62] The bottom line, assuming Robinson has basically got it right, is that even if absolutely everything about Poe's parents and about Poe's genes could be known today, the current (and possibly future) state of science could not possibly predict (or rather retrodict) his genius (assuming such, of course).

Another of Robinson's chapters is devoted to the topic of the unconscious mind. The central matter of interest here, to my mind, is the question of whether "unconscious processing" (called "unconscious scientific creativity" and "unconscious problem solving" by Smith in the section above) really exists, the question of whether during the so-called period of incubation, which extends from the problem situation with typically failed attempts to the moment of sudden illumination, is really unconscious processing. After all, if unconscious processing exists, it cannot be observed, either subjectively (otherwise it would not be unconscious) or objectively (otherwise it would not be mental). Instead it is a kind of black box, with only input and output capable of being observed. A possibility other than unconscious processing is *cryptomnesia*, (which Robinson refers to all but in name,) "Generally used for ideas and thoughts (often apparently creative and novel) that are really memories of past experiences and events that the individual does not (consciously) recall."[63] Perhaps eureka moments, then, are not really eureka moments but simply cases of

62. Robinson (note 61), 23, 24.

63. Arthur S. Reber (1985). *The Penguin Dictionary of Psychology*. New York: Viking Penguin, 169.

bad memory, of forgetting the steps and conclusions one (or someone else) went through before, and then later, sometimes much later, all of a sudden having the memory of the conclusion pop into one's conscious mind, so that it seems completely new. After all, says Robinson, "Poincaré in 1908, like Kekulé in 1890, was recalling events a quarter of a century after they occurred. Maybe he, too, forgot some preparatory steps in his thinking and misremembered his precise circumstances." Hence some researchers, such as Robert Weisberg, are "sceptical about the reality of unconscious processing." Robinson himself definitely leans toward the reality of unconscious processing, but unfortunately his argument rests on his key piece of evidence, the eureka moment of an experimental procedure by Otto Loewi in 1920 that led to his discovery of chemical neurotransmitters, (for which he was awarded a Nobel Prize,) a eureka moment that was "carefully described [by Loewi] in 1960."[64]

This is an important problem. (It boils down to Dupin versus Holmes!) I should say that while probably *some* cases of eureka moments can legitimately be dismissed as cases of cryptomnesia, *not all* can be dismissed so easily, that some even positively resist that interpretation. A prime example is Crick's discovery of mRNA discussed in the section above. This is because he had been working on the problem along with others, not alone. Had his "sudden flash of enlightenment" on Good Friday 1960 been simply a case of cryptomnesia, surely his colleagues would have reminded him of what he had forgotten. In other words, the cryptomnesia explanation can in some cases be tested, or exposed to testing by circumstantial evidence, and ruled out beyond a reasonable doubt, which I believe is the case with Crick. Moreover, the way a eureka moment, as we have seen in the previous section over and over again, in case after case, *illuminates* a problem like never before, suggests, in repetition, that what came to the conscious mind in the eureka moment had not been in that conscious mind ever before. Recall Crick's words: "One could kick oneself for not having the idea earlier, it now seems so obvious. Yet before, everything was in a fog." If the fog had been cleared previously but he simply forgot, then surely the peak experience

64. Robinson (note 61), 33, 35, 38.

the first time around would have made a memorable impression, rather than a prematurely buried one. Finally, there is a feature of cryptomnesia that weighs against the dismissal of eureka moments as cryptomnesic. According to Carl Jung, in *verified* cases of cryptomnesia, concerning stories, while "there are cases where the reproduction occurs unconsciously, almost word for word," the "peculiarities of cryptomnesia" are such that "a quite unimportant detail which only deserves to be forgotten as quickly as possible is suddenly reproduced with almost literal fidelity, while the main point of the story is, one cannot say modified, but re-created in an individual manner."[65] If one applies this to eureka moments in science one would have to say that *in many cases* the original idea, in being forgotten and then remembered later as an original idea, has subsequently been modified, and in an important way—but then that is all but to openly admit the reality of unconscious processing, of unconscious problem solving! (Berenice, after all, is not the same before her burial as after, evidence alone the box from which "there rolled out some instruments of dental surgery, intermingled with thirty-two small, white and ivory-looking substances that were scattered to and fro about the floor.")

Another of Robinson's chapters is entitled "The Lunatic, the Lover, and the Poet," which is devoted to (i) so-called *idiot savants*, people with low IQ but extraordinary talent in something, such as memory recall or mathematical calculation; (ii) *autism*, which involves extreme "narrow focus" and "mind-blindness," the latter a deficiency in attributing mind to others; and (iii) *synesthesia*, an umbrella term for "cross-talk" between sensory modes, such as seeing sounds in color, or smelling sounds, or tasting visual shapes. In examining cases of individuals and scientific studies on these topics, Robinson finds suggestive evidence for connections with creativity and genius, but nothing solid. Autism, for example, can possibly be conducive to genius by confining mental energy to minutiae and a "rage to master," and to originality by not being affected by social pressure or stereotypes, and yet there are no significant examples of eureka moments in art and science that

65. C.G. Jung (1902). "On the Psychology and Pathology of So-Called Occult Phenomena." R.F.C. Hull, trans. In William McGuire, ed. (1977). *C.G. Jung: Psychology and the Occult*. Princeton, NJ: Princeton University Press, 84, 86.

come from autistic savants. Whatever their originality, it fails to be "an advance on existing thinking," but is "simply bizarre or maladaptive."[66]

Much more interesting, however, is synesthesia. Although there is conflicting evidence, "According to one study," Robinson tells us, "synaethesia is seven times more common in creative people than in the general population," with Poincaré an example of someone who "saw the letters of the alphabet in colour."[67] First formally studied in the 1880s by Darwin's younger half-cousin, Francis Galton, it should be noticed that there are synesthetic elements in Poe's writings. In his poem "Al Aaraaf" (1829), for example, we find "the murmur of the grey twilight," "the darkness coming as a cloud," and "its form—its voice—most palpable and loud," to which Poe adds in a footnote, "I have often thought I could distinctly hear the sound of darkness as it stole over the horizon." Another example is Poe's prose dialogue "The Colloquy of Monos and Una" (1841), in which Monos describes what it was like just after he died. He was still sentient, his senses "assuming often each other's functions at random. The taste and the smell were inextricably confounded, and became one sentiment, abnormal and intense." And the light rays that hit the side of his eyeballs produced "a more vivid effect" than the rays that struck the front, in that "I appreciated it only as *sound*—sound sweet or discordant as the matters presenting themselves at my side were light or dark in shade."[68]

In a chapter devoted to family relations, Robinson doesn't find much to go on that is solid, with the possible exception that creative geniuses are more likely to have lost a parent while a child, especially the mother, are more likely to have been deprived of love and support while a child, (that would seem to follow simply from the preceding,) and are more likely to spend time alone. The latter, of course, while possibly contributing to creative genius,

66. Robinson (note 61), 46.

67. Robinson (note 61), 51.

68. *P&T*, 45, 453–454. For discussions on how these passages relate to synesthesia as it is known today, see Lawrence E. Marks (1978). *The Unity of the Senses: Interrelations among the Modalities*. New York: Academic Press, 229–234. Cretien van Campen (2008). *The Hidden Sense: Synesthesia in Art and Science*. Cambridge: MIT Press, 105–106.

might also simply be an effect of it, as a need. At any rate, the Poe lover can hardly fail to notice that both of Poe's biological parents died shortly before he turned three. And while he was raised by the Allans apparently not suffering from want, whatever affection he received from his stern foster father was already in decline by the time Poe hit puberty, with his loving foster mother suffering from illness, beginning early in his second lustrum, and dying shortly after his fourth. It should also be noticed that the life of an orphan, no matter how well provided for, must surely be attended with a powerful sense of insecurity at some level, in the very least an indefinite feeling that what was once taken away can easily be taken away again. And then there's Poe's poetry, which speaks over and again of aloneness, such as "I dwelt alone / In a world of moan," or much earlier, "From childhood's hour I have not been / As others were—I have not seen / As others saw—I could not bring / My passions from a common spring— / . . . And all I lov'd—*I* lov'd alone."[69] As Robinson puts it, "Genius does not sit well on committees."[70]

On the topic of formal education, especially public schooling, creative geniuses tend not to have enjoyed it much. What stands out far more importantly is being an autodidact, a self-learner, and following one's interests. Regimentation, learning by rote, and the like, tend to produce nothing but human robots. When it comes to the natural sciences, of course, the postgraduate formal setting has become indispensable for mastering the necessary background knowledge and experimental or observational techniques. And yet that produces a trade-off in terms of creative imagination, given that the former produces constraints on the latter. Robinson is surely right, in speaking like a Kuhnian, that "academia has an inherent tendency to ignore or reject highly original work that does not fit the existing paradigm."[71] And he is also surely right, following the psychologist Dean Keith Simonton, in concluding that "What is generally accepted . . . is that the huge growth in size and competitiveness of higher education in the second half

69. "Eulalie—A Song" (1844), "Alone" (1829). *P&T*, 80, 60.

70. Robinson (note 61), 265.

71. Robinson (note 61), 274.

of the twentieth century and after did not increase the number of exceptionally creative scientists."[72]

It may be, however, that the more mature and advanced a scientific field becomes the less opportunities there are for creative discoveries in the sense of Poe's "leaps," as there is less of nature to know in the theoretical sense, less gaps in our knowledge, (leaving the future more to advances in technology,) which could explain the conclusion above. If this is indeed the case, then Poe's picture of science appears better suited to, or more accurately captures, the earlier stages of a science than the later ones. Poe himself, of course, if we look back at Chapter 4, was an autodidact when it came to the science of his day, which, combined with his passion, energy, imagination, and intellect, allowed him to produce the results we examined in Chapter 5. But it seems far less likely that the equivalent of a Poe could happen in science today.

And yet there remains much to be said for Poe's picture. For even in the highly specialized world of modern science, there is still a place for imagination, even for unconscious scientific creativity, if not anymore for the putting together of big picture scenarios, (though still for recognition of it by others,) then for the more particular problems over which scientists rack their brains, like Crick's discovery of mRNA. Someone like Poe can no longer acquire the necessary background knowledge and training that only formal institutions can provide, to do something like Crick did. But Crick himself provides a wonderful testament to the fact that training and knowledge and institutions and teams and financial resources are not enough. What is needed is still something more, much more, and Poe nailed it back in 1848. It is not method but what Einstein called "inward freedom," which he says "consists in the independence of thought from the restrictions of authoritarian and social prejudices as well as from unphilosophical routinizing and habit in general." It is a necessary requirement, he insists, for the "development of science and of the creative activities of the spirit as well."[73] Surely it would be folly to lose this lesson in our modern age of specialization and professionalization—it

72. Robinson (note 61), 276.

73. Albert Einstein (1940). "On Freedom." In Sonja Bargmann, trans. (1954). *Ideas and Opinions by Albert Einstein*. New York: Crown Publishers, 32.

would be to lose art and science to technology, the spiritual and imaginative man to the machine man.

Next, does the scientific genius have its typical personality type, like introvert or extrovert, agreeable or uncooperative, selfish or unselfish, and so on? The answer is *no*. As Robinson puts it, "At present, personality psychology has nothing very useful to say about genius." And yet his educated guess is that "the more creative a person is, the more multifarious is his or her personality," that creative geniuses have a "chameleon personality," appearing to others, and possibly even to themselves, as a "bundle of contradictions."[74] Certainly this is the impression one gets about Poe if one consults a number of biographies of the man, with one biographer referring to "Poe's abrasive and often bellicose personality," while another, who knew Poe during "the last three months of his life," as well as many of his friends and acquaintances, shared the view that "Everybody could see what his faults were, but few gave him credit for his good qualities—his generous nature and kindly and affectionate disposition, especially as exemplified in the harmony always existing between himself and his wife and mother-in-law."[75] At any rate, apart from the matter of a chameleon nature, the only certainty as a common denominator is that scientific geniuses, as with creative geniuses in general, "share a personality that is highly motivated to work and determined to succeed in their field."[76] And given what we have seen in Chapter 4, that surely describes Poe.

Finally, there is the so-called *ten-year rule*, discovered (or created) by the psychologist John Hayes in 1989 and followed by a number of others since then, such as the psychologists Howard Gardner, Robert Weisberg, and Keith Sawyer. "The only widely respected 'law' of creativity," Robinson puts it as follows: "it states that a person must persevere with learning and practising a craft or discipline for about ten years before he or she can make a breakthrough." The rule, of course, admits of exceptions, (as do most rules,) and Robinson himself attempts to improve on the rule by distinguishing three versions: *weak*, ("a breakthrough requires

74. Robinson (note 61), 295, 296.

75. *Hutchisson*, 151; *Weiss*, 8, 102.

76. Robinson (note 61), 303.

a minimum of ten years of hard work and practice in a relevant domain,") *medium*, ("a breakthrough requires a minimum of ten years of hard work and practice focused on the particular problem solved by the breakthrough,") and *strong*, ("a breakthrough requires about ten years—no less and no more—of hard work and practice focused on the particular problem solved by the breakthrough"). Robinson claims that of the ten creative geniuses studied in his book, (each getting their own chapter,) all of the ten "obey the weak version," eight "obey the medium version," and five (including Darwin and Einstein) "obey the strong version."[77]

My problem with this is that it is reminiscent of the fallacy of *presentism* repeatedly cautioned by historians, the fallacy of reading present ideas into the past without sufficient evidence, (universal human rights in the Bible or the Koran, for example,) and also of what psychologists call *confirmation bias*, which is when one selects, interprets, or recalls evidence only in accordance with one's pre-existing beliefs or theories (you see this in racism, for example). Both involve, one might say, a mixture of projection and selection. And both are similar to an experience Freud wrote about in a letter to Jung, dated April 16, 1909. "A few years ago," he says, "I took it into my head that I would die between the ages of 61 and 62, . . . Shortly afterward I made a trip to Greece with my brother, and it was absolutely uncanny to see how the number 61, or 60 in conjunction with 1 and 2, kept cropping up on anything that had a number, especially on vehicles. I consciously noted down three occasions. . . . At our hotel [in Athens] we were assigned rooms on the second floor, . . . it turned out that my room was No. 31 (which, with fatalistic license, I regarded as after all half of 61–62). . . . From that day until very recently the number 31 remained faithful to me, with a 2 all too readily associated with it." Freud attributes the experience to two factors, the first "an enormously intensified alertness on the part of the unconscious," and the second "the undeniable 'co-operation of chance.'"[78]

I suggest that something like the above is happening with the ten-year rule, or at least that it poses a very real danger. In the case of Darwin, Robinson sees the strong version of the ten-year

77. Robinson (note 61), xxxiii, 322–323.

78. Quoted in Jung (note 56), 333, 334.

rule as applying from 1828, when Darwin's study of natural history "became serious at Cambridge University," to 1838, "when he made his breakthrough."[79] But as I believe I have shown near the end of Chapter 6, the fundamental problem that Darwin faced didn't really strike him until midway through his *Beagle* voyage, around 1832 after reading the first two volumes of Lyell's *Principles of Geology*, which was magnified by his own observations of flora and fauna as naturalist aboard the *Beagle*. That would make the gestation period of Darwin's theory of evolution by natural selection, from when he first genuinely started working on the problem until he had his breakthrough, no more than six years, not ten. Or one could date the beginning of Darwin's problem situation to his tutelage under and friendship with Robert Grant during his second year at the University of Edinburgh, which began in the fall of 1826, at which time Grant (who also got Darwin doing a lot of natural history) shocked Darwin by advocating the evolutionary theories of Lamarck and Darwin's own grandfather, Erasmus. Of this episode Darwin wrote in his autobiography, "I listened in silent astonishment, and as far as I can judge, without any effect on my mind."[80] In this case, the gestation period would be close to twelve years.[81] We should also bear in mind an important insight noted by Einstein: "It is never easy to talk about how I got to the theory of relativity because there would be various concealed complexities to motivate human thinking and because they worked with different weights."[82] Surely the same would apply to Darwin's road to evolution by natural selection, and it creates a serious problem for the ten-year rule.

Turning now to Einstein, Robinson tells us that he "studied some physics at school in Germany, but did not begin to address the subject seriously until he dropped out of school in 1894 and spent many months at home in Milan in 1895, when he probably made his thought experiment with a light ray. From then on, he

79. Robinson (note 61), 324.

80. Barlow (note 19), 49.

81. Or it could be even more. According to Browne, Darwin on Grant was "Far too disingenuous in his *Autobiography*," for "he was in truth well prepared to understand the importance both of Lamarck and of his grandfather." Janet Browne (1995). *Charles Darwin: Voyaging*. New York: Alfred A. Knopf, 83.

82. Quoted in Pais (note 41), 132.

devoted most of his time and energy to reading, thinking, and talking about mathematics and physics. His breakthrough with special relativity came ten years later, in 1905."[83] It is true that in his autobiographical "obituary," published six years before his death, Einstein states that "After ten years of reflection such a principle [of relativity] resulted from a paradox upon which I had already hit at the age of sixteen: If I pursue a beam of light with the velocity *c* . . . . One sees that in this paradox the germ of the special relativity theory is already contained."[84]

The paradox or light beam thought experiment, Einstein tells us, occurred sometime while Einstein was attending school in Aarau, Switzerland, which was from late October 1895 to early fall 1896,[85] (Einstein turned seventeen on March 14, 1896,) so that the ten years is now possibly reduced to a little more than nine, given that Einstein did not provide a specific date for the thought experiment and that the final insights for his special theory of relativity did not occur until five or six weeks before he wrote his 1905 paper, which was completed near the end of June.[86] Moreover, the roughly ten-year period of gestation (or closer to nine) does not span the period "of hard work and practice focused on the particular problem solved by the breakthrough," as Robinson defines his strong version of the ten-year rule. This is because the light beam thought experiment was not the beginning of the particular problem that special relativity solved. The particular problem began when Einstein was struck by a serious problem about the existence of the ether, a "nineteenth century edifice,"[87] (eliminated in his own theory of the Universe, I might add, by Poe in 1848, as we have seen in Chapter 5). And Einstein was struck by that problem apparently from his reading of a monograph by

83. Robinson (note 61), 324.

84. Albert Einstein (1949). "Autobiographical Notes." In Paul Arthur Schilpp, ed. (1949). *Albert Einstein: Philosopher-Scientist*. La Salle, IL: Open Court, 53. This is also the conclusion of the extended reconstruction of the events by Pais, that "special relativity was born after a decade of gestation." Pais (note 41), 139. I suggest, however, that the ten might function here merely as a convenient shorthand (and with unconscious influences).

85. Pais (note 41), 131, 521.

86. Pais (note 41), 139, 522.

87. Pais (note 41), 117.

Hendrik Lorentz published in 1895, what Einstein deigned "the decisive investigation."[88] The monograph discussed at length the failed experiment of Fizeau in 1851 to measure the effect of the ether on a split beam of light traveling in opposite directions through water, the failed experiment by Michelson in 1881, and then by Michelson and Morley in 1887, to measure the effect of the "ether wind" on the motion of the Earth by measuring the speed of a split beam of light traveling at right angles but the same distance, and Lorentz's proposed solution to the failed attempts, which like the others still retained the belief in the ether. Einstein didn't think Lorentz's solution worked and that was the beginning of the problem that ultimately led to his relativity theory.

It is unclear, however, when Einstein actually first read Lorentz's 1895 monograph. The most we get are his chronologically unclear statements provided in his address at Kyoto University in 1922 (which was recorded by hand by a Japanese physicist in attendance, then published in Japanese in 1923, and then translated back into English and published in 1982): "As a student I got acquainted with the unaccountable [failed] result of the Michelson experiment and then realized intuitively that it might be our incorrect thinking to take account of the motion of the earth relative to the aether, if we recognized the experimental result as a fact. In effect, this is the first route that led me to what is now called the special principles [*sic*] of relativity. . . . I had just a chance to read Lorentz's 1895 monograph, . . . In this connection I took into consideration Fizeau's experiment."[89]

So does this at least provide some support for the strong version of the ten-year rule, from 1895 to 1905, given that "a breakthrough requires about ten years—no less and no more—of hard work and practice focused on the particular problem solved by the breakthrough," as Robinson defines it? Not necessarily, as Einstein never clearly stated when he first read Lorentz's 1895 monograph, only that it was sometime before 1905.[90] Of course, there might have been a different source that got Einstein onto the problem of the ether during his school days, a source that included

88. Quoted in Pais (note 41), 117.

89. Quoted verbatim from Pais (note 41), 118–119.

90. Pais (note 41), 116–117.

descriptions of the failed experiments of Fizeau, Michelson, Morley and possibly others. Walter Isaacson refers to a paper by Wilhelm Wien that Einstein read in 1899, which was "probably" referred to him by Heinrich Weber, his physics professor at the Zurich Polytechnic.[91] Earlier than that there is darkness. At any rate, it should be noted that in 1924 Einstein stated that "After seven years of reflection in vain (1898–1905), the solution came to me suddenly, . . . I thus arrived at the special theory of relativity."[92] In 1932 he referred to "a state of psychic tension in me that, after seven years of vain searching, was resolved by relativizing the concepts of time and length."[93] And in October 1952, two-and-a-half years before he died, again in reference to the special theory, Einstein referred to "the seven years that relativity had been my life."[94]

That one sees ten years in all of this, then, might possibly have nothing more to it than the obvious fact that typically we each have ten fingers and ten toes. If we instead were to have seven, (figure that one out for yourself,) maybe psychologists and biographers would be emphasizing the *seven-year rule*.

Now if I may add Poe to the scene, there's nothing at all to indicate ten years of hard work in the "relevant domain" preceding *Eureka*, (weak version), or a minimum ten years of hard work from the particular problem to the breakthrough, (medium version,) or ten years of hard work, no more and no less, (strong version). There's absolutely nothing to indicate ten years, not even if we turn to Poe's tale "The Fall of the House of Usher," first published in September of 1839,[95] which as I argued in the second-last section of Chapter 1 is not (contrary to Peeples) an anticipation of *Eureka* and cannot be maintained as dealing with the same problem, the meaning of the Universe, except by an enormous stretch of exegesis. Instead, as I argued, we do not see a real beginning until 1844–1845, in works that contain many of

91. Walter Isaacson (2007). *Einstein: His Life and Universe*. New York: Simon & Schuster, 115–116.

92. Quoted in Stachel (note 40), 112.

93. Quoted in Stachel (note 40), 111.

94. Quoted in Pais (note 41), 117.

95. *Poe Log*, 267.

the elements of Poe's theory of the Universe found in *Eureka*. That period was followed by a steep decline extending through 1846 due to the advanced illness of Virginia, their poverty, and Poe's correspondingly poor mental and physical health. Then followed an even worse year, Poe's least productive of all, which began with the death of Virginia on January 30, 1847, and during which he suffered almost a year of mental depression and physical illness. Toward the end of 1847, however, Poe experienced a burst of creative activity, composing most notably his poem "Ulalume" in October and then his lecture entitled "The Universe," which he gave on February 3, 1848, and which was published as *Eureka* on or around July 11 of the same year.[96]

At any rate, it surely remains a fact, as Robinson puts it, that in spite of their periodic breaks "geniuses work habitually and continually," that "'sudden genius' is a myth" while "genius is not a myth," and that "Genius is the name we give to a quality of work that transcends fashion, fame, and reputation: the opposite of a period piece."[97]

I want to finish this section by taking a brief look at a book on creative genius by the psychologist Mihaly Csikszentmihalyi, entitled *Creativity*. The book is the culmination of interviews, recorded by videotape using a set of questions, of "ninety-one exceptional individuals" in the worlds of science and medicine, art and the humanities, technology, business, and politics, with twelve of the interviewees having received a Nobel Prize (eight of them in science).[98] I cannot possibly do justice to this remarkable book in a few pages, or even in an entire chapter. But let me highlight the conclusions in the book that seem to me especially salient.

First and foremost is the phenomenon of eureka moments. The relevant question put to each interviewee was, "How important is rationality versus intuition in your work? . . . Have you ever had a useful idea while lying in bed, or in a dream?" Csikszentmihalyi tells us that "Most of the people in our sample—but not all—recall with great intensity and precision a particular moment when some major problem crystallized in their minds in such a way that a

96. *Poe Log*, 705, 720, 714, 742.

97. Robinson (note 61), 317, 329, 315.

98. Csikszentmihalyi (note 47), 12, 107, 13.

solution became all but inevitable, requiring only a matter of time and hard work. For presented problems, the insight might even include the particulars of the solution." In the previous section we have seen how Freeman Dyson answered, who was one of the interviewees. But one has to wonder if Csikszentmihalyi would have garnered an even stronger response from the rest, given the seemingly restrictive dichotomy at the end of the second question above, which should have been preceded by a "for example" or "such as." At any rate, one of the interviewees, a writer named Madeleine L'Engle, answered that "Your intuition and your intellect should be working together . . . making love. That's how it works best." This adds an interesting angle to C.G. Jung's view, which he called *individuation* and which we shall examine near the end of the neuroscience section below. What is interesting here is the theory that Csikszentmihalyi, a psychologist who takes evolution seriously, adds as an explanation: "The insight presumably occurs when a subconscious connection between ideas fits so well that it is forced to pop out into awareness, like a cork held underwater breaking out into the air after it is released."[99]

A related matter concerns cross-domains. As Csikszentmihalyi puts it,

> Yet it is important to keep in mind that most breakthroughs are based on linking information that usually is not thought of as related. Integration, synthesis both across and within domains, is the norm rather than the exception. Madeleine L'Engle is inspired by molecular biology to write her stories; Ravi Shankar finds ways of harmonizing the music of India and Europe; and almost all scientists cross and recross the boundaries of physics, chemistry, and biology in the work that turns out to be creative.[100]

One has to be careful with the "almost all scientists," of course, as it refers to the scientists among his interviewees, not to "normal scientists." At any rate, what Csikszentmihalyi states here is

99. Csikszentmihalyi (note 47), 396, 239, 104.

100. Csikszentmihalyi (note 47), 329.

in perfect accord with a theme that recurs throughout the book before you, which is Aristotle's insight about genius and metaphor, about "an intuitive perception of the similarity in dissimilars." Once again Poe is in very good company.

And what about Poe? Can Csikszentmihalyi's book help us to understand Poe even better? Not only is the answer *yes*, but it can also help us to appreciate him more than ever before.

For a start, there is the matter of formal education. On the one hand, we are told that "Performance in school matters more in some domains than in others. In mathematics and the sciences, the exposure one gets in high school is necessary for further advancement." On the other hand, we are told that "Young artists, especially visual artists, are notoriously uninterested in academic subjects, and their scholastic records usually reflect this."[101] Poe was therefore unique even among the unique, for as we have seen in Chapter 4, he loved school and excelled at academics, all the while being an artist, including a visual artist. In a way, then, Poe transcended C.P. Snow's "two cultures," the sciences and humanities.

Then there is the matter of age and achievement. According to Csikszentmihalyi, "Recent studies suggest that not only quantity but quality is retained with age, and some of the most memorable work in a person's career is done in later years."[102] Poe died at forty, and we can only pause to wonder if the world has been denied his greatest work.

There is also the matter of convergence in circumstances. "Most of us deep down believe that a person who is creative will prevail regardless of the environment." And Csikszentmihalyi is surely right about this. "But the reality appears to be different," he tells us. "Favorable convergences in time and place open up a brief window of opportunity for the person who, having the proper qualifications, happens to be in the right place at the right time."[103] Georges Lemaître, as we have seen in Chapter 5, had the proper qualifications and was in the right place at the right time to found the Big Bang theory, according to the official story, but

101. Csikszentmihalyi (note 47), 178.

102. Csikszentmihalyi (note 47), 210.

103. Csikszentmihalyi (note 47), 330.

so were many others. What is surely more interesting is that Poe beat him to it by a full eighty-three years, and in far less favorable circumstances along with a very different definition of "proper qualifications."

This leads us to the topic of determinism. In looking at what shapes creative lives, Csikszentmihalyi is astonished by "the great varieties of paths that led to eminence." Many people evidently had the genes and talent, but failed to achieve, others seemed to be average in that department but got the lucky breaks, some had normal childhoods and some not, and so on. What seems to matter most is the will of creative geniuses, that "they shaped events to suit their purposes." But even here, says Csikszentmihalyi, "a creative life is still determined, but what determines it is a will moving across time—the fierce determination to succeed, to make sense of the world, to use whatever means to unravel some of the mysteries of the world."[104] I wonder if Csikszentmihalyi is a Poe lover in disguise, for (determinism or not) he has given us here the perfect encapsulation of Poe!

And then there is the imperfect encapsulation, provided most notably by T.S. Eliot, who as we have seen in the Prologue accused Poe and Poe lovers of never having grown up, of being immature. One should read his words again. According to Csikszentmihalyi, on the other hand, and we have seen Schweitzer and Einstein say much the same thing, "the first step toward a more creative life is the cultivation of curiosity and interest, that is, the allocation of attention to things for their own sake. On this score, children tend to have the advantage over adults; . . . With age most of us lose the sense of wonder, the feeling of awe in confronting the majesty and variety of the world. . . . Creative individuals are childlike in that their curiosity remains fresh even at ninety years of age."[105]

For all of Csikszentmihalyi's insights, however, there remains a profound problem with his book, which is what he did *not* ask his interviewees and what he *should* have.

Perhaps the biggest oversight was about birth order. Although Csikszentmihalyi asked his interviewees a lot of questions about family matters, he did not ask them whether they were the oldest

104. Csikszentmihalyi (note 47), 181, 182.

105. Csikszentmihalyi (note 47), 346.

child in the family (biologically or functionally). The question is important, given that the historian Frank Sulloway, in his massive statistical analysis of scientific revolutions that took him twenty-six years to complete, found that birth order best explains the place of a scientist during a revolutionary divide. Scientists resistant to a revolution tend to be firstborns, functionally if not also biologically, while those defending the revolutionary science tend to be laterborns, so much so that, in the case of the Darwinian revolution, "80-year-old laterborns were as open to this theory as were 25-year-old firstborns. . . . being laterborn was equivalent to a 55-year dose of the open-mindedness that typically resides in youth."[106] The basic idea is that being a firstborn tends to make a person more traditional- and closed-minded, more conservative, while being a laterborn tends to make one more rebellious and innovative, and that our functional birth order triggers one or the other strategy as they both evolved in our genes over eons. Firstborn scientists are still often innovative, but, as Sulloway claims, they "are much more likely to make scientific breakthroughs of a technical nature than they are to pioneer radical revolutions," much more likely to accomplish "clever puzzle solving" within a paradigm rather than to change the paradigm, so that in science "Firstborns and laterborns tend to become eminent in different ways."[107] If this is true, then we can view birth order as placing, statistically, a filter on the kinds of eureka moments produced by one's unconscious mind. In looking at the scientists examined in the previous section, Darwin, Wallace, and Trivers (the latter a self-styled rebel) were typical laterborns, given that their eureka moments implied revolutionary changes in biology. Einstein, on the other hand, was a firstborn, and as we have seen in Chapter 6 he did not consider his relativity theories as revolutionary at all, but instead as further developments of classical physics. Moreover, in typical firstborn fashion he was initially resistant to the idea of a Big Bang universe, inventing even the "cosmological constant" to make it so, and never accepted the realist interpretation of quantum physics, which was truly revolutionary with its statistical

106. Frank J. Sulloway (1996). *Born to Rebel: Birth Order, Family Dynamics, and Creative Lives*. New York: Pantheon Books, 36.

107. Sulloway (note 106), 331, 356, 357n.

laws, its indeterminism, and its ontology of genuine chance. Poe, for his own part, although biologically a laterborn, (a secondborn, to be exact,) was raised as a singleton—a kind of poetic justice in itself, given especially *Eureka*.

Another oversight is that Csikszentmihalyi did not ask his interviewees if they were fluent in more than one language. It will be recalled from Chapter 4 that Northrop Frye claimed, approvingly, that "the humanists have always insisted that you don't learn to think wholly from one language: you learn to think better from linguistic conflict, from bouncing one language off another."[108] Einstein was fluent in more than one language, as was Poe, and one has to wonder whether Csikszentmihalyi's interviewees exceeded the average on this matter.

Csikszentmihalyi also didn't ask them about love of pets, which affords a prime opportunity, especially in youth, for the development of analogical and metaphorical thinking, of the "intuitive perception of the similarity in dissimilars." Darwin loved dogs, which he called a lifelong "passion," while Poe not only claimed to "like" cats but often wrote at night with Catterina sitting on his knee or shoulder and was "notorious for taking in stray cats."[109] It would be interesting to know how many of Csikszentmihalyi's interviewees shared similar sentiments.

Similarly, we are not told how many played musical instruments, regularly and with a respectable degree of proficiency, especially in childhood. Einstein, for example, began playing the violin at the age of six and played it excellently for most of his life, turning to the piano in his later years.[110] Poe himself professed to have not only a love of music but also a musical ear, the "accuracy" of which he developed by "some odd chromatic experiments," and of course his own practice of poetry was a kind of music, the "*Rhythmical Creation of Beauty*" using words, as he

108. Northrop Frye (1963). *The Educated Imagination*. Toronto: CBC Enterprises. Republished (2002). Toronto: House of Anansi Press, 72.

109. Barlow (note 19), 27; *Weiss*, 89, 72; *Evermore*, 71. Poe also apparently regarded the killing of a cat as murder. See Clark T. Moreland and Karime Rodriguez (2015). "Fuseli's *The Nightmare* and Collapsing Masculinity in Poe's 'The Black Cat.'" *The Edgar Allan Poe Review* 16 (2), 209–210.

110. Philip Glass (2005). "Einstein and Music." In Andrew Robinson (2005). *Einstein: A Hundred Years of Relativity*. New York: Harry N. Abrams, 153-155.

put it in "The Poetic Principle" (1850).[111] Perhaps music, since it is mainly a right-brain activity, (given *amusia*, which results mainly from damage to the right brain,[112]) contributes to a greater development of the corpus callosum, which connects our left brain and right brain both anatomically and functionally in terms of information transfer (and also the inhibition of that transfer). There is in fact evidence from brain imaging that musicality contributes to this greater development, especially if the person began playing before the age of seven,[113] and certainly Einstein's brain fit the pattern.[114] As we shall see in the neuroscience section below, the significance is that greater functional capabilities in the hemispheric connection is quite possibly at the center of the eureka experience in scientific imagination.

Finally, there is the topic of depression followed by creative insights, or related states that might serve as a surrogate, such as delirium in the case of Wallace. This is arguably related to "writer's block," which concerns "the domain of the word," a common phenomenon that Csikszentmihalyi naturally found in the writers among his interviewees.[115] Mayer and Tesla, as we have seen in the section above, were subject to depression, a bout of which apparently preceded each of their eureka moments. Similarly, in the madness section of the previous chapter we noted the case of Robert Trivers, whose bouts of depression seemed sometimes to precede his scientific problem solving. We also noted that Chambers' *Vestiges* was written immediately following an extended period of depression, as was Poe's *Eureka*. These and related cases suggest

111. *Letters*, 115; *E&R*, 78. For an excellent discussion on the unique musicality in Poe's poetry, see Sławomir Studniarz (2014). "Poetry as 'an inferior or less capable Music': Sound and Meaning in 'The Conqueror Worm' and 'To One in Paradise.'" *The Edgar Allan Poe Review* 15 (1), 59–81.

112. Sally P. Springer and Georg Deutsch (1998). *Left Brain, Right Brain: Perspectives from Cognitive Neuroscience*. 5th ed. New York: Freeman, 18, 220–224.

113. Springer and Deutsch (note 112), 224. See also Norman Doidge (2007). *The Brain That Changes Itself: Stories of Personal Triumph from the Frontiers of Brain Science*. London: Penguin, 290, 395n.

114. Weiwei Men et al. (2013). "The Corpus Callosum of Albert Einstein's Brain: Another Clue To His High Intelligence?" *Brain* 136 (Sept.), e1–e8.

115. Csikszentmihalyi (note 47), 242, 263.

the possibility of an evolved mechanism for creative depression that is unconsciously activated, but I want to save this for the evolution section below.

And on that note, it is time now to go yet deeper into the abyss, into a darker level of the mind.

Cognitive Science

What is *cognitive science*? Perhaps surprisingly, it is not the name of an actual science at all. Instead, it is an umbrella term for the interdisciplinary approach to the study of how the mind thinks, which traditionally includes mainly linguistics, (inspired early on by Chomsky's "language organ,") philosophy of mind, cognitive psychology, and computer science (especially work on artificial intelligence). According to John Searle, one of its critics, it is the view that "Thinking is processing information, but information processing is just symbol manipulation. Computers do symbol manipulation. So the best way to study thinking (or as they prefer to call it, 'cognition') is to study computational symbol-manipulating programs, whether they are in computers or in brains."[116] Searle is famous for his Chinese Room thought experiment, which concludes that you cannot get semantics from syntax, that you cannot get understanding from following rules, and since all a computer does is apply rules to data, computers cannot have mind. What we need, instead, he says, is to study the brain. Similarly, Francis Crick characterizes cognitive science as "part of the rebellion against behaviorism. . . . Cognitive scientists . . . think it important to make explicit models of mental processes, especially those of humans. . . . There is no great enthusiasm, however, for looking into the actual brain itself. Many cognitive scientists tend to regard the brain as a 'black box,' better left unopened. In fact, some people *define* cognitive science as studies that take no account of such things as nerve cells. In cognitive science the usual procedure is to isolate some psychological phenomenon, make a theoretical model of the postulated mental processes, and then test

116. John Searle (1984). *Minds, Brains and Science*. Cambridge, MA: Harvard University Press, 43. 117. Crick (note 6), 149–150.

the model, by computer simulation, to make sure it works as its author thought it would."[117] Although Crick is open to a multilevel approach, his criticism of cognitive science is not only "If you want to understand function, study structure," but also that cognitive science ignores the way nature works. Modelling in cognitive science is largely guided by "elegance and simplicity," given the underdetermination of theory by evidence, but biological evolution did not and does not operate in accordance with epistemic virtues and values. That is because evolution does not have a mind and accordingly works on what already exists, rather than from scratch. It is "descent with modification," to use Darwin's phrase. Consequently, as Crick so rightly warns, elegance and simplicity, in trying to figure out how the mind works, "are dangerous guides to the right answer."[118] What this means for the study of eureka moments in particular should be obvious: eureka moments in the human brain might not at all function the way computer scientists think they do or should.

Searle's and Crick's criticisms were made in the 1980s. In the years following, given mainly the rise of neuroimaging techniques, cognitive science increasingly turned to the brain, so that now one routinely finds neuroscience included in the definition of cognitive science. And yet the fundamental premise of cognitive science remains, that thinking is a form of information processing, so that the penchant for computer modelling is greater than ever before along with the belief that thinking can be simulated (weak artificial intelligence) and possibly even realized (strong artificial intelligence) in computers. Nevertheless, I want to reserve neuroscience on its own for the next section, to examine what it has to contribute in addition to, or separate from, cognitive science as more recently conceived. This is because Crick has a fundamental point about evolution that fits with a central theme in Chapter 6. Science is a process, done by scientists, whose brains were evolved by evolution, not created by computer scientists. If we want to understand scientific imagination, then, we need not only to view eureka moments as central, as per Poe, we also need to investigate how they are actually produced by the human brain, very much

117. Crick (note 6), 149–150.

118. Crick (note 6), 150, 141.

like wanting to know how mutations are produced in evolution. Anything less won't do.

There is no question that cognitive science recognizes the existence of the unconscious mind. So let us begin with a question that was asked in a special issue of a psychology journal: "Is the unconscious smart or dumb?" The authors of the article, as well as the editors and authors contributing to the issue, concluded that the unconscious is dumb. That was in 1992.[119] And it is what happens when interests remain parochial, when one is not interested in everything, for the history of eureka moments alone, sampled in the second section above, proves beyond a reasonable doubt that the unconscious is capable of *profound* intelligence, absolutely remarkable *brilliance*, and in the relatively few, *genius*. Many years later, two different psychologists reviewed the topic, noted that no consensus definition yet exists for the unconscious, and argued that the earlier researchers used a much too narrow concept of the unconscious, that "cognitive psychology" still operates with a "conscious-centric bias" that equates "*unconscious* with *subliminal*," according to which the subliminal is usually restricted to either (i) "the preconscious analysis of stimuli prior to the products of the analysis being furnished to conscious awareness," like the primitive associative learning of Pavlov's dogs, the so-called "New Look research," according to which "higher mental processes" must be conscious, or (ii) "the gain in efficiency of processes with practice over time until they become subconscious," like driving a car absentmindedly, the so-called "skill-acquisition research,"—either of which leads to the conclusion that the unconscious is "dumb as dirt." In the view of these more recent authors, instead, "the true power and scope of the unconscious" becomes apparent when one looks "outside of psychology," to the "natural sciences, especially evolutionary biology," which involves a change in the definition of the unconscious to "the influences or effects of stimulus processing of which one is not aware," a definition which "affords the opinion that it [the unconscious] is highly intelligent and adaptive." This is certainly much better, and it is good that the authors turn to the field of evolutionary biology,

119. Elizabeth F. Loftus and Mark R. Klinger (1992). "Is the Unconscious Smart or Dumb?" *American Psychologist* 47 (6), 761–765.

to "widespread discoveries of sophisticated unconscious behavior guidance systems," and to the view that "In nature, the 'unconscious mind' is the rule, not the exception."[120] And yet they leave us with no mention of eureka moments at all, either in word or in concept. That was in 2008.

All of a sudden it is feeling like Chapter 6 again, though now with psychologists rather than with philosophers and historians of science.

I want to close this section with what I consider to be a prime example of research in cognitive science devoted explicitly and exclusively to the topic of eureka moments, co-authored by a philosopher of science-turned-cognitive scientist and an artificial intelligence cognitive scientist. In a paper devoted to what they call "the treasured *Aha!* experience," in which they do take into account some neuroscience and apply it to scientific discovery, Paul Thagard and Terrence Stewart provide a model based on what they call *convolution*. A mental representation, in their view, which is common in cognitive science including neuroscience, consists of, or is carried in, a *neural network*, excitatory links between neurons. Convolution, at its simplest, is the combination of two representations, two neural networks, into a novel one, with "emergent properties," they tell us, features that are not the sum of their parts but are genuinely novel. This alone, of course, will not produce a eureka experience. What is needed is the convolution of two more components, "two aspects of emotional processing—cognitive appraisal and physiological perception," each of them involving "patterns of neural activity," the former referring to the accomplishment of the researcher's goals, the latter to a dopamine kick in the brain or something like it. Moreover, since the operations of the brain are "highly parallel," their model is not one of a "serial process," but rather "all these processes take place concurrently, . . . The *Aha!* experience seems mysterious because we have no conscious access to any of these processes or their integration."[121] There is a lot more to their model, of course, but I

120. John A. Bargh and Ezequiel Morsella (2008). "The Unconscious Mind." *Perspectives on Psychological Science* 3, 73–79.

121. Paul Thagard and Terrence C. Stewart (2011). "The *Aha!* Experience: Creativity through Emergent Binding in Neural Networks." *Cognitive Science* 35, 1–33. Reprinted in Paul Thagard (2012). *The Cognitive Science of*

believe that my neurons and written memes have accurately represented its basic form. In a follow-up paper, Thagard analyzes two hundred "very important" scientific discoveries and technological innovations and makes "The stronger claim that *all* creativity requires novel combination of representations."[122]

Thagard and Stewart are well aware of the limitations of their model and discuss them freely. One that seems to me quite important is that "we have not produced a new model of analogical processing."[123] I think Poe would have made something more out of this, which is that they have not actually provided a model of analogical reasoning at all. It will be recalled from Chapters 4 and 7 that Poe made much of great scientists who were also poets. He also made his personification of the scientific mind, Auguste Dupin, a poet. In line with this is Aristotle's insight into genius and metaphor, "an intuitive perception of the similarity in dissimilars." Convolution does not seem to, and perhaps cannot, account for this kind of perception, so key to great discoveries in science, such as Darwin's discovery of evolution by selection as fundamentally the same process in nature as in manmade breeding. Seeing the process of selection in nature, analogous to the process of selection in the breeding of plants and animals by man, does not seem to fit the definition of an "emergent property" in a convolution, the "properties not possessed by (or simple aggregates of) either of the two vectors [neural representations] out of which it is combined," as our authors put it.[124] Seeing the similarity in dissimilars seems to be something different. Andrew Robinson puts it this way in his biography of Champollion:

Science: Explanation, Discovery, and Conceptual Change. Cambridge: MIT Press, 108, 117, 128, 118, 120.

122. Thagard and Stewart (note 121), 108; Paul Thagard (2012). "Creative Combination of Representations: Scientific Discovery and Technological Invention." In Robert W. Proctor and E.J. Capaldi, eds. (2012). *Psychology of Science: Implicit and Explicit Processes*. Oxford: Oxford University Press, 389–405. Reprinted in Thagard (note 121), 141–157. Thagard's sources for his analysis are Kendall Haven's *100 Greatest Science Discoveries of All Time* (2007) and Tom Philbin's *100 Greatest Inventions of All Time* (2003).

123. Thagard and Stewart (note 121), 139.

124. Thagard and Stewart (note 121), 117.

> Young [Thomas Young, Champollion's main competitor] brought to the problem one other extremely valuable and relatively uncommon ability. He had trained himself to sift, compare, contrast, retain and reject large amounts of visual linguistic data in his mind. This ability has been a *sine qua non* for serious decipherers ever since Young and Champollion, . . . Although outsiders to decipherment often like to imagine that, in today's world, computers could be programmed to accomplish such sifting, in reality the human factor remains all-important—mainly because a human being can spot that two signs that objectively look somewhat different are in fact variants of the same sign.[125]

Thagard and Stewart, for their model of "representation combination," emphasize that "Importantly, this is not a selective process: all of the original patterns are preserved," and although their "four component" model of convolution does "not preclude the presence of many reentry (feedback) loops," their main emphasis is on hierarchical, two-way "causal interactions" between social, psychological, neural, and molecular levels,[126] not the horizontal operations on representations that Robinson emphasizes. And maybe "sifting," after all, is not the necessary or common denominator to major eureka moments. Nevertheless, the key to a eureka moment in science does not seem to be an emergent property in a novel combination of representations, but rather, and at least often, the perception of a profound similarity in dissimilars, one that solves a mystery in a holistic way.

In line with this, I think Poe would have made a further criticism, which is that the convolution model does not account for another key feature of the eureka experience, which Poe took to be a criterion of truth and to be connected with the experience of beauty, namely, as we have seen in Chapter 4, what he called "the majestic highway of the *Consistent*." What is it that makes ideas put together in a new way productive of the eureka experience? It is that the new way *makes perfect (or near perfect) consistency out of the material in the domain of the problem situation*, data that

125. Robinson (note 18), 85.

126. Thagard and Stewart (note 121), 127, 129, 134.

did not previously fit together so that an explanation was wanting,—which is the very stuff of wonder if not necessity. Recall how Crick described his eureka experience in the second section above: "it cleared away so many difficulties. . . . I woke up that morning with only a set of confused ideas about the overall control of protein synthesis. When I went to bed all our difficulties had resolved and the shining answers stood clearly before us. Of course, it would take months and years of work to establish these new ideas, but we no longer felt lost in the jungle. We could survey the open plain and clearly see the mountains in the distance." This feature of major eureka experiences in science, of "*a perfect consistency [that] can be nothing but an absolute truth*," which is how Poe put it hyperbolically, as we have seen in Chapter 4, with Kepler as his model, points to something that is missing in the convolution approach to scientific discovery. Combine this now with a further key feature of the eureka experiences examined in the second section above, which is that the solution came to the scientist either in a dream, or right after awakening, or when they were awake but were doing something mundane, or the solution was preceded by a state of depression or something like it, or the conscious mind was not given the solution but was given, instead, a major clue, like Darwin's sudden reading of Malthus "for amusement," or Trivers' dream about Ernst Mayr.

All of this requires that we take a much more serious look at the human brain than normally afforded by cognitive science, which is to say we must now descend even deeper into the abyss, into the domain of neuroscience, the study of the brain.

Neuroscience

The brain is composed mainly of nerve cells called *neurons*, support cells called *glial cells*, and blood vessels called *blood vessels*. Glial cells are the glue that holds the brain together (*glia* is from the medieval Greek word for "glue") and they also provide metabolic support to neurons, part of which comes from digesting dead neurons. The ratio of glial cells to nerve cells in the brain is roughly 1:1 (the ratio of 1:10 or more is a dying myth). The human brain has upwards of 100 billion neurons, (depending on the individual,) each of which has upwards of 10,000 synaptic connections with other neurons. In a staggering display of complexity, these connections, with their electrochemical nerve signals that travel up to 250 mph, are the way neurons "communicate" with one another, allowing for seemingly infinite neural circuits and networks. The brain, furthermore, is divided into two halves called *hemispheres*, known also as the *left brain* and the *right brain*. Remarkably, the left brain controls the right half of the body, and the right brain controls the left half of the body. The two hemispheres are connected by what is called the *corpus callosum*, sometimes called the *great cerebral commissure*, (from the Latin *commissura*, meaning "a joining together,") a band of 200 to 800 million nerve fibers through which the two hemispheres communicate with each other. The outer layer of each hemisphere is known as the *cerebral cortex* (*cortex* means "bark"). In a mouse, for example, the cerebral cortex is almost perfectly smooth, while in higher animals it has folds and fissures as a way of accommodating the evolution of more functions in the limited space of the cranium. The cerebral cortex of humans is extremely bulbous and overshadows the rest of the human brain. Not surprisingly it is where our higher mental functions occur, namely, language, abstract thought, perceptual awareness, and consciousness. Over eons there has evolved an asymmetry of functions between the two halves of the brain, moreso in humans than in any other species of animal (with the possible exception of dolphins). Biology is statistical, but it is a genetic fact that 90 to 95% of the human species is left-brain dominant, meaning that they are right-handed and that their main language centers, both for comprehension and for speech, are in the left brain. Gender changes the asymmetries only a little. (For the sake of verbal economy, it should be kept in mind that throughout this section I shall confine myself to the statistical norm in hemispheric asymmetry.)

The above describes, in short order, the scene of the mystery, the interior of the room. It is a mystery that takes us back to Poe, with the "Bi-Part Soul" and his "double Dupin," to the nature of scientific imagination in the grand scale, as glorified in *Eureka*. According to Margaret Boden only a few years ago, creativity, defined as "the ability to generate ideas/artifacts that are new, surprising, and valuable," is "an unsolved puzzle for neuroscientists, not an ineluctable mystery essentially beyond their grasp," albeit "a puzzle that will be very hard to solve," so that in "common parlance . . . it's a 'mystery.'"[127] Poe would have loved this language, and most assuredly would have wanted to succeed in the solution of the mystery, with his ultimate object only the truth, though without making the mistakes which arise in the path of Reason through her propensity for seeking truth *in detail*.

What we shall be doing in this section is to follow a set of evidence from neuroscience, putting together the clues if you will, (some of them more controversial than others,) clues that ultimately lead us to a neuroscientific picture of Poe's "double Dupin" examined in the previous chapter,—of the "creative" and the "resolvent," the former the domain of the flight of the educated imagination, the *context of discovery*, the latter the domain of the creeping and crawling methods of deduction and induction and testing, the *context of justification*. In all of this I shall harbor no pretension to having unpuzzled the hitherto "unsolved puzzle for neuroscientists" of how scientific imagination in the grand sense actually works, the big picture cases of unconscious scientific problem solving, the kind we examined in the second section above, either the unconscious solutions themselves or the unconscious promptings that led to them and that subsequently entered consciousness as eureka moments. My goal, instead, since the debate in neuroscience is more alive now than ever before and has finally come of age, is much more modest, which is to follow a line of evidence in a manner that is, so far, a plausible and scientifically respectable unriddling of the riddle of scientific imagination begun with Poe's "double Dupin" and that is true to the spirit of Poe. In the very least, the inquiry will afford us amusement. Having accomplished this task, which is to fulfil my duty as a philosopher of science inspired by Poe, I shall then leave the mystery to others to see where it goes,—with the hope, of course, of seeing Poe vindicated before I lose my individuality to the Conqueror Worm.

---

127. Margaret Boden (2013). "Creativity as a Neuroscientific Mystery." In Oshin Vartanian, Adam S. Bristol, and James C. Kaufman, eds. (2013). *Neuroscience of Creativity*. Cambridge: MIT Press, 3.

Before we begin, it is to be noted that the asymmetries in function between the left and right brain have been discovered mainly by observation of the cognitive functions of people with diseased or damaged brains, and also from surgery, most importantly (i) *callosotomy* or *commissurotomy*, synonyms for when the corpus callosum is severed in order to reduce the severity of the patient's epileptic seizures, resulting in what is known as a *split-brain patient*, and (ii) *hemispherectomy*, the removal of one hemisphere of the brain, performed when the damaged hemisphere threatens the functioning of the undamaged hemisphere, which in adults "is almost never performed on the dominant sphere," and then usually only on the non-dominant cortex, "because of the very severe consequences."[128] More recently, brain scanning and imaging techniques have become widely used for understanding how the brain works, but although they "may have much to tell us about creativity," as two researchers recently put it, "in their current state of development they focus at too high a level to explain the magic of creativity."[129] Much the same conclusion was arrived at by two other researchers, who point out that brain image acquisition is correlated with task demands, such that "the inability to localize a network of underlying creativity may have as much to do with methodological vagaries related to task and acquisition techniques as with construct problems. Structural and lesion studies avoid task demand problems."[130] What this means is that we are more likely to be on the right track if we focus mainly on evidence from brain structure and brain damage or surgery.

Clue #1: Here is how the neuropsychologist Rhawn Joseph (well published in his field and not to be discredited simply because of his heterodox views on evolution and cosmology) describes the differences in function of the two hemispheres. The left brain has "linguistic consciousness," its domains of information processing include "math," "grammar," "logical and analytical reasoning," "perception of details," "seeing the trees." The right

128. Springer and Deutsch (note 112), 175.

129. Liane Gabora and Apara Ranjan (2013). "How Insight Emerges in a Distributed, Content-Addressable Memory." In Vartanian et al. (note 127), 19.

130. Rex E. Jung and Richard J. Haier (2013). "Creativity and Intelligence: Brain Networks That Link and Differentiate the Expression of Genius." In Vartanian et al. (note 127), 242.

brain has "unconscious awareness," "comprehension of music, emotion," "visual, emotional, musical, creative, and geometrical thinking," "seeing the forest," "reading between the lines," "perceiving the overall 'big picture,'" "gestalt formation."[131] Similarly, the neuropsychologist Joan Borod states that "Left-hemisphere strategies are described as analytic, linear, serial, detailed, and temporal, and as involving abstract, logical, sequential reasoning," while she adds that "emotional processing involves strategies and functions for which the right hemisphere is dominant: strategies termed nonverbal, synthetic, holistic, and Gestalt, and functions such as pattern perception, visuospatial organization, and visual imaging."[132] More succinctly, the neuroscientist John Eccles, a Nobel laureate, describes the left brain as "analytical and sequential," "an arithmetical hemisphere," while the right brain has "a strongly developed pictorial and pattern sense" and is "a geometric hemisphere."[133]

In all of this, it is remarkable how these descriptions of the left brain resemble Poe's description of the police detectives in the Dupin mysteries, examined in the previous chapter, including the Prefect of the Parisian police, who proceed by "the book," who "make a vast parade of measures," who "impaired his vision by holding the object too close" and "necessarily, lost sight of the matter as a whole," who "have no variation of principle in their investigations," who "extend or exaggerate their old modes of *practice*, without touching their principles," who regard their kind of reasoning as "*the* reason *par excellence*," and who accordingly reject educated guesses as merely "lucky."[134] And it is especially remarkable how the descriptions of the left brain *combined* with the descriptions of the right brain resemble Poe's description of Dupin and also the Minister D—, as mathematicians *and* poets, who because of both "reason well," which involves observation, to

131. R. Joseph (1992). *The Right Brain and the Unconscious: Discovering the Stranger Within*. Cambridge, MA: Perseus Publishing, 116.

132. Joan C. Borod (1992). "Interhemispheric and Intrahemispheric Control of Emotion: A Focus on Unilateral Brain Damage." *Journal of Consulting and Clinical Psychology* 60 (3), 339, 340–341.

133. John C. Eccles (1989). *Evolution of the Brain: Creation of the Self*. London: Routledge, 208, 209.

134. *P&T*, 398, 412, 690, 691, 689.

be sure, as a "species of necessity," and also "educated thought," which involves book learning, with books even as the "sole luxuries," but also a "wild fervor" and "vivid freshness of . . . imagination," glancing at a matter "in a side-long way," with "deviations from the plane of the ordinary" to "prominences above the plane of the ordinary," regard for "collateral or circumstantial events" and "the seemingly irrelevant," which includes "metaphor, or simile" and accordingly "guesses," with "some principle of guessing."[135] This is painting in brush strokes, but the picture should be clear enough.

Clue #2: It has long been known that the two main language centers are in the left hemisphere, Broca's area for speech expression, whether verbal or written, and Wernicke's area for speech comprehension. In infants the capacity for development seems equal in both hemispheres, though in most people the left brain acquires the dominance in speech by age four or five, while the right hemisphere "retains some competence in understanding."[136] Here it seems safe to say that ontogeny probably recapitulates phylogeny, given not only that the above is a cross-cultural occurrence but also that primates are typically right-handed and therefore left-brain dominant. What the right hemisphere seems to add to the understanding of language in humans is the emotional content, which includes the "prosodic elements of speech."[137] At any rate, what is extremely significant is that in a split-brain patient, although there is some subcortical transfer of information, in the absence of external clues and cues the left brain literally does not know what the right brain is *thinking*, and vice versa.

Clue #3: Because of the above, it is sometimes claimed that there are fundamentally two conscious minds in the human brain, a double consciousness proper in the case of split-brain patients and an integrated double consciousness in the rest of us. Roger Sperry, for example, who received a Nobel Prize for his pioneering work with split-brain patients in the 1950s and '60s, concluded not

135. *P&T*, 690, 403, 412, 400, 412, 414, 520, 534, 425, 694, 689.

136. Eccles (note 133), 81.

137. Isabelle George (2010). "Hemispheric Asymmetry of Songbirds." In Kenneth Hugdahl and René Westerhausen, eds. (2010). *The Two Halves of the Brain: Information Processing in the Cerebral Hemispheres*. Cambridge: MIT Press, 92. See also Springer and Deutsch (note 112), 234.

only that "In many respects each disconnected hemisphere appears to have a separate 'mind of its own,'" but that "Everything we have seen so far indicates that the surgery has left these people with two separate minds, that is, two separate spheres of consciousness. . . . This mental dimension has been demonstrated in regard to perception, cognition, volition, learning, and memory."[138] Others, however, such as Eccles, have concluded that consciousness fully resides only in the left hemisphere. He allows that many nonhuman animals have a kind of consciousness, evidenced alone in "the play of young animals," but since the right brain in humans did not develop linguistically, and "the conscious being or self" in the left hemisphere "is recognizably the same person as before the operation [commissurotomy]," it follows for Eccles that the consciousness in the right hemisphere "is of a limited kind and would not qualify the right hemisphere to have personhood," so that "the commissurotomy has split a fragment off from the self-conscious mind, but the person remains apparently unscathed, with mental unity intact in its now exclusive left hemisphere association."[139]

Perhaps the debate is better conceived along a different line. According to the neurophilosopher Patricia Churchland, one of the defining features of consciousness is that it is single focused, that "You cannot follow two conversations at once, you cannot at the same time do mental long division and watch for dangerous eddies in a fast-moving river," so that when we think we are multitasking we are probably only "shifting attention back and forth between two or possibly three tasks."[140] For basically the same reason Carl Jung rejects the concept of "universal consciousness" advocated by yogis and meditative gurus. "They do not realize," he says, "that a 'universal consciousness' is a contradiction in terms, since exclusion, selection, and discrimination are at the root and essence of everything that lays claim to the name 'consciousness.'" As one expands one's consciousness it loses "clarity of detail" and becomes "nebulous," he adds, so that it is "logically

138. Quoted in Springer and Deutsch (note 112), 52, 334.

139. Eccles (note 133), 174, 207, 210.

140. Patricia S. Churchland (2013). *Touching a Nerve: The Self as Brain*. New York: W.W. Norton, 240. Note the word "eddies" in the quotation. We shall return to it below.

identical with unconsciousness."[141] It would therefore seem that the light of consciousness is best thought of like a flashlight. But the unconscious arguably has its own light, too, as we have seen in the second section above. Recall especially Dyson's "Suddenly the whole picture became clear" and Crick's "we no longer felt lost in the jungle. We could survey the open plain and clearly see the mountains in the distance." The light of the unconscious, then, would seem best thought of like a ceiling light, (at least sometimes,) illuminating the entire room. So perhaps Joseph's "unconscious awareness" is a better label for what goes on in the right brain. As we shall soon see, there is more that sustains the analogy.

Clue #4: We have seen in the second section of this chapter that many eureka moments were experienced by scientists in a dream, or just after waking from a dream, or during a daydream or similar mental state. It is therefore important to ask whether there is any difference in the hemispheres in terms of dreams. According to Springer and Deutsch, "some investigators have suggested that dreaming is part of the realm of the right hemisphere. A few have proposed that the right hemisphere does all of the dreaming; others proposed that the dream state allows the right hemisphere to express itself more freely than usual because the left hemisphere does not dominate or interfere."[142] This is of obvious relevance to eureka moments. However, the investigators were speculating based mainly on the already known functional asymmetries in the hemispheres. In more recent years, a veritable mountain of research has been done on laterality and sleep using brain scan and neuroimaging techniques. In the case of EEGs, however, which measure brain waves, although it is clear that both hemispheres dream, studies of functional asymmetry in dreams have produced "mixed results" overall, "contradictory findings," "very conflicting results," as the neuropsychologist Maria Casagrande puts it in her recent review of the literature. Other physiological techniques, such as PET and fMRI, similarly "do not allow us to reach any clear-cut conclusions," although an interesting

141. C.G. Jung (1939). "Conscious, Unconscious, and Individuation." In Herbert Read et al., eds. (1969). *C.G. Jung: The Archetypes and the Collective Unconscious*. 2nd ed. Princeton, NJ: Princeton University Press, 287–288.

142. Springer and Deutsch (note 112), 343.

behavioral asymmetry that has long been established is that right-handed people during sleep will use their left hand in a defensive move when their nose is tickled, indicating right-brain dominance during sleep. Casagrande's ultimate conclusion is nonetheless significant, which is not only that there does seem to be a shift in "mentation" from left brain to right brain with the onset of sleep, from left-brain dominance to right-brain dominance, accounted for mainly by "a relative absence of external input and the voluntary control of the stream of thought," but that this "could enable a hierarchical reorganization of cognitive mechanisms, functional to a mental activity that is different from the one typical of the wakeful condition. That is, a change in hierarchical integration from left hemisphere to right hemisphere could be functional to some internally oriented operations that are more prominent when the brain is drowsy or sleeping."[143] Big picture thinking, of course, along with Aristotle's "an intuitive perception of the similarity in dissimilars," are mental activities "different from the one typical of the wakeful condition," as revealed by eureka moments.

Split-brain research also strongly supports the above. For example, Springer and Deutsch refer to a study published in 1977 of twelve split-brain patients in which it was reported by the researcher that "patients after commissurotomy reveal a paucity of dreams, fantasies, and symbols. Their dreams lack the characteristics of dream work; their fantasies are unimaginative, utilitarian, and tied to reality; their symbolization is concretistic, discursive, and rigid."[144] Although Springer and Deutsch caution that these patients suffered from epilepsy beforehand, it is surely significant that it was only the left brain that got to describe its dreams to the researcher, and that there were twelve of them.

Clue #5: A further line of research highly relevant here but overlooked in Casagrande's review is hemisphere asymmetry in brain activity correlated with "the subjective experience of a flash of insight, or '*aha*!,'" the "*self-rated* originality of ideas," as the neuropsychologists Andreas Fink and Mathias Benedek put

143. Maria Casagrande (2010). "Laterality and Sleep." In Hugdahl and Westerhausen (note 137), 315, 316, 317, 319, 328, 331.

144. Sally P. Springer and Georg Deutsch (1993). *Left Brain, Right Brain*. 4th ed. New York: Freeman, 280.

it. Building on two earlier studies, conducted in 2004 and 2006, respectively, what they found is that "the production of ideas that were subjectively rated as more original was reflected in a different activity pattern of the brain than the production of less original ideas. Analyses revealed that the production of more original ideas exhibited a larger right-hemispheric ERS in the lower alpha band . . . than the production of less original ideas, whereas in the left hemisphere no ERS differences in relation to self-rated originality of ideas were found."[145] ERS means "event related synchronization," and alpha waves are "the electrophysiological background of a relaxed eye-closed waking state," which are replaced by faster beta waves when we are attentive, and then by much slower delta and theta waves when we are asleep, although "the alpha rhythm has also been recorded during sleep—above all, during REM."[146] Rapid eye movement (REM) is most closely connected with vivid dreaming, but it is not exclusive to the dream state per se. The conclusion of Fink and Benedek was later replicated comparing individuals rated higher in originality than others, with the result that "the higher original individuals exhibited a comparatively strong hemispheric asymmetry with respect to alpha activity, with a stronger task-related alpha synchronization in the right than in the left hemisphere, while in less original individuals no hemispheric differences with respect to alpha activity emerged." The authors are well aware that the originality tasks given to their subjects were relatively simple, like finding the association between three seemingly unrelated words such as "rat, blue, cottage,"[147] and that the subjects were hooked up to electrodes or lying supine in an fMRI scanner, so that it is difficult to generalize to "ecologically valid 'real-life' creativity tasks."[148] And yet their

145. Andreas Fink and Mathias Benedek (2013). "The Creative Brain: Brain Correlates Underlying the Generation of Original Ideas." In Vartanian (note 127), 213.

146. Casagrande (note 143), 318.

147. Did you think "_ _ _ _ _ _"? *Aha!*

148. Fink and Benedek (note 145), 214, 223. For an empirical test of the *objective accuracy* of relatively simple eureka moments, see Carola Salvi et al. (2016). "Insight Solutions Are Correct More Often Than Analytic Solutions." *Thinking & Reasoning* (August 4), 1–18.

research is in the middle lane of what Poe in *Eureka* calls "the great thoroughfare—the majestic highway of the *Consistent*."[149]

So is the highly-related subjective experience of the person who invented and developed the world's first safe and effective polio vaccine, thereby putting an end to one of the world's worst plagues, namely, the virologist Jonas Salk, (he was the inventor, not the plague,) as reported among the interviews collected by Csikszentmihalyi:

> Salk's best ideas often come to him at night when he suddenly wakes up and after about five minutes of visualizing problems he had thought about the day before he begins "to see an unfolding, as if a poem or a painting or a story or a concept begins to take form." Sometimes when such associations of ideas begin to occur in his mind, Salk claims he feels a palpable physiological response which indicates to him that the right side of the brain has become active. At this point he either falls into a deep sleep, or he sits up in bed, turns on the light, and writes down the thoughts that have occurred to him, . . .[150]

Clue #6: What sends us skidding off "the great thoroughfare," possibly even into a multiple rollover with teeth chattering and eyes starting from their sockets, is what the neuroscientist Vilayanur Ramachandran—in a book appropriately entitled for our purposes—calls "metaphor blindness." Fully recognizing that "the use of metaphor and our ability to uncover analogies is the basis of all creative thought," and admitting that "we know almost nothing about why metaphors are so evocative and how they are represented in the brain," he brings to the fore the fact, which he calls "compelling," that "there are hints that people with damage to the left inferior parietal lobule (IPL) often lose the ability to interpret metaphors and become extremely literal minded." For example, when asked for the meaning of "All that glitters is not gold," even when told that it is a proverb, a patient with a stroke in his left IPL ultimately replied, "it means you have to be very careful when you go to buy jewellery." Although formerly a prominent

149. *P&T*, 1269.

150. Csikszentmihalyi (note 47), 287.

physician, the patient was now extremely literal minded. Ramachandran found metaphor blindness in a number of his patients with damage to their left IPL, specifically to the angular gyrus portion, but of course he admits that a larger sample of patients is needed in order to draw firm conclusions.[151]

Only recently discovered in terms of its operation, the left IPL is situated right beside Wernicke's area (language comprehension) and is connected by large bundles of nerve fibers with both Wernicke's area and Broca's area, (language expression,) both of which are in the left brain. It is also one of the last structures to have evolved in the human brain, and accordingly is one of the last of the brain structures to mature in children. In fact, even though it is more prominent in primates than in lower mammals, and more prominent in the great apes compared with other primates, only in humans did the left IPL evolve two parts, the angular gyrus and the supramarginal gyrus. Does this mean, then, not only that apes taught (allegedly) ASL, such as Koko the gorilla, can never be poets, but also that the right brain in humans is not involved in metaphorical thinking, in Aristotle's "an intuitive perception of the similarity in dissimilars"?

At least the latter does not follow. First, just because damage to a particular part of the brain seems to knock out metaphorical thinking, it doesn't automatically mean that that part of the brain is the "metaphor center," as other parts of the brain can contribute to the same function (token or type). There is an analogy here with genes and their physical expression. As Richard Dawkins puts it, "to speak of a gene 'for' something only ever means that a *change* in the gene causes a *change* in the something."[152]

Second, the left IPL has a homologue in the right brain, appropriately called the right IPL and which is also subdivided like the left IPL. Moreover, not only are these two lobules connected by the corpus callosum, but there is evidence that the right IPL is predominant when it comes to auditory spatial processing,[153]

151. V.S. Ramachandran (2011). *The Tell-Tale Brain: A Neuroscientist's Quest for What Makes Us Human*. New York: W.W. Norton, 105, 7, 130–131, 180.

152. Richard Dawkins (1989). *The Selfish Gene*. 2nd ed. Oxford: Oxford University Press, 281.

153. Robert A. Weeks et al. (1999). "A PET Study of Human Auditory Spatial Processing." *Neuroscience Letters* 3, 155–158.

to temporal order processing in the visual field,[154] and that both hemispheres play a complementary role in visual object recognition while only the right IPL is integral to visual space perception, including both 2D and 3D representations.[155] Given that the left and right IPLs are homologous, what this suggests is that the perception of similarity in dissimilars, as we all should know, and certainly as Einstein emphasized, need not be linguistic at all.

Third and finally, and most important as a matter of evidence, in a recent meta-analysis of neuroimaging studies on metaphor, (the first of its kind, actually,) in which studies were distinguished between those that used novel metaphors and those that used familiar, conventional metaphors, (which are more likely to be confined to memory retrieval,) the authors conclude that "the RH [right hemisphere] is involved in processing mainly novel, unconventional concept combinations, such as novel metaphors" and "only novel, but not conventional metaphors report robust RH activation in the RIFG [right inferior frontal gyrus] and right ACC [anterior cingulate cortex]."[156] Ramachandran's few cases of metaphor blindness, then, might simply involve cases of left-brain incapacitation for metaphorical interpretations of known metaphors. At any rate, the implication of novel metaphors for scientific discovery should be obvious, whatever the brain mechanisms in the right hemisphere.

Clue #7: Perceiving "the similarity in dissimilars" is a kind of abstract thinking. But how does that work? The neuroscientist Rodrigo Quiroga, wanting to understand how abstraction works in the human brain, claims that abstraction works not by making a comparison with an exemplar (the view of Berkeley and Hume in the eighteenth century), but by overlapping neural representations and *keeping what is common while forgetting what is not common* (presaged by Locke in the seventeenth century). Interestingly, "this abstraction, this loss of detail," he says, "reaches

154. Lorella Battelli et al. (2007). "The 'When' Pathway of the Right Parietal Lobe." *Trends in Cognitive Science* 11 (50), 204–210.

155. Selene Schintu et al. (2014). "Object and Space Perception—Is It a Matter of Hemisphere?" *Cortex* (57), 244–253.

156. Isabel C. Bohrn et al. (2012). "Looking at the Brains Behind Figurative Language—A Quantitative Meta-Analysis of Neuroimaging Studies on Metaphor, Idiom, and Irony Processing." *Neuropsychologia* 50, 2681.

its pinnacle in the hippocampus," and he claims that he and his colleagues discovered neurons in the hippocampus that do this, that "encode concepts in the abstract form in which they will be used by memory functions," which is "indeed ideal for associating concepts, for generating relationships, for thinking."[157] The hippocampus is a seahorse-looking structure, doubled in mammals with a left and right version, both located immediately below the cerebral cortex and with their own connecting fibers called the *hippocampal commissure*, which means the hippocampus should have a neural information connection with the left and right brains even in patients with a severed corpus callosum. Long believed to play an important role in memory and spatial navigation, it extends far back in evolutionary history, and given the "pinnacle" claim of Quiroga I would think it plays a central role in animals in recognition and in stereotyping. At any rate, whether Quiroga is right about the hippocampus and abstraction, the role of memory is surely central in whatever the process of scientific imagination, which leads us to our next clue.

Clue #8: Memory in brains does not work at all like memory in computers. In computers, what you retrieve is exactly what you stored. People naturally think that human memory works the same way, or more simply like playing back a tape recording, or rereading something written down. This has led to some bad philosophy of science, even given that human memory can be remarkably accurate. In brains, as Liane Gabora and Apara Ranjan put it, (a neuroscientist and a cognitive scientist, respectively,) "an item in memory is never re-experienced in exactly the form it was first experienced, but is colored, however subtly, by what has been experienced in the meantime, and reassembled spontaneously in a way that relates to the task at hand (this is one reason eye-witness accounts cannot always be trusted)."[158] All of this is because memories are stored in networks of neurons, networks moreover that are not individually dedicated to a particular memory but overlap with other networks, such that a given neuron can play a role, more or less, in thousands of memories, perhaps even more. (The

157. Rodrigo Quian Quiroga (2012). *Borges and Memory*. Juan Pablo Fernández, trans. Cambridge: MIT Press, 189, 190.

158. Gabora and Ranjan (note 129), 24.

same would have to be true of the neural representations of concepts, that they overlap.)

This undermines the view that scientific imagination proceeds in the manner of a rapid process of "guesses," that scientific imagination, as we have seen Whewell put in Chapter 6, is "a constant invention and activity, a perpetual creating and selecting power . . . of which the last results only are exhibited to us," or that it is a process of "blind trials" and the "natural selection of hypotheses," as Popper put it, or similarly a "blind-variation-and-selective-retention process," as Campbell put it, which we have seen later in that same chapter. Given that "There are potentially infinitely many ways of tweaking what we know to come up with something new," as Gabora and Ranjan put it, which connects with the problem of underdetermination in philosophy of science, what follows from the overlapping way that neurons represent is that "this enables one to go beyond what one knows without resorting to trial and error," that it "enables the brain to accomplish creative acts *without* recourse to a 'search and selection' type explanation."[159] Their answer, in short, is that in overlapping representations the "something new" of creative insight is "*implicitly* present" in those overlapping representations, that solutions are problem-sensitive and are the more creative when they come from an overlap in neuronal distributions that is wide rather than narrow, in "associative thought" and its "broad activations," when the "particular distributed set of neurons has never been activated before as an ensemble," thus accounting for the "*magic of creativity*" which "isn't really magic after all," what they call "the ability to pull out of memory something new and appropriate that was never explicitly stored there."[160] What is missing from this account, of course, is what we have already seen, which is the revelatory insight of explanatory consistency, (the lights going on,) the accompanying emotion of the eureka experience, and the asymmetric roles of the brain hemispheres, all of which were emphasized by Poe in 1848 except for the latter (at least as far as we know). Nevertheless, the neurobiology of memory has given us another important clue.

159. Gabora and Ranjan (note 129), 21, 24.

160. Gabora and Ranjan (note 129), 21, 31, 33, 19, 36.

Clue #9: Any investigation into the divided brain cannot afford to ignore what is known about the corpus callosum, the thick and complex band of nerve fibers that connects the left and right brain, cortical homologue to cortical homologue rather than in a crisscross fashion. From mainly the study of split-brain patients, it became evident that the corpus callosum doesn't just hold the two hemispheres together but also serves a communicative role, integrating the left and right brain into a functional whole. There is no question about this anymore. But there is an interesting twist that has come out in more recent years, which is that the corpus callosum not only transmits nerve impulses between the hemispheres—*it also functions to inhibit them*. As the neuropsychologists Juliana Bloom and George Hynd sum the situation in their review of the literature, "the corpus callosum may be inhibitory at times and excitatory at other times, depending on the task. In some instances, it might be more efficient for the hemispheres to operate independently and in some instances interhemispheric communication might be necessary to complete the task."[161]

What this means is that the corpus callosum can no longer be viewed as passive, but must instead be viewed as active. It also means that even while we have a fully intact corpus callosum, our left and right brains do not necessarily function as one brain, as an integrated brain, but sometimes are functionally *dis*integrated to a substantial degree. And, finally, it means that a larger than normal corpus callosum, in any of its regions, cannot automatically be interpreted as a greater degree of integration and cooperation between the hemispheres, since it could also indicate a greater degree of inhibition and independence between them.[162]

161. Juliana S. Bloom and George W. Hynd (2005). "The Role of the Corpus Callosum in Interhemispheric Transfer of Information: Excitation or Inhibition?" *Neuropsychology Review* 15 (2), 68.

162. This is an important implication. As Bloom and Hynd point out, "one cannot assume that a larger callosum necessarily indicates a greater number of nerve fibers." Bloom and Hynd (note 161), 60. This is because "Callosal size may vary due to the differences in the number and size of nerve fibers, glial cells and blood vessels." Suganthy Rabi et al. (2007). "Quantitative Analysis of the Human Corpus Callosum Under Light Microscopy." *European Journal of Anatomy* 11 (2), 95. I would take this brain matter one step further: *even a greater number of nerve fibers in the corpus callosum does not necessarily mean greater interhemispheric communication and*

How interesting, then, that we have a double brain, which in turn is connected by a double function, one that allows the two brains not only to collaborate directly with each other but also to function simultaneously and independently, one doing something while the other is doing something else.

Summing up the evidence from Clues #1 to #9, the model of scientific imagination that we are left with, especially for scientific imagination in the grand sense, is that (using the language of Joseph) the "language-dependent conscious mind," which resides in the left brain, is faced with problems in its environment that it tries to solve, collecting as much information as it can, but it is limited by its single focus analytical, syntactical, and serial nature,

---

cooperation. This is because it is not yet known, as Bloom and Hynd themselves point out, "if both excitation and inhibition can occur at the same time." Bloom and Hynd (note 161), 69. What I take this to mean is that some nerve fibers in the corpus callosum *might possibly* serve a primarily or exclusively excitatory (communicative) function and others a primarily or exclusively inhibitory function, so that a thicker corpus callosum, if thicker because of a greater number of nerve fibers, might be thicker because of a greater *proportion* of the inhibitory compared with the communicative nerve fibers. Greater callosal size might also be the result not of more nerve fibers but of the requirements of the mechanisms that control whether nerve fibers excite or inhibit, given that we are told that "inhibition or excitation can occur depending on the neurotransmitters, receptors, and interneurons involved." Bloom and Hynd (note 161), 69. Either way, given that a larger corpus callosum does *not* automatically mean greater interhemispheric communication and cooperation, this brings us back to Einstein's brain. A recent analysis has shown that it has "an extraordinary prefrontal cortex" and very unusual "parietal lobes," which the authors say "may have provided underpinnings for some of his extraordinary cognitive abilities, including his productive use of thought experiments." Dean Falk et al. (2013). "The Cerebral Cortex of Albert Einstein: A Description and Preliminary Analysis of Unpublished Photographs." *Brain* 136 (April), 1304, 1325. From this and the fact that Einstein's corpus callosum was thicker than usual compared with other males, it *cannot* be concluded that "connectivity between the two hemispheres was generally enhanced in Einstein compared with controls," that "The results of our study suggest that Einstein's intellectual gifts were not only related to specializations of cortical folding and cytoarchitecture in certain brain regions, but also involved coordinated communication between the cerebral hemispheres." Men et al. (note 114), e7. Given Einstein's eureka moments and his repeated emphasis on imagination and intuition, we *could* also conclude, perhaps even more strongly, that his intellectual gifts involved *enhanced coordinated inhibition* between the cerebral hemispheres!

while the unconscious mind in the right brain, with its synthetical, holistical, and metaphorical abilities, which are language independent, attempts (preferably independently of the left brain) to solve the problem as a big picture, in terms of overall consistency. How "unconscious awareness" does this is still largely a mystery, although it is apparently not a sifting but some sort of abstractive process motivated by total consistency, involving representations and ideas embedded in overlapping networks of neurons. The left brain, with its linguistic kind of consciousness, can work on the problem for hours or days or even years without success, but once the right brain has the solution to the problem, (if it does,) and the left brain is in a low-active receptive state, such as upon awakening or while doing something relaxed, the right brain sends the solution via the corpus callosum to the left brain, such that "the creative solutions suddenly burst into the conscious mind,"[163] to linguistic consciousness, where the "intuition" can be processed in terms of language and be expressed to others and worked out in detail, including testing.

In an important sense, then, Poe's "double Dupin" turns out to be, after all, the double brain, the right brain the throne of the "creative," the left brain the throne of the "resolvent," mind. And it is a model not just for the scientific genius but for us all—each of us mentally, in the very least, a two in one.[164]

163. Joseph (note 131), 117, 115, 275.

164. It is also possibly a model for the West vis-à-vis the East. In an interesting view of the matter, the psychologist Iain McGilchrist designates the right brain as properly the "master" and the left brain as properly the "emissary," which he applies to "the cultural history of the West" over roughly the past five hundred years, such that "the emissary became contemptuous of his master," "the master was usurped, the people were duped, the domain became a tyranny,"—again, "the Master, the one whose wisdom gave the people peace and security, is led away in chains. . . . betrayed by his emissary,"—again, though the emissary "is ultimately dependent on, one might almost say parasitic on," the master, the emissary "is filled with an alarming self-confidence" and is, albeit "gifted," "an ambitious regional bureaucrat with his own interests at heart." The upshot for McGilchrist is that if the "power struggle" in our "bihemispheric" brains is not resolved with its "need to cooperate," then the West is doomed, although "the awareness of the situation may enable us to change course before it is too late." Iain McGilchrist (2009). *The Master and his Emissary: The Divided Brain and the Making of the Western World.* New Haven: Yale University Press, 14, 6,

Before we move on to evolution, the question we must now ask ourselves is the proverbial and seemingly timeless one: "Who, or what, am I?" According to the neurophilosopher Patricia Churchland, "Lately, I think about my brain in more intimate terms—as *me*." She sums her reasons in her book as follows: "In death, brain cells quickly degenerate, with massive loss of information. Without the living neurons that embody information, memories perish, personalities change, skills vanish, motives dissipate. Is there anything left of me to exist in an afterlife? What would such a thing be? Something without memories or personality, without motives and feelings? That is no kind of *me*."[165] The extreme of the brain view is that the self is *ontologically* nothing more than the brain. As the philosopher of science Alex Rosenberg, an advocate of this view, puts it, "The mind is the brain." More exactly, "There is no self, soul, person. Scientism must firmly deny its existence. The self, as conveyed to us by introspection, is a fiction. It doesn't exist."[166] Similarly Quiroga, that Descartes was wrong with his

3, 6. Whatever one makes of McGilchrist's application of the divided brain to the rise (and possible fall) of the West, the focus of the second half of his book, (I myself am highly skeptical, given that the ancient history of the East seems dominated by "tyranny" rather than genuine "peace and security,") there is no doubt that the first half of his book, devoted to the brain itself, is a veritable *tour de force*.

165. Churchland (note 140), 11, 12. Churchland, by the way, is not entirely hardcore in her reductionism. Mental reality "depends on a neural reality," "reductionism is essentially about *explanation*," and "Of course, science is not the be-all and end-all." After all, "Last week, unaccountably, a raven flew in the back door and perched on the rocking chair. She was utterly statuesque and stunningly huge. All beauty. I talked to her briefly, and then she found her way out again. Balance, dear Aristotle, balance." Churchland (note 140), 263, 264, 261. The reference to Aristotle is a nod to him on the importance of striking a balance between believing too easily and disbelieving too easily. But we have to wonder about the raven episode. Is Churchland serious or had she been reading Poe and is playing with us? She did, after all, as we have seen above, also use the word "eddies." My bet is that cryptic references to Poe, whether conscious or unconscious, are afoot here. In fact I would bet my afterlife on it!

166. Alex Rosenberg (2011). *The Atheist's Guide to Reality: Enjoying Life without Illusions*. New York: W.W. Norton, 195, 223.

"distinction between mind and body," as "mind and neurons are one and the same thing."[167]

It is impossible to say how first the idea entered my brain; but once conceived, it haunted me day and night. Oh God! what *could* I do? What I call my brain, in this view, is the real David N. Stamos, perhaps four pounds in weight (much less if only the neurons) and the tell-tale author of the book before you. But as such it is a *co-authored* book, given that me, my brain, is a left–right double brain, (each half warranting only half the royalties, perhaps,) with only the book itself enjoying (hopefully) a kind of afterlife.

But now we are back to Descartes. The brain, single or double, is objective. It has extension. It can be scanned, dissected, measured, and weighed. But in addition to the brain there is the *subjective* self. It is accessible to no one but *me*. It does not have extension. It cannot be dissected, measured, or weighed. And yet it is surely something, not nothing. Perhaps this is why the philosopher-turned-neuroscientist Sam Harris identifies the self not with the brain but with the conscious self. In his book against free will, for example, he tells us not only that "We are conscious of only a tiny fraction of the information that our brains process in each moment," but that given any so-called choice of mine it "was made for me by events in my brain," events "that I, as the conscious witness of my thoughts and actions, could not inspect or influence." Even more clearly, he quotes someone who characterizes his view as "the conscious self . . . as constituting the *real* self," to which he adds that the author "perfectly articulates" his view. In this view, then, the real me is my conscious self, not my brain, even given what Harris calls "the unconscious origins of the conscious mind."[168] One could also take something of a mediated

167. Quiroga (note 157), 92, 172.

168. Sam Harris (2012). *Free Will*. New York: Free Press, 7–8, 21, 22, 5. Harris' mention of information raises an interesting problem: Rosenberg's physicalism is so extreme that he not only believes that computers do not contain information but that brains ("neurons working together") lack *aboutness*, since "nothing physical can be *about* anything." Rosenberg (note 166), 187, 191, 193. The self-referential paradox is that his book is literally *about* nothing, so that at $25.95 US it is highly overpriced. The real problem with Rosenberg's view is that information itself is not physical but instead

view, such as that of the ethicist Peter Singer, who for the legal and moral purpose of defining human death does not draw the line at brain death, since some parts of the brain can be dead but not others, but at the death of the part of the brain responsible for consciousness, at "the permanent cessation of function of the cerebral cortex, not of the whole brain," since what we "ought to care about—is the *person* rather than the body."[169]

But no matter, never mind. The emphasis on the conscious self means that when "I" go to sleep at night—"I" as my conscious self—"I" cease to exist, (ignore dreaming for the moment, and think, notably, of sedation by the prophet Propofol,) and that when "I" awake in the morning (or the middle of the night) it is not the same "I," not *numerically* the same *me*, but *qualitatively* sufficiently similar to evoke the mistake (subjectively in me and objectively in others) that it is numerically the same *me*. And so the author of this book before you is not one, but many. (In fact, "My name is Legion," each of us inside this man not synchronically but diachronically.) That hardly feels right, but if it is right, I should think the fact ought to make its way into the royalty contract, perhaps calculated by the number of days (roughly equal to the number of conscious selves) it took to write this book.

At any rate, if "I" am an internal projection of my brain, (which is apparently the most common view in cognitive science and neuropsychology,) then this certainly presents us with an interesting twist to Poe's lament, "Is *all* that we see or seem / But a dream within a dream?"[170] If the plot of my life, centered on this brain of mine, is something like a dream of Poe's God, then as an

---

supervenes on a disjunctive base of physical mediums (a token of the book before you, for example, can exist on paper, a CD, a magnetic tape, and so on, but the book itself is none of those). Information is therefore not reducible to the physical, (even though it depends on it,) which opens the door to the reality of mind. See, e.g., my (2008). *Evolution and the Big Questions: Sex, Race, Religion, and Other Matters*. Malden, MA: Blackwell, 246, 56, 58, 104.

169. Peter Singer (1995). "Is the Sanctity of Life Ethic Terminally Ill?" *Bioethics* 9 (3–4), 307–343. Reprinted in Helga Kuhse and Peter Singer, eds. (1999). *Bioethics: An Anthology*. Malden, MA: Blackwell, 296.

170. "A Dream within a Dream" (1849). *P&T*, 97. In line with Chapter 3, the role of "Hope" in the poem should not go unnoticed.

internal projection I am like a dream of this brain, and so yes, I *am*, in a sense, a "dream within a dream."

But here is an even more disturbing thought. If I am my conscious self, then really I am not only an internal projection of what I call my brain, but obviously an internal projection with *subjectivity*. And if what I call my brain can do that on a daily basis, and even in some cases with more than one personality, (known as multiple personality disorder,) there would then seem no reason why the people and pets in my dreams (other than myself) could not also have subjectivity. After all, it is not only while awake that I am above the "threshold intensity" (Jung's phrase) for being conscious, but also (I submit) while I am dreaming, for in the very least I have subjectivity and perspectivity in my dreams, too. So why not also the others in my dreams? While awake we naturally assume, those of us of a scientific mind, that those people and pets in our dreams were produced by whatever is the dream mechanism in each of our brains, that they were *not* really real, *not* really persons, that they only *seemed* so. But we need to think this thought through to the end. For if the brain can produce *one* (single or serial) conscious self at a time, (and even flip between two or more over time,) there would then seem no good reason why the very same brain could not produce *many* conscious selves,—*simultaneously*,—whether in hallucinations covered in books and chapters on hallucinations—*Reynolds!*—or in by far the most common hallucination of them all, the land known as "dreamland." Dream persons would then be *just as real* as you or me.

When I first thought this thought my nerves became thoroughly unstrung, and I fell a prey to perpetual horror. In the very least I was terrified of falling asleep, for in waking from a dream the consequence is possibly a mass dying, like the flames of multiple candles snuffed out with no more thought or conscience than that of the morning breeze. Never mind, then, "what dreams may come" in what Hamlet calls "that sleep of death," for "To sleep! perchance to dream"—that's the "rub" that comes before all else!

But now what of the *unconscious* mind? Is it not a *self*, too, perhaps even the *real* self? And if so, should it not be given the credit, or at least half the credit, for the authorship of the discoveries chronicled in the second section of this chapter and other discoveries like them? Robert Trivers jokingly suggested to Ernst

Mayr that he should cite "Ernst Mayr, personal communication" in his paper, given that Mayr figured in his dream with the key to the solution.[171] Perhaps, more accurately, Trivers should have included his own unconscious as the co-author instead.

If we turn to Freud for the answer, he divided the human mind into conscious, preconscious, and unconscious, in addition to *das Es*, *das Ich*, and *das Über-Ich*, ("the It," "the I," and "the Over-I,") rendered as *id*, *ego*, and *superego* using Latin. While *ego* means "I," it would certainly be a mistake to think Freud thought of the ego as one's true self, given that it is a derivative entity in psychoanalysis, that it is not equated with consciousness but is partly unconscious, that it can be "split" as a defense mechanism, and that consciousness itself is filled with what Freud called "gaps." But it would be no less mistaken to think of the Freudian unconscious as the true self, even though Freud considered it the vast majority of our mind or psyche and called the id "the core of our being." This is because *id* means "it" and the Freudian unconscious consists of the instinctual forces of the id along with much else, including much that is antagonistic within itself and with the ego itself, to the point that Freud called the id "the internal enemy." According to the philosopher Alex Watson, Freud in fact freed us from the "narcissism" of the concept of self. "It was part of the genius of Freud," he says, "that he was able to see through this concept. He did not accept the existence of any single entity that could be put forward as an answer to the question, 'Who am I' or 'What am I?' We neither are nor contain anything that remains identical over time. Even at one moment of time, we are not one thing. Rather we are a multiplicity of interacting systems and processes."[172]

Jung, on the other hand, who unlike Freud seemed to be interested in everything, (which showed in his approach to the unconscious, in his remarkably wide-ranging search for "the similarity in dissimilars,") provided a very different answer. Jung divided the mind into the conscious, the personal unconscious, and the collective unconscious: the personal unconscious consisting of elements acquired in one's lifetime and therefore unique to the individual,

171. Trivers (note 52), 161.

172. Alex Watson (2014). "Who Am I? The Self/Subject According to Psychoanalytic Theory." *SAGE Open* (July–September), 1–6.

and the collective unconscious consisting of instinctual archetypes inherited as a member of the human species. Jung clearly thought of the unconscious as fundamental. In his autobiography, for example, he writes of consciousness as "phylogenetically and ontogenetically a secondary phenomenon," that "Consciousness began its evolution from an animal-like state which seems to us unconscious, and the same process of differentiation is repeated in every child," that "Throughout life the ego is sustained by this base," that the unconscious is "the generator of the empirical personality," and he even suggests, speaking for the unconscious, that "in the opinion of the 'other side,' our unconscious existence is the real one and our conscious world a kind of illusion, an apparent reality constructed for a specific purpose, like a dream which seems a reality as long as we are in it."[173]

The "other side" sounds somewhat like an *angry* unconscious speaking, a *rejected* unconscious. But in addition to the above is Jung's theory of what he calls "the central concept of my psychology: *the process of individuation*."[174] Not to be confused with *individualism*, which is egocentric and somewhat selfish, Jung's concept of individuation has, at its core, the idea of psychic integration. To not be integrated is to deny the existence of one's unconscious outright, or to think that consciousness is one's true self, one's "I," and the unconscious something stupid and irrational, in the very least as something beneath one, (as implied by the word "subconscious,") or that it is something else altogether, something foreign, an alien or stranger, something evil even, a monster or enemy within, a Goliath to be battled and defeated for one's homeland, or it is simply to ignore the influence of the unconscious on the conscious, such as dreams and intuitive promptings. It is also to invite mental problems. As Jung puts it in an earlier work, "the unconscious is life and this life turns against us if suppressed," which he repeatedly warns can lead to neuroses and psychoses. The process of individuation, then, is when "a person becomes a psychological 'in-dividual,' that is, a separate, indivisible unity or whole," a unity characterized by "open conflict and open collaboration at once," the conscious and unconscious

173. Jung (note 56), 320, 321, 299–300.

174. Jung (note 56), 200.

having "equal rights," a "harmonizing" of "two incongruous halves," like the "inner polarity" of the Heraclitean lyre or bow.[175]

Essential to this working relationship, moreover, is what Jung calls "active imagination." "The creative activity of imagination frees man," he says, "from his bondage to the 'nothing but' and raises him to the status of one who plays. As Schiller says, man is completely human only when he is at play." "All the works of man," Jung even goes so far as to say, "have their origin in creative imagination. What right, then, have we to disparage fantasy?"[176] (There is much more to Jung's concept of "active imagination," but how Poe would have loved this!)

The upshot is that with Jung's concept of individuation the "I" returns anew, and authorship ceases to be a problem.[177] The

175. Jung (note 141), 288, 275, 288, 289, 287; Jung (note 56), 319. Did Jung, then, think of the unconscious as another self, (or selves,) in the sense of another ego or "I," not just a subject but a person with subjectivity, so that the individuated self is still something of a "double consciousness"? The quick answer is *no*, as the unconscious, to which we have no direct access, manifests itself to us with personas that are "decentralized" and "fragmentary." Jung (note 141), 276, 281, 283–284, 286. However, Jung was certainly open to the possibility of one or more subjective selves residing in the unconscious. Jung (note 56), 298, 310. Meier, in turn, definitely leans closer toward *yes* for the unconscious as a whole. Taking the view that creative actions involve "a positive *attitude* to the unconscious" and that a negative attitude produces *parapraxes*, (faulty actions caused by the unconscious,) "a conception of this kind," he says, "involves something approaching an anthropomorphism of the unconscious, since it seems as if the unconscious behaves well when we are friendly toward it—as in fact we expect civilized human beings to do. But analytical experience has shown that this hypothesis stands the test of practice." Meier (note 16), 39. (Recall Poincaré.) At any rate, the theory of individuation should make one look anew at the meaning and value of prayer and meditation.

176. C.G. Jung (1931). "The Aims of Psychotherapy." In Joan Chodorow, ed. (1997). *Jung on Active Imagination*. Princeton, NJ: Princeton University Press, 91.

177. The problem now is self-inflation. As Meir points out, the ancient Greeks and Romans invoked the gods and then thanked them for their moments of inspiration, in particular the Muses. This served to curb hubris, or megalomania. Meir (note 16), 10. With modern concepts of the self, however, such as Jung's concept of individuation, the problem arises anew. As a useful antidote, I suggest a muted version of Sartre's claim that you are what you do, that man is "nothing else but the sum of his actions," such that "there is no

individuated self is not simply the brain or part of the brain or the conscious functioning of the brain, but the integrated self, involving the conscious self in collaboration with the undiscovered, deeper self, a conscious–unconscious *double*, a one out of two, a mystery-solving duo working together and *supervening* on—beyond that I am not prepared to say—the neurons and actions of one brain, however divided, so that we may therefore freely speak, with a good conscience, of *Darwin*'s theory of evolution, of *Einstein*'s theory of relativity, and so on, each work written by a *time-slice* of its author.

And indeed, for my own part, if I may speak for myself as a "whole," in writing the book before you, from the beginning to the end, I could not help but think and feel that it was not so much my conscious self that wrote it but moreso my unconscious self, that the latter directed my conscious research and readings, brought to conscious memory time after time what hitherto I did not consciously imagine was there, and got my conscious self to work incessantly on section after section and line after line, often interrupting my sleep to change something, to add something, or to move something, to rewrite even the simplest line, to look here and there for something new, for something missing, and submitting to my conscious self an idea or even a solution while that self was doing something else, not just sleeping, but sometimes just relaxing, even while watching a movie, at times a line in a movie prompting a thought of something that needs to be done in the book before you. And of course my unconscious self often gave my conscious self leave of Poe, for a day or two and sometimes more, to work on and take care of other matters.

At any rate, speaking now only as my conscious self, I look back and I marvel at it all. How infinitely more to be a Darwin or an Einstein, then, or a Da Vinci, to name another major example

---

genius other than that which is expressed in works of art." Sartre (note 8), 358, 359. I take this to mean that one does not deserve to be called a genius, whether in art or science or anything else, until one has produced a bona fide work of genius. Until then, to call or fancy oneself a genius is nothing more than mere arrogance and inflated self-esteem. Otherwise we are all geniuses and the term ceases to have meaning, like every child getting a trophy—political correctness and Maslow's fourth level arm-in-arm, blissfully running amok.

of genius, with their moments of sudden inspiration and their use of personal notebooks—I can only begin to imagine, as most of us can only begin to imagine.

We *truly* begin to imagine when we start to take notice of moments of creative flashes, problem-solving flashes, no matter how small or seemingly trivial, including dreams, and when we begin to learn how to use them for our conscious purposes. When we do this, when we *actively strive* for individuation, (which is something we must never stop doing, according to Jung,) the results will be life changing. To give a simple example of how this might work, in recent years I've told my students to read over the entire exam before they start writing, and when they write to not get stuck on a question for which they cannot remember the answer or cannot figure out the solution, that they should first answer the questions of which they feel confident, for their unconscious is meanwhile working on the others that gave them trouble, and that when they come back to them later in the exam they may well discover that they can now remember what they could not before or can now see the solution that had eluded them.[178]

178. A striking example of individuation is what happened to me personally just before I gave my talk on Poe's "double Dupin" at the Fourth International Edgar Allan Poe Conference held in Manhattan. I arrived at the Roosevelt Hotel on the Friday at around 6:00 p.m. and checked in. I then went for a walk outside looking for a restaurant to have supper and also for a place to have breakfast the next morning. It was very cold, very dark, I hadn't eaten since noon, and I got lost. It didn't take long for my hereditary temper to render me more impatient of control. And then it happened—the area under my left eye swelled up like a balloon. I could *feel* it swell, most palpable and loud. *Freak* was how I would be looking the next morning at 10:30 for my talk. *Ruined* was now the special day I had been planning for months. *Hope* for surcease of swelling there was none. I continued walking, up and down the streets of Manhattan, lost and in a rage. The fury of a demon instantly possessed me. I knew myself no longer. I gnashed my teeth. I foamed at the mouth. I glared at anyone who passed me with an expression of maniacal rage. *I foamed—I raved—I swore!* At length I found my way back to Madison Avenue at 45th, and proceeded two lefts around the corner to where I *knew* there was a pizza restaurant. I entered and looked round upon the company with a half-insane stare. Surprisingly, amazingly, they seated me and took my order. Waiting, overwaiting, *fuming*, suddenly three officers of the police entered and I braced myself, for I was certain they had come for *me*. (Unless you think me mad, I also thought I could not blame the manager for having called them.) But the gentlemen got their food and left. At

In many situations, of course, one needs not only to give the unconscious the problem that needs to be solved, (and with passion and gravity,) one also needs to feed the unconscious with as much information as one can and to listen carefully to its promptings. That, if I may say so, is the Jungian concept of individuation taken to its logical conclusion, and it is also—not accidentally—the

---

last I ate, at 7:30, and returned to my hotel room with the rest of my food. *Hideous* is how I looked in the bathroom mirror, and I knew it would be a few days for the abomination to be gone. Angry still as all hell, reading is how I passed the rest of the evening, *viva voce*. At 11:30, even though tired, I had the distinct feeling that I would not be able to sleep, no matter how far into the night. I knew the feeling well. Many a night, over many a year, when all the world slept, I had experienced the very same feeling. Having brought with me my bottle of prescription sleeping pills, along with a vial of laudanum, I took out three of the pills (knowing this would be a personal record) and downed them with an ounce of the medicine using my official Edgar Allan Poe shooter glass that I had bought months earlier by mail order from the Poe Museum. (Surely a fitting way to die for a Poe lover, I thought.) The light of my Soul was out at about midnight. Strangely, I awoke at 5:00 a.m., but rather than take another pill, (which would have left me dopey for 10:30,) I decided to read, and mercifully shortly afterward I sank back into slumber. I next awoke at precisely 8:40 a.m., (I knew for the first thing I did was look at the clock,)—and from the strangest dream. I was giving my talk, but I brought the wrong papers. What I was reading had nothing to do with Poe. People in the audience were looking at me and at one another with a puzzled expression, but they were also polite. Realizing my mistake, I then tried to give my talk from memory. But I couldn't remember anything that I had prepared. So I went back to reading from the papers I had brought, which again, had nothing to do with Poe. Then I saw a man to the right of me make some gestures to the Chair, who was to the left of me. He pointed twice to his wristwatch and then drew his finger along his neck. The Chair then looked at me and said, "Thank you, Dr. Stamos, your time is up." I looked at her and said, "Thank you," and quietly sat down. At that precise moment I awoke. My former Self would have been furious! Why would my unconscious give me such a stupid dream hours before my talk? But the Jungian in me instantly intervened and asked, "What would be the purpose of my unconscious giving me this dream at this time?" The answer was immediately obvious: to make me laugh. And it worked! The dream was so ridiculous and surreal that I could not help but laugh, and really laugh the more I thought about it. Then, in an instant, the demon gave way to the better angels of our nature—out of Evil proceeded Good, Ge-Henna returned to Hinnon, the Night's Plutonian shore became the distant Aidenn, the vexatious Evil Eye was now the Discord in Music, and I went on to have one of the happiest and most fulfilling days of my life.

educated imagination, what Poe called the "poetic intellect," the very stuff of scientific imagination.

But enough of these meditations on the self. Let us now proceed even deeper, not deeper down into the well, not deeper in space, but in time, to the water in the well and the river of time, to where we are to find, from a scientific view, the ultimate explanation of scientific imagination.

Evolution

For a start, we need an important distinction from evolutionary biology. As Ernst Mayr puts it, "biology can be divided into the study of proximate causes, the subject of the physiological sciences (broadly conceived), and into the study of ultimate (evolutionary) causes."[179] To use his example, a proximate explanation for why a species of bird migrates south would focus on the physiology of the brains of the species, combined with the environmental triggers. An evolutionary explanation would focus on the reasons why the species evolved the migratory instinct over thousands or millions of years, such as the fact that the species only feeds on insects, which are available in the north in the summer but only in the south in the winter. In other words, then, the migratory instinct is an *adaptation*, evolved because it increased survival and the opportunity to reproduce, (otherwise the species would go extinct,) the bottom line of which is passing on the genes (natural selection involves a kind of sifting) responsible for the beneficial trait.

So now why the evolution of the *double* brain? Except for two points that should be of special interest to Poe lovers, I want to provide only a few brushstrokes here, as the topic is too extensive for a just treatment in this section—it properly requires a book. First, evolution, in general, proceeds from the simple to the complex, with greater brain functions requiring more neurons and more complex circuitry between them. Hence, as a general rule when applied to species, (not to individuals within a species,) the

179. Ernst Mayr (1982). *The Growth of Biological Thought: Diversity, Evolution, and Inheritance*. Cambridge, MA: Harvard University Press, 67.

more intelligent animals have larger brains in proportion to their body size. (There is a qualification known as the *encephalization quotient*, but there is no need to get into it here.) The cerebral cortex is the most evolved layer in the mammalian brain, and in humans (with the possible exception of dolphins) it is the most complex of all animal species.[180] Accordingly, it is the throne (doubled) from which we rule the world.

Second, a double-hemispheric brain is not a biological necessity in animals, as there are many bilaterally symmetrical animals that have a symmetrical brain with mirrored symmetries but not a double hemisphere, animals as dissimilar as water-bears, death's-head moths, and gold-bugs, exemplified by *Milnesium tardigradum degrasseæ*, *Acherontia styx lectera*, and *Scarabæus caput hominis* respectively. The double brain, with bilateral hemispheres, is unique to vertebrates. So whatever the original cause of its evolution,[181] (it could have began as an accident, after all, possibly as a chromosomal mutation caused by gamma rays,) once the double brain became established as a basic plan it also became a developmental constraint, a restriction on the future evolution of vertebrate brains, one that could not be eliminated but that did allow for the evolution of certain advantages, most notably an asymmetry in hemispheric functions and specializations, as if to say, "two brains are better than one." It should therefore not be surprising that humans are not alone in having hemispheric asymmetry, both structural and functional. As one researcher puts it, "evidence is accumulating that cerebral asymmetry is widespread in the animal kingdom,"[182] which leads to the conclusion, as another researcher points out, of "an early common origin of lateralization in vertebrates."[183] This gradual evolution of a "division of labor" in the double brain reached its peak (so far as we know) in our own species, with "manual dexterity and language"

180. Daniel H. Geschwind and Pasko Rakic (2013). "Cortical Evolution: Judge the Brain by Its Cover." *Neuron* 80 (3), 633–647.

181. R. Glenn Northcutt (2002). "Understanding Vertebrate Brain Evolution." *Integrative & Comparative Biology* 42, 743–756.

182. Michael C. Corballis (2010). "Handedness and Cerebral Asymmetry: An Evolutionary Perspective." In Hugdahl and Westerhausen (note 137), 65.

183. George (note 137), 91.

generally considered as our two unique functions.[184] The ultimate point for our purposes, given everything we have seen so far in this section, is that our laterality in brain functions allows for our brains to carry out "simultaneous, parallel processing,"[185] especially given callosal inhibition.

Given the bilateral brain asymmetries outlined in the previous section, and their connection with the phenomenon of eureka moments in science, which all bring us back to Poe's *Eureka* and even earlier to his "double Dupin," one cannot help but notice the metaphors that scientists often use for eureka revelations. Our evolutionary past as humans, it must be noted, was characterized by a nomadic, hunting-gathering lifestyle. We should also note that we expanded beyond Africa roughly 100,000 years ago and spread throughout the globe. The enormous complexity of situations faced by our ancestors cannot be overstated, which included mentally mapping the geography of their environment, learning about the behaviors of predators and prey, mostly avoiding the former and tracking and hunting the latter, making tools and weapons, learning about the properties of plants for food and medicine, social relations, including mating and social hierarchy within the group, and competition with rival groups, the latter often violent. What I find particularly interesting is that scientists commonly use geographical metaphors for big picture thinking in scientific discoveries, or discoveries that lead to big picture thinking in science. Francis Crick, for example, as we have seen in the second section of this chapter, states that "When I went to bed all our difficulties had resolved and the shining answers stood clearly before us. Of course, it would take months and years of work to establish these new ideas, but we no longer felt lost in the jungle. We could survey the open plain and clearly see the mountains in

184. Hugdahl and Westerhausen (note 137), 8. It should be obvious, by the way, that what is called *brain plasticity*, the remarkable ability of the brain to "rewire" itself, including after damage, (which accounts for left-handedness in humans, for example,) does not undermine anything in this evolutionary perspective, for the plasticity itself is yet another amazing example of the many evolved adaptations that constitute the wonder of our being. See, e.g., Springer and Deutsch (note 112), ch. 10.

185. Hugdahl and Westerhausen (note 137), 9.

the distance."[186] Similarly, Einstein refers to "Planck's trailblazing work" and to his own work as "alternations of confidence and exhaustion and the final emergence into the light."[187]

Equally interesting is the possibility of an evolved role for depression. We naturally think of depression as a malfunction, as we can "slide" into depression and as people can commit suicide out of depression. And given its varied causes, no doubt depression is often a malfunctioning of the brain/mind, one that needs to be treated. But that does not mean it is *always* a malfunctioning. A single effect can have multiple sufficient causes. In evolutionary psychiatry there are two standard theories for the evolution of depression as an adaptive strategy, namely, *deprivation depression* and *defeat depression*, the former having evolved because it helps us to get over an attachment to a deceased loved one, which allows us to get on in the struggle for survival and reproduction, and the latter having evolved because it helps to increase social cohesion, since depression curbs violence in those who have lost rank in the social hierarchy.[188]

But creativity seems another matter altogether. Given the "crippling and disabling" effect of depression and the "self-harm and suicide" of the "disorder," the "implications" of the idea that depression can be creative "feel vulgar and sound wrong" to some psychologists, (and graduate students,) so that "creativity science should invest more heavily in understanding the basic processes of normal creativity."[189] The problem with this sentiment is that depression might have a creative side that *is* "normal," or at least common enough to be called "somewhat normal." How an idea *feels* or *sounds* is synesthesia, not science.

And in fact our case studies in scientific imagination—Mayer, Tesla, Trivers, and much earlier Chambers and Poe—suggest an

186. Crick (note 6), 120.

187. Einstein (note 84), 53. "Notes on the Origin of the General Theory of Relativity" (1934). In Bargmann (note 73), 289–290.

188. Anthony Stevens and John Price (1996). *Evolutionary Psychiatry: A New Beginning*. New York: Routledge, 63–71.

189. Emily C. Nusbaum, Roger E. Beaty, and Paul J. Silvia (2014). "Ruminating About Mental Illness and Creativity." In James C. Kaufman, ed. (2014). *Creativity and Mental Illness*. Cambridge: Cambridge University Press, 398, 400.

evolved function for depression in addition to the two already discussed, a function that is *for* scientific imagination, a function that more generally may be called *creative depression*. This is a novel insight, so let us investigate more closely into how it might work.

For a start, depression is typically characterized by pronounced sadness, negativity about one's own life or about life in general, and lack of motivation, but also by enhanced inward reflection on the possible causes and consequences of one's distress, known in psychology as *rumination*. All of this is a description at the conscious level, however. What comes out over and over again in the literature on depression in neuroscience, based primarily on EEG and fMRI studies, is that during depression the left hemisphere is typically characterized by lower activity than usual ("hypoactive," "under-activation") and the right hemisphere by higher activity than usual ("hyperactive," "over-activation"). This is not to say the left hemisphere consumes less energy than normal in the depressed state while the right hemisphere consumes more, given that the human brain, which represents usually only 2% of the body's total weight, already consumes approximately 20% of the body's total caloric intake in the restive state (which is why it is called "metabolically greedy") and evidence from fMRI and PET scans supports the counterintuitive conclusion that "This high rate of metabolism [of the human brain] is remarkably constant despite widely varying mental and motoric activity."[190] It only means, instead, that during depression more is going on than normal in the right hemisphere and less is going on than normal in the left hemisphere.

It might simply be the case that the right brain takes advantage of this situation, so that many cases of unconscious problem solving, including many cases of eureka moments in science, are merely preceded by depression, rather than having depression caused for their own sake. But "simpler" does not necessarily mean "more likely" in the logic of evolution, which builds not from scratch but from what already is, and basically because it increases survival and reproduction. What is suggestive of an evolutionary

190. Marcus E. Raichle and Debra A. Gusnard (2002). "Appraising the Brain's Energy Budget." *Proceedings of the National Academy of Sciences of the United States of America* 99 (16), 10237. See also Ferris Jabr (2012). "Does Thinking Really Hard Burn More Calories?" *Scientific American* (July 18).

mechanism is that "depression with its RH [right hemisphere] hyperactivity is very often associated with hypercortisolemia [high amounts of circulating cortisol, the stress hormone]" and that the secretion of cortisol is "mainly under excitatory control of the RH." "Accordingly," as the neuropsychologist David Hecht continues with the point, "the link between depression and RH hyperactivity can be described in both ways, as a cause and as an effect. Continuous fear, anxiety, stress as well as other negative emotions, activating primarily the RH, can cause depression. Alternatively, . . ."[191] What this means, I should think, is not only that the stress of a problem situation can cause depression, but that *the right hemisphere might even secrete cortisol to bring on the depressed state.* And this suggests a possible evolutionary mechanism. If the secretion of cortisol by the right hemisphere occurs as an *instinctive* reaction to a problem in the environment best suited for solution by the right hemisphere, a "forest" as opposed to a "tree" kind of problem, then we have an evolutionary *product*, a mechanism for depression evolved by natural selection because it increased the ability for unconscious problem solving in certain situations. Another possibility is that the right hemisphere might secrete cortisol as a *willful* act, for the purpose of inducing depression. This is because *the right hemisphere has a mind, too.* It might intentionally do this, thereby causing lower activity in the left hemisphere, because this situation allows the right hemisphere to work harder on solving the stressful problem than it could in the normal situation, somewhat like the need "to work alone" and "to work in quiet." In this case we would have a *byproduct*, not a product of evolution.

Either way, this possible role for depression, for the purpose of unconscious problem solving, of which eureka moments are the epitome, is an intriguing possibility that gains support from an entirely different line of evidence discussed by the psychologist Dean Simonton. In 1994 the psychiatrist Arnold Ludwig conducted a massive histriometric study on the "creativity–psychopathology relationship" in the biographies of 1,004 creative geniuses who lived during the late nineteenth and/or early

191. David Hecht (2010). "Depression and the Hyperactive Right-Hemisphere." *Neuroscience Research* (68), 79.

twentieth centuries. The result was that "the highest rates of any lifetime disorder were exhibited by the poets (87%), followed by the fiction writers (77%), dramatists (74%), visual artists (73%), non-fiction writers (72%), musical composers (60%), and architects (52%). In contrast, the eminent natural scientists had a rate less than half as high as the composers did (28%)." Moreover, the kind of psychopathology varied between groups, such that poets "dominated with respect to psychosis (17%), suicide rate (20%), and depression (77%)." Moreover still, the members with mental illness in these groups were more likely to have parents and siblings with mental illness than the "normal" population, especially for alcohol abuse, drug abuse, and depression.[192] These conclusions indicate not just environmental causes but also genetic causes and have been corroborated by a substantial number of other studies.[193]

Ludwig himself, in a follow-up study, concluded that, as Simonton puts it, "creative geniuses in domains 'that require more logical, objective, and formal forms of expression' tend to feature less psychopathology than those active in domains 'that require more intuitive, subjective, and emotive forms.'"[194] Other researchers, however, have suggested that "rumination, which figures so persistently and painfully in the depression episode, fundamentally reflects an adaptive mode of analytic thinking that could also be a most useful cognitive tool in dealing with complex problems. So one effect of sad or depressed mood could also be to promote an analytical reasoning style, where greater attention is devoted to detail, and where information is processed more systematically."[195] But why use the word "analytical" when they mean "logical," "relies on logical rules of inference"?[196] Poets, after all, use

192. Dean Keith Simonton (2014). "The Mad (Creative) Genius: What Do We Know After a Century of Histriometric Research?" In Kaufman (note 189), 30.

193. Neus Barrantes-Vidal (2014). "Creativity and the Spectrum of Affective and Schizophrenic Psychoses." In Kaufman (note 189), 175–177.

194. Simonton (note 192), 31.

195. Geir Kaufmann and Astrid Kaufmann (2014). "When Good is Bad and Bad is Good: Mood, Bipolarity, and Creativity." In Kaufman (note 189), 212.

196. Kaufmann and Kaufmann (note 195), 221.

metaphors, and as Aristotle well knew, metaphors are not arrived at by following "logical rules."

And that's fine for poets. But but but our topic is *scientific* imagination. If poets are the most depressed of creative geniuses and scientists the least, does this not undermine the idea of an adaptive role for depression in scientific imagination? The answer *yes* would seem obvious, until one recalls from Chapter 6 that there are scientists and then there are scientists. Simonton cites "a fascinating historiometric study of 82 scientific geniuses in the natural (paradigmatic) sciences" by Ko and Kim (2008). The scientists were divided into two groups à la Kuhn, the normal scientists and the revolutionary scientists. As Simonton puts it, "The results were striking: For paradigm-preserving contributors eminence was negatively related to the level of mental illness, but for paradigm-rejecting contributors the eminence–pathology relation became positive—a classic 'cross-over' interaction effect."[197] Poe would put it that the revolutionary scientists *were* poets, with a "poetic intellect," and we should keep in mind the evidence from neuroscience, examined in the previous section, in particular the recent neuroimaging meta-analysis in which it was found that "conventional metaphors" are processed in the left brain and "novel metaphors" are processed in the right brain. Now add the evidence we have just seen from Ko and Kim, on normal scientists and revolutionary scientists. The former are characterized by common metaphors, left-brain thinking, and relatively low levels of mental illness, while the latter are characterized by novel metaphors, right-brain thinking, and relatively high levels of mental illness. This is an exaggerated way of putting it, but it makes the point.

Instead of "metaphor" we could also use the word "non-linear," which is what James and John Swain do, a neuropsychologist and a physicist respectively. They note not only that "creativity involves arriving at a thought (brain state or state of mind) that is not close to that reached by most people from the same starting point" and that it "might be achieved as a sudden 'gestalt' by what one could term a 'catastrophic' route," but also that "much of mental illness" seems "quite similar." The main difference, they

197. Simonton (note 192), 32.

suggest, "between creativity and mental illness is very much one of how much control one has over the non-linear wanderings of the mind to places—either via a chaotic route drifting farther and farther away, or a catastrophic route, making sudden transitions to new states," of which, for the latter, they provide the examples of Edison, Tesla, Kekulé, Einstein, and Ramanujan.[198] And from fiction we could add Poe's Dupin and Legrand.

Granting that non-linearity plays "an essential role" in scientific creativity, there is certainly much more to it than control (Dupin) versus non-control (Bedlamite). But at this point it is all a circle that ever returneth in to the self-same spot, with much of Madness, and more of Sin, and Horror the soul of the plot. It really is quite amazing how it all fits together, and circles round right back to Poe, as a kind of *poe*tic justice. But of course as they say in science, "more research is needed"—and I shall leave it at that.[199]

---

198. James E. Swain and John D. Swain (2014). "Non-linearity in Creativity and Mental Illness: The Mixed Blessings of Chaos, Catastrophe, and Noise in Brain and Behavior." In Kaufman (note 189), 138, 139–140.

199. I should like to add, however, given all of the above, that for a scientist suffering from depression in the context of a scientific problem best suited for the right brain it might be advisable, in my medical opinion as a doctor, to refrain from taking antidepressants, that it might not always be wise to take the pills. This is because "Patients who respond to antidepressants are characterized by a relatively higher left hemisphere activity in comparison to non-responders, and successful treatment with antidepressants increases left hemisphere activity," while "Right-hemispheric skills are less sensitive to antidepressants and are not restored in the process of drug treatment," that "right hemisphere functions are not restored by antidepressants and remain distorted even after the successful treatment." V.S. Rotenberg (2008). "Functional Brain Asymmetry as a Determinative Factor in the Treatment of Depression: Theoretical Implications." *Progress in Neuro-Psychopharmacology & Biological Psychiatry* 32, 1772, 1775. Rotenberg accepts that "The function of the right hemisphere is a simultaneous capture of an infinite number of connections as the formation of an integral polysemantic context that is relevant to the polydimensional world," that its function is "integrative creativity," and he takes it that "the core reason for depression is the inability of the right hemisphere to correspond to the demands of the polydimensional environment," that "A subject's inability to create a polysemantic context that is relevant to the polydimensional and complicated reality, especially to the reality of the emotional interpersonal environment, puts the subject in a state of stress, where he/she feels unequipped to master

I want to finish this section with a *finale*, one that returns us to the left and right brain, to "simultaneous, parallel processing" especially,—and to the role of terror. In an interview on his famous taxicab scene with Marlon Brando in *On the Waterfront*, Rod Steiger reflects on, as he puts it, "one of the greatest sources of creativity, which is terror." I recently watched that interview anew, and it got me thinking. We can readily imagine many different scenarios in our evolutionary past that favored the evolution of simultaneous left-brain analytical thinking and right-brain intuitive thinking in single individuals. But if ever there were *immediate* situations in which this was needed, *immediate* situations in which life and death hung in the balance on the interposition of a single thought into consciousness, these would have been situations that were filled with terror. And who better to give us an illustrative example than the terror master himself?

In "The Pit and the Pendulum" (1842), arguably a case of literary method acting, Poe blends the "creativity of terror" with "simultaneous, parallel processing" into a literary paradigm. Imprisoned in a dungeon of the Inquisition, the narrator, after a sleep "like that of death," finds himself lying supine on "a species of low framework of wood" and "bound by a long strap resembling a surcingle," a strap that "passed in many convolutions" around his limbs and body, leaving only his head and his left arm with some liberty of movement. Above him was "a huge pendulum, such as we see on antique clocks," but unlike the pendulums of clocks the edge of this pendulum was "a crescent of glittering steel, about a foot in length from horn to horn . . . the under edge evidently as keen as a razor." With every sweep of the pendulum "the whole *hissed* as it swung through the air," and with every hiss

a demanding environment." Rotenberg (this note), 1772, 1775, 1772, 1775. But he fails to see that the "stress" of the right hemisphere in failing to achieve a "simultaneous" capture of "an integral polysemantic context that is relevant to the polydimensional world" *might sometimes be a good thing, one not to be alleviated*, (except perhaps when the problem is "the reality of the emotional interpersonal environment,") since the right hemisphere, rather than being functionally "distorted," *might* instead be functionally *normal* but in need of a stretch of time and relative independence in order to solve the problem, to accomplish "integrative creativity," and accordingly would *benefit* from depression in the left hemisphere,—a situation that provides the ideal conditions for a eureka moment.

it inched closer to his chest at a right angle. What to do? He grew "frantically mad" and tried to struggle free. Then he "fell suddenly calm," smiling at the instrument of his death "as a child at some rare bauble." There was "an interval of utter insensibility." Recovering from that interval he tried to reach a morsel of food by painfully outstretching his arm to the "earthen dish" beside him on the floor, "for the human nature craved food." Touching the food to his lips, there rushed to his mind "a half-formed thought of joy—of hope." But what in this situation had he to do with hope? The thought "perished in its formation" and he struggled in vain to regain it. "The long suffering," he says, "had nearly annihilated all my ordinary powers of mind. I was an imbecile—an idiot." With every sweep of the pendulum, "some thirty feet or more," its gradual descent now meant that he was only "several minutes" away from death. Taking "a frenzied pleasure in contrasting its downward with its lateral descent," he "alternatively laughed and howled, as the one or the other idea grew predominant." With the blade now only a few inches from his chest, he "gasped and struggled at each vibration," and "shrunk convulsively at its every sweep." And yet hope resurfaced—"the *hope* that prompted the nerve to quiver"—"the hope that triumphs on the rack—that whispers to the death-condemned even in the dungeons of the Inquisition." Then something remarkable happened: "there suddenly came over my spirit all the keen, collected calmness of despair. For the first time during many hours—or perhaps days—I *thought*." What "now occurred" to him was that he was bound by a *single* "bandage, or surcingle," and that if it could be cut at any one point he could then unravel himself to freedom with his left hand. But how to cut it? All of a sudden "there flashed upon my mind what I cannot better describe than as the unformed half of that idea of deliverance." Earlier, with the dish of food beside him, he had managed to reach the dish with his left arm, scaring away the rats, who got most of the food in spite of his efforts, and then had managed to get a morsel and raise it to his "burning lips." But now the half-thought which he had at that time came to him "entire"—"the whole thought was now present." Since the dungeon was "literally swarming with rats," *voracious* rats, eagerly waiting for him to die so they could eat his flesh and drink his blood, he ceased the habitual waving of his arm, put his hand in the dish, and rubbed "the oily and spicy viand which now remained" into the bandage, "wherever I could reach it." He then "lay breathlessly still," a total "cessation of movement," at which signal the rats descended on their prey. They "leaped in hundreds upon my person," they "writhed upon my throat; their cold lips sought my own," and avoiding the strokes of the pendulum they also "busied themselves with the anointed bandage." All the while the narrator "lay *still*," with "a more than human resolution." Knowing not what they do, the ambitious rats chewed through his strap sufficiently enough that the narrator "felt that I was *free*," just in time as the blade pierced the linen of his robe and "a sharp sense of pain shot through every nerve." He then waved his hand to scare away his "deliverers" and "slid from the embrace of the bandage and beyond the reach of the scimitar." "For the moment," then, he says, "*I was free*."[200]

What happened afterword is something you'll have to find out for yourself. But what Poe has given us here, in the most dramatic and thrilling fashion, is a striking example of *individuation* in action, the very stuff of the evolution of the "double Dupin"—the "Bi-Part Soul" of you and me.

Adieu

Edgar Poe, ill-fated and mysterious man! Entering this cold, dreary world in 1813, or 1811, or 1809, in the midst of cares and vexations of all kinds, himself the son of an actress, her brief career one of beauty and of genius. Orphaned in the second year of my age, I myself never knew my mother, and never knew the affection of a father. Though of an ancient & noble family, Mr. John Allan adopted me, and with his appetite for system and regularity, and not very often sober and mistaken parsimony, treated me with what I considered indignity. His wife, whom I had always regarded as a mother, the love she used to have for me,—my dearest love to Ma,—and yours the slanderous tongue, that did to death the innocence, that died and died so young, the funeral song be sung. Being young and dipt in folly, I fell in love with

200. *P&T*, 498–503.

melancholy. My first mental development had in it much of the uncommon,—even much of the *outré*,—the Reverend Dr. Bransby, with robes so glossy, the economic ingenuity, the Draconian laws of the academy, and my natural rights of self-agency, denied so pertinaciously, so insultingly! Reason had at once hurled superstition from her throne, in a strange city lying alone, on account of the dampness of the catacombs, the old rampart of bones. My own height, about five feet eight inches, with a profusion of curling, black hair, from which a forehead of unusual breadth gleamed forth, examined by several phrenologists—all of whom spoke of me in a species of extravaganza, my sinciput a Spurzheimite *principia*. Even George Bush and Cassius Clay, for certain Sartain and John Kennedy, my head a model for statuary. I wore no garb of *vraisemblance*, no snuffy habiliments *par excellence*. My life was no "exercise," or experiment, in the plausible or verisimilar style. I was a man whose eye reads the heavens while his feet step firmly on the ground and his hands are strong and dextrous in the use of human instruments, a man to whom this world is no mere spectacle or fleeting shadow. With a *habit* of intense and continual thought, books, indeed, were my sole luxuries, and observation became with me, of late, a species of necessity. Well educated, with unusual powers of mind, but infected with misanthropy, it was doubled with the wild fervor, and the vivid freshness of my imagination, a palace of imagination, combined with an excessive nervous agitation, a morbid acuteness of the senses, and an imaginative and easily excitable temperament. By strange alchemy of brain, the effect of wine on my excitable brain, my pleasures always turn'd to pain. But for Sissy, Muddy, and Catterina too, I dwelt alone, in a world of moan, my (double) marriage ill-omened. Soft may the worms about her creep! While I weep—while I weep! The soul is a cypher, in the sense of a cryptograph; and the shorter a cryptograph is, the more difficulty there is in its comprehension. A phial labelled "laudanum," and emptied, was found near him. Oh, I loved with a love that was more than love. But in the end, in my lonesome latter years, all I loved—*I* lov'd alone.

Death was in that poisonous wave, the pitiless wave, and in its gulf a fitting grave. How mysteriously did it act as a check to human bliss—saying unto it "Thus far, and no farther! For the hours of thy happiness are over; and joy is not gathered twice in a life." As in that fleeting, shadowy, misty strife, thy soul shall find itself alone, 'mid dark thoughts of the gray tomb-stone. And as they lie so composedly, now in their bed, who ever *really* saw anything, but horror in the smile of the dead?—in that suspiciously lingering smile upon the lip, which is so terrible in the dead? And throw open the graves of all mankind, is it not—oh, God! is it *not* a very pitiful sight? Their sad and solemn slumbers with the worm, through the realms of the subterrene Night? A blood-red thing that writhes from out, amid the mimic rout. It writhes!—it writhes!—in human gore imbued, the mimes become its food. And "*blood*" too, that word of all words—so rife at all times with mystery, and suffering, and terror. Besprinkled with the scarlet horror, the figure was tall and gaunt, with corpse-like mask which concealed the visage, and shrouded with grave-cerements. Untenanted by any tangible form, his vesture was dabbled in *blood*, the redness and the horror of blood, the Avatar and the seal of the Conqueror Worm. O God! O Divine Father!—shall this Conqueror be not once conquered? Are we not part and parcel in Thee? Is *all* that we see or seem, but a dream within a dream? Virginia's cough, a Chevalier's cryptograph, Conqueror Worm or Butterfly, requiring a most knowing eye.

Genius irritabile, at its epochs of most intense tranquility, *liberality* is *economy*, *extravagance* is *parsimony*. An artist *is* an artist only by dint of his exquisite sense of Beauty, but at the same time implying, or involving, an equally exquisite sense of Deformity. In lauding Beauty, Genius merely evinces a filial affection. To Genius Beauty gives life—reaping often a reward in Immortality. The *highest* genius acts upon individuals, as well upon the mass, by a species of magnetism incomprehensible but irresistible and *never resisted*, by that intuitive feeling for the forcible and the true, which is the *sixth sense* of the man of genius, and is the result of generally large mental power existing in a state of *absolute proportion*—so that no one faculty has undue prominence. The true genius, then, is necessarily, if not universal in its manifestation, at least capable of universality. That *fictitious* "genius"—that "genius" in the popular sense—which is but the manifestation of the abnormal predominance of some one faculty over all the others—and, of course, at the expense and to the detriment of all the others—is a result of mental disease, of organic malformation of mind. Another error, that industry, in great measure, was genius itself, for perseverance is one thing and genius quite another, the former here, and there, and every where, while true genius is rare. And then there's the small geniuses—the literary Titmice—animalculae which judge of merit solely by *result*. The worth of a work is most accurately estimated, they assure us, by the number of those who peruse it; and "does it sell?"—the critical and popular taste, the most odious and insupportable despotism that ever was heard of upon the face of the Earth. A book, of course, may be readily sold, may be universally read, for the sake of some half or two-thirds of its matter, which half or two-thirds may be susceptible of popular appreciation, while the one-half or one-third remaining may be the delight of the highest intellect and genius, and absolute *caviare* to the rabble, whose thing is only the whizz!—fizz!—of *style*. There lives (and died) no man feeling a deeper reverence for genius than ourself.

A consideration of intense interest: chaste, vigorous, and glorious *imagination*. How exquisitely artistic is the *management of imagination*. The bard that paints imagination's powers, and him whose song revives departed hours, *all* that we see or seem, my words the music of a dream. That we can imagine a griffin, thought's dominion, or an enchanted garden, up to Heaven, each soul is, *in part*, its own God—its own Creator,—the less within the greater, a glimpse of things supernal and eternal.

Romance, who loves to nod and sing, with drowsy head and folded wing, is opposed to a work of poetry, in my opinion, by having for its object a *definite* instead of an *indefinite* pleasure, turning naivete to wild desire, wit to love and wine to fire, and after-drunkenness of soul, succeeds the glories of the bowl, presenting perceptible images with definite, poetry with *in*definite sensations, with beauty even exceeding our anticipations, to which end music is *not* an essential, (let alone spectacles,) for a *magnetic* sympathy of soul for soul.

And, through all, your true poor-devil author, forced into the service of the magazine prison-house, struggling with despair in the shape of a ghastly poverty, which had no alleviation, I—*wrote*. Through joy and through sorrow, I—*wrote*. Through hunger and through thirst, I—*wrote*. Through sunshine and through moonshine, I—*wrote*. How I labored—how I toiled—how I wrote! The sickness—the nausea—the pitiless pain, I wrote with the fever called "Living" that burned in my brain. By day I adhered to my desk, and at night,—I consumed the midnight oil, in the blackness of fright. My fancy grew charnal. I talked "of graves, of worms, and epitaphs," and of the more abstruse cryptographs.

Literature is the most noble of professions—there was no seducing me from the path. My first object (as usual) was originality, *the absoluteness of novelty*, with variety one of my chief aims. Having a novel effect, then a subject, I rejected every word not having a tendency to develope the effect. And now my actions—my works—clear and forcible—the *physical power of my words*—are the property of mankind. Forgive these egotisms, the inflation,—you should have seen *me*,—but it was what I was born to be, the sole unquestionable aristocracy—that of intellect, along the majestic highway of the *Consistent*.

Love—as in infancy was mine—her young heart the shrine. The angels, whispering to one another, can find, among their burning terms of love, none so devotional as that of "Mother." The unselfish and self-sacrificing love of a brute, nothing but a truth absolute. A love that hesitated was not a love for me, the Venus Aphrodite. For Heaven to them no hope imparts, who hear not for the beating of their hearts. But I could not love except where Death, was mingling his with Beauty's breath. And so I loved as no man ever loved before, with a love that was more than love, each love a sweet Lenore, a quaint and curious volume of forgotten lore,—weep now or never more! And be gone thou grim, ungainly, ghastly, gaunt, and ominous bird of yore.

A poem, in my opinion, is *not* opposed to a work of science, when the province of the latter involves the sentiment of Poesy, the sense of the beautiful, of the sublime, and of the mystical, of that Beauty which, I maintain, is the excitement, or pleasurable elevation, of the soul. Perfection of rhyme is very imperfectly understood, unexpectedness must arise from expectedness, without mere *arbitrariness*, as there is no exquisite beauty without some *strangeness*, and as evil cannot exist without good. The poetic sentiment implies an abnormally keen appreciation of poetic excellence, with an *unconscious assimilation* of it into the poetic entity, with a totality, or unity, of effect and a perfect consistency. The true character, the august aims, the supreme majesty and dignity of the poetic sentiment, lies in the creation of novel forms of beauty. The true poetical effect the ambrosia which nourishes his soul, delicacy is the poet's own kingdom, his *El Dorado*,—the bird beat the bug, all hollow. The origin of Poetry lies in a thirst for a wider Beauty than Earth supplies, for Beauty (which is Hope) from out the Future cries, the Beauty beyond the grave, far from the pitiless wave. Poetry itself is that imperfect effort to quench this immortal thirst, for the napthaline river of Passion accurst, the desire of the moth for the star, thro' storm and night so trembled from afar. Poetry, for this Virginian, formerly a Bostonian, was therefore not a purpose, but a passion, held in reverence, without an eye to the paltry compensations, or the more paltry commendations, of mankind, and, under happier circumstances, would have been the field of my choice. Most palpable and loud, is not its form—its voice? Rather than too much, my poetry wearies in having done too little, 'tis a token and a symbol. Even so, with my rhythmical creation of beauty using words, I put the Poe into Poesy, and banished poetasters from the know.

Nemo me impune lacessit. Of the innumerable paragraphs of personal and literary abuse, I foamed—I raved—I swore! A huge human foot d'or. The evident toadyism: "*monstrum, horrendum, informe, ingens, cui lumen ademptum*." English, Hopkins, and Griswold,—the yelpings of my foes, to each I hold a glass, and let him see himself, each a proper ass! Bonaparte, Hoffman, and Weekes,—who would be men by imitating apes, low ruffians and boobies, with a chorus of turkeys, the ludicrous heightened into the grotesque, better to have Outis. Eliot and Bloom anon,—a tale told by an idiot and the King of Donkey-dom; small geniuses—the literary Titmice—animalculae, puffery and corruption, with no positive *genius* for *adaptation*. Oh, horror upon horror! The eight corpses swung in their chains,—the rhythmical creation of beauty, a discordant melody, with unity or totality of *effect*, the most poetical topic,—a fetid, blackened, hideous, and indistinguishable mass, plagiarism, immorality, inanity, and bombast. *For the love of God, Montresor!* No more—no more—no more. *In pace requiescat*.

"Prophet!" said I, "thing of evil!—prophet still, if bird or devil!" It was night, in the lonesome October, the skies they were ashen and sober. Lo! 'twas a gala night, within the lonesome latter years, bedight in veils, and drowned in tears, my life a play of hopes and fears, a circle that ever returneth in, to the self-same spot, with much of Madness, and more of Sin, and Horror the soul of the plot. Nothing is more clear than that every plot, worth the name, must be elaborated to its *dénouement* before any thing be attempted with the pen. And when, finally, I sank into slumber, it was only to rush at once into a world of phantasms, above which, with vast, sable, overshadowing wings, hovered, predominant, the one sepulchral Idea, that Death has reared himself a throne, in this strange city, with me lying alone, while the angels uprising, unveiling, affirm, that my life is a tragedy, and its hero the Conqueror Worm. Then, a deep sleep fell upon me—a sleep like that of death. But, alas! with a degree of appalling and intolerable horror, my spirit flickered up, as a lamp which is near its death-hour. Then a brief re-sinking into non-entity; then a sudden recovery. "Scoundrel! Imposter! Accursed villain!" Words are impotent to convey any justness of the fierceness of resistance with which I wrestled with the Shadow. At length I struggled no more, now in my bed, that any beholder might fancy me dead. Again my spirit flickered up as the flame within the socket of the lamp. But now, with the madness of a memory which busies itself among forbidden things, the agony of my soul found vent in one loud, long, and final scream of despair—"*Reynolds!*" Then in a singular whisper, "Lord help my poor soul." An icy hand upon my forehead, thinking me dead, here I opened wide the door;—Darkness there and nothing more. The light of life is o'er. Never again to find my lost Lenore. Poe the man is nevermore—his death a mystery of mysteries. And then, amid all my infinite miseries, came the sweet cherub Hope. A dull shock like that of electricity pervaded my frame. In an instant I seemed to rise from the ground. But I had no bodily, no visible, audible, or palpable presence. Surcease of sorrow, nevermore to wish the morrow, a wider Beauty than Earth supplies, I saw no Heaven—but in her eyes. A dream within a dream, a wreath upon my brow, my soul is filled with Beauty, and I am happy now.

Of all conceivable subjects I approached the reader with the most solemn—the most comprehensive—the most difficult—the most august. With the desire of the moth for the star, an object of unceasing pursuit,—with a thirst unquenchable, the thirst for supernal Beauty,—with a science that preyed not upon the poet's heart, but soared with an undaunted wing, and with a breath that dimmed not the mirror of our joy,—with ardent imagination and unthought-like thoughts that are the souls of thought, dreaming dreams no mortal ever dared to dream before,—with a rapid whirling on my heel from the top of Ætna, comprehending the panorama in the sublimity of its *oneness*, the one sepulchral Idea,—with glimpses of eternity bringing us upon the verge of the great secret, through a circle that ever returneth in, to the selfsame spot,—the riddle, so far, was now unriddled, a mystery of mysteries! And so my legacy was saved, from the pitiless wave, of a surf-tormented shore, as I hold within my hand, grains of the golden sand, my Book of Truths, an Art-Product alone, the Beauty that abounds in its Truth, is ours—for evermore. All was rigorously self-consistent. But a prophet has no honor in his own land.

Eureka, I could accomplish nothing more, given the Night's Plutonian shore. All Nature speaks, and ev'n ideal things, flap shadowy sounds from visionary wings. To love and deem this poet wise, is to seek for treasure in the jewelled skies. *Perfection of plot*, it shall not be forgot. Combining the radical and chasmal differences of Poetry and Truth, over a gorge that is spanned by a rope, *Eureka* is my *E pluribus unum*—the reduction of many into one—and the dawn of a more exciting *hope*.

Index